THE FRONTIERS COLLECTION

The books in this collection are devoted to challenging and open problems at the forefront of modern science and scholarship, including related philosophical debates. In contrast to typical research monographs, however, they strive to present their topics in a manner accessible also to scientifically literate non-specialists wishing to gain insight into the deeper implications and fascinating questions involved. Taken as a whole, the series reflects the need for a fundamental and interdisciplinary approach to modern science and research. Furthermore, it is intended to encourage active academics in all fields to ponder over important and perhaps controversial issues beyond their own speciality. Extending from quantum physics and relativity to entropy, consciousness, language and complex systems—the Frontiers Collection will inspire readers to push back the frontiers of their own knowledge.

More information about this series at http://www.springer.com/series/5342

Shyam Wuppuluri · Newton da Costa
Editors

WITTGENSTEINIAN (*adj.*)

Looking at the World from the Viewpoint of Wittgenstein's Philosophy

Foreword by A. C. Grayling

Springer

Editors
Shyam Wuppuluri
JVPD Scheme
Yoga Housing Society
Mumbai, India

Newton da Costa
Fuzzy Sets Laboratory
HCTE
Rio de Janeiro, Brazil

ISSN 1612-3018 ISSN 2197-6619 (electronic)
The Frontiers Collection
ISBN 978-3-030-27568-6 ISBN 978-3-030-27569-3 (eBook)
https://doi.org/10.1007/978-3-030-27569-3

This Springer imprint is published by the registered company Springer Nature Switzerland AG
The registered company address is: Gewerbestrasse 11, 6330 Cham, Switzerland

Foreword

Let us compare and contrast Spinoza and Wittgenstein. The former was ambitious to set out his views with thoroughgoing explicitness and clarity, following a plan that was fully articulated in his mind and which he wished to express with the exactitude of a deductive mathematical exposition: *more geometrico*. The respects in which his work is difficult have to do with the subject-matter of his work, not his method of presenting it.

Wiggenstein attempted a methodology similar to Spinoza's in his only published work, the *Tractatus Logico-Philosophicus*. He came to think that the method as well as the project of the *Tractatus* was misconceived, and therefore he changed tack entirely: his later efforts at expounding his views took the form of notes, questions, suggestions, hints, assertions which sometimes might or might not have been experiments rather than commitments: they form an unfinished and open-textured set of explorations.

Whereas the *Tractatus* has a Spartan economy as a text, its principal doctrines and its motivations are clear. Matters are not quite the same with Wittgenstein's later writings. The suggestive and rough-hewn character of the unsorted notes which posthumous editors ordered and put into book form have, by contrast, a protean nature: there are different and not seldom competing ways of interpreting them.

From that kind of *nachlass* students and scholars can build many different philosophical stories. This collection of papers is testament to that fact. It raises the question of distinctions to be drawn among three different enterprises: (a) interpretations of Wittgenstein's views, (b) applications of his views to subject-matters that he did not address (for example, AI), and (c) views which philosophers develop under the inspiration of their reading of Wittgenstein—where Wittgenstein's views and *dicta* are the occasion rather than the substance of their own thoughts. This collection of papers, again, exhibits examples of all three.

Of course, some of what is written in the name of Wittgenstein is a combination of more than one of (a) to (c). In particular what constitutes what has been called *The New Wittgenstein*, following a celebrated volume of papers by that name, exemplifies the combination of (a) and (c). This is perhaps now the common form of engagement with Wittgensteinian themes.

For anyone who comes new to the study of Wittgenstein and Wittgensteinian themes, it might be of help to have a sketch of what was once, at least, an orthodox view of the story of Wittgenstein's 'two philosophies'—the 'early' and the 'later' (a distinction challenged by the 'new Wittgenstein' paradigm). Having a background knowledge of this former orthodoxy makes more sense of the interpretations and inspirations that revision of the orthodoxy prompts. I shall now give such a sketch. After giving it I offer two remarks about the former orthodoxy thus sketched which, more by way of asides than anything else, suggest that it retains more value than later reinterpretations claim.

One of Wittgenstein's overarching aims, common both to his 'early' and his 'later' thought, was to show that the problems of philosophy can be solved by understanding how language works. In the early philosophy, as set out in the *Tractatus*, he argues that we are to do this by understanding what Frege and Russell had sought to identify as 'the logic of our language'. His later philosophy is based on a rejection of this approach, and a completely different view of language. In both phases, the problems of philosophy are non-problems arising from misunderstanding of language; a correct account of how language works therefore lays an axe to the root of philosophy's problems.

The central idea in the *Tractatus*, acquired from Russell, is that language has an underlying structure, inspection of which shows what can be meaningfully said. For Wittgenstein what can be *said* is the same as what can be *thought*, so once one has shown what can meaningfully be said, one has shown the limits of thought. Beyond these limits, language and thought are meaningless. It is in this outer zone, he argued, that traditional philosophical problems arise, as the result of trying to say the unsayable and think the unthinkable. At the beginning of the *Tractatus*, he accordingly writes, 'What can be said at all can be said clearly, and what we cannot talk about, we must consign to silence'; likewise the very last sentence of the *Tractatus* is the famous remark, 'Whereof we cannot speak, thereof we must be silent'. In a letter to Russell explaining these remarks Wittgenstein wrote, 'The main point [of the *Tractatus*] is the theory of what can be expressed by propositions —i.e. by language—(and, which comes to the same thing, what can be thought), and what cannot be expressed by propositions, but only shown; which, I believe, is the cardinal problem of philosophy'.

The argument is as follows. Both language and the world have structure. Language consists of propositions, which are compounds of 'elementary propositions', which in turn are combinations of 'names'. Names are the ultimate constituents of language. Correspondingly, the word consists of the totality of facts, which are compounded out of 'states of affairs', which in turn are combinations of 'objects'. The 'names' and 'objects', and the 'elementary propositions' and 'states of affairs', are *logically* specified levels of structure; what they correspond to in reality is a question for a quite different enquiry to establish. This logically specified structure is purely abstract; it stands for whatever there is at the more elemental levels of the related complexes which are language and world.

Each level of structure in the world is matched by a level of structure in language. Names denote objects, the combination of names constituting elementary propositions correspond to states of affairs, and each of these in their turn combines to form, respectively, propositions and facts. The arrangement of names at the most fundamental level of language structure 'mirrors' or 'pictures' the arrangement of objects at the most fundamental level of the world's structure. This is the basis of the 'picture theory of meaning' central to the *Tractatus*.

The elementary propositions are logically independent of each other. Therefore, we have to say which are true and which false to give a complete account of reality. This says that reality consists of all possible states of affairs, whether actual or not; reality is everything that is and is not the case. Propositions are built out of elementary propositions by the logical connectives 'and', 'or' and the rest (actually, by one primitive connective from which the others can be defined), and therefore the truth-value of propositions is a function of the truth-values of the elementary propositions composing them. This does not apply, of course, in the case of tautologies, which are always true, and contradictions, which are always false. The true propositions of logic and mathematics are tautologies (analytic truths). They do not say anything about the world because their truth-value is consistent with any way the world could be (with the existence or non-existence of *any* states of affairs.)

When a string of signs fails to express a proposition, it is nonsense. It is not false; it cannot be, because it says nothing capable of being either true or false. The propositions of philosophy belong to this class—including the propositions in which Wittgenstein tells us this. He says, 'My propositions serve as elucidations in the following way: anyone who understands me eventually recognises them as nonsensical, when he has used them—as steps—to climb up beyond them. (He must, so to speak, throw away the ladder after he has climbed up it.)' He borrows the image of the discardable ladder from Schopenhauer.

And of course the propositions of ethics, aesthetics, religion and the 'problems of life' belong to the same class as philosophical propositions; they are not pictures of actual or possible facts, and therefore they are meaningless.

The *Tractatus* sets out these themes in seven numbered propositions, each of which has subordinate propositions appended to it—with the subordinate comments themselves having subordinate comments, and so on—in an elaborate system of decimal notation. The propositions and sub-propositions are highly compressed, but they make out in detail the nature of the structural levels and how they relate. We learn that it is of the essence of objects to be possible constituents of states of affairs, so that if we knew all the objects there are, we would know all the states of affairs there could possibly be. Objects themselves do not undergo change; it is only the combinations they enter into that do so. When they combine, the states of affairs they constitute are determinate. This is why a complete analysis of a proposition will terminate in the specification of a determinate combination of objects, given that there is only one complete analysis of a proposition, and that each of its constituent names denotes an object. Correlatively, the states of affairs that exist thereby settle the question of which states of affairs do not exist.

The 'picture theory of meaning' is based on these considerations. 'A proposition is a picture of reality' and 'A logical picture of facts is a thought', thoughts being what are expressed by propositions. The totality of propositions is language, and 'the totality of true propositions is science'. Wittgenstein had got his idea for the picture theory from reading of a Paris court case about a traffic accident, in which toys were used to portray what had happened. The relationship between the toy cars and their arrangements and the real cars was one of correspondence; the mock-up in the Paris courtroom was a model of the actual scene. 'Pictorial form is the possibility that things are related to one another in the same way as the elements of a picture … there must be something identical in a picture and what it depicts, to enable one to be a picture of the other'. He gives an even better analogy later: 'A gramophone record, the musical idea, the written notes, the sound waves, all stand to one another in the same internal relation of depicting that holds between language and the world'.

Wittgenstein thought that these views, by delimiting the possibility of meaningful discourse, had shown that the problems of philosophy are spurious—which is the same thing, in effect, as solving them. But he was persuaded by Moritz Schick and Friedrich Waismann of the Vienna Circle that he had not in fact succeeded in this aim. As a result, he returned to Cambridge in 1929 and thereafter wrote copiously. His views became known in philosophical circles, partly by word of mouth, partly through discussions with colleagues and visitors, and partly through the circulation of manuscripts, such as the *Blue* and *Brown Books* containing notes from his lectures in Cambridge in the years 1933–4 and 1934–5, respectively. These notebooks were later published as *Preliminary Studies for the Philosophical Investigations*, referring to the principal work that emerged from this period and which appeared after Wittgenstein's death, the *Philosophical Investigations* (1953).

Wittgenstein begins the *Philosophical Investigations* by saying that it ought to be read alongside the *Tractatus* so that a comparison between them can be made, the implication being that this will illustrate the respects in which the views of the earlier work are wrong. In one significant respect, there is continuity between the early and late philosophy: in both phases, Wittgenstein's claim was that philosophical problems arise because we misunderstand the way language works. What changed was his conception of how language works.

In the *Philosophical Investigations*, the governing idea is that language, instead of having a single underlying essence exemplifiable as its 'logical form', consists of many different practices—Wittgenstein called them 'language-games'—within each of which the meaning of expressions consists in the uses made of them in the context (the 'form of life') which is the language-game's setting. He now argued that the austerely systematic account attempted in the *Tractatus* is the wrong way to show this. Instead of devising theories to solve philosophical problems we should see our task as a 'therapeutic' one of 'dissolving' those problems by showing that they arise from misuse of language. The metaphor of therapy is applied with perfect seriousness: 'The philosopher's treatment of a question is like the treatment of an illness'. He varies the metaphor: 'What is your aim in philosophy? To show the fly the way out of the fly-bottle'.

A chief source of misunderstanding arises when we take an expression from one context of use and apply it in another where it does not belong. If we look at the proper use of expressions, we will resist the temptation—the 'urge'—to misapply them. We begin—as the *Tractatus* had implicitly done—by accepting the beguiling view that we learn the names of objects by associating them with objects. Wittgenstein quotes St Augustine describing how his elders pointed at things and said their names in order to teach him to speak. On this model, says Wittgenstein, we commit ourselves to think that 'every word has a meaning. The meaning is correlated with the word. It is the object for which the word stands'. And we then proceed to assume that sentences are combinations of names. That is the *Tractatus* view. It leads us to search for the *essence* of language, as something hidden: the 'logical form' underlying the surface.

But in fact how language works, he now argues, lies open to view. As soon as we look in the right place we see that language is not *one uniform thing* but many different activities. We describe, report, inform, affirm, deny, speculate, give commands, ask questions, tell stories, make jokes, sing, greet, curse, pray, reminisce, play-act, thank, warn, complain and much besides. Each of these activities is a language game: 'the term "language-game" is meant to bring into prominence the fact that the *speaking* of language is part of an activity, or form of life'. The allusion to 'games' does not imply something unserious; Wittgenstein is trading on the fact that there is no one thing that is common to all games, no single essence or defining property that makes everything that is a game a game, and differentiates all games from non-games. 'Consider … the proceedings we call "games". I mean card games, board games, Olympic games, and so on. What is common to them all? Don't say: there must be something, but look and *see* whether there is something common to all, for if you look at them you will not see something that is in common to all, but similarities, relationships, and a whole series of them at that … I can think of no better expression than to characterise these similarities than "family resemblances"; for the various resemblances between members of a family: build, features, colour of eyes, gait, temperament, etc. etc., overlap and criss-cross in the same way.—And I shall say: "games" form a family'.

In these games, expressions have uses, functions, purposes, offices, roles, employment—Wittgenstein uses all these words to talk of what is done with them —and the general idea is that the meaning of expressions consists in the parts they play in the language-games they belong to. This is sometimes described as a 'use theory of meaning', which over-simplifies it, but is not inaccurate, providing one remembers Wittgenstein's insistence that 'knowing the meaning' of an expression is not an inner mental state of 'understanding' but is a *technique*: to know the meaning of an expression is to be master of a technique for following the rules for its use. And not just the usage-rules for that particular expression, but of language as such: to be a speaker is, holistically, to be the speaker of a whole language.

Wittgenstein insists that the rules followed in 'following a rule' must not be understood as constituting a structure coercively determining in advance what their correct application must produce, as, for example, in arithmetic the rules of multiplication do. What makes a rule a *rule* in language is the collective use that speakers of the language make of it. So although the rules guide and provide a measure of correctness for each individual language user, they are not like immutable railway tracks, but instead are a product of the forms of language-using life themselves. They record customs, or practices; they are like the sign-posts that guide one on a footpath. What justifies or provides grounds for them? They do it themselves: 'Giving grounds … comes to an end … the end … is our acting, which lies at the bottom of the language-game'.

We acquire the ability to follow the use of expressions by our training as members of our language community. The training occurs in the 'form of life' shared by teachers and learners alike. The shared form of life is the frame of reference and the ground of the practice of speaking the language. A corollary of this is that language is *essentially public*; there cannot be a language known to and spoken by only one speaker. There can be a *contingently* private language, like the code Samuel Pepys invented to keep secret what he wrote in his diary: but such a code merely disguises a public language. A 'logically private language' could not be learned in the first place; and in any case if being able to speak a language is having mastery of a rule-following technique, then there has to be a distinction between following a rule and only thinking that one is doing so, and the check on usage can only be provided by others in one's linguistic community.

A corollary of the argument in the *Philosophical Investigations* is that examining language to guard against philosophy-generating misunderstandings 'leaves everything as it is', that is, does not change anything about how we view the world or ourselves—other than, of course, to liberate us from the temptation to engage in philosophy and therefore, like the fly, to get ourselves trapped in the fly bottle.

This might not have been Wittgenstein's final view of the role of philosophy, however. The notes he was working on at the end of his life were edited by Elizabeth Anscombe into the volume entitled *On Certainty*. It is the draft of an argument to the effect that we have to have some indubitable beliefs in order for us to be able to do anything at all. It starts from Moore's 'here is one hand' argument, and sets out the view that discourses depend upon there being some propositions which are immune to question; they serve as the 'hinges' on which the discourse turns, or (to vary the metaphor) they are the river-bed and banks through which the water of discourse flows. Like the banks of a river, they can be eroded over time, but that has to happen slowly enough for the public meanings of expressions in the discourse to remain relatively stable so that the aims of communication can be fulfilled.

What is interesting about *On Certainty* is that it is an exercise in standard epistemology, a discussion of a central and traditional philosophical problem: how to deal with sceptical challenge about the possibility of knowledge. It does not treat this as a pseudo-problem that will disappear of its own accord if we pay attention to the ordinary surface forms of language. Perhaps Wittgenstein was at that point moving beyond what he had inspired others to accept, in imitation of his approach: the idea that, by itself, attention to ordinary language will solve philosophical problems.

Here I now offer the two remarks mentioned earlier. One is that it is not a straightforward matter to uphold the revisionist interpretation of Wittgenstein's projects without taking into account the overwhelming evidence presented by Brian McGuiness in his *Young Ludwig* (Oxford: Clarendon Press, 2005)—an important work in this respect—and also Allan Janik's and Stephen Toulmin's *Wittgenstein's Vienna* (New York, 1973), showing that Wittgenstein's ideas were richly anticipated in the philosophical and cultural milieu of the *fin de siècle* and early twentieth-century Vienna in which Wittgenstein grew up. A surprising aspect of Wittgenstein scholarship is that this background has not been more thoroughly researched, both for its intrinsic interest and for the light it throws on the emergence of the philosophy of language generally—to say nothing of the increment of understanding it offers about Wittgenstein's own ideas. There is much material there for further and wider illuminations in all these respects.

More immediately to the present point: I said above that '*one* of Wittgenstein's overarching aims … was to show that the problems of philosophy can be solved by understanding how language works'. The *other* overarching aim, explicit in the early philosophy and implicit in the later, was to defend matters of value and faith from what he saw as the reductive and corrosive encroachments of natural and social science. We see this as follows.

In the *Tractatus*, Wittgenstein argues that the proper task of philosophy is 'to say nothing except what can be said, i.e. the propositions of natural science—i.e. something that has nothing to do with philosophy—and then, whenever someone wanted to say something metaphysical, to demonstrate to him that he had failed to give a meaning to certain signs in his propositions'. Does this rule out ethics, aesthetics, religion and the 'problems of life' as nonsensical? No: on the contrary. Wittgenstein wishes to establish that it is only the attempt to *say* anything about them which is so. 'There are, indeed, things which cannot be put into words; they make themselves manifest. They are what is mystical'. Here 'showing' rather than 'saying' is all that is possible. Wittgenstein accordingly alludes to the 'more important unwritten second half of the *Tractatus*', meaning that the *Tractatus* shows from within the limits of language what is important. In another letter, he wrote, 'For the Ethical is delimited from within, as it were, by my book … All of what *many* are *babbling* today, I have defined in my book by remaining silent about it'.

The natural sciences, including psychology, and the social sciences, including philology, archaeology, anthropology and history, were closing in on all three of ethics, aesthetics and religion in reductive and debunking ways, explaining them in naturalistic and empirical terms. Psychologists described religious experiences as hallucinations or delusions; textual critics and historians argued that the Bible was the work of many people over large stretches of time, full of inconsistency and error; anthropologists pointed out that different cultures think about ethical questions in very different ways. The thesis of the *Tractatus* protects ethics and religion from these encroachments by placing them outside the realm of what can be talked about, restricting meaningful language to propositions that are actual or possible pictures of facts in the world. This is what makes talk of subjects other than science meaningless—but not, in Wittgenstein's view, therefore unimportant. Nothing can be said about them, and if efforts are made to say something about them the result will be nonsense. That is the fate of philosophy itself, he says, even on the charitable view that the propositions of philosophy can be viewed as eventually disposable elucidations.

In the later philosophy, ethical and religious discourses are their own justification; they constitute the meanings of their own key terms by the role they play in the language-game and the form of life in which it is played. Once again there is a kind of incommensurability thesis at work—by which I mean the irreducibility of the discourses of value and faith to those of science, and the untestability of the former by means of the latter—very different in expression from the parallel thesis in the *Tractatus*, but the same in effect. And this connects to the point about how it is, as a result of the form taken by Wittgenstein's later writings, that they have proved so fertile in allowing alternative interpretations, and likewise allowing the development of novel ideas from his themes. This collection of papers is a further testament to that fact.

2019

A. C. Grayling
New College of the Humanities
London, UK

Preface

"Tell me", Wittgenstein once asked a friend, "why do people always say, it was natural for man to assume that the sun went round the earth rather than that the earth was rotating?" His friend replied, "Well, obviously because it just looks as though the Sun is going round the Earth". Wittgenstein replied, "Well, what would it have looked like if it had looked as though the Earth was rotating?"
—G. E. M. Anscombe, *An Introduction to Wittgenstein's Tractatus* (2nd ed., 1963), Chap. 12: Knowledge and Certainty.

'Tell them, I had a wonderful life!' These were supposedly the last words uttered by Ludwig Wittgenstein, to his friend Joan and her husband, Edward Bevan, who was also his doctor. It takes something for a man to speak these words. Enduring the moments of sheer despair and depression, intellectual struggles with his peers and with his own self, Wittgenstein perhaps attained an intellectual ataraxia towards the end of his life. What do we know about his life? And his work? Ludwig Wittgenstein is undoubtedly one of the outstanding philosophers of all time. He is peculiar in so many different ways that any description or category that we try to impose on him rapidly proves inadequate. He shares at most a family resemblance (to use his own phrase) with other 'philosophers', in the true spirit of that word. In this volume, some of the world's leading thinkers come together to expound upon the *Wittgensteinian* ways of looking both at the sciences and at the world as a whole.

Though Wittgenstein did not write or publish a lot, in the little he did manage to write, he said it all in the most concise way possible. Not a word more, not a word less. Such was his precision. His work has had a far-reaching impact and has permeated a vast number of fields within science; it has infiltrated even the cognitive tools we use to think about things. The difficulty in dealing with a multifaceted personality like Wittgenstein is that his readers are liable to experience literary and ideational pareidolia: We often try to put into the mouth of Wittgenstein, things that he didn't say or mean; indeed a fear of being misunderstood is something that haunted him frequently during his life. Great care has been taken by the outstanding authors of this volume to remain immune to anachronistically 'reading back' and interpreting the texts in the manner that suits some formalism, or to favour a certain schism. The views presented in these pages truly, or very closely, reflect those that Wittgenstein held (or did not hold). Of course, no one can be sure what views Wittgenstein would have held about Quantum Mechanics or about things like Artificial Intelligence. But the authors have preserved the Wittgensteinian lens and carefully outlined the landscapes that Wittgenstein's known views engender. We took efforts to put these articles into various baskets ranging from Philosophy to Theoretical sciences and beyond, even though they don't always adhere neatly to such classifications. All authors have attempted to make their articles as readable as possible so that a passionate layperson can easily skim through the book and understand a good deal of it. My hope is that, through this book, readers will be able to appreciate and understand Wittgensteinian views—as well as how and why they can help us obtain deeper insights into the pressing intellectual issues within the sciences.

This book wouldn't have been possible without the constant editorial support I have received from Dr. Angela Lahee at Springer. Having helped us in innumerable ways, she remains humble and leaves no trace of her involvement. Like a silent spring that comes, does its job and leaves. I am immensely grateful to Prof. Newton Da Costa, for his endless support, simple humility and kindness. An intellectual giant, his shoulders enabled me to see farther than I otherwise could have done. I also express my utmost gratitude to Prof. A. C. Grayling. As a leading intellectual figure of our times, he has been nothing but kind towards me and supportive of this volume. I am also very thankful to all the authors, without whom this book would not have been possible. I would like to thank Freeman Dyson, for his support and kindness during our interaction. In the context of typesetting the book, I would like to acknowledge and thank the wonderful editorial support I have received from Dharaneeswaran Sundaramurthy, Priyadharshini Subramani, Antje Endemann and their team. In times of adversity, it is our loved ones, most of all, who stand by and help us see the positive side of the world. I would like to thank my family and friends for always being there. Especially Wu Dali; I couldn't have endured the hardships without her gracious unconditional support, which is a balm to my soul.

Mumbai, India Shyam Wuppuluri

Contents

Contributors

Jonathan Beale Researcher-in-Residence, Tony Little Centre for Innovation and Research in Learning, Eton College, Windsor, Berkshire, UK

David R. Cerbone Department of Philosophy, West Virginia University, Morgantown, WV, USA

Annalisa Coliva Department of Philosophy, UC Irvine, Irvine, USA

Freeman Dyson Institute for Advanced Study, School of Natural Sciences, Princeton, NJ, USA

Susan Edwards-Mckie Darwin College, Cambridge, UK

Lars Hertzberg Department of Philosophy, Åbo Akademi University, Turku, Finland

Paul Horwich Department of Philosophy, New York University, New York, NY, USA

Oskari Kuusela School of Politics, Philosophy, Language and Communication Studies, University of East Anglia, Norwich, Norfolk, UK

Gregory Landini Department of Philosophy, University of Iowa, Iowa City, IA, USA

Nana Last The University of Virginia, Charlottesville, VA, USA

Jakub Mácha Masaryk University, Brno, Czech Republic

Yoshihiro Maruyama The Hakubi Centre for Advanced Research, Kyoto University, Kyoto, Japan

Nikolay Milkov Philosophy Department, Universität Paderborn, Paderborn, Germany

Danièle Moyal-Sharrock University of Hertfordshire, Hatfield, Hertfordshire, UK

Michael Nedo Wittgenstein Archive, Cambridge, UK

Christopher Norris School of English, Communication and Philosophy, Cardiff University, Cardiff, UK

Carlo Penco Philosophy of Language, Department of Philosophy, University of Genova, Genoa, Italy

Duncan Pritchard University of California, Irvine, USA;
University of Edinburgh, Edinburgh, Scotland, UK

Rupert Read School of Politics, Philosophy, Language and Communication Studies, University of East Anglia, Norwich, Norfolk, UK

Beth Savickey Philosophy Department, University of Winnipeg, Winnipeg, MB, Canada

Paul F. Snowdon Department of Philosophy, University College London, London, UK

Ilse Somavilla Universität Innsbruck, Innsbruck, Austria

Susan G. Sterrett Department of Philosophy, Wichita State University, Wichita, KS, USA

Chon Tejedor Philosophy Faculty, Universitat de València, Valencia, Spain

Michel ter Hark Faculty of Humanities, Vrije Universiteit Amsterdam, Amsterdam, The Netherlands

Edward Witherspoon Department of Philosophy, Colgate University, Hamilton, NY, USA

Pascal Zambito University of Cambridge, Cambridge, UK

Abbreviations

BB	The Blue and Brown Books
BT	The Big Typescript (TS 213)
CE	Cause and Effect: Intuitive Awareness
CV	Culture and Value
DB	Denkbewegungen: Tagebücher 1930–1932, 1936–1937
EPB	Eine Philosophische Betrachtung (Revision of parts of The Brown Book)
GT	Geheime Tagebücher
LC	Lectures and Conversations on Aesthetics, Psychology and Religious Belief
LE	(Wittgenstein's) Lecture on Ethics
LWPP I	Last Writings on the Philosophy of Psychology, Vol. 1
LWPP II	Last Writings on the Philosophy of Psychology, Vol. 2
NB	Notebooks 1914–1916
NFL	Notes for Lectures on "Private Experience" and "Sense Data"
OC	On Certainty
P	'Philosophy' (in BT/TS 213)
PG	Philosophical Grammar
PI	Philosophical Investigations (Part two is referred to as PI II, pg. nr.)
PO	Philosophical Occasions
PPO	Private and Public Occasions
PR	Philosophical Remarks
PTLP	Prototractatus: An Early Version of Tractatus Logico-Philosophicus
RC	Remarks on Colour
RFGB	Remarks on Frazer's Golden Bough
RFM	Remarks on the Foundations of Mathematics
RPP I	Remarks on the Philosophy of Psychology, Vol. 1
RPP II	Remarks on the Philosophy of Psychology, Vol. 2

TLP	Tractatus Logico-Philosophicus
WLPP	Wittgenstein's Lectures on the Philosophy of Psychology 1946–1947
Z	Zettel

Letters

CL	Ludwig Wittgenstein: Cambridge Letters, ed. Brian McGuinness and G. H. von Wright
EL	Letters from Ludwig Wittgenstein–With a Memoir, by Paul Engelmann, trans. L. Furtmüller, ed. Brian McGuinness
FB	Familienbriefe, ed. M. Ascher, B. McGuinness, and O. Pfersmann
GB	Gesamtbriefwechsel, ed. B. McGuinness, M. Seekircher, and A. Unterkircher
LO	Letters to Ogden, ed. Charles K. Ogden and G. H. von Wright
LRKM	Letters to Russell, Keynes and Moore, ed. G. H. von Wright
WC	Wittgenstein in Cambridge: Letters and Documents 1911–1951, ed. Brian McGuinness
WE	Wittgenstein–Engelmann: Briefe, Begegnungen, Erinnerungen

Notes Taken by Others

AWL	Wittgenstein's Lectures, Cambridge 1932–3, ed. Alice Ambrose
LC	Lectures and Conversations on Aesthetics, Psychology and Religious Belief
LFM	Wittgenstein's Lectures on the Foundations of Mathematics, Cambridge 1939, ed. Cora Diamond
LSD	'The Language of Sense Data and Private Experience' (p. xiii) MA Unpublished: Wittgenstein's lectures 1932–3 from the notes of G. E. Moore, Cambridge University Library G. E. Moore Archive, ref: ADD 8875, 10/7/7
MWL	'Wittgenstein's Lectures 1930–33', G. E. Moore's notes from Wittgenstein's lectures
NTW	'Notes on Talks with Wittgenstein'
VW	The Voices of Wittgenstein: The Vienna Circle, Gordon Baker ed.
WL	Wittgenstein's Lectures, Cambridge 1930–32, ed. Desmond Lee
WLPP	Wittgenstein's Lectures on the Philosophy of Psychology 1946–47, ed. Peter Geach
WVC	Wittgenstein and the Vienna Circle: Conversations recorded by Friedrich Waismann, ed. Brian McGuinness

Part I
Philosophy

Chapter 1
Showings (Wittgenstein): A Double Sestina

Christopher Norris

This inseparableness of everything in the world from language has intrigued modern thinkers, most notably Wittgenstein. If its limits—that is, the precise point at which sense becomes nonsense—could somehow be defined, then speakers would not attempt to express the inexpressible. Therefore, said Wittgenstein, do not put too great a burden upon language.

Peter Farb, *Word-Play*

If a person tells me he has been to the worst places I have no reason to judge him; but if he tells me it was his superior wisdom that enabled him to go there, then I know he is a fraud.

Ludwig Wittgenstein, *Personal Recollections* (ed. Rush Rhees)

The real discovery is the one which enables me to stop doing philosophy when I want to. The one that gives philosophy peace, so that it is no longer tormented by questions which bring *itself* into question.

Wittgenstein, *Philosophical Investigations*

The world is everything that is the case.
All that's the case is all that we can say.
Some things cannot be said but may be shown.
These are the most important things in life.
A change in them will be a change of world.
Let silence show where saying leads astray.

So many ways we can be led astray!
Delinquent speech is not the only case,
Though certain evils may infect our world
Through word-abuse. Believing we can say
What matters most, in language or in life,
Is Russell's error. This much can be shown.

C. Norris (✉)
School of English, Communication and Philosophy,
Cardiff University, Cardiff CF10 3AT, UK
e-mail: norrisc@cardiff.ac.uk

S. Wuppuluri and N. da Costa (eds.), *WITTGENSTEINIAN (adj.)*,
The Frontiers Collection, https://doi.org/10.1007/978-3-030-27569-3_1

That's why my faithful few won't have it shown
How moral compass-points can swing astray
Even with such ascetic forms of life
Or utterance as mine. Count it a case
Of things-gone-wrong that nobody could say
Belonged exclusively to word or world.

Russell and Moore: they were my Cambridge world
Back then although, despite some kindness shown,
They failed to grasp how using words to say
Those things unsayable led sense astray.
Their verdict on me: genius, but a case
Of life screwed up by mind and mind by life.

'Just tell them that it's been a wonderful life.'
My dying words, and spoken from a world
So distant, now, from all that is the case
With their world that what's said by them, or shown,
Will likely lead my auditors astray
As much as anything I've had to say.

Yet there's some truth in what the others say,
My critics, who'd regard a tortured life
Like mine as leading and as led astray
Since formed within the solipsistic world
Of my obsessions. That's the sole thing shown,
They'd say, by such a cautionary case.

I keep my life a closed book just in case
Some rogue biographer should have his say
And seek, for no good cause, to have it shown
That there were certain chapters in that life
Kept secret from the academic world
Lest scandal lead my acolytes astray.

Yet could it be some young men went astray
Because I'd cruise the Prater and then case
The gay joints in my craving for a world
As far removed as possible from, say,
The wealth and privilege of my old life,
Or the mixed spite and condescension shown

By Moore and his Apostles? If I've shown
A seamy side, a will to go astray
In quest of what they'll call 'his other life',

It's not (the vulgar-Freudian view) a case
Of my abject desire that they should say
Harsh things that show me up before the world

For what I am. Rather, I deem that world
Of theirs a world in need of being shown
Such truths as neither they nor I can say
Since, in the saying, sense would go astray
And make me out a monster or a case
For some corrective treatment. It's my life,

Not anything I've written, but my life
As lived that bears sole witness to the world
Concerning just those matters in the case
Of Ludwig Wittgenstein that should be shown,
Not said, since uttering them sends words astray
And has them mimic what they fail to say.

And yet I ask: why think of 'show' and 'say'
In such bi-polar terms unless your life,
Like mine, has gone unspeakably astray
And left you stranded in an alien world
Where your 'condition' can at most be shown,
Not talked about or stated, just in case.

A modest claim: to say, not save, the world,
Yet still too statement-bound, as life has shown.
What was it went astray with what's the case?

No world exists that logothetes might say
'Here's all we've shown: that words bring worlds to life'.
What if 'the case' just is what goes astray?

Chapter 2
Wittgenstein's Ways

Nikolay Milkov

2.1 What Are the Ways?

Aristotle explored different modes (*qua*) of being. Unfortunately, in modern literature, discussion on this concept has been largely neglected. A possible exception was Roman Ingarden, who investigated it in his *Time and Modes of Being* (1964).

Even more neglected is the concept of *ways* as a specific type of mode. Interest in this concept has increased only in recent years. Some authors explore it in connection with the ontological status of universals and particulars. For example, one claims that universals (for example, the shape of my house) are the different ways in which the building blocks (the particulars) of the object (my house) are ordered (Berman 2008, 220). The trope theory of ways argues in a similar vein (Simons 1994).

Closer to our approach but developed in another direction is the discussion on constitutional relation by Barker (2000) and Zangwill (2012). They consider constitutional relation as a primitive and unanalysed notion. It cannot be reduced to other, more fundamental relations. Unfortunately, these two authors fail to suggest a conception that convincingly explains the emergence of new ontological orders.

In this paper, we will discuss the ontology of ways that employs the concept of construction (constitution) as well as of modal entailment. It defends hierarchical organization of the being with many levels of ontological dependence. To be more explicit, we will explore Wittgenstein's conception of the ways of arranging a set of elements of one ontological level that produces a new ontological level. Its main advantage is that it sees the emergence of new entities like matter, space, life, mind, works of art, etc., as the construction of new ontological levels from the old ones.

To be sure, not every arrangement of the elements of one ontological level leads to the emergence of a new ontological order; it only takes place in specific cases. Moreover, most often, one cannot produce a new ontological level according to plan.

N. Milkov (✉)
Philosophy Department, Universität Paderborn, Warburger Straße 100,
33098 Paderborn, Germany
e-mail: nikolay.milkov@upb.de

S. Wuppuluri and N. da Costa (eds.), *WITTGENSTEINIAN (adj.)*,
The Frontiers Collection, https://doi.org/10.1007/978-3-030-27569-3_2

However, the fact that we cannot predict the exact way of arranging individuals of one ontological level, which will produce a new ontological level, is not a valid argument against the constructivist ontology of ways. Scientific knowledge of human anatomy also fails to predict how life emerges from the analysed structure of the body. Nevertheless, it is a useful discipline that helps the most in our efforts to learn more about how our body functions.

Wittgenstein's ontology of ways is not reductivist but deductivist, in the sense that he rejects the view that entities of higher ontological orders are reducible to entities of lower order. However, Wittgenstein supports the view that the former are *deducible* from the latter, with which they stay in contact. This happens because they are ontologically dependent. Furthermore, Wittgenstein does not reject the emergence of higher ontological orders. He denies that the emergence is a product of an occult process, though (see § 2.6 (iv)). Moreover, Wittgenstein explores how exactly the new ontological order comes into existence.

2.2 Frege, Husserl, and Russell

The early analytic philosophers, Frege, Russell, and Husserl were the first to intensively discuss the constructivist ontology of ways.[1] Frege maintained that the sense of the sign is determined by the way (*die Art*) of its presenting (Frege 1892, 158). Though he did not explore the sense of the proposition in similar terms (as we will see in § 2.3, [ii], only Wittgenstein did this), we can construct it accordingly as the way in which the elements of the proposition are combined.

Husserl held that different types of mental acts (*noeses*) are the specific ways (*die Weise*) in which the human mind (the 'Ego') is directed to its objects (*noemata*) (Mayer 2009, 70–71; Husserl 1900/1901, ii, 761). Since mental acts have a constitutive function in Husserl's ontology, the objects of the external world are only given by the way in which they are grasped.

However, the first person to explicitly and most innovatively explore the concept of 'way' in the constructivist sense was Bertrand Russell. In his paper 'What is Logic?' Russell defined that discipline (logic) as the study of the forms of complexes. Russell further claimed that these forms are nothing but 'the way the constituents of complexes are put together' (1912, 55). The constituents of a proposition are united neither by their 'organic unity' (Moore) nor by the 'context principle' (Frege), but by the way in which they are ordered.

Russell had great hopes from the introduction of the concept of ways into his ontology. He believed that it could help eliminate superfluous entities, such as propositions and logical constants. Importantly enough, Russell's programme was not reductive

[1] Many historians of philosophy agree now that the early Husserl was an analytic philosopher.

but merely eliminative.[2] As we will see shortly, it massively influenced Wittgenstein's ontology of ways.

The aforementioned steps of elaborating the concept of way presented by Frege, Husserl, and Russell suggest that the interest in that concept was a product of extensive exploration of the problem of analysis. That such exploration can be fruitful by discussing the problems of non-reductive metaphysics is well known from the fact that G. E. Moore is today unanimously credited with the introduction of the concept of supervenience.[3] The motivating intuition behind the present paper is that important ideas of the early analytic philosophers remained unexplored by the upcoming philosophical generations. Our objective here is to rationally reconstruct perhaps the most fruitful of them: Wittgenstein's constructivist ontology of ways.

2.3 The Early Wittgenstein: Six Ways

A full-fledged ontology of ways (*Art und Weise*) as a means of producing ontological stages of a different order was first elaborated in Wittgenstein's *Tractatus*. Like Russell, Wittgenstein hoped that with its help he could reach the most austere (the most parsimonious) ontology that eliminates much philosophical confusion. Regrettably, Wittgenstein's ontology of ways remained largely overlooked. Also the newly awakened interest in the concept of ways among the ontologists (see §1) did not change the situation. One reason for this was that Wittgenstein's ontology of ways was only implicitly articulated. Moreover, as we will see in §4, this implicitness had its deep theoretical reasons. The task of this paper is to make it explicit and show how resourceful this ontology is.

Wittgenstein developed his Tractarian ontology of ways in (at least) six steps:

(i) *Constructing States of Affairs*. In the *Tractatus*, Wittgenstein maintained that 'in a state of affairs [in atomic facts], objects stand in a determinate relation [*Art und Weise*] to one another' (1922, 2.031). 'They fit into one another like the links of a chain' (2.03). In other words, states of affairs are nothing but 'the determinate way_1 in which objects are connected in [it]' (2.032).

By this connection, we have a topologically tight 'fitting' (*passen*) of the objects (of their boundaries) of the states of affairs of one and another, in a specific arrangement (in a specific way_1). So, according to this conception, the elements (the objects) in the state of affairs are not connected with the help of a *third element* (a mortar)[4]; they stick together through the topology of their boundaries alone.[5] In other words, states of affairs are nothing but collections of invariant items (objects) ordered in a

[2]On the difference between reductionism and eliminativism by Russell, see Landini (2003) and Beyney (2016).

[3]Cf. Moore (1922, 261).

[4]'The elements [of the state of affairs] are not connected with one another by anything' (Wittgenstein 1930, 252).

[5]The building blocks of the walls of many Inca constructions in Machu Picchu have similar topological forms of cohesion. These centuries-old buildings were built with stones having boundaries

specific way_1. Moreover, the new way_1 in which they are tied together gives birth to the new ontological level: the states of affairs.

Apparently, exactly this kind of ontology urged Wittgenstein to introduce the concept of state of affairs in his logical philosophy. To be sure, the original German word for this term—*Sachverhalt*—underlines that states of affairs are unities composed of objects, or things (*Sachen*), which are interrelated (*sich zueinander verhalten*; cf. Mulligan 1985) in a certain way_1. Moreover, the tight fitting of the boundaries of the elements (things) to one another and, as already noted, not some third element between them, is the cement (the glue) that ties these things together into a whole so that they are not merely loose complexes. Importantly enough, in contrast to Frege and G. E. Moore, Wittgenstein held that the unity of the states of affairs is not 'organic' but achieved because of their topology alone.

(ii) *Constructing Propositions via Projecting States of Affairs*. Wittgenstein's next claim was that in the act of copying (modelling) states of affairs through propositions (pictures), the structure of states of affairs is articulated in a new, strictly determined way_2. 'What a proposition expresses it expresses in a determinate manner [way_2] which can be set out clearly' (3.251). Way_2 is thus the new element that the act of modelling (articulating, projecting) states of affairs introduces. It can also be called the *form* of picturing ('pictorial form'), or of projecting, or simply the way_2 in which the picture is articulated in time and space.

As it can be expected, the relation between the piece of reality (state of affairs) and its picture is a topological relation of the tight fitting of the elements of the two structures. By its correlation to the state of affairs, the picture 'touches' the state of affairs with its 'feelers' (2.1515). In fact, only the 'end-points [of the feelers ...] actually *touch*' the state of affairs (2.15121). Here again, the touch occurs without the help of a third element.

The 'hinge' that connects the two elements—state of affairs and its picture—is their 'logical multiplicity'. It is 'what a picture must have in common with reality, in order to be able to depict it—correctly or incorrectly—in the $way_{[2]}$ it does' (2.17).

Apparently, Wittgenstein held that the shared multiplicity (the hinge) between the two structures—picture and state of affairs—(ontologically) depends on the way_1 in which the two are composed. Indeed, he maintained that 'what constitutes a picture [what makes a picture *picture*] is that its elements are related to one another in a determinate $way_{[1]}$' (2.14). To be more exact, the picture takes shape because 'the things are related to one another in the same $way_{[1]}$ as the elements of the picture' (2.151).

Similar is the construction of the propositional sign, or of 'the [sensibly] perceptible sign of a proposition' (3.11): 'What constitutes a propositional sign is that in it its elements (the words) stand in a determinate relation to one another' (3.14)—in a way_1. The way_1 in which the elements of the propositional sign are connected shows 'how [the way_1] things stand' (4.022, 4.5) in the state of affairs.

---

that exactly fitted one another. This fitting together is precisely what made them stay together in good order for centuries.

Among other things, this conception shows that way_2 is ontologically dependent on way_1.[6] To be more exact, a proposition can project a state of affairs in a certain way_2 so that it can model the state of affairs only because the way_1 in which the elements of the picture are ordered is the same as the way_1 in which the elements of the state of affairs are built. In fact, exactly the identity of the way_1 of arranging the elements of both the state of affairs and its picture fixes their shared multiplicity.

That way_2 ontologically depends on way_1 betrays an implicit principle of the *Tractatus* that deserves attention. This principle maintains that there is a hierarchy of ontological dependencies of ways. As we will see in the lines to come, however, the sequence of these ontological dependencies has more of a character of fractals than of a Porphyrian tree. Indeed, the relation between way_3 and way_2 is not the same as that between way_2 and way_1, and even more so the relation between way_4 and way_3 that we are going to discuss now.

(iii) *Constructing Thoughts (Application of Propositional Signs).* The formed pictures/propositions, resp. propositional signs can be applied. To be more exact, 'a positional sign, applied and thought out, is a thought' (3.5). It goes without saying that the hinge between the propositional sign and its application—the thought—is their identical logical multiplicity.

This is Wittgenstein's answer to the question: What is thought? Thought is not an occult process[7]—it is not a process at all[8]—but an element (a state of affairs) of reality, (logically) projected (transformed) in specific way_2 by the propositional sign that is applied (transformed) in a specific way_3. In a sense, thought is the reverse side of picturing states of affairs, or of the way_2 of their articulation. This, however, does not mean that way_3 is reducible to way_2; it is only ontologically dependent on it. This indicates that entities constructed by $ways_3$ are deducible from entities constructed by $ways_2$.

(iv) *Truth.* Truth and falsity are two different $ways_4$ of relating propositions to states of affairs: '"*p*" signified in the true $way_{[4]}$ what "~*p*" signified in the false way_4' (4.061). It follows that Frege's conception of truth and falsity of propositions as objects is false. They are only forms ($ways_2$) of relatedness of propositions to states of affairs.

(v) *Propositions of Logic (Tautologies).* An important point in Wittgenstein's Tractatus is that human beings copy states of affairs spontaneously, without following acts of their will. It is true of thinking, too. We do not think because we decide to do so. Speaking and thinking are simply parts of human natural history. Alternatively, as Wittgenstein put it, 'man possesses the ability to construct languages capable of expressing every sense [and to think about every sense], without having any idea how each word has meaning or what its meaning is' (4.002).

The situation is quite different in logic. Indeed, one constructs propositions of logic (tautologies) on purpose. Moreover, in Wittgenstein's *Tractatus*, tautologies

---

[6]Similar conception defends also Peter Simons: 'States of affairs are *existentially dependent* on its constituent objects' (2014, 413).

[7]Wittgenstein further developed this claim in his *Philosophical Investigations*. Cf. §2.6 (iv) below.

[8]This position results from Wittgenstein's non-emergentist stance.

are something of an optical instrument that helps us to see—it shows[9]—the form of propositions of science and ordinary language: 'The propositions of logic demonstrate the logical properties of propositions by combining them to form propositions that say nothing' (6.121). Exactly this 'combining' shows the way_5 the logical properties of the propositions are hooked to each other in the *logical knot* they build in order to form their sense. Wittgenstein also called this kind of showing a *method*; in fact, the whole of mathematics is nothing but a logical method that employs equations.

The use of the term 'method' here is no accident. Often, when Wittgenstein spoke about active operation with symbols, he used the term 'method' instead of 'way'. As we will see in § 2.5, the later Wittgenstein frequently spoke about *method*—about the method of verification, for example—instead of way. Without any doubt, this growing use of the term 'method' is connected with the increased role of action in Wittgenstein's later philosophy. Ontologically, however, the two terms mean the same.

(vi) *Contemplation beyond Time*. Works of art and the good life are also products of seeing elements of reality in a specific way_6: in the way_{6a} we see a single object (the works of art) *sub specie aeternitatis*, or in the way_{6b} we see the whole world (the good life) *sub specie aeternitatis* (1961, 83). The same is also true of the mystical way_{6c} of contemplating the world. It makes it possible 'to view [to feel] the world *sub specie aeterni*...as a limited whole' (6.45).

2.4 Concluding Remarks on Wittgenstein's Tractarian Ways

We have seen that there are at least six different 'ways' in Wittgenstein's Tractarian constructivist ontology. By way of taking stock of our study so far, we would like to say that the *Tractatus* introduced a full-fledged ontology of ways that helps one to construct entities of at least the following ontological levels: states of affairs, propositions, propositional signs, thoughts, tautologies, truths/falsehoods, works of art/the good life/the mystical. For obvious reasons, however, this paper will not analyse them in detail. Our objective here is more humble. It is to present in brief Wittgenstein's concept of ways and, at the end of the paper, to reveal their explanatory power in naturalistic metaphysics.

We have already seen (in § 2.3, [v]) that the tautologies of logic and mathematics show the logical form of the propositions of science and ordinary language. Wittgenstein insisted that 'a proposition *shows* its sense. A proposition *shows* [the way_2] how things stand [*wie es sich verhält*]' (4.022). This point betrays that Wittgenstein's ontological ways are ineffable. We can only notice them and point at them; we can grasp them or show them. But we cannot define them, or theoretically analyse them. We can only conceive of them intuitively. Among other things, this point (at least partly)

[9]We will say more about Wittgenstein's concept of showing in §2.4.

explains why theoreticians and historians of analytic philosophy badly overlooked Wittgenstein's many concepts of way.

We would also like to emphasize that Wittgenstein's Tractarian ontology of ways remained essentially the ontology of one world.[10] It thus opposed the ontology of many subordinated worlds of his teachers, Frege (the author of the conception of three worlds) and Russell (the author of the Ramified Theory of Types).[11]

Moreover, Wittgenstein's Tractarian ontology was not emergentist. The transition from one ontological level to a higher one does not mean a transition from one world to a higher one. Wittgenstein's ontological levels are only different stages of one and the same world. In a sense, Wittgenstein's Tractarian ways are simply 'bridges' to new, higher ontological levels.

Perhaps this is the main advantage of Wittgenstein's constructivist ontology of ways: It dispenses with the accepting of many ontologically subordinate worlds. At the same time, the Tractarian ontology of ways is not reductionist. The level of propositions and of propositional signs, for example, is not reducible to their elements. Without the specific way in which the elements are ordered, the proposition/propositional sign would not come into existence.

Yet another point is that Wittgenstein's different concepts of ways ($ways_{1-6}$) form a family with the members bearing close resemblance to each other. They, however, have no 'general form' and no general structure. This point is especially well demonstrated by the fact that the relation between different ways is divergent. Indeed, whereas way_2 is ontologically dependent on way_1, way_6, for example, is not based on way_5 (cf. § 2.3, [ii]).

2.5 The Later Wittgenstein: Four Further Ways

As it is well known today, the focus of Wittgenstein's attention changed in the early 1930s. While the *Tractatus* was mainly concerned with the acts of modelling (articulating) parts of the world (states of affairs) through language (propositions) and thoughts, at the beginning of the 1930s, Wittgenstein deepened the exploration of the relation between the action of language-speaking and the action of language-learning (between the actions of demonstrating how to compute and of learning how to compute). In this connection, he also critically discussed the concept of *rule*—the supposed principle that apparently guides us when we use a language or when we calculate. The whole of Wittgenstein's later philosophy can be advantageously seen from this perspective.

Importantly enough, with the change in his conception, starting with his 'Theses' (1930), Wittgenstein gradually abandoned the *term* of 'ways' and spoke more often

---

[10]Incidentally, the very term 'hinge' betrays that we have here not two worlds but two planes of one and the same world.

[11]At the same time, Wittgenstein's *Tractatus* can be also seen to maintain the existence of many possible worlds (cf. Bradley 1992)—without hierarchy between them, though.

about 'method' and 'use'.[12] However, the concept he employed remained the same. From this change in Wittgenstein's position, four further types of ways emerged in his ontology:

(vii) *Propositions' Meaning.* Around 1930, Wittgenstein started claiming that '*the meaning of a proposition is the way*$_{[7]}$ *it is verified*' (1930, 244). In other words, it is the way$_7$ in which we can apply it—or linguistically act with it. In his later work, this thesis was explicated in the doctrine that 'the meaning of a word is its use in the language' (1953, §43). The use here means nothing but the way$_7$ in which the word is applied, or in which we can linguistically act with its help.

(viii) *Language/Calculating Learning.* The change in the theoretical interest of the later Wittgenstein urged him to put at the centre of his attention the problem of language/calculus learning. He held that two persons, teacher and student, teach/learn a language sentence by sentence, by which one action (the speaking of the sentence by the pupil) models another action (the demonstration of the sentence-command by the teacher). To be more exact, the pupil is 'drilled' (*abgerichtet*) for this purpose. The same is also true about teaching and learning how to calculate.

Importantly enough, this kind of action-modelling proceeds in a form similar to that in which pictures model states of affairs in the *Tractatus*. The hinge between the two levels here is again the 'logical multiplicity'—this time, the logical multiplicity of the action of teaching and learning language/calculus.

Correspondingly, the action of the pupil, his/her getting command of language/calculus, can be seen as a *way*$_8$ of performing (replicating) the action of the teacher—the demonstrated command of the language/calculus taught.

(ix) *Mind.* In *Philosophical Investigations*, Wittgenstein held that the human mind (*Geist*) is only an instrumental concept—it is a way$_9$ of *doing* something. Following Descartes, we usually believe that the mind is a substance, but only because we think about the mind in physical terms. Concepts like 'understanding', 'meaning', etc., are thus typically interpreted as a kind of thin *stuff*. Wittgenstein's main point now was that 'where our language [misleadingly] suggests a body [a stuff ...] there is none: there, we should say, is a *mind* [*Geist*]' (1953, §36). To be more exact, the concept of mind denotes nothing but the way$_9$ in which we mentally *do* something: speak a language, calculate, etc.

(x) *Action.* Wittgenstein had already discussed the problem of connecting language with action, including the action of operating with language (action of language producing and conceiving) in the *Tractatus*, wherein he claimed that 'if a sign is *useless* [*wird nicht gebraucht*], it is meaningless' (3.328).[13] Only in the *Philosophical Investigations*, however, he clearly articulated the doctrine that we can correctly use language only in terms of actions.

---

[12]In fact, Wittgenstein already spoke about the application of the propositional sign as the 'method' of its application in the *Tractatus* (cf. §2.3, [iv]).

[13]The illustrious commentator of Wittgenstein's philosophy, B. F. McGuinness has correctly noticed that 'we ought not to contrast the *Tractatus*, with its notion of *Bedeutung*, and the *Philosophical Investigations*, with its notion that naming is the use. Use determines reference in the *Tractatus* also' (1981, 66).

The explicit change in Wittgenstein's understanding can be tracked to the following remark made in the 'Theses' by Friedrich Waismann:

> A propositional sign must have as many distinguishable elements as the corresponding situation. Both must have the same *multiplicity*.
>
> A very good illustration of that is provided by taking the proposition of language as the instruction for doing something. By using words, I can, for instance, guide somebody around my room by saying, 'Take three steps straight ahead! Now two to the left! Now stretch out your arm! Etc.' Hence, it is evident that language must have the same multiplicity as those movements. (1930, 236)

In *Philosophical Remarks*, Wittgenstein claimed that language has the multiplicity of a signal box that produces actions that correspond to different sentences of that language (1964, §10). This idea held a central place in *Philosophical Investigations,* wherein Wittgenstein compared (in §11) the different roles words play in our language with the different functions of the tools in a toolbox. Now, Wittgenstein maintained that an action, following an instruction (i.e. action that is not radically spontaneous and free), is nothing but a way_{10} of carrying out the instructions for acting.

* * *

It is time now to sum up the changes that Wittgenstein introduced in his conception of ways in the 1930s and 1940s. In these years, he concentrated his efforts on the different forms of connecting language/calculus action-instructions with language/calculus action-acquisition. He also explored the problems of the relation between mind and action in general. Importantly enough, in this second period of Wittgenstein's philosophical development, his ontology remained non-reductionist and non-emergentist. The world of the language/calculus instructor is not different in kind from the world of the instructed. They are just its different levels. The 'bridge' (the 'hinge') between the two world levels is the logical multiplicity of the mental actions.

2.6 Applications of Wittgenstein's Ontology of Ways in the Naturalistic Metaphysics

In this last section, we will demonstrate that Wittgenstein's ontology of ways can help treat difficult philosophical problems of the emergence of mind, science, and nature. Its main advantage is that it dispenses with ontologies of different worlds of increasing complexity without elapsing into reductionism.

(i) *Technology*. It is a common place that many technical devices employ the constructive principle of ways. Here is a trivial example, well known for more than 150 years. The telephone transforms—in a certain way_{11}—human speech into electrical impulses to transport them to another point in space. (The recorder transports them to another point in time.) In the next step, the electrical impulses are transformed again—following the same way$_{11}$ but activated in reverse—into human speech. In the latter phase, electrical impulses construct an entity of a higher level (but not of a higher world): human speech.

(ii) *Philosophy of Biology*. Similar considerations can be also adduced in the realm of the emergence of life. Life on Earth emerged as a result of the specific way$_{12}$ in which some of its elements (chemical substances) were ordered. Standard theories of the origin of life, like that of Oparin–Haldane, can be interpreted in this sense. No final cause, no Aristotelian ἐντελέχια, no Kantian teleology, no Bergsonian *élan vital*, no *vitalism* in Hans Driesch's sense, is needed for this purpose.

This approach puts more stress on the ontological side of the biological diversity than on the genesis of new species. To be sure, ontology is interested in the being, not in genealogy. Importantly enough, this remark is not a criticism of Darwin's theory of evolution. Even less is it a defence of the design argument in biology. It only underlines that Darwin's theory of natural selection, realized through the 'struggle for existence', is primarily genealogical (historical), not ontological. An alternative, ontological (structural) conception of biological evolution was suggested, for example, by Thompson (1917).[14]

(iii) *Anthropology* . Wittgenstein's ontology of ways can also help explain how human beings emerged out of the animal world. A crowd of creationists claims that a convincing answer to this question requires referring to God's design. In contrast, the ontology of ways we have exposed and defended in the lines above supports the view that human form of life is only another way$_{13}$ of arranging the building blocks of the basic forms of life of other species—more so since, as we have already seen in §3, (ii), (iii), and also in §5, (x), language, thought, and action can be conceived of as ways of arranging lower levels of being.

(iv) *Philosophy of Mind.* Wittgenstein's ontology of ways is especially effective in attempting to solve the problem of how the mind relates to the brain. Thereby, it dispenses with all 'occult processes' (1953, §38) that are usually associated with this relation. In contrast, the 'new mysterians', Thomas Nagel and Colin McGinn, maintain that human reason cannot understand this relation in principle (cf. McGinn 1991)—or, as Emil du Bois-Reymond put it more than a century ago, *ignoramus et ignorabimus*.[15]

In contrast, the consequential Wittgensteinians support the view that qualia depend ontologically on the ways$_{14}$ in which the brain cells and the connections between them are ordered.[16] Their opponents (cf. Davidson 1970; Putnam 1975), in contrast, maintain that even if two brains are identical, their psychological product will be different. In his theory of 'neutral monism', Donald Davidson, for example, claims that whereas mental events can be token-identical with the physical events, they are not

---

[14]See also Milkov (2002), Rescher (2011) and ch. 25 of this volume.

[15]'We don't know and will never know.'

[16]Arguments first formulated by Wittgenstein were successfully used against the ignorabimus principle in the philosophy of mind (cf. Bennett and Hacker 2013).

type-identical with them. In particular, propositional attitudes are not ontologically dependent on the brain.

In fact, however, what Davidson's argument really indicates is only that besides the brain events, a second physical factor interacts with the mental events: parts of the external world to which the mind is directed. In other words, mental events are ontologically dependent, in different proportions, both on the physical events of the brain, and on the events of the external world. There is nothing intrinsically 'anomalous' in this completely monistic conception of the relation between mind and matter.

(v) *Philosophical Cosmology.* Wittgenstein's ontology of ways also shakes hands with the latest discoveries in astrophysics. According to relativistic quantum theory, particles are to be understood as specific $ways_{15}$ of arrangement of quantum fields. Certain arrangements of the fields (particularly their arrangement in our universe) correspond to the idea of 24 particles, and others to 276 particles.

Taking off from there, some authors (Krauss 2012) conclude that the universe—the stuff of which is composed of 24 particles—is practically constructed from nothing. To be more exact, the stuff—the particles—are simply arrangements out of quantum fields. No energy is needed to this purpose. Correspondingly, the fact that 'there is something [stuff, particles] rather than nothing' is only due to the specific way_{15} of arranging quantum fields.

This conception was severely criticized by Albert (2012) with the argument that quantum fields are not 'nothing'. Be this as it may, this point is irrelevant to our ontological study. What is important to it is that the external world can be seen as being composed of the specific $ways_{15}$ in which the quantum fields are arranged.

(vi) *Style* . While the early Wittgenstein was above all interested in the contemplative side of art—in the art consuming (cf. §2.3, [vi])—the later Wittgenstein, who, as already seen, experienced an action-turn, paid more attention to art creating. Accordingly, in the mid-1930s, Wittgenstein started maintaining that 'style is the general necessity seen *sub specie aeterninatis*' (1997, 27).[17] Of course, Wittgenstein meant here the style in works of art, not in science where it is of less importance.[18] Indeed, since art is radically expressive, the style in it plays a key role. Propositions and theories of science and of everyday life are clearly different: they picture (model) parts of the world (states of affairs). That is why the style in them plays a minor role.

The above-mentioned claim of Wittgenstein can be well interpreted in the sense of his ontology of ways. According to this interpretation, what Wittgenstein meant here was that style is constitutive by creating works of art: It is the 'engine'[19] by creating the work of art and the 'rule' of its interpretation after it is produced. Most importantly, this rule can also be seen as the way_{15} in which the artefact is *constructed*

[17] See ch. 27.

[18] The fact that in these lines Wittgenstein spoke about necessity 'seen *sub specie aeterninatis*'—a term he used when he discussed works of art in the *Tractatus* (cf. §2.3 [vi])—clearly suggests that here he meant the style in works of art.

[19] We speak here about an 'engine'—which is a physical object—only in a metaphorical sense. As we have seen, Wittgenstein strongly opposed using physical terms when speaking about the human mind.

(not just *seen*, as it was with way_6). Different interpretations of the work of art are only its instantiations that follow its rule.

2.7 Epilogue

We have seen that Wittgenstein suggested a sophisticated ontological model (method) that explains the emergence of new levels of reality of increasing complexity. Over recent years, this topic has been extensively investigated using alternative methods (Clayton 2004; Deacon 2012; Kaufman 2000). At the end of our study, we would like to express our conviction that the theory we followed in it—the ontology of Wittgenstein's ways—is more promising and resourceful for this purpose.

References

Albert, David. 2012. 'On the Origin of Everything', *New York Times Book Review*, March 25, 2012.

Barker, Lynne. 2000. *Persons and Bodies: A Constitution View*, Cambridge: Cambridge University Press.

Beaney, Michael. 2016. 'The Analytic Revolution', *Royal Institute of Philosophy Supplement* 78:227-249.

Bennett, Max, and Hacker, Peter. 2003. *Philosophical Foundations of Neuroscience*, Oxford: Blackwell.

Berman, Scott. 2008. 'Universals: Ways or Things?', *Metaphysica* 9: 219–234.

Bradley, Raymond. 1992. *The Nature of All Being: A Study of Wittgenstein's Modal Atomism*, Oxford: Oxford University Press.

Clayton, Philip. 2004. *Mind & Emergence: From Quantum to Consciousness*. Oxford: Oxford University Press.

Davidson, Donald. 1970. 'Mental Events', in: idem, *Essays on Actions and Events*, Oxford: Oxford University Press, 1980, pp. 207–25.

Deacon, Terrence. 2012. *Incomplete Nature: How Mind Emerged from Matter*, New York: W. W. Norton.

Frege, Gottlob. 1892. 'On Sense and Reference', in: *Collected Papers on Mathematics, Logic, and Philosophy*, Oxford: Basil Blackwell, 1985, pp. 157–177.

Husserl, Edmund. 1900/1901. *Logische Untersuchungen*, 2 vols. Halle: Niemeyer.

Ingarden, Roman. 1964. *Time and Modes of Being*, trans. by Helen R. Michejda, Springfield, Illinois: Charles C. Thomas.

Kauffman. Stuart. 2000. *Investigations*, New York: Oxford University Press.

Krauss, Lawrence. 2012. *A Universe From Nothing: Why There Is Something Rather Than Nothing*, New York: Free Press.

Landini, Gregory. 2003. 'Wittgenstein's Tractarian Apprenticeship', *Russell* 23: 101–130.

Mayer, Verena. 2009. *Edmund Husserl*, München: Beck.

McGinn, Colin. 1991. *The Problem of Consciousness*, Oxford: Blackwell.

McGuinness, Brian. 1981. 'The So-called Realism of Wittgenstein's *Tractatus*', in: I. Block (ed.), *Perspectives on the Philosophy of Wittgenstein*, Oxford: Blackwell, pp. 60–73.

Milkov, Nikolay. 2002. 'Logical Forms of Biological Objects', *Analecta Husserliana* 77: 13–28.

Moore, G. E. 1922. *Philosophical Studies*, London: Kegan Paul.

Mulligan, Kevin. 1985: '"Wie die Sachen sich zueinander verhalten' inside and outside the *Tractatus*', *Teoria* 5:2, *Wittgenstein and Contemporary Philosophy*, eds. B. McGuinness and A. Gargani, Pisa: ETS.

Putnam, Hilary. 1975. 'The Meaning of "Meaning"', *Minnesota Studies in the Philosophy of Science*, 7: 131–193.

Rescher, Nicholas. 2011. *Productive Evolution: On Reconciling Evolution with Intelligent Design*, Frankfurt: Ontos.

Russell, Bertrand. 1912. 'What is Logic?', in: *The Collected Papers of Bertrand Russell*, vol. 6, London: Routledge, 1993, pp. 54–56.

Simons, Peter. 1994. 'Ways', *Kriterion* 4: 12–15.

Simons, Peter. 2014. 'Winnowing Wittgenstein: What's Worth Salving from the Wreck of the *Tractatus*', in: Anne Reboul (ed.), *Philosophical Papers Dedicated to Kevin Mulligan*, vol. 1, Berlin: Springer, pp. 407–21.

Thompson, D'Arcy Wentworth. 1917. *On Growth and Form*, Cambridge: Cambridge University Press.

Wittgenstein, Ludwig. 1922. *Tractatus logico-philosophicus*, London: Routldege, 1962.

Wittgenstein, Ludwig. 1930. *Ludwig Wittgenstein and the Vienna Circle*, ed. by B. F. McGuinness, trans. Joachim Schulte and B. F. McGuinness, Oxford: Basil Blackwell, 1979.

Wittgenstein, Ludwig. 1953. *Philosophical Investigations*, ed. by G. E. M. Anscombe and G. H. von Wright, Oxford: Basil Blackwell.

Wittgenstein, Ludwig. 1961. *Notebooks 1914–1916*, ed. by G. E. M. Anscombe and G. H. von Wright, Oxford: Basil Blackwell, 1979.

Wittgenstein, Ludwig. 1964. *Philosophical Remarks*, ed. by Rush Rhees, Oxford: Basil Blackwell.

Wittgenstein, Ludwig. 1997. *Geheime Tagebücher 1914–1916*, Hg. von Wilhelm Baum, Wien: Turia & Kant.

Zangwill, Nick. 2012. 'Constitution and Causation', *Metaphysica* 13: 1–6.

Chapter 3
Philosophy as Education in Thinking: Why Getting the Reader to Think Matters to Wittgenstein

Oskari Kuusela

Wittgenstein writes in the preface to the *Philosophical Investigations*: 'I should not like my writing to spare other people the trouble of thinking. But if possible, to stimulate someone to thoughts of his own.' In the following I argue that this indicates something essential about Wittgenstein's approach. In order to remain true to his conception of philosophy without theses (PI §§109, 128) he could not, instead, for example, aim to instruct his reader about grammar by means of theses or put forward prescriptions about grammar, logic or language use. Thus, there is an essential connection between aim of stimulating the reader to thoughts of their own, and philosophizing without theses.

Addressing this issue will require me to discuss certain features of Wittgenstein's philosophical approach that one might not immediately connect with the quoted statement. These include his rejection of philosophical foundations in a sense that implies a hierarchical organization of philosophy, such as assumed in the *Tractatus*. As I explain, this rejection of a hierarchical organization underlies Wittgenstein's conception of philosophy as addressing particular problems as opposed to solving a single fundamental problem whose solution would constitute the basis for the solution to all others. Another relevant issue is the notion of agreement in Wittgenstein's philosophy. In the final section, I take up certain remarks from Wittgenstein's *Nachlass* where he speaks about how the *Philosophical Investigations* ought to be written and comments on his way of writing. The points made in these remarks, which seem to have informed the composition of the *Investigations*, may be taken to lend further support to my suggestion that getting the reader to think is essential to Wittgenstein's approach. If so, Wittgenstein's philosophy can be described as education in thinking, as opposed to philosophy as establishing philosophical theses. Relatedly his book can be seen a manual for developing one's thinking.

O. Kuusela (✉)
School of Politics, Philosophy, Language and Communication Studies,
University of East Anglia, Norwich Research Park, Norwich, Norfolk NR4 7TJ, UK
e-mail: o.kuusela@uea.ac.uk

S. Wuppuluri and N. da Costa (eds.), *WITTGENSTEINIAN (adj.)*,
The Frontiers Collection, https://doi.org/10.1007/978-3-030-27569-3_3

I begin from the *Tractatus'* attempt to abandon philosophical theses and how this ultimately fails. As I explain, this failure informs in important ways Wittgenstein's later view of philosophy without theses.

3.1 The *Tractatus'* Rejection of Philosophical Theses and Logical Prescriptions

Regarding the view that the task of philosophy, logic or grammatical investigation is to prescribe how we ought to think, Wittgenstein rejected this view of philosophy and logic already in the *Tractatus*. In this respect his early philosophy of logic stands in contrast with, for example, Frege's view of logic as a normative science and his characterization of the laws of logic as prescriptions for making judgments (PW, 128). As Frege explains his view regarding the principles of logic as the laws of truth, 'We can also think of these as prescriptions for making judgments; we must comply with them in our judgments, if we are not to fail of the truth.' (PW, 145) Similarly Wittgenstein's early philosophy of logic stands in contrast with Russell's view that the task of the theory of types is to prescribe how we can use our symbols. As Russell explains in a letter to Wittgenstein after reading the manuscript of the *Tractatus*: '3.331 The theory of types, in my view, is a theory of correct symbolism: (a) a simple symbol must not be used to express anything complex; (b) more generally, as symbol must have the same structure as its meaning' (CL, 122; cf. Russell 2010, xli). To this Wittgenstein replies: 'That's exactly what one can't say. You cannot prescribe to a symbol what it may be used to express. All that a symbol CAN express, it MAY express. This is a short answer but it is true!' (CL, 125; cf. TLP 3.33–3.334). The point that logic does not prescribe is then also made in the *Tractatus*: 'Logic must take care of itself. \\ A *possible* sign must also be able to signify. Everything that is possible in logic is also permitted. [...] In a certain sense, we cannot make mistakes in logic' (TLP 5.473; cf. 5.473–5.4733). In the pre-Tractarian *Notebooks* this principle is characterised as 'an extremely profound and important insight', when first introduced (MS 101, 13r/NB, 2).

As regards the early Wittgenstein's reasons for rejecting philosophical theses, they are intimately connected with, even if not exhausted by, the point that logic takes care of itself, and that there are no prescriptions in logic. To outline this connection, on the *Tractatus'* view a comprehension of the principles of logic is always already assumed when thinking or speaking (This is the also the sense in which logic is a priori, according to the *Tractatus*. TLP 5.4731). Being a speaker or a thinker involves a comprehension of the principles of logic, including a comprehension of what counts as a proposition as opposed to a nonsensical string of signs. This is not something that logicians could inform speakers or thinkers about, and then take as the basis of logical prescriptions, as one can be informed of scientific findings and use them to justify medical prescriptions, for example. What counts as nonsense, and what follows from a proposition depends on the propositions themselves, not on the rules of logic laid out by logicians (TLP 5.11–5.12, 5.131–5.132). There is therefore nothing to prescribe. Instead, the task of the logician is to clarify what speakers already know by virtue of being speakers. 'Logic takes care of itself; all we have to do is to look and see how it does it' (MS 101, 39r/NB, 11; cf. 43).

Of course the view that we cannot 'in a certain sense' make mistakes in logic does not mean that speakers or thinkers would never get confused about logic. Indeed, Wittgenstein describes philosophy as being full of such logical confusions, that is, confusions relating to the modes of signification or logical function of our words (TLP 3.323). There is thus a need for logic as a clarificatory or explicatory discipline which clarifies or explicates to speakers what they already know implicitly in the capacity of thinkers or language users. This is how Wittgenstein then characterizes the role of logic or the kind of logical philosophy that he inherits from Frege and Russell, and remodels in the *Tractatus*. Philosophy is an activity of clarification and its results are not philosophical propositions (TLP 4.112, 6.53).

Besides the point that there is no need for theses as the basis of logical prescriptions, another reason for Wittgenstein's rejection of theses in the *Tractatus* is the failure of theses to give proper expression to logical exceptionless necessity and their failure as expressions of essential as opposed to accidental generality. This is exemplified by how theses about the essence or general form of propositions, such as the *Tractatus* seems to put forward (4.5, 5, 5.471ff.), would ultimately leave it unclear whether the thesis (that all propositions have such and such a form) really holds for every possible proposition. When presented with such a thesis an interlocutor can reasonably wonder whether the thesis really is true, that is, whether it really covers every possible proposition. Moreover, even if every possible proposition did have this form, could this not be accidental? That a reader of the *Tractatus* may reasonably raise questions about both issues indicates that the matter has not been fully clarified. As theses always seem to leave open the possibility of such questions, exceptionless necessity and generality cannot be properly clarified by means of theses.

The preceding is a very important consideration, I believe, behind Wittgenstein's accepting, and developing further in a revised form, Russell's idea that the proper way to express logical exceptionless generality and necessity is a variable. 'The expression for a formal concept is a propositional variable in which this distinctive feature alone is constant.' (TLP 4.126; cf. 4.127–4.1272; for the relation between Wittgenstein and Russell, see Kuusela, forthcoming.) This is also how Wittgenstein characterizes the notion of general propositional form. 'The general propositional form is a variable' (TLP 4.53). In the correct logical notation, such as the *Tractatus* seeks to outline, the essence of propositions (their having a general form) would then be clarified by making it plain that there is no other way to construct a proposition except as having the general propositional form, i.e. as a substitution instance of the propositional variable. In this way this notation is meant to make evident that every possible proposition does indeed possess the general propositional form. There simply is no other way to construe a proposition in this notation. Thus, if we did accept Wittgenstein's notation as the correct one, it would make the point about all propositions sharing a general form as obvious as it obvious to anyone possessing the concept of a triangle that triangles always have three angles. (Since the general propositional form is meant to be a rule for the use of signs, that is, a rule for the construction of propositions, the comparison can be taken quite literally.) Accordingly, we find Wittgenstein making this comment in 1929: ‘The notation is the last expression of a philosophical view.’ (MS 105, 12; for a detailed discussion of this remark in connection with the *Tractatus*, see Kuusela 2011, 2019.)

3.2 The *Tractatus’* Relapse to Theses and Prescriptions

The preceding sums up some key considerations behind Wittgenstein’s rejection of philosophical theses in the *Tractatus*—or his attempt to do so. For in the end the attempt fails. However, as regards this very interesting attempt to develop a philosophy without theses, the preceding puts us in a position to see that his rejection of theses is not a consequence of the *Tractatus’* so-called picture theory. There is no need to think so, contrary to how interpreters have traditionally read the *Tractatus* (for example Hacker 1986). Problematically, interpretations along these lines lead to the infamous paradox of the *Tractatus*: if the book contains a theory or an argument it cannot be nonsense; if it is nonsense it cannot contain a theory or an argument. This, it seems, makes it impossible to understand what Wittgenstein wanted to achieve with his book, incurring in this sense a high cost on the traditional readings. (They now need to explain how nonsensical propositions or unthinkable thoughts can express or convey logical insights. No good answer has been given to this question, however, and this makes the traditional readings quite uncharitable to Wittgenstein.) It is also noteworthy that Wittgenstein himself never mentions a paradox when speaking about problems with his early philosophy. He explains its failure quite differently. (I discuss the *Tractatus’* failure in detail in Kuusela 2008a, Chap. 3.2, and will only outline some key points below insofar as they are relevant to the present discussion. For an explanation of how the alleged paradox is dissolved, or why there is no paradox in the *Tractatus*, see Kuusela 2019a.)

Note also that if Wittgenstein’s rejection of theses were based on his picture theory, this would constitute a dogmatic argumentation strategy against Frege’s and Russell’s philosophies of logic according to which logic is a science, i.e. a discipline that puts forward true theses or propositions about logic, and whose basis are axioms considered to be self-evident truths. The argumentation strategy would be dogmatic in the sense in which Kant speaks about dogmatism: it would be a matter of arguing against Frege’s and Russell’s accounts of logic from within Wittgenstein’s own theory of propositions which neither target of his criticism subscribes to. Instead, on the account just outlined we can understand Wittgenstein as criticizing Frege and Russell on the basis of tensions within their accounts, that is, from within their accounts of logic. (This is what Kant recommends as a non-dogmatic way of proceeding. See Kuusela 2008b for discussion.) A relevant kind of tension is evident, for example, in Russell’s view that comprehension of logical forms is always part of understanding discourse *and* that logic is a science that informs us about logical truths, or that logic results in knowledge (true/false knowledge claims) about logic. As Russell writes, '[…] some kind of knowledge of logical forms, though with most people it is not explicit, is involved in all understanding of discourse. It is the business of philosophical logic to extract this knowledge from its concrete integuments, and to render it explicit and pure' (OKEW, 53). Clearly, Russell cannot consistently maintain both that comprehension of logic is already assumed in understanding any discourse and that the science of logic informs speakers about the principles of logic in the way in which science informs us about truths we did not know.[1] As the tension within Russell's philosophy of logic might be expressed, logic cannot both be an object of knowledge and something already assumed by the possibility of comprehending knowledge claims. (This emphasizes the importance of the idea that logic takes care of itself for the early Wittgenstein. For a detailed discussion, see Kuusela 2019, Chap. 2.)

Wittgenstein's response to Frege and Russell in the *Tractatus* is therefore not meant to be prescriptive in the sense of an attempt to make them accept prescriptions about logic based on his theory of propositions, contrary to what the prescription looking propositions of the *Tractatus* might suggest, according to which logic, logical forms or formal concepts cannot be the object of propositions (TLP 4.12ff.).[2] Rather, Wittgenstein's intention in the *Tractatus*, as I read him, is merely to introduce a logical method (a method of logical analysis) to be used in philosophy that will enable us to philosophize without making any statements about logical necessity and possibility. Instead of putting forward theses and prescriptions based on them, philosophical analysis clarifies the logical features of what is said with the help of a logically perspicuous notation. This is a matter of translating relevant statements into the notation which makes clear the logical features of the expressions in question, and clarifies what they show in the sense of the *Tractatus'* distinction between saying and showing (TLP 4.122ff., 6.53).

The preceding makes a long story very short. Arguably, however, the gist of the *Tractatus* is to introduce an improved version of a Fregean-Russellian concept-script (logical language or notation) which provides us with a framework for carrying out logical analyses, and thus puts us in a position to philosophize without making any statements about exceptionless necessity, avoiding the problems with such statements outlined in Sect. 1. (TLP 3.323-325; Kuusela 2019 provides a proper exegetical argument.) The aspirations of the *Tractatus* are therefore methodological. Firstly, rather than seeking to lay down a foundation for logical analysis or a formal philosophy by means of theses, when introducing his notation Wittgenstein only intends to rely on the readers' pre-theoretical comprehension of logic which they have as virtue of being speakers or thinkers. Secondly, rather than any alleged Tractarian theses, the proper expression for Wittgenstein's logical insights is the notation which the Tractarian sentences introduce; his logical insights are encoded into the structure or

---

[1] Although Russell does not explicitly speak about informing anyone in the quote, Russellian philosophical logic is meant to result in true propositions regarding logical forms, thus placing the statements of logic in the same category as those of science (OKEW, 66–67).

[2] For an account of the function of the sentences of the *Tractatus* as intended to introduce logical concepts and principles constitutive of a logical notation, see Kuusela 2019, Chaps. 2 and 3.

rules of this notation. To emphasize, to introduce a notation is not the same as putting forward theses. By introducing a logical calculus (its key concepts and principles) one has not yet made a true/false claim about anything.

Ultimately the *Tractatus* fails to reach this goal of philosophy devoid of theses, however. This failure is due to Wittgenstein's assumption that his logical-philosophical method would put us in a position to solve any possible philosophical problem or that, when this method has been introduced, all philosophical problems have been solved 'in essentials', as he says in the book's Preface. (Given that the *Tractatus* does not discuss all possible philosophical problems, Wittgenstein's point must be that they have been solved in outline through the introduction of Wittgenstein's method.) Importantly, however, not only does Wittgenstein commit himself to a thesis about the nature of philosophy and philosophical problems through this programmatic thesis about philosophical method. The method also commits him to a thesis about language. For the method introduced in the *Tractatus* assumes that all sensible propositions are analysable into truth-functions of elementary propositions which in turn consist of logically simple names. On the proposed reading Wittgenstein's so-called picture theory of propositions or language thus is partly constitutive of Wittgenstein's framework for the logical analysis of propositions. Although his purpose, as such, was not to put forward any theses of claims about language, nevertheless, similarly to Frege's and Russell's logical languages, Wittgenstein's logical language is built around the notion of a true/false judgment or proposition which constitutes its core. In this way a thesis about the nature of language as a totality of propositions that constitute true/false pictures or representations entered the *Tractatus* underhandedly. A metaphysical thesis about nature of language, and of what all propositions must be, hid itself in Wittgenstein's claim about the method of philosophy.

We are now in a position to see how philosophy assumes a hierarchical structure in the *Tractatus*. Underlying Wittgenstein's approach there is a fundamental question about philosophical method and language whose answer is meant to constitute the basis for solving all other problems relating to particular philosophical issues. For, in order for the Tractarian method of logical analysis to make possible the resolution of any philosophical problems arising from logical or linguistic unclarities, this method must presumably be underpinned by a correct account of proposition, judgement or language. Only if the view of language which the method of analysis involves can accommodate all possible forms of sensible language use, can its applicability be guaranteed in all instances of philosophical clarification. In this way the question of the essence of language then constitutes *the* fundamental problem of philosophy for the early Wittgenstein. Accordingly, the *Tractatus* can be seen as an attempt to dissolve all problems of philosophy by solving what Wittgenstein perceives as *the* fundamental problem regarding the essence of language. As he remarks in his pre-*Tractatus* notebooks: 'My *whole* task consists in explaining/clarifying [erklären] the nature of the proposition' (MS 102, 63r-64r/NB, 39).

3.3 Wittgenstein's Rejection of a Hierarchical Organization of Philosophy

The later Wittgenstein questions the hierarchical organization of philosophy assumed in the *Tractatus*, coming to regard it as constituting a vulnerability for philosophy. According to Wittgenstein, this hierarchical way of structuring philosophy prevents philosophy from reaching its goal, which described as, 'Thoughts at peace. This is the goal that someone who philosophizes longs for.' (CV, 50/MS 127, 41v; 1944) Relevant points are expressed in the *Investigations* as follows:

> The real discovery is the one that makes me capable of stopping doing philosophy when I want to.–The one that gives philosophy peace, so that it is no longer tormented by questions which bring *itself* in question.–Instead, we now demonstrate a method by examples, and the series of examples can be broken off.–Problems are solved (difficulties eliminated), not a *single* problem. (PI §133)

This condensed remark requires unpacking. A key point is that a hierarchical organization, such as assumed in the *Tractatus,* exposes philosophy to criticisms and attacks directed at its foundations. Given how the *Tractatus'* approach makes the solutions to any particular philosophical problem dependent on the solution to the fundamental problem concerning philosophical method and, relating to this, the nature of language, all solutions to particular problems can be questioned by questioning the solution to the fundamental problem. In other words, should the *Tractatus'* method of analysis reveal itself as problematic—for example, due to involving a simplistic conception of language, as Wittgenstein later believes—all results achieved by means of this method emerge as potentially problematic too. Thus, as he says in §133, philosophy is at risk of being 'tormented by questions which bring *itself* in question' (see Kuusela 2008a, Chap. 1.5 for discussion).

Wittgenstein's later response to this problem is to reconceive philosophy in such a way that solutions to particular problems no longer depend on the solution to a fundamental problem. Consequently, philosophy can be understood as genuinely concerned with particular questions or problems, rather than with a single fundamental question or problem that is assumed to contain a solution to them all. As §133 states: 'Problems are solved (difficulties eliminated), not a *single* problem.' Helpfully, the point is stated with a few additional explanatory words in the *Big Typescript* (1932–33), where Wittgenstein uses an early version of §133 to begin a new subchapter entitled 'The methods of philosophy. Possibility of peaceful progress.' (TS 213, 431) Here the possibility of peaceful progress, finding peace in philosophy, and being able to stop when one wants to, is explicitly explained in terms of a transition from addressing a single problem to addressing a plurality of problems. As Wittgenstein writes, clarifying the first three sentences that subsequently form the beginning of §133: 'But more correctly one would say: Problems are solved (disquietudes//difficulties eliminated), not a single problem' (TS 213, 431).

The later Wittgenstein therefore rejects philosophical foundations in the sense of an abstract general justification concerning the philosophical method, achieved by solving a fundamental problem. An important point here is that such an abstract justification is possible only, if an overarching characterization of the nature of philosophical problems and their solution can be provided. Only if this is so, can it be claimed that a certain method makes it possible to solve all philosophical problems. The later Wittgenstein, however, abandons the assumption that philosophical problems could be delimited in such a way without running the risk of encountering problems to which the method claimed to be able to solve all problems is not applicable. Rather, a variety of different methods may be required to address different philosophical problems. For, even if philosophy is broadly conceived as a logical or conceptual clarification, such problems may relate to different types of concepts that require different treatment, and cannot all be clarified by means of one and the same method. For example, it seems that the concept of freedom cannot be helpfully clarified through an analysis in terms of a logical calculus. Instead, it seems better approached by means of Wittgenstein's later method of language-games which involves clarifying the roles that the concept plays in human life. (See 'Lectures on the Freedom of the Will'; for the method of language-games, see Kuusela 2019, Chap. 5.) Or as Wittgenstein remarks in connection with problems relating to colours: 'In philosophy we must always ask: "How must we look at this problem in order for it to become solvable?"' (RC II, §11).

As explained, the *Tractatus'* clarifications regarding the logical principles governing thought and language were not intended as a theory but only as reminders concerning what speakers and thinkers already know. Nevertheless, when combined with Wittgenstein's claim that his book has solved philosophical problems 'in essentials', his clarifications assume the role of a thesis that provides a foundation for all subsequent philosophical clarifications (see Sect. 2). In this sense theses have not been left behind. By contrast, in the *Investigations* Wittgenstein seeks to introduce his methods in a different way. Instead of being introduced by means of a foundational thesis about method and/or language, '[…] we now demonstrate a method by examples […]' (PI §133). Accordingly, a variety of actual philosophical problems that are by no means mere toy-examples are discussed in the *Investigations*, with Wittgenstein stopping to explicitly comment on his method here and there.

More specifically, from the point of view of the later Wittgenstein, philosophical methods and clarifications in their terms are to be justified with reference to their capacity to resolve philosophical problems or to contribute to their resolution. As he also notes, in order for something to count as a clarification it must actually clarify (MS 123, 18r; quoted in Sect. 4). The introduction of methods through examples is therefore not *merely advantageous* over Wittgenstein's early method in that it enables the reader to better evaluate whether the proposed methods really enables one to solve philosophical problems. Rather, insofar as clarificatory capacity can only be determined with reference to actual problems, this way of demonstrating methods is *essential* for their justification. As the point might be put, methods cannot be introduced and justified abstractly or hypothetically by means of a general argument, as if by outlining solutions without actually spelling them out. There can be no solution to a philosophical problem from a distance, so to speak, before addressing relevant issues in required detail, and applying the methods to actual cases. Here it can be no substitute to postulate ideal cases to which a method is applicable as illustration, because they do not constitute examples of the solution any actual problems. Wittgenstein writes relatedly in draft version of the *Investigations* (1936) about the Tractarian calculus and its employment:

> It is of the greatest importance that we always think of examples to which a calculus of logic is really applicable; that we don't give examples and say: these are not the ideal ones of which the calculus is really valid which we don't have yet. This is a sign of the wrong conception. If I can employ the calculus at all, this is the ideal application and the application that matters. […] The mistake lies in promising an employment for it in a foggy distance. (TS 220, 84; cf. MS 111, 118; MS 115, 55–56; MS 157b, 13r)

Problems pertaining to the possibility of giving an abstract justification for a philosophical method thus constitute the main reason for Wittgenstein's demonstration of his methods by examples. A related notable feature of his way of introducing methods is the open-endedness of such demonstrations. A series of examples can in principle be continued in more than one way, and accordingly a set of methods introduced in this way will be open to further development and modification. In this sense the later Wittgenstein is no longer committed to any claims or theses about *the* correct method of philosophy. His later way of introducing his methods does not involve a philosophical thesis about the nature of philosophical problems or philosophical methods which is put forward as exclusively true. (Neither does it involve a thesis about the nature of language, as I explain presently.) In this regard it is important that §133 presents demonstration by examples as an *alternative* to justifying a philosophical approach by means of a foundational thesis. This is indicated by the word 'Instead' in §133 which contrasts demonstration by examples with being concerned with a single fundamental problem.

The sense in which Wittgenstein's later methods do not involve any foundational theses about language can be explained with reference to his rejection of what he calls 'metalogic' in the early 1930s. Although Wittgenstein does not later use this locution, he retains the key point expressed in its terms, which is put in somewhat different terms in *Investigations* §121 (see below).[3] To use the *Tractatus* as a foil again, its failure to abandon philosophical theses can be described by saying that its method of analysis hid within itself a metaphysical thesis regarding the essence of language, a modal claim about what all propositions *must* be. Thus, metaphysics disguised itself as methodology in the *Tractatus*. By contrast, the later Wittgenstein seeks to chase metaphysics out of this last hiding place. He writes (in 1931):

> Why do we perceive the investigation of grammar as fundamental?
>
> The word "fundamental" cannot mean anything metalogical or philosophical, wherever it has any meaning at all.
>
> We chase metaphysics out of all its hiding places.
>
> The investigation of grammar is fundamental in the same sense as we can call language fundamental – roughly, its own foundation. (MS 110, 194/TS 213, 412–413/TS 211, 245;

[3]Possibly the reason why Wittgenstein changed his terminology was simply that around this time, in the early 1930s, Carnap and other logically oriented philosophers started using the term 'metalogic' in a different way, as referring to concepts and principles articulated in a metalanguage. However, I am not aware of any comments by Wittgenstein on this issue.

the third paragraph occurs in this place only in MS 110, but is included elsewhere in 211 and revisions for 213.)

By 'metalogic' Wittgenstein means here considerations that cannot be articulated from within a logical system or a method, such as the Tractarian method of analysis or Wittgenstein's later clarifications in terms of grammatical rules. Rather, metalogical considerations pertain to something that the possibility of the system or method seems to presuppose. (I return to the fundamentality of grammatical investigation shortly.) In this sense he rejects, more specifically, accounts on which concepts such as picturing, understanding, meaning, agreement with reality, rule, or game would be metalogical concepts or, as he later calls them, super-concepts (*Über-Begriffe*). (MS 110, 189, MS 114, 2, 15, 27, 83; MS 115, 85, MS 116, 2, 16, 21) Rather, Wittgenstein emphasizes, relevant words have an ordinary non-metaphysical use. As he puts the point in the *Investigations*, commenting on the fundamental role of concepts such as world, proposition and language in the *Tractatus*, or that there would be 'a super-order between—so to speak—super-concepts', '[…] if the words "language", "experience", "world" have a use, it must be as humble as that of the words "table", "lamp", "door".' (PI §97) Thus, concepts that might seem to have a metalogical significance, or to function as super-concepts with foundational significance, are concepts among others, not metaphysically more fundamental. Such concepts, or any corresponding philosophical accounts of meaning, understanding, rules, and so on, do not establish a foundation for philosophy.

In this sense Wittgenstein's later method of the description of language use by means of grammatical rules, for example, does not presuppose a foundational thesis or rest on a 'super-rule' about language use as rule-governed. While this method does indeed assume that uses of language can be described by means of rules, or that language can be compared with games according to rules (see PI §§54, 81–83), this method supports itself, and justifies itself by its clarificatory capacity. (See Kuusela 2008a for further discussion, and 2019, Chap. 5.5 for the status of descriptions in terms of rules.) Accordingly, Wittgenstein rejects the idea that the method of describing language use by means of rules would require him to provide a definition of the notion of a rule, committing him in this sense to 'super-rules' or to a 'metaphilosophy' regarding the notion of a rule or the rule-governedness of language (MS 114, 104/PG, 115–116). He refers to this as something that could be presented as a leading principle of his philosophy:

> […] those rules are not super-rules. But if [philosophy] was really concerned with the concept of calculus, and thus with the concept of the calculus of all calculi, there would be such a thing as metaphilosophy. (But there is not. We might so present all that we have to say that this would appear as a leading principle.) (MS 114, 104/PG, 115–116; my square brackets)

A grammatical investigation therefore constitutes its own foundation, justifying itself by its clarificatory capacity. Although any notion relevant to the investigation can be clarified, such clarifications do not constitute a hierarchy, whereby a 'super-clarification' provides a foundation for all the rest of clarifications. A related remark regarding philosophical hierarchies, which connects this issue with the possibility of stopping philosophy without everything being brought into question, similarly to §133, enables us to connect the preceding points with *Investigations* §121:

> Thoughts are to be arranged in such a way that the investigation can be interrupted at any point without the sequel being able to put into question what was said up to that point. Here we come again to the thought that spelling the word 'spelling' is not higher order spelling. (MS 163, 40v-40r)

In the *Investigations* Wittgenstein's rejection of metaphilosophy is explained in terms of this analogy with spelling: 'One might think: if philosophy speaks of the use of the word "philosophy" there must be a second-order philosophy. But that's not the way it is; it is, rather, like the case of orthography, which deals with the word "orthography" among others without then being second-order.' (PI §121) In other words, like questions relating to the spelling of the word 'spelling' do not play a special fundamental role in orthography, but the question of the spelling of this word is like any other, similarly it is not the case in philosophy that certain notions constitute its foundation, with determinations of relevant notions making up a metaphilosophy.

Accordingly, philosophical clarifications in Wittgenstein's the later sense no longer presuppose any single overarching conception of language assumed as a foundation or framework. Instead, the later Wittgenstein makes use of a variety of conceptions of language connected with his methods of clarification: The two most important ones, described in general terms, are: (1) The conception of language use as a rule-governed activity, a game according to rules or a calculus according to simple and exact rules. This conception is connected with the method of clarification by means of grammatical rules and the variants of the language-game method that regard language as a game according to rules. (2) Language as embedded in non-linguistic activities, for example, as an extension of natural pre-linguistic behaviour (such as pain behaviour). This conception is connected with quasi-anthropological variants of the language-game method that envisage language as arising from non-linguistic modes of action and behaviour, and as in natural historical facts about humans and their environment.[4]

The outlined non-hierarchical structure then puts Wittgenstein into a position to avoid the problem with his early philosophy of a metaphysics having disguised itself as methodology. Structured in the outlined way, Wittgenstein's later philosophy is not exposed to the problem of attacks directed to foundations. Because solutions to particular problems no longer rest on a solution to a single fundamental problem, they cannot be undermined by problems with this solution either. As explained, philosophical clarifications and the methods employed are justified insofar as they can actually clarify whatever they purport to clarify. Consequently, philosophy is released from questions that could bring philosophy *itself* into question. In this way Wittgenstein's shift from addressing *the* fundamental problem to particular problems

[4]These two conceptions of language correspond to the senses in which language is, according to Wittgenstein, both arbitrary and non-arbitrary, as explained in PI §492. Different variants of Wittgenstein's methods are discussed in Kuusela (2019, Chaps. 4–6).

makes possible finding peace and a peaceful progress in philosophy. This transformation of the approach is then what §133 calls 'the real discovery'. Now it is possible to stop philosophizing when one wants to, because there is no foundational thesis that one must defend indefinitely due to the worry that if it collapses, every particular result achieved by its means collapses too. It is now possible to deal with particular problems, and then rest, making peaceful, piecemeal progress.

Next, let us look more closely at what Wittgenstein means by addressing particular problems and the clarification of such problems. This brings us to the notion of agreement in Wittgenstein's philosophy, the last issue that needs to be discussed before I am in a position to answer the question concerning the sense in which it is essential to Wittgenstein's philosophical approach to get the reader to think for themselves.

3.4 The Notion of Agreement in Wittgenstein

On Wittgenstein's view, whether a statement counts as clarification depends ultimately on its clarificatory effect. As he explains: 'An explanation of words has clarificatory value for the person to whom it clarifies something, upon whom it has a clarifying effect. Independently of *that* it is not an explanation.' (MS 123, 18r; 1940) A clarification therefore counts as such only insofar as it actually clarifies something; what does not clarify, despite intentions to do so, is only an attempt at clarification. This has important implications for what can qualify as a philosophical statement, provided that the goal of philosophy is clarification.

Wittgenstein remarks on this in the *Investigations:* 'If one tried to advance *theses* in philosophy, it would never be possible to debate them, because everyone would agree to them.' (PI §128/MS 110, 259; cf. WVC, 183) This can be understood as follows. Insofar as a philosophical clarification must be agreed upon as clarifying something, theses in the sense of controversial true/false statements do not qualify as philosophy. Being controversial, they are not statements agreed upon as clarifying. Accordingly, as Wittgenstein remarks in the same manuscript where §128 is first drafted: 'only when I say what is self-evident is it philosophy.' (MS 110, 214/TS 211, 256; 1931; a similar point is made in MS 136, 41b from 1948) Numerous statements can also be found in Wittgenstein's lectures according to which he is willing put to the side anything not be agreed upon. For instance: 'I won't say anything that anyone can dispute. Or if anyone does dispute it, I will let that point drop and pass on to say something else' (LFM 22; see 55, 102, 103; AWL, 97).

Wittgenstein therefore clearly does not intend to force his interlocutor to agree on anything that she feels she cannot agree upon. In this sense his later approach to philosophy leaves no room for prescriptions, not even only underhandedly as in the *Tractatus*. But can one really take seriously the claim that Wittgenstein would drop any point not agreed upon? I believe one can, but this requires us to distinguish Wittgensteinian clarificatory statements from philosophical, logical or grammatical theses, and comprehending the former as devices of clarification. More specifically, by such devices of clarification I mean models to be used as objects of comparison, as Wittgenstein explains the point in *Investigations* §§130–131. Models in this sense are modes of representing the objects of investigation with which their actual instances are compared for both similarity but and difference, rather than asserting that all relevant cases must fit the model.

Wittgenstein explains this point with reference to simple language-games as models for language use: 'Our clear and simple language-games are not preliminary studies for a future regimentation of language […]. Rather the language-games stand there as *objects of comparison* which, through similarities and dissimilarities, are meant to throw light on features of our language.' (PI §130) And: 'For we can avoid unfairness or vacuity in our assertions only by presenting the model as what it is, as an object of comparison – as a sort of yardstick; not as a preconception to which reality *must* correspond. […]' (PI §131) His point can be explained by saying that we ought to see our philosophical accounts as modes of representing reality, including actual uses of language, by means of which we seek to render comprehensible the targets of our study, and bring order into relevant facts for the purpose of clarification (PI §132). But we should not forget what our models are. They are just that, models developed by us with the purpose of understanding how things are or how our concepts function. Here we must then be careful, as Wittgenstein keeps warning, not to falsify reality by means of our models. As he also remarks, a philosophical model '[…] is supposed to be a picture with which we compare reality, through which we represent how things are. Not a picture by which we falsify reality' (MS 183, 163–164/PPO 97; from 1937; see Kuusela 2008a, 121–122). We run the risk of falsification, however, if we expect our objects of investigation to neatly correspond to our models and project those models on to reality, or onto our interlocutors' thinking, in the capacity of claims or theses about necessities pertaining to her concepts or thinking.

By contrast, as long as grammatical remarks and reminders are conceived on the model of statements of 'grammatical truths' comparable to philosophical theses as the statement of truths (such as those that in Frege and Russell serve as the basis of their prescriptions) a feeling will persists that Wittgenstein's statement about dropping anything not agreed upon is merely rhetorical and less than honest. Secretly he knows, even if he does not say it but pretends the opposite, that insofar as the interlocutor uses language in the usual way, she will *have* to acknowledge, on pain of inconsistency, the grammatical rules that Wittgenstein articulates for the purpose of clarifying those concepts or relevant uses of language. (See Glock 1991, 79, 84–86 for a latter kind of interpretation.)

Let me briefly illustrate the difficulty that arises here of making sense of what Wittgenstein says. Similarly to Glock, Peter Hacker writes about Wittgenstein's views such as the conception of an inner process as standing in need of outer criteria and the conception of mathematical propositions as rules: 'If challenged, they can be withdrawn, and painstaking description of the use of words resumed until it becomes clear that the way we use psychological predicates in the third person involves behavioural criteria for their application, and that the characteristic employment of mathematical propositions is normative' (Hacker 2001, 334). According to this interpretation, what needs to be agreed upon, is already known by the clarifier in advance of the discussion.

The discussion will continue until agreement about the grammatical clarifications has been reached—or the attempt to clarify matters is given up.

We can now connect this with the issue of stimulating the reader into thinking. If the preceding Glock-Hacker view were Wittgenstein's, getting the reader or interlocutor to think for themselves would not be essential to Wittgenstein's philosophy. Rather, on the Glock-Hacker view we can know what the right clarification is quite independently of the interlocutor coming to agree that what is offered as a clarification actually clarifies matters. By contrast, Wittgenstein insists in his lectures that he has no right to make such 'clarificatory claims' (compare such claims with what §128 above says about philosophical theses). As he remarks in his lectures about his view of mathematical propositions as rules of grammar: 'I have no right to want you to say that mathematical propositions are rules of grammar' (LFM, 55). Wittgenstein does indeed propose, in this context, to investigate the matter with the interlocutor. But unlike Hacker, Wittgenstein does not suggest that the outcome of the investigation, and what is to be said about the matter, is already known. Rather, in order for the clarification to succeed it is essential to get the interlocutor to think for themselves, and hopefully to agree to clarifications offered. Part of the issue, as I will explain, is that the clarification must really address what the interlocutor—not Wittgenstein or other philosophers—finds problematic about the matter at hand.

Here it is important that from Wittgenstein's point of view, dropping a grammatical statement is not the same as abandoning a 'grammatical truth'. Rather, dropping grammatical statements is better understood in analogy with exchanging a tool for another better suited for the particular job at hand. An important point here is that what clarificatory accounts or devices one ought to employ in tackling a problem depends on what exactly those to whom the clarification is offered find problematic about the issue. This also means that in some cases one might not be able to employ certain clarificatory devices, because one's interlocutors regard them as problematic. In such a case, the clarifier then has to find other means to address the problem. In this sense, when selecting instruments for clarification, one must take into account the interlocutors' philosophical views that provide a background for the problem and constitute a context in which it is to be resolved. Readiness to drop clarificatory statements, if they are not agreed upon, then simply means being ready to respond to a clarificatory task as required by its specific circumstances. This is not the same as holding no views about grammar or the matters under investigation, or having to give up an account because it cannot clarify matters to someone else.

The issue can be further elucidated as follows. What needs to be said about a philosophical issue is whatever is relevant from the point of view of solving the particular problems that one is concerned with, and different things may be relevant from the point of view of different problems, even if those problems relate to one and the same object of investigation. For example, even if it may be helpful in some contexts to regard language as used according to rules, this does not seem very helpful if one's object of interest is poetic language use specifically. As this illustrates, quite different things might need to be said about an object of investigation in different contexts of discussion. But of course this does not mean that just anything could be helpfully said. Nor does it mean that there is not a way that things are, as if the facts of reality, including facts about language use, depended on what one finds problematic or interesting about an issue. In this sense Wittgensteinian pluralism of clarifications is perfectly compatible with objective knowledge and truth, and this kind of pluralism has no relativistic implications.

The preceding implies that it is indeed essential to philosophy in Wittgenstein's sense to engage his reader or interlocutor and to get them to think for themselves. Only in this way it is possible to elicit a meaningful agreement from the interlocutor that a clarification has helped them with their philosophical problem or some specific aspect thereof. As the point might be put, one cannot clarify anything to another without the other person's engagement with the clarification. Unlike theses, clarifications cannot be passively received and accepted. Moreover, the success of clarification depends on their taking into account what exactly the person's in question find problematic about the matter at hand. Philosophy thus conceived could not offer generic solutions in the form of philosophical theses, because it could not achieve its clarificatory aims by so doing. Therefore, to answer my original question, the preceding considerations explain Wittgenstein's desire not to spare his reader from thinking but to stimulate them to thoughts of their own. On this account, stimulating the reader into thinking is essential to philosophy. No theses or prescriptions regarding logic or grammar can substitute for this. Philosophy cannot be passively received.[5]

3.5 Wittgenstein on How the *Investigations* Ought to Be Written

A number of remarks can be found in Wittgenstein's *Nachlass* on how he thinks his book ought to be written. Moreover, it seems that ideas expressed in these remarks were in fact taken into account in the composition of the *Investigations*. I conclude with a brief discussion of relevant remarks. They suggest a picture of the book as a sort of manual for philosophical self-education, a practice book for philosophical thinking. In the light of the preceding, discussion this emerges as an essential characteristic of Wittgenstein's philosophical approach.

Wittgenstein writes in a pocket notebook (estimated to be from 1941–44): 'When this book is correctly written it should contain many examination questions' (MS 164, 161). As well known, the *Investigations* does indeed contain many questions, more than are answered in it. This suggests that Wittgenstein did take into account the idea in the remark. But for what purpose should the book contain such questions? We must ask: what is the purpose of an exam question? It is not that the question needs answering as such; those who set the questions are not trying to find out about the matter in this way. (Exams are not part of the research activities of universities, whereby students are used as informants.) Their purpose, rather, is to find out about the student's knowledge or understanding in order to assess it. However, Wittgenstein

[5]Wittgenstein's view seems in this respect reminiscent of Kant who famously said that it is not possible to teach philosophy (in the sense of philosophical doctrines), only philosophizing.

can be hardly have literally wished to examine his reader. In this case he would also need to learn about their answers, but no arrangement has been set in place to do this. The purpose therefore must be for the reader to test their own understanding or ability to deal with relevant issues. Why? This can be understood as serving the goal of self-education in that becoming aware of the shortcomings of one's understanding is necessary for being able to develop it further. In this sense it might then not be exaggerated to regard Wittgenstein's book as a kind of training manual in philosophical thinking.

Consider also the following two remarks in another pocket notebook from 1938: 'When this book is written as it should, everything that I say must be comprehensible, even trivial, but [it will be] hard to understand why I say it' (MS 160, 4r; my square brackets). And: '"You give with one hand and take with the other!" Certainly, for I do not want to give anything to you, but only demonstrate giving and taking.' (MS 160, 4r-v) Regarding the first remark, again it seems that it may well have informed the composition of Wittgenstein's book. It can certainly be quite hard to know, when reading the *Investigations*, why Wittgenstein is saying what he is saying. Indeed, this seems to be a characteristic experience reading it, at least in the beginning. But in what way would this be desirable? Presumably the point is to make the reader to think harder. When the *Investigations* makes one ask why Wittgenstein says what he is saying, what the issue really is that Wittgenstein addresses, or what the core of the matter is, and what point he wishes to make about it, the book is making one think harder. This might therefore be seen as a literary technique for stimulating the readers' thinking. By contrast, domesticating the book for oneself with the help of commentaries may likely have the opposite effect. This might therefore well be what Wittgenstein wishes one would not do. Once again this indicates that it is impossible to be a passive receiver of Wittgenstein's philosophy or of his presumed theses in the sense that one would learn about those theses from a commentary, and then see the *Investigations* merely as their more authoritative statement. Moreover, as the second remark can be interpreted to say, there are no such theses to be found in the *Investigations*. Wittgenstein's purpose there is not to offer any philosophical theses to the reader. Rather, his purpose is to give examples of the philosophical dialectic of giving and taking, whereby an answer is offered to a question and subsequently brought into question again.

Finally consider a remark from 1948 where Wittgenstein comments on his way of writing, after he had stopped working on (what is known as Part I of) the *Investigations*. 'The difference between my philosophizing and that of others is really: what I write down as *preparation* for philosophical clarification they regard as a philosophical clarification' (MS 136, 41b). This, too, seems to become comprehensible, if we regard clarification as necessarily requiring active engagement from the interlocutor, and that she should think for herself. What Wittgenstein writes down is only preparation for clarification in that it is only intended as something that his reader would engage with, when trying to work her way towards clarity.

Acknowledgements This paper was presented in a philosophy of education seminar at the University College London, Institute of Education in December 2017. I would like to thank those present for their questions and comments.

References

Frege, Gottlob. *Posthumous Writings*. Hermes, H., Kambartel F., Kaulbach, F. eds. Oxford: Basil Blackwell, 1979 (PW).

Glock, Hans-Johan. "*Philosophical Investigations* Section 128: 'Theses in Philosophy' and Undogmatic Procedure". In *Wittgenstein's* Philosophical Investigations, *Text and Context*, eds. Arrington, R. L. & Glock, H.-J. London: Routledge, 1991.

Hacker, Peter. 'Philosophy'. In Glock, H.-J. ed., *Wittgenstein: A Critical Reader*. Oxford: Blackwell, 2001.

Kuusela, Oskari. 'Wittgenstein's Critical Inheritance and Development of Russell's Logic', in Klement, K. *The Oxford Handbook of Russell*. Oxford: Oxford University Press, forthcoming.

Kuusela, Oskari. *The Struggle against Dogmatism: Wittgenstein and the Concept of Philosophy*. Cambridge, MA: Harvard University Press, 2008a.

Kuusela, Oskari. 'Transcendental Arguments and the Problem of Dogmatism'. *International Journal of Philosophical Studies*, Vol. 16, No. 1, pp. 57–75, 2008b.

Kuusela, Oskari. 'The Development of Wittgenstein's Philosophy', in Kuusela O. and McGinn M. eds., *The Oxford Handbook of Wittgenstein*. Oxford: Oxford University Press, 2011.

Kuusela, Oskari. *Wittgenstein on Logic as the Method of Philosophy: Re-examining the Roots and Development of Analytic Philosophy*. Oxford: Oxford University Press, 2019.

Kuusela, Oskari. 'On Wittgenstein's and Carnap's Conceptions of the Dissolution of Philosophical Problems, and against a Therapeutic Mix: How to Solve the Paradox of the Tractatus', *Philosophical Investigations*, 2019a.

Russell, Bertrand. *Our Knowledge of the External World*. London: George Allen and Unwin, 1926 (OKEW).

Russell, Bertrand. *The Principles of Mathematics*. London: Routledge, 2010.

Wittgenstein, Ludwig. *Tractatus Logico-Philosophicus*. London: Routledge & Kegan Paul, 1951 (TLP).

Wittgenstein, Ludwig. *Lectures on the Foundations of Mathematics, Cambridge 1939*. Diamond, Cora ed. Hassock, Sussex: Harvester Press, 1976 (LFM).

Wittgenstein, Ludwig. *Wittgenstein's Lectures, Cambridge 1932–35*. Ambrose, Alice ed. Oxford: Basil Blackwell, 1979a (AWL).

Wittgenstein, Ludwig. *Notebooks 1914–16*. Oxford: Blackwell, 1979b (NB).

Wittgenstein, Ludwig. *Remarks on Colour*. Oxford: Blackwell, 1979c (RC).

Wittgenstein, Ludwig. 'Lectures on the Freedom of the Will'. In Klagge, James and Nordmann, Alfred eds., *Philosophical Occasions 1912–1951*. Indianapolis: Hackett, 1993.

Wittgenstein, Ludwig. *Ludwig Wittgenstein: Cambridge Letters, Correspondence with Russell, Keynes, Moore, Ramsey and Straffa*. G.H. Von Wright and Brian McGuinness eds. Oxford: Blackwell, 1997 (CL).

Wittgenstein, Ludwig. *Culture and Value*. Revised Edition. Oxford: Blackwell, 1998 (CV).

Wittgenstein, Ludwig. *Wittgenstein's Nachlass: The Bergen Electronic Edition*. Edited by The Wittgenstein Archives at the University of Bergen. Oxford: Oxford University Press, 2000 (References by manuscript or typescript number).

Wittgenstein, Ludwig. *Philosophical Investigations*. Oxford: Blackwell-Wiley, 2009 (PI).

Chapter 4
Wittgenstein's Grammar: Through Thick and Thin

Danièle Moyal-Sharrock

I would say that one of the continuous tracks of Wittgenstein's philosophy is the discernment and elucidation of grammar—its nature and its limits. This paper traces Wittgenstein's evolving notion of grammar from the *Tractatus* to *On Certainty*. We can distinguish in Wittgenstein what I will call a 'thin grammar'—a grammar that governs our use of words independently of facts about the world—from a 'thick grammar'—a grammar that is 'reality-soaked'[1] or fact dependent. It seems to me that Wittgenstein's thick grammar grows increasingly thick; so much so that there occurs in 'the third Wittgenstein'[2] what I call *a grammaticalization of experience*. This is particularly notable in his notion of 'patterns of life' in *Last Writings* and in his concept of 'hinge' certainties in *On Certainty*. This reflects his growing realisation that grammar can be, as it were, anthropo-*logical*; and that it can manifest itself as *a way of acting*. In moves that exceed anything in *Philosophical Investigations*, the third Wittgenstein makes grammar *enactive*. However, we shall see that his unrelenting, albeit hesitant, connection of grammar to the stream of life in no way infringes on the 'autonomy of grammar'.

4.1 Defining Grammar

> By *grammatical* rule I understand every rule that relates to the use of a language.
> (VOW 303)

[1] I borrow the term 'reality-soaked' from Bernard Harrison, who uses it to speak of a 'reality-soaked language' (1991, 58).

[2] An expression I coined to denote the post-*Investigations* corpus. See Moyal-Sharrock (2004).

D. Moyal-Sharrock (✉)
University of Hertfordshire, Hatfield, Hertfordshire, UK
e-mail: d.moyal-sharrock@herts.ac.uk

S. Wuppuluri and N. da Costa (eds.), *WITTGENSTEINIAN (adj.)*,
The Frontiers Collection, https://doi.org/10.1007/978-3-030-27569-3_4

The limits of what makes sense and what does not; what can be said and what cannot, is a leitmotif of Wittgenstein's philosophy. But what determines those limits? Wittgenstein's astonishing answer, already inscribed in the *Tractatus*, is: grammar. For him, language—any language—is rule-governed (RC 303); that is, governed by rules of grammar. What Wittgenstein means by 'grammar' is both similar to and different from what we usually mean by grammar. Ray Monk recounts the following (due to Rush Rhees):

> Moore, who attended Wittgenstein's lectures, insisted that "Wittgenstein was using the word 'grammar' in a rather odd sense. ... Thus, he argued, the sentence: 'Three men was working' is incontrovertibly a misuse of grammar, but it is not clear that: 'Different colours cannot be in the same place in a visual field at the same time' commits a similar transgression. If this latter is also called a misuse of grammar, then 'grammar' must mean something different in each case." No, replied Wittgenstein. 'The right expression is "It does not have sense to say ..."' Both kinds of rules were rules in the same sense. 'It is just that some have been the subject of philosophical discussion and some have not.' (1991, 322–323)

And what Monk importantly adds here is that grammatical mistakes made by philosophers are more 'pernicious' than ordinary grammatical mistakes. Wittgenstein, then, merely expands our ordinary understanding of grammar rather than altering it: he does not see grammar as comprised merely of syntactic rules, but of any rule that governs 'the way we are going to talk' (MWL 72): 'By *grammatical* rule I understand every rule that relates to the use of a language' (VOW 303). For him, grammar is 'a preparation for description, just as fixing a unit of length is a preparation for measuring'; so that 'A rod has a length' is as much a preparation for description (e.g. 'This rod is 3 feet long') as the grammatical rule to use 'were' and not 'was' in some cases is a preparation for our intelligible use of language. Wittgenstein is simply more liberal than grammarians as to what he will count as grammar:

> Everything that's required for comparing the proposition with the facts belongs to grammar. That is, all the requirements for understanding. (All the requirements for sense.) (BT 38)

Another way of putting this is that grammar consists of the conditions of intelligibility of a language. It is the conventionally-established basis on which we can *make sense*: 'Grammar consists of conventions' (PG 138), keeping in mind that conventions here are not due to a concerted consensus, but to an unconcerted agreement in practice.

Now if grammar includes '[a]ll the requirements for sense', it must then also include rules such as 'There exist people other than myself'. For isn't that a requisite underpinning of sense—a preparation for such descriptions as 'There are twenty of us in this room' or 'Vietnam's population is 96.5 million'? Moreover, following Wittgenstein's criterion for the misuse of grammar in his reply to Moore above ('It does not have sense to say ...'), it has at least as little sense to say 'I can't be sure that anyone exists but me' as to say 'Three men was working'. In fact, people are more likely *not* to understand what you are saying in the first case than in the second. In both cases, they'll understand all the words, but as Monk noted, violations of grammar can be more or less pernicious—so that whereas 'Three men was working' is laughable at worst; 'I can't be sure that anyone exists but me' smacks of the pathological. We'll come back to this.

Grammar, then, is a normatively sanctioned system or method of representation/description; it allows us to use words in order to *intelligibly* represent, describe, express, misrepresent, misdescribe, imagine, pretend, lie about, etc. how things are. I will now briefly retrace Wittgenstein's drawing of the limits of language in the *Tractatus* as it relates to nonsense and ineffability; for it remains, in this, essentially unchanged and informs what the later Wittgenstein will call grammar.

4.2 Drawing the Limits of Language

In 'On Heidegger on Being and Dread', written in 1929, Wittgenstein writes:

> Man feels the urge to run up against the limits of language. Think for example of the astonishment that anything at all exists. This astonishment cannot be expressed in the form of a question, and there is also no answer whatsoever. Anything we might say is a priori bound to be nonsense. Nevertheless we do run up against the limits of language. … This running-up against the limits of language is *ethics*. (WVC 68)

Why is anything we might say in explanation of the astonishment that anything at all exists, nonsense? Why would such an attempt constitute a 'running-up against the limits of language? And how is *that* ethics? In his 'Lecture on Ethics' (written in the same year), Wittgenstein writes:

> I see now that these nonsensical expressions were not nonsensical because I had not yet found the correct expressions, but that their nonsensicality was their very essence. For all I wanted to do with them was just to go beyond the world and that is to say beyond significant language. My whole tendency and, I believe, the tendency of all men who ever tried to write or talk Ethics or Religion was to run against the boundaries of language. This running against the walls of our cage is perfectly, absolutely hopeless. Ethics so far as it springs from the desire to say something about the ultimate meaning of life, the absolute good, the absolute valuable, can be no science. What it says does not add to our knowledge in any sense. (LE 44)

At this early period of his thought, Wittgenstein viewed as nonsensical any expression that did not 'add to our knowledge'[3]—that was not a proposition of natural science (6.53). The nonsensical included ethics and aesthetics (6.421), the mystical (6.522), and his own Tractarian sentences (6.54). None of these have sense—none are bipolar propositions susceptible of truth and falsity—and cannot therefore add to our knowledge. Indeed, even his Tractarian sentences do not *inform*; they *elucidate* (6.54), which is the rightful task of philosophy (4.112). It is *their not adding to knowledge* that makes Tractarian *Sätze* technically nonsensical, devoid of sense.

---

[3]For my present purpose, it is not necessary to mark the distinction made in the *Tractatus* between 'nonsense' and 'senseless', which dissolves in the later Wittgenstein. See, for example, PI 247 and pp. 175, 221 for cases where *sinnlos* is used for reasons which would have, according to the *Tractatus*, required the use of *unsinnig*. But this indiscriminate use of the terms is already present in *Philosophical Grammar* (1931–1934), e.g. PG 129.

It is clear then that the *Tractatus* advances a *non-derogatory*, indeed *positive*, understanding of the nonsensical,[4] which it was a mistake on the part of New Wittgensteinians to reject.[5] For, preferring a monochrome, 'austere', reading of nonsense as exclusively gibberish meant having to view Tractarian sentences as gibberish—a consequence they embraced, with no enduring success. Already the first sentence in the 'Lecture on Ethics' passage shows Wittgenstein alluding to different uses of nonsense ('I see now that these nonsensical expressions were not nonsensical because I had not yet found the correct expressions, but that their nonsensicality was their very essence'); but he was also later to make this clearer: '... the word 'nonsense' is used to exclude certain things [...] for different reasons' (AWL 64). By the time of the *Investigations*, Wittgenstein uses the terms 'nonsense', 'senseless', 'has no sense' indiscriminately to refer to combinations of words that are excluded from the language, 'withdrawn from circulation' (PI 500), and insists that this exclusion may be for different reasons:

> To say 'This combination of words makes no sense (*hat keinen Sinn*)' excludes it from the sphere of language and thereby bounds the domain of language. But when one draws a boundary it may be for various kinds of reason. (PI 499)

But the *Tractatus*, I'd have thought, was clear enough in its distinctions of different kinds of nonsense. Unlike nonsense which (1) results from a *violation* of sense, say, when categorial boundaries are misread and allowed to overlap, (as in the question: 'Is the good more or less identical than the beautiful?' (4.003) or in '2 + 2 at 3o'clock equals 4' (4.1272)), there is nonsense (call it 'important nonsense') that does not make sense either because (2) it is *impossible to put into words* (e.g. the mystical, ethics and aesthetics[6]) or because (3) it *enables* or *regulates* sense (e.g. 'There is only one 1' (4.1272)) or because (4) it *elucidates* sense—and this includes Tractarian remarks. Nonsense that regulates sense is one of the early manifestations of what Wittgenstein will later call 'grammar'.

Moreover, inasmuch as the Tractarian Wittgenstein takes only truth-conditional utterances to be sayable (6.53),[7] any string of words that does not express a truth-conditional proposition is not, technically-speaking, *sayable*. On that count, all nonsense is *ineffable*. However, as regards important nonsense, whereas (2) the mystical, ethics and aesthetics cannot even be put into words, (3) regulative and (4) elucidatory nonsense, though not *sayable* strictly speaking, can be meaningfully formulated for heuristic purposes. That is, they can be formulated to serve as steps towards a clearer understanding of the conditions of sense or 'limit to thought' (TLP Preface). Tractarian remarks must be passed over in silence—in that they are not hypothetical

---

[4]For a full-blown argument, see Moyal-Sharrock (2007a).

[5]See Crary and Read (2000) *passim* in which so-called 'Ineffabilists' (philosophers who, like Peter Hacker, view some nonsense in the *Tractatus* as 'illuminating') are rebuked for 'chickening out', for not being 'resolute' enough to recognise that Wittgenstein viewed all nonsense as 'plain nonsense'; i.e. gibberish.

[6]Ethics, aesthetics and the mystical 'cannot be put into words' (TLP 6.421; 6.522).

[7]'... what can be said; i.e. propositions of natural science—i.e. something that has nothing to do with philosophy' (TLP 6.53).

propositions, but the 'steps' or 'ladder' to intelligibility or perspicuity (6.54). Once used, the ladder must be thrown away (6.54), for these heuristic aids do not belong to the sphere of language but to its delimitation. That is, they belong to what Wittgenstein will later call the *scaffolding of thought* (OC 211). As to elucidatory nonsense, he later makes clear that 'nonsense is produced by trying to express by the use of language what ought to be embodied in the grammar' (MWL 103).

If important nonsense is by definition ineffable, ineffability, too, appears validated. This sheds light on the value for Wittgenstein of 'what we must pass over in silence' (TLP 7); as does this, from Paul Engelmann: 'Wittgenstein passionately believes that all that really matters in human life is precisely what, in his view, we must be silent about'[8] (EL 97). But silence here is not deadening: what really matters cannot be said but it isn't imperceptible; it can *show* itself, make itself manifest, to us. And it often does this *through* language: 'the unutterable will be—unutterably—*contained* in what has been uttered' (EL 7), and we can apprehend it. Our apprehension of significance, as non-propositionally conveyed in a work of literature is of that kind—it resembles our grasp of wit. But grammar, too, *shows* itself in our use of language.

The later Wittgenstein will extend the list of the *sayable* to include non-truth-conditional uses of language (e.g. spontaneous utterances, questions, imperatives),[9] but he will never give up the idea that some things cannot be *said* in the sphere of language—that is, 'in the flow of the language-game'; or the idea that some things cannot be put into words *at all* but can only show themselves *through* words (and, he will add, through deeds).[10] In fact, he will add *certainties* to the list of the ineffable—the grammatical ineffable. Like regulative nonsense, certainties cannot be *said* because they constitute the scaffolding of sense, not its object.[11] Basic certainties (e.g. 'There exist people other than myself', 'I have a body', 'Human beings need nourishment') are 'removed from the traffic' (OC 210); they cannot meaningfully be said in the flow of the language-game as if they were open for discussion because they are bounds of sense (rules of grammar), not objects of sense.

In fact, the *Tractatus* sets the stage for what Wittgenstein will later call 'grammar': grammar is that which enables or regulates sense (and so is itself nonsensical) and cannot meaningfully be said in the flow of the language-game but only heuristically articulated. What is also incipient in the *Tractatus* is the idea of philosophy as conceptual or grammatical elucidation:

> Philosophy aims at the logical clarification of thoughts. … A philosophical work consists essentially of elucidations. [Indeed, this is what he says his Tractarian propositions are: elucidations (6.54)] Philosophy does not result in 'philosophical propositions', but in the

[8]Note also Wittgenstein's acknowledgment, in a letter to Fricker, that the important part of the *Tractatus* was the silent part: 'My work consists of two parts: the one presented here plus all that I have *not* written. And it is precisely this second part that is the important one' (EL 143).

[9]See also PI 23.

[10]For a discussion of the presence of the saying/showing dichotomy as late as *On Certainty*, see Moyal-Sharrock (2007b), 94–97, and passim.

[11]For further discussion on the ineffability of basic certainty see Moyal-Sharrock (2007b) 65–71; 94–99.

> clarification of propositions. Without philosophy thoughts are, as it were, cloudy and indistinct: its task is to make them clear and to give them sharp boundaries. (TLP 4.112)

Here again, Wittgenstein will never retract his Tractarian view that conceptual elucidation is the task of philosophy; that: '[t]he philosophical problem is an awareness of disorder in our concepts, and can be solved by ordering them' (BT 309). This emphasis on grammatical elucidation has given Wittgenstein a bad name in mainstream philosophy: it seems as if he reduces everything to language, which would make him a linguistic idealist. Does Wittgenstein think philosophers only play games with words? Language-*games*? What philosophers should be primarily interested with is life, not language. Language is for linguists, no? No. Grammatical elucidation is important for life because grammar is inextricable from life. Let's see in what way it is inextricable.

4.3 Grammar Delimits the World

> Whenever we say that something *must* be the case we are using a norm of expression.' (AWL 16)

The connection between language and reality has long perplexed philosophers. Is human language the result of our attempts to translate 'nature's own language'? But how can nature have a language? More plausibly, is human language the result of our attempts to facilitate and enhance our relationship with nature, and each other? 'The connection between "language and reality" is made by definitions of words, and these belong to grammar', writes Wittgenstein[12] (PG 97).

There is no such thing in the world, or nature—or however we want to name what Paul Boghossian calls: the 'basic worldly dough' (2006, 35)—as an outline, a system or a concept. The system does not reside in the nature of things (Z 357). Nature is conceptually unmarked; it is we who, with our language, cut paths or inroads of salience and understanding in order to harness the contingent in ways that produce and govern sense for us. For Wittgenstein, this tracing is not metaphysical, but grammatical: 'One thinks that one is tracing the outline of a thing's nature over and over again, and one is merely tracing round the frame through which we look at it' (PI 114). Boghossian is right: of course, 'we have no choice but to recognize that there must be some objective, mind-independent facts' (2006, 35); however, as linguistic beings, we come to experience and grapple the world *always already* with language, and there is no getting out of language to compare or measure our outline against bare particulars. At the *conceptual* basis of our confrontation with experience are not bare particulars, but grammar: it is *grammar* that tells us what *kind* of object

[12]Note that I will be using the terms 'reality', 'nature' and 'the world' interchangeably to refer to unconceptualized or raw reality—what Wittgenstein refers to as the 'reality lying behind the notation' (PI 562); where I mean a conceptualized world or reality, I will speak of 'my world', 'our world' or 'human reality'.

anything is (PI 373). Wittgenstein's realisation that the demarcation is not empirical but grammatical—that the connection between language and world is a grammatical or logical connection (these are synonymous for the later Wittgenstein)—is, as far as I know, unprecedented in the history of philosophy.

Wittgenstein, however, is no linguistic idealist[13]; he does not reduce reality, 'the world' or even 'our world', to language. Though we are responsible for its conceptual outline, the world is not our invention. Is not even the Tractarian Wittgenstein a linguistic idealist? I think not. 'The limits of my language mean the limits of my world' may be read as expressing the view that 'my world'—i.e. my perception of the world—is concept-laden. The later Wittgenstein fleshed this out in his many remarks on how enculturation conditions us to perceive the world through a conceptual frame.[14] But what I want to stress here is that our perception being (from a very early age) concept-laden does not make it *judgment*-laden: though I must possess the concept of 'grief' to see grief on someone's face, judgment need not come into it.[15] In most cases, I see emotion in someone's face immediately: no judgment, interpretation or inference is needed; and therefore no so-called 'Theory of mind'.[16] Of course, in describing the face as 'sad', I am using a concept, but I am not thereby necessarily making a judgment. That much perception is concept-laden does not make it theory-laden. This is important in that it shows the nonempirical and nonepistemic nature of our *basic* conceptual interactions with our world.

When Peter Hacker writes that the post-Tractarian Wittgenstein sees the harmony between language and reality as no longer orchestrated between language and reality but *within* language—that is, by grammar—and that far from reflecting the logical structure of the world, grammar is 'arbitrary', it owes no homage to reality' (2000, 9–19), he should be wary of throwing out the baby with the bathwater. For Wittgenstein, grammar's demarcations, though not accountable to reality,[17] do owe some homage

[13]For further discussion, see Moyal-Sharrock (2016).

[14]For example: 'How can I be taught to recognize these patterns [in the weave of life]? I am shown simple examples, and then complicated ones of both kinds. It is almost the way I learn to distinguish the styles of two composers' (LW II 42-3); 'When he first learns the names of colours—what is taught him? Well, he learns e.g. to call out "red" on seeing something red—But is that the right description; or ought it to have gone: "He learns to call 'red' *what we too* call 'red'"?—Both descriptions are right. … What I teach him however must be a *capacity*. So he *can* now bring something red at an order; or arrange objects according to colour' (Z 421); 'As children we learn concepts and what one does with them simultaneously' (LW II 43).

[15]As Wittgenstein makes clear: 'We do not see facial contortions and *make the inference* that he is feeling joy, grief, boredom. We describe a face immediately as sad, radiant, bored, even when we are unable to give any other description of the features—Grief, one would like to say, is personified in the face. This is essential to what we call 'emotion' (RPP II, 570).

[16]The mind is often *directly observable*, thus obviating the need for a Theory of Mind to attribute mental states—thoughts, perceptions, desires, intentions, feelings—to others. The notion of the mind as 'something inner', accessible only to oneself, has been seriously undermined by Wittgenstein. For an excellent discussion, see Vaaja (2013).

[17]'Grammar is not accountable to any reality. It is grammatical rules that determine meaning (constitute it) and so they themselves are not answerable to any meaning and to that extent are arbitrary' (PG 184).

to it. Language (or rather grammar[18]) constitutes *our* world in that it conceptually demarcates *the* world; but that demarcation is itself conditioned by the world.

4.4 The World's Impact on Grammar: A 'reality-Soaked' Grammar

... This fact is fused into the foundations of our language-game. (OC 558)

Certainly, Wittgenstein is clear that grammar is not—in fact, *cannot be*—accountable to reality: 'Grammatical conventions cannot be justified by describing what is represented. Any such description already presupposes the grammatical rules' (PR 67). But he is also clear that reality has to impact our grammar or concepts[19]: 'The rule we lay down is the one most strongly suggested by the facts of experience' (AWL 84). Though we don't read off our concepts from nature, our concepts are impacted by natural *facts* which are fundamental or salient for us: 'What we have to mention in order to explain the significance, I mean the importance, of a concept, are often extremely general facts of nature: such facts as are hardly ever mentioned because of their great generality' (PI, p. 56). But what is the nature of this correspondence? 'Would it be correct to say our concepts reflect our life?' he asks; to which he replies: 'They stand in the middle of it' (LW II 72). This means that our concepts are immersed in our life, intertwined with it in a dynamic interaction: the fabric of life provides the milieu in which our concepts are formed; in turn, our concepts order the fabric of life. Does this mean that the grammar of our language is *justified* by facts?

---

[18]Note that though one of Wittgenstein's most persistent philosophical concerns is to make the distinction ever clearer between grammar and the use of language (BT 38), he does not always mark that distinction. Where he is more preoccupied with distinguishing word and world rather than grammar from propositions, he often speaks of language where he means grammar [e.g. 'If I want to tell someone what colour some material is to be, I send him a sample, and obviously this sample belongs to language' (PR 38)]; or speaks of them quasi-interchangeably [e.g. 'The calculus is as it were autonomous—Language must speak for itself' (PG 63)]. But the distinction between language and grammar remains a logical distinction: grammatical rules *govern* our use of language; they are not language, but are essential to language; make language possible.

[19]These—'grammar' and 'concepts'—are often used interchangeably by Wittgenstein. A conceptual elucidation is a grammatical elucidation.

Wittgenstein asks himself this question again and again, from at least 1930.[20] Here is one late formulation of it:

> If we can find a ground [by which Wittgenstein means a 'justification': he uses the German *begründen*] for the structures of concepts among the facts of nature (psychological and physical), then isn't the description of the structure of our concepts really disguised natural science; ought we not in that case to concern ourselves not with grammar, but with what lies at the bottom of grammar in nature?
>
> Indeed the correspondence between our grammar and general (seldom mentioned) facts of nature does concern us. But our interest does not fall back on these *possible* causes. We are not pursuing a natural science ... Nor natural history .... (RPP I, 46; cf. PI p. 230)

This[21] makes it clear that when Wittgenstein speaks of the correspondence between concepts and nature, he is talking about the correspondence between the *structures* of concepts—that is, our grammatical rules—our grammar—and facts of nature. Take the concept of pain, some of the 'structures' of that concept can be expressed in grammatical rules such as: 'Human beings are normally susceptible to pain'; 'Tables and chairs don't feel pain'; 'There is psychological as well as physical pain', etc. In these passages, then, Wittgenstein is saying that of course we are interested in the correspondence between our grammar and very general facts of nature, but not in the way natural scientists or historians are interested in this correspondence. That is, we are not interested in any empirical justification or historical account for our having the grammatical rules we do. Let me give an example.

There are cases where a rule of grammar may have its historical root in an empirical discovery; for example, the realization that men have something to do with the reproductive process (it was long thought that women singly procreated[22]) may well be at the root of the grammatical rule: 'Every human being has two biological parents'. However, such correspondence or justification is of no interest to philosophers; what philosophers are interested in is what *logically* demarcates sense from nonsense, and that cannot be due to justification: 'The essence of logical possibility is what is laid down in language. What is laid down depends on facts, but is not made true or

---

[20]See Malcolm: 'The notes for 1930–1932, edited by Desmond Lee, exhibit at first a striking continuity with the *Tractatus*: language consists of propositions; a proposition is a picture of reality; a proposition must have the same logical multiplicity as the fact which it describes; thought must have the logical form of reality. But new concerns soon appear. What is the relation of the logical grammar of language to reality? The application of grammar to reality is not shown by the grammar; a picture does not contain its own application. "In all language," Wittgenstein says, "there is a bridge between the sign and its application. No one can make this for us; we have to bridge the gap ourselves. No explanation ever saves the jump, because any further explanation will itself need a jump." 'Can grammar be justified? Can we say why we use just these rules of grammar and not other ones? Is the logic of our language to be justified on the ground that it fits the nature of reality? No. "Our justification could only take the form of saying 'As reality is so and so, the rules must be such and such.' But this presupposes that I could say 'If reality were otherwise, then the rules of grammar would be otherwise.' But in order to describe a reality in which grammar was otherwise I would have to use the very combinations which grammar forbids. The rules of grammar distinguish sense and nonsense and if I use the forbidden combinations I talk nonsense' (1980, online).

[21]PI Part II (1946–49) is roughly contemporaneous with RPP I (1946–47).

[22]With the help of fertility goddesses.

false by them' (AWL 162). Our grammatical rules are not the result of reasoning or justification. So how are facts connected to grammar?

In contrast to what I will call a '*thin* grammar', whose rules are *not* conditioned by facts (e.g. '2 + 2 = 4'; 'A rod has a length'; 'This is (what we call) a table') and are normatively engendered and sanctioned, the rules of '*thick* grammar' (e.g. 'Human beings have bodies, need nourishment, sleep'; 'Human beings can go to the moon') *are* rooted in facts and may be either experientially or normatively engendered and sanctioned. Wittgenstein opposes the terms: 'cause' and 'ground' to distinguish a nonratiocinated rootedness in facts and experience from a justified rootedness. Would it be right to say that my having a body is a bound of sense for me *because* previous experience has taught me so? No, our lifelong experience of having a body may be the 'cause' of it being a bound of sense, but it is not its 'ground' (cf. OC 429; 474). I have not come to the conclusion that I have a body 'by following a particular line of thought' (OC 103). Our *experience* of having a body or of mountains not sprouting up in an hour is not empirical or epistemic; it is either instinctive (as is the case of having a body) or acquired via repeated exposure or conditioning, and so second nature (as is the case with mountains being there immemorially). It may help here to distinguish between 'experiential' (*embedded in* experience) and 'empirical' (*inferred from* experience). Our lifelong experience of ourselves as embodied; our experience of the world as populated by people other than ourselves; our experience of mountains as geological structures that do not sprout up in an hour[23] make such experience part of the scaffolding or background of sense for us. These facts are 'fused into the foundations of our language-game' (OC 558)—that is, they are part of our grammar. And where some thick rules of grammar are originally grounded in fact and articulated as news-breaking propositions (e.g. 'Human beings can go to the moon'), it is only with time and repetition, once they have hardened into a rule, that they belong to the scaffolding of thought.[24]

4.5 'The Common Behaviour of Mankind'

While acknowledging the buzzing indeterminacy, spontaneity and irreducibility of human life, Wittgenstein also keeps reminding us of certainties and regularities. Amongst these are our *shared* natural reactions; what he calls 'the common behaviour of mankind' (PI 206): reactions such as crying when in pain or sad; smiling when glad; jumping when startled; gasping or screaming when afraid; reacting to someone's suffering. He calls these instinctive common reactions or action patterns 'prototype[s] of thought' (RPP I, 916). These prototypes or action patterns are the necessary starting

[23]'[A child] doesn't learn *at all* that that mountain has existed for a long time: that is, the question whether it is so doesn't arise at all. It swallows this consequence down, so to speak, together with *what* it learns' (OC 143). 237. 'If I say "an hour ago this table didn't exist" I probably meant that it was only made later on. … If I say "this mountain didn't exist half an hour ago", that is such a strange statement that it is not clear what I mean' (OC 237).

[24]For more detailed discussion, see Moyal-Sharrock (2007b).

points of language: 'it is characteristic of our language that the foundation on which it grows consists in *steady ways of living, regular ways of acting*' (CE 397; my emphasis). Without certain constantly recurring patterns, our concepts would have no grip; so that our acquiring concepts, such as pain, requires that we have normal prototypical human reactions: 'If a child looked radiant when it was hurt, and shrieked for no apparent reason, one couldn't teach him to use the word "pain"' (LPP 37).

In his last writings on the philosophy of psychology, Wittgenstein delves deeper into these action patterns as they typically, and often internally, relate to feelings and emotions. He introduces the expressions 'patterns of life' and 'patterns of experience'[25] to denote the 'constant repetition' (RPP II, 626), the regular and tell-tale characteristics of our various psychological expressions and behaviours (e.g. those of pain, joy, grief, hope, but also of *simulated* pain, joy grief, hope etc.)—going as far as to suggest that typical physiognomies and constitutive rhythm and tempo attach to them:

> 'Grief' describes a pattern which recurs, with different variations, in the weave of our life. If a man's bodily expression of sorrow and of joy alternated, say with the ticking of a clock, here we should not have the characteristic formation of the pattern of sorrow or of the pattern of joy. (PI, p. 174; cf. LW I, 406)

> For pretence is a (certain) pattern within the weave of life. It is repeated in an infinite number of variations. (LW I, 862)

> Someone smiles and his further reactions fit neither a genuine nor a simulated joy. We might say 'I don't know my way around with him. It is neither the picture (pattern) of genuine nor of pretended joy.' (LW II, 61)

'Patterns of life' refer to recurring—mostly behavioral and facial—expressions characteristic of psychological concepts. There is not only one, or even a handful of 'occasions' that we might call 'grief', but innumerable ones that are interwoven with a thousand other patterns (cf. LW I, 966). And this is so for *all* our psychological concepts, because the 'natural foundation' for the way they are formed 'is the complex nature and the variety of human contingencies' (RPP II, 614). As a result the concepts themselves lack determinacy and have a kind of elasticity; but where most philosophers attempt to tame or reduce the indeterminacy, Wittgenstein wants to capture it: 'I do not want to reduce unsharpness to sharpness; but to capture unsharpness conceptually' (MS 1367, 64). Yet this unsharpness does not mean that our concepts are so elastic as to lack a hard core. Indeed Wittgenstein's depiction of psychological indeterminacy is everywhere bounded not by rules, but by certain *regularities*: an order or pattern emerges from obstinate, though constantly varied, repetition; the evidence has tell-tale *characteristics*, our feelings and behaviours are informed by *typical physiognomies*. Of course, 'there are simple and more complicated cases; and that is important for the concept' (LWI, 967), for it is the simple cases that give the concept its solid centre, its unambiguous core' (RPP II, 614): 'There is an *unmistakable* expression of joy and its opposite' (LW II, 32; original emphasis). The point here is not to eradicate indeterminacy, but to recognize that there are *basic* regularities in the 'hurly burly of human action' (Z 567), and that these

[25] Wittgenstein uses the German word *Muster*, which can be translated both as *model* or *pattern*.

are what shape our psychological bedrock or psychological grammar. This, without losing sight of the fact that 'simple language-games [...] are poles of a description, not the ground-floor of a theory' (RPP I, 633). The point is that, though there is an *indefiniteness* or *indeterminacy* essential to the kind of *repetition* in question—for it is a repetition that is embedded in life—there *is* a repetition, a pattern; and our psychological grammar is conditioned by such patterns:

> Seeing life as a weave, this pattern (pretence, say) is not always complete and is varied in a multiplicity of ways. But we, in our conceptual world, keep on seeing the same, recurring with variations. That is how our concepts take it. For concepts are not for use on a single occasion. (Z 568)

4.6 The Grammaticalisation of Experience

The 'hinge' certainties of *On Certainty* and the 'patterns of life/experience' of *Last Writings* attest to the fact that one of the leitmotifs in the work of the third Wittgenstein is the *grammaticalisation of experience*. Thick grammar—inasmuch as it is conditioned by our human form of life[26]—is a grammaticalisation of experience[27]; one might say it is anthropo-*logical*: 'The basic concepts are interwoven so closely with what is most fundamental in our way of living that they are therefore unassailable' (LW II, 43–44). Unassailable means impregnable and infallible. So that grammar, though impacted by facts, remains autonomous.

The grammaticalisation of experience did not come easily to Wittgenstein. As early as 1930, he writes: 'I will count any fact whose obtaining is a presupposition of a proposition's making sense as belonging *to language*' (PR 45)' (by which he means grammar[28]). And indeed in PI, he had gone as far as to count objects such as samples as belonging to grammar.[29] However, in his last notes: *Remarks on Colour* and *On Certainty*, we find him hesitant to apply this to contingent facts: 'Here it could now

---

[26]There are also culture-bound, or 'local' thick grammars—conditioned by the different forms of human life. For more on this distinction, see Moyal-Sharrock (2015).

[27]To consider the thick rules that are reality-soaked but not strictly speaking *experiential* [e.g. 'Napoleon existed' (cf. OC 185)] as part of the 'grammaticalisation of experience', we would need to either understand the term 'experience' in this phrase in a broad (culture-inclusive) sense, or say that not all of thick grammar is part of Wittgenstein's grammaticalisation of experience. I would opt for the first.

[28]See note 18.

[29]'If I want to tell someone what colour some material is to be, I send him a sample, and obviously this sample belongs to language; and equally the memory or image of a colour that I conjure by a word, belongs to language' (PR 38). 'We can put it like this: This sample is an instrument of the language used in ascriptions of colour. In this language-game it is not something that is represented, but is a means of representation. ... And to say "If it did not *exist*, it could have no name" is to say as much and as little as: if this thing did not exist, we could not use it in our language-game—What looks as if it *had* to exist, is part of the language. It is a paradigm in our language-game; something with which comparison is made. And this may be an important observation; but it is none the less an observation concerning our language-game—our method of representation' (PI 50). Although

be asked what I really want, to what extent I want to deal with grammar' (RC 309); and what he'd asserted as a 'correspondence between concepts and very general facts of nature' in PI, is now put in the form of a question, albeit a rhetorical one, in *On Certainty*: 'Indeed, doesn't it seem obvious that the possibility of a language-game is conditioned by certain facts?' (OC 617).

Though comfortable with the idea that some contingent facts condition our grammar, Wittgenstein was uneasy with assigning the expression of these contingent facts a grammatical status—indeed, with recognising seemingly empirical propositions to be grammatical rules. So much so that, as late as OC, he even contemplates a compromise: 'Is it that rule and empirical proposition merge into one another?' (OC 309). However, his '[inclination] to believe that not everything that has the form of an empirical proposition *is* one' (OC 308) wins the day; in *On Certainty*, he comes to see that the contingent or empirical nature of these propositions is invalidated by their very unassailability or indubitability. In fact, he could have reminded himself of this passage from the *Remarks on the Foundations of Mathematics*: 'To accept a proposition as unshakably certain … means to use it as a grammatical rule' (RFM 170). Be that as it may, we can safely say that in *On Certainty*, he is finally reconciled with the fact that some apparent empirical propositions 'form the foundation of all operating with thoughts (with language)' (OC 401). Wittgenstein has understood that grammar can be 'reality-soaked' without being empirical, but the journey was laborious, confirming that 'Not empiricism and yet realism in philosophy' is indeed 'the hardest thing' (RFM 325).

Our concepts delimit the world; but the world has its impact. Realising the full extent to which—as Cora Diamond puts it: '… grammar is to be seen in how we live' (1989, 20), Wittgenstein expands our conception of the logical. He gives grammar/logic an anthropological twist, and thereby redefines its limits. Logic is seen to be 'reality-soaked', but without falling into the traps abhorred by Frege: it is flawed neither by subjectivity (psychologism) nor fallibility (empiricism). This is crucially important for philosophy; it redefines the limits of sense or possibility.

4.7 Redefining the Limits of Possibility

Wittgenstein's conception of the logical is internally linked to our human form of life and this does not sit well with traditional conceptions of logical necessity. Stanley Cavell points out this apparent shortcoming:

> Wittgenstein's view of necessity is … internal to his view of what philosophy is. His philosophy provides, one might say, an anthropological, or even anthropomorphic, view of necessity: and that can be disappointing; as if it is not really *necessity* which he has given an anthropological view of. As though if the a priori has a history it cannot really be the a priori in question. (1979, 118–119)

Wittgenstein speaks of samples as being part of, or belonging to, 'language', strictly speaking, he means to 'grammar'. See note 18.

On the standard philosophical view, the logical must encompass *all possible worlds.* Most philosophers think, like Bertrand Russell, that 'No logical absurdity results from the hypothesis that the world consists of myself and my thoughts and feelings and sensations, and that everything else is mere fancy' (1912, 10). This, the Tractarian Wittgenstein would have endorsed: 'A thought contains the possibility of the situation of which it is the thought. What is thinkable is possible too' (TLP 3.02). But, on Wittgenstein's anthropo-*logical* view of logic (which would include the grammatical rule: 'There exist people other than myself'), logical absurdity *would* result from such hypotheses or thoughts as 'The world consists of myself alone'—unless meant figuratively or embedded in a fictional context. Conceivability can be equal to possibility in fictional worlds, but in our world, possibility is dependent on the conditions of sense anchored in our human form of life. A thought that has lost its human mooring and runs wild on the uncharted tracks of the imagination is not a possibility; it's a thought. It takes anthropo-*logical* bounds of sense to enable us to discern the *humanly* possible from the merely imaginable. Of course, these reality-soaked bounds of sense are, by definition, *basic*—but it only takes something as basic as the grammatical rule: 'There exist people other than myself' to render senseless the alleged possibility that 'The world consists of myself alone'.

Philosophical speculation should not be deemed sufficiently confined by conceivability; it must be constrained by logical parameters specific to our world. Rai Gaita echoes Cavell's denunciation of traditional epistemology and logic as a 'denial of the human' because the concepts informing them are 'essentially unconditioned by the fact that they are concepts deployed by human beings'—and, he adds, echoing Winch this time, that 'we cannot purify our concepts of their embeddedness in human life … without being left with only a shadow play of the grammar of serious judgment' (1990, xi; xii). Wittgenstein decried this purification of our concepts, or as he put it, the 'crystalline purity of logic' (PI 107). For logic to stop being 'empty', it must '[g]et back to the rough ground'! (ibid.).

When the only constraints on logical necessity are a few logical laws pared of any *human* specification, nothing impedes the so-called 'possibilities' that: 'There may exist no one in the world but myself'; 'The world may be five minutes old' and 'Human beings can switch bodies'.[30] To any ordinary person, these are inanities unless uttered in sci-fi contexts; and philosophers shouldn't be in the business of propagating them as possibilities. Hume is right that '[n]othing is more dangerous to reason than the flights of the imagination, and nothing has been the occasion of more mistakes among philosophers[31]' (*Treatise* 1.4.7).

Wittgenstein had a second look at logical necessity and saw that, as Diamond puts it, it has a *human* face (1991, 6, 13). The logical limit of possibility in our form of life has to be an anthropo-*logical* limit. Wittgenstein's extension of logic to incorporate

[30]David Lewis sees 'the mere possibility' that a person might switch bodies as real or serious enough to require refutation (1971, 47).

[31]And, I would add, nothing more time-wasting for scientists to engage in experiments to try and (dis)prove philosophers' time-wasting thought experiments: http://www.pbs.org/wgbh/nova/next/physics/physicists-confirm-that-were-not-living-in-a-computer-simulation/.

specifically human bounds of sense prevents—or ought to prevent—philosophers from envisaging life-size absurdities as legitimate possibilities.

4.8 Grammar as Enactive

> 'How is the word used?' and 'What is the grammar of the word?' I shall take as being the same question.
>
> (AWL 3)

We have seen that grammar can manifest itself in action. All grammar—thin or thick—deploys itself as a technique, a know-how, not a set of principles that we learn and apply rationally: 'To understand a language means to be master of a technique' (PI 199). In mastering different language-games, a child masters the *grammar* of words; their use in the language. Wittgenstein's notion of grammar is not more complicated than this. Rules of grammar are simply expressions of the norms of sense that grow out of, and in tandem with, our natural ways of acting and our socio-cultural practices. Grammar does not generate language; nor does it exist independently of language or action; it is embedded and enacted in what we say and do. The proper use of the word 'pain', for example is manifested in our ways of speaking and acting, and so the child assimilates grammatical rules or norms as it assimilates the language—through exposure to, and guidance in, correct usage. That is, children are habituated into standards of correctness of the practice in question, and thereby formed to act and react in particular ways; they are trained to master a technique through the inculcation of a normative attitude. The assimilation of rules—be they linguistic or mathematical—does not have the features of a *learning that*, but of a *learning how*—of a training or a conditioning, a way of acting.

Wittgenstein's account of the emergence, transmission and practice of language is an enactive account. In teaching a child to replace the natural expression of pain with the word 'pain', adults teach the child 'new pain-behaviour' (PI 244). Language is, in such cases, an extension of an underlying action pattern, a more sophisticated way of acting. So that, in investigating our grammar—thick or thin—the philosopher examines and reflects on the practices, forms of life and patterns of life in which it is inextricably embedded. *Pace* its detractors, grammatical or conceptual investigation is not a philosophical method that ignores reality, but a way of elucidating *our* bounds of sense.

References

Boghossian, P. (2006) *Fear of Knowledge: Against Relativism and Constructivism* (Oxford: Clarendon Press).

Crary, A., & Read, R. (Eds.). (2000). *The New Wittgenstein* (London: Routledge, 2000).

Diamond, C. (1989) "Rules: looking in the right place" in Phillips, D.Z. (ed.) *Wittgenstein: Attention to Particulars: Essays in Honour of Rush Rhees*. L, Macmillan, 12–34.

Gaita, R. (1990) Introduction to *Value & Understanding: Essays for Peter Winch* (New York: Routledge).

Hacker, P.M.S. (2000) 'Was he trying to whistle it?' in Crary, A. & Read, R. *The New Wittgenstein* (London: Routledge), 353–388.

Harrison, B. (1991) *Inconvenient Fictions: Literature and the Limits of Theory* (New Haven: Yale University Press).

Lewis, D. (1971) 'Counterparts of Persons and Their Bodies' in *Philosophical Papers*: *Vol. 1.* (Oxford, Oxford University Press, 1983), 47–54.

Monk, R. (1991) *Ludwig Wittgenstein: The Duty of Genius* (London: Vintage).

Moyal-Sharrock, D. (2004) (ed.) *The Third Wittgenstein: the post-Investigations works* (Routledge, 2004).

Moyal-Sharrock, D. (2007a) 'The Good Sense of Nonsense: A Reading of Wittgenstein's *Tractatus* as Nonself-repudiating' in *Philosophy* 82:1, 147–177.

Moyal-Sharrock, D. (2007b) *Understanding Wittgenstein's On Certainty* (Basingstoke: Palgrave Macmillan).

Moyal-Sharrock, D. (2015) Wittgenstein's Forms of Life, Patterns of Life and Ways of Living in *Nordic Wittgenstein Review*, October, 21–42. http://www.nordicwittgensteinreview.com/issue/view/NWR%20Special%20Issue%202015.

Moyal-Sharrock, D. (2016) 'Wittgenstein, No Linguistic Idealist' in *Wittgenstein and the Creativity of Language*, eds S. Greve and J. Mácha. (Basingstoke, UK: Palgrave Macmillan), 117–140.

Vaaja, T. (2013) 'Wittgenstein's "Inner and Outer": Overcoming Epistemic Asymmetry' *Nordic Wittgenstein Review* (2:1), 107–129. http://wab.uib.no/nwr/index.php/nwr/article/view/1012/pdf.

Chapter 5
Playing Language Games

Beth Savickey

Wittgenstein plays with language throughout his later philosophy. Imaginary scenes and invented languages turn metaphysical concerns into conceptual play. He sometimes describes his activities as five finger exercises in thinking—exercises that enable us to think *differently*. They range in form from individual games to open-ended grammatical play and Wittgenstein refers to them as *Sprachspiele*; a term that has traditionally been translated as 'language games.' However, activities characterized as *Spiele* are broader than those characterized as games, so any activity that can be characterized as playing, like throwing a ball in the air or ring-a-ring-a-roses, can be called a *Spiel*. Wittgenstein compares his language games and conceptual play to the games by means of which children learn their native language, and he draws attention to their entertaining character. However, his philosophical play is not a form of language acquisition. Rather, it offers insight and clarification if and when we become conceptually lost or confused.

While playing with language may sound trivial, its philosophical implications are profound. Problems arising through a misinterpretation of forms of language are conceptual and deep: 'They are deep disquietudes; their roots are as deep in us as the forms of our language and their significance is as great as the importance of our language.' However, they are also playful. 'Let's ask ourselves: why do we feel a grammatical joke to be *deep*? (And that is what the depth of philosophy is.)' (PI 111) Wittgenstein suggests that metaphysical questions express confusion concerning the grammar or use of words in the form of scientific questions (BB 35). In other words, metaphysical assertions look empirical but are really conceptual. Thus, he investigates metaphysical concerns through linguistic play. In so doing, he challenges a variety of philosophical theories including realism, idealism, and solipsism. He claims that in such disputes, one party attacks the normal form of expression as if they were attacking a statement, while the others defend it as if they were stating facts recognized by every reasonable human being (PI 402). In

B. Savickey (✉)
Philosophy Department, University of Winnipeg, Winnipeg, MB, Canada
e-mail: b.savickey@uwinnipeg.ca

S. Wuppuluri and N. da Costa (eds.), *WITTGENSTEINIAN (adj.)*,
The Frontiers Collection, https://doi.org/10.1007/978-3-030-27569-3_5

response, Wittgenstein attempts to destroy the outward similarity between conceptual and empirical propositions, while also trying to find a form of expression that fulfills a certain craving on the part of the metaphysician which our ordinary language does not fulfill and which produces puzzlement (BB 55). Simple language games and imaginative grammatical play challenge our most fundamental philosophical pictures and ideas, and enable us to think differently without recourse to theory or argumentation. As such, they also challenge us to think differently about the role of theory and argumentation in philosophy itself.

Wittgenstein plays with language for several reasons. First, we are often confused about language itself. He explains:

> Frege ridiculed the formalist conception of mathematics by saying that the formalists confused the unimportant thing, the sign, with the important, the meaning. Surely, one wishes to say, mathematics does not treat of dashes on a bit of paper. Frege's idea could be expressed thus: the propositions of mathematics, if they were just complexes of dashes would be dead and utterly uninteresting, whereas they obviously have a kind of life. And the same, of course, could be said of any proposition. Without a sense, or without the thought, a proposition would be an utterly dead and trivial thing. And further it seems that no adding of inorganic signs can make the proposition live. And the conclusion which one draws from this is that what must be added to the dead signs in order to make a live proposition is something immaterial, with properties different from all signs.
>
> But if we had to name anything which is the life of the sign, we should have to say that it was its *use*. (BB 4)

In a humorous twist on this insight, Wittgenstein presents a language game in which A gives an order to B through written signs consisting of dots and dashes, and B executes them by doing a figure in dancing with a particular step. 'Thus the order "__." is to be carried out by taking a step and a hop alternately; the order "… __ __ __" by alternately taking two hops and three steps' (BB 99). These dots and dashes immediately call to mind Morse code; a code in which a series of dots and dashes indicate letters of the alphabet. In Morse code '__.' is the letter 'N'. What makes a dash or a dot into a dance move, a letter of the alphabet, or part of our ordinary punctuation? Their distinct and varied uses. Wittgenstein's language game, with its allusion to Morse code, plays with Frege's idea and reminds us that language, notations, and systems of communication can be visual, sonic, or tactile. His language games include not only dots and dashes, but arrows, squiggles, doodles, coloured squares, geometric figures, smiling faces, cardinal numbers, individual letters, ordinary words, and everyday sentences. Wittgenstein draws attention to innumerable types of language and language games, and reminds us that their multiplicity is not something fixed or given once for all. Rather, new types of language and new language games come into existence and others become obsolete and get forgotten. The simple language game of dots, dashes, and dance moves enables Wittgenstein to investigate the use of signs, as well as concepts such as training, executing an order, being guided, reading, and being able to make a move. In so doing, he helps clarify our understanding of language.

Second, if the life of a sign is its use, then it is necessary to bring words and signs to life by setting them in motion:

> Logicians use examples which no one would ever think of using in any other connection. Whoever says 'Socrates is a man'? I am not criticizing this because it does not occur in practical life. What I am criticizing is the fact that logicians do not give these examples any life. We must invent a surrounding for our examples…a game in which its use is a move. It does not matter whether in practice the word has a place in a game, but what matters is that we have a game, that a life is given for it. (AWL 124)

To imagine a language is to imagine a form of life. And the term language *game* is meant to bring into prominence the fact that speaking a language is part of an activity or of a form of life (PI 23). Conceptual or grammatical play is one method of exhibiting such activity. The purpose of invented languages and imaginary scenes is to study the phenomena of language in primitive kinds of application in order to command a clear view of the aim and functioning of words. By using words or playing grammatical games, language comes alive.

Third, invented languages and imaginary scenes are presented as models or objects of comparison that are meant to throw light on the facts of our language by means of similarities and differences (PI 130). In other words, the purpose of grammatical play is not to demonstrate how language has evolved or to prepare for the future regimentation of language. Wittgenstein acknowledges:

> We seem to be asking about the natural history of human beings. Yet you know that in some obvious sense we are not interested in natural history…it is not natural history to invent languages of our own, as I have done…We are interested in language only insofar as it gives us trouble. I only describe the actual use of a word if this is necessary to remove some trouble we want to get rid of. (AWL 97)

Grammatical play allows Wittgenstein to avoid ineptness or emptiness in his assertions by presenting language games as objects for consideration, rather than preconceived ideas to which reality must correspond. He warns against the latter as a form of philosophical dogmatism.

Wittgenstein's language games and grammatical play are simple but not easy. Although his investigations often begin with a simple statement such as 'This is my body,' they quickly lead to the investigation of complex metaphysical claims such as 'I know that *I* feel pain but not that anyone else does' or 'Only my own experience is real.' He writes, for example:

> There is a temptation for me to say that only my own experience is real: "I know that *I* see, hear, feel pains, etc., but not that anyone else does. I can't know this, because I am I and they are they."
>
> On the other hand I feel ashamed to say to anyone that my experience is the only real one; and I know that he will reply that he could say exactly the same thing about his experience. (BB 46)

For Wittgenstein, both the temptation and its apparent absurdity are expressions of philosophical puzzlement and conceptual confusion. However, he acknowledges that it is often difficult to recognize that metaphysical assertions express discontentment with the grammar or use of words, because the words in question can also be used to state facts of experience (BB 56–57):

Consider how the following questions can be applied, and how settled:

(1) "Are these books *my* books?"
(2) "Is this foot *my* foot?"
(3) "Is this body *my* body?"
(4) "Is this sensation *my* sensation?"

Each of these questions has practical (non-philosophical) applications.

(1) Think of cases in which my foot is anaesthetized or paralyzed. Under certain [other] circumstances the question could be settled by determining whether I can feel pain in this foot.
(2) Here one might be pointing to a mirror-image. Under certain circumstances, however, one might touch a body and ask the question. In others it means the same as: "Does my body look like *that*?"
(3) Which sensation does one mean by 'this' one? That is: how is one using the demonstrative pronoun here? Certainly otherwise than in, say, the first example! Here confusion occurs because one imagines that by directing one's attention to a sensation one is pointing to it (PI 411).

For Wittgenstein, one way to respond to metaphysical confusion is to look at how words are actually used in our language. He writes: 'What *we* do is to bring words back from their metaphysical to their everyday use' (PI 116). However, it is important to note that looking at how we use words is itself an act of imagination, for we are reminding ourselves of the *kinds of statements* we make about phenomena. In other words, 'grammatical characteristics must characterize what it is that we talk about, as opposed to what is said about them' (LPP 293). These investigations are conceptual. As noted earlier, Wittgenstein only describes the actual use of words if this is necessary to remove some trouble we want to get rid of. Each of the numbered questions in §411 invokes a variety of imagined scenes. For each question, we can imagine circumstances that turn the use of these words into a move in a game such that they lose everything that is philosophically astonishing. This is not meant to pre-empt conceptual confusion or dismiss philosophical concerns as pseudo-problems. Rather, such imaginative play serves several purposes. First, this exercise reminds us of the diverse uses of words, as in cases (1) and (2). In so doing, it reveals that we may be confusing two or more uses of a word, or that their different applications may not be clearly before our eyes. Second, it demonstrates that a given word or expression may not *yet* have a clear meaning, although one might be given for it, as in case (3) and perhaps case (4). Although 'This is *my* body' may be used to state certain facts of experience in the examples imagined above, these are not the examples that philosophers generally invoke. Third, looking at how we actually use words demonstrates that a particular meaning may be unclear although we are using simple or familiar words, such as case (4). And finally, such a method reminds us that we are not always tempted to make metaphysical assertions, and that we need to become aware of when and under what circumstances such temptations arise. These imagined scenes enable us to develop our own conceptual thinking in greater depth and complexity.

If we do not keep specific examples or circumstances in mind, we may be tempted to draw metaphysical conclusions. For example, when sensations are distinguished from bodies and books, it is easy to think that something important has been discovered. Wittgenstein refers to this as the apparent queerness of the grammar of a word, which results from the temptation to use it in several different ways. The grammars of 'this' and 'my' differ with reference to books, bodies, and sensations, and we are tempted to see important metaphysical implications in these differences, not only for our understanding of such objects but for understanding ourselves. However, the importance of these distinctions rest, not on the discovery of new facts, the articulation of new theories, or the significance of personal experiences, but in the recovery of what is lost when we fail to recognize or acknowledge such differences. For example, in philosophy we often focus on, or attempt to distinguish between, different kinds of 'objects,' and we think we are making statements such as 'A railway train, a railway station and a railway car are different kinds of objects,' whereas our statements are analogous to 'A railway train, a railway accident, and a railway law are different kinds of objects' (BB 64). Wittgenstein's conceptual play attempts to demonstrate such differences. To say that books and human bodies are different kinds of objects, or that our knowledge or identification of them differs by degree, fails to do justice to the concept of being human.

The question 'Is this body *my* body?' appears philosophically unproblematic because we can imagine numerous examples in which the question can be asked and answered empirically. Wittgenstein acknowledges that the question has practical, non-philosophical applications. However, when abstracted from such applications, we continue to think that the question remains meaningful. For example, in Moore's 'Defence of Common Sense,' the first proposition he claims to know with certainty to be true is: 'There exists at present a living human body, which is *my* body' (Moore 1959: 33). Wittgenstein asks, 'What is the criterion for 'This is *my* body'? In other words, how might we settle or verify such a claim? As he did in §411, he begins with reference to a particular part of his body. In this case, it is his nose rather than his foot. He suggests that a criterion for 'This is my nose' might be that it is attached to a particular body or *possessed* by it (AWL 24). Then he asks, what is the criterion for 'This is my body'? He anticipates and articulates the temptation to say that there is a soul to which a body belongs, and that my body is the body that belongs to me.

He challenges this temptation by imagining the following scene:

> Suppose that all bodies were seen in a mirror so that all were on the same level. I could talk of A's nose and *my* nose in the same way. But if I singled out a body as mine, the grammar changes. Pointing to a mirror and saying 'This is my body' does not assert the same relation of possession between me and my body as is asserted by 'This is A's nose' between 'A's body and A's nose'. (AWL 24)

Wittgenstein imagines looking in a mirror for several reasons. First, looking in a mirror is one way in which we observe ourselves and our bodies, and it is easy to imagine such a scene. Second, it is possible *not* to recognize oneself in a mirror or not to recognize a particular reflection as one's own. Wittgenstein asks: 'Isn't it *imaginable* that I lived all my life looking in a mirror, where I saw faces and did not know which was my face, nor how my [nose] was distinguished from anyone else's?' (AWL 17) Third, to *observe* oneself as one observes others is not as simple, easy, or obvious a task as it might at first appear. Wittgenstein acknowledges: 'I can't see myself ordinarily; and I can't observe myself unobserved, even if I do watch myself in a mirror' (LPP 95). Fourth, looking in a mirror allows Wittgenstein to observe his body and those of others on the same level (in the mirror) and in real time (in the present). It is a simple, elegant, and low tech act of imagination that today might be augmented through experiences in Virtual Reality that enable us to leave our bodies, change bodies, or look at them from other locations. And finally, *reflections* in a mirror are often used by Moore and others in Cambridge while discussing the external world and things internal or external to our minds (Moore 1959: 136). Only Wittgenstein incorporates a *mirror* into his investigations.

When all bodies are seen in a mirror, A's nose is possessed by A's body in the same way that Wittgenstein's nose is possessed by his body. Thus, 'A's nose' and 'my nose' [when said by Wittgenstein] and 'A's body' and 'my body' [when said by Wittgenstein] play a grammatically similar role. However, if Wittgenstein *points* to the mirror and says 'This is *my* body' the grammar changes. Although it still looks like an empirical statement, its meaning is not yet clear. Think of the examples provided in §411 above. Wittgenstein imagines specific cases in which he might not recognize his body in a mirror. But when he stands in front of a mirror while doing philosophy and points to his reflection saying 'This is *my* body,' what does it mean?

Try the exercise: Imagine or actually stand in front of a mirror and say 'This is *my* body' while pointing to the mirror. It is unclear what one is pointing to and what the word 'this' refers to. When Wittgenstein points to the mirror, he points to his reflection in the mirror not to his physical body. The gesture of pointing is significant and its meaning changes with each imagined scene. Is it possible to point directly to one's own body? Ordinary, we are more apt to put a hand on our chest and say 'This is *my* body.' However, as Wittgenstein notes, this is a gesture of emphasis not identification. We often become confused when doing philosophy because we think we are pointing to something when we are concentrating our attention on it. However, we might imagine saying 'This is *my* body' when asserting autonomy in the face of physical or verbal violence. But in such cases, reference to one's body is an affirmation of one's autonomy, dignity, or humanity. In this exercise, pointing to a mirror image and saying 'This is *my* body' is not *yet* a meaningful act of identification.

Wittgenstein continues by imagining a scene in which 'This is *my* body' *becomes* an act of identification. He imagines that bodies become transparent but their mirror images remain. Thus, we no longer see bodies—we see through them. In such circumstances, mirror images might be used as a means of identification, and saying 'This is *my* body' might become an act of identification because we no longer see bodies directly. However, Wittgenstein observes:

> If all bodies are seen in a mirror and the bodies themselves become transparent but the mirror images remain, my body will be where the mirror image is. And the criterion for something being my nose will be very different from its belonging to the body to which it is attached. (AWL 24)

If we imagine that Wittgenstein's body *becomes* transparent while he points to himself in the mirror, two interesting things happen. First, his reflection points out of the mirror at a transparent body. Thus, it points to nothing at all. Now, what does 'this' refer to? Pointing to a particular body or its location will only be meaningful with reference to the mirror world. This example also changes the earlier criterion that noses belong to the bodies to which they are *attached*, for we will no longer speak of the reflection of a nose being *attached* to the reflection of a body in the same way as we would if referring to bodies that are not in the mirror world. And what about the *possession* of one's own body? Further, we may own or possess a mirror, but do we own or possess the reflections in it? (Z 670)

Wittgenstein then asks: 'In the mirror world, will deciding which body is mine be like deciding which body is A's?' (AWL 24) He leaves this question open for it requires further investigation. How might such identification take place when bodies are transparent? Wittgenstein imagines that A's body is identified with reference to a voice called 'A' which is correlated to the body, and his body is identified with reference to a voice called Wittgenstein. If bodies become transparent, it is easy to imagine relying on voices and shifting identification and verification from sight to sound. But, as Wittgenstein notes, it will not make sense for him to ask 'Which is *my* voice?' (AWL 24). This is particularly funny if he asks the question aloud. In case we are tempted to think that sound is a more appropriate mode of identification than sight, Wittgenstein asks us to imagine a variation on this scene in which all voices are alike. In such a case, it will not make sense to identify bodies by their voices. He also asks us to imagine a case in which sound comes from a loudspeaker when people speak and all voices are alike. In such a case, not only will voices no longer be used for the purpose of identification, but the word 'I' [or 'my'] might no longer have any use at all. 'Were all human bodies seen in a mirror, with a loudspeaker making the sounds when mouths moved, the idea of an ego speaking and seeing would become very different' (AWL 23). In such a case, the personal pronoun 'I' might only be used with proper names.

Wittgenstein imagines a final method of identification during this investigation. He asks, 'What is the criterion for one of the bodies [in the mirror world] being mine?' and answers: 'It might be said that the body which moved when I had a certain feeling will be mine' (AWL 24). The temptation to invoke feelings or sensations as a means of identification is a recognizable philosophical move. In the *Investigations*, for example, an interlocutor states: 'My kinaesthetic sensations advise me of the movement and position of my limbs' (PI 158e). The investigation of this temptation simultaneously concludes Wittgenstein's investigation of the proposition 'This is *my* body' and begins an extended investigation of sensation.

Wittgenstein presents a series of exercises in which he imagines feeling pain in the body of another (BB 49). While we may think that such exercises are unimaginable, he challenges us to try. He never presents imaginary scenes in order to demonstrate—by default—that it is impossible to imagine something. On the contrary, he asserts that:

> the grammatical difficulty which we are in we shall only see clearly if we get familiar with the idea of feeling pain in another person's body. For otherwise, in puzzling about this problem, we shall be liable to confuse our metaphysical proposition "I can't feel his pain" with the experiential one "We can't have (haven't as a rule) pains in another person's [body]". (BB 49)

The temptation to use feelings or sensations as a means of identification takes for granted that we cannot feel pain in the body of another. But according to Wittgenstein, we are confusing conceptual and empirical claims. He suggests that it is not only possible to imagine feeling pain in another person's body, but that it is easy to do so. For example, he asks us to imagine that the hand which he sees and takes to be his own is connected to the body of a person standing beside him. He writes:

> Suppose I feel a pain which on the evidence of the pain alone, e.g., with closed eyes, I should call a pain in my left hand. Someone asks me to touch the painful spot with my right hand. I do so and looking round perceive that I am touching my neighbour's hand (meaning the hand connected to my neighbour's torso). (BB 50)

Or, alternately, consider a wireless connection between two bodies in which Wittgenstein feels pain when another person is injured. If we object that these pains are still Wittgenstein's own because he feels them in his own body, he offers a further example:

> Suppose I and someone else had a part of our bodies in common, say a hand. Imagine the nerves and tendons of my arm and A's connected to this hand by an operation. Now imagine the hand stung by a wasp. Both of us cry, contort our faces, give the same description of the pain, etc. Now are we to say we have the same pain or different ones? (BB 54)
>
> If in such a case you say: "We feel pain in the same place, in the same body, our descriptions tally, but still my pain can't be his," I suppose as a reason you will be inclined to say: "because my pain is my pain and his pain is his pain." And here you are making a grammatical statement about the use of such a phrase as "the same pain". (BB 54)[1]

A reluctance or refusal to use the phrase 'the same pain' will exclude certain other phrases such as 'he has my pain' or 'we both have the same pain.' But it might include phrases such as 'his pain is exactly like mine.' The point is that we may use the phrase 'the same pain' or not in this case, because we are not making a metaphysical or empirical discovery but a grammatical or conceptual statement about pain. Wittgenstein writes:

> In so far as it makes *sense* to say that my pain is the same as his, it is also possible for us both to have the same pain. (And it would be imaginable for two people to feel pain in the same – not just the corresponding – place. That might be the case with [conjoined] twins, for instance.)
>
> I have seen a person in a discussion on this subject strike himself on the breast and say: 'But surely another person can't have THIS pain!" – The answer to this is that one does not define

---

[1] For further discussion see BB 46–48. Many contemporary experiments in Virtual Reality exemplify and enact Wittgenstein's exercises in thinking. They include the experience of moving from body to body and seeing through the eyes of another. As scientific experiments, they do not *validate* Wittgenstein's conceptual play or present new information or discoveries. Rather, they lessen the philosophical task of imagining possibilities and raise shared conceptual concerns. See, for example, Rothman 2018; Yin 2018, and TheMachinetobeanother.org.

> a criterion of identity by emphatic stressing of the word "this." Rather, what the emphasis does is to suggest the case in which we are conversant with such a criterion of identity, but have to be reminded of it (PI 253).

It may appear as if a person striking himself on the breast were expressing a state of affairs within himself, which he alone can directly see. The idea is that the self is within and is to be observed by introspection. But there is no question of direct or indirect observation, because there is no question of observation at all:

> A *primitive language game*: If "I've got a pain" replaces, as it does, a moan, then *knowing* [or observing] I've got a pain doesn't come in: Nor does doubting. You can say "I can't doubt that I have a pain" if that equals something like "I can't score a goal in chess." In this case there is no knowing and no doubting. (LPP 328)

Paradoxically, this also means that it makes sense to say 'I can know that someone else is in pain, but not that I am.' Further, the statement 'Only I can know whether I am really in pain; another person can only surmise it' is either wrong or nonsensical, because others often know when we are in pain, and we do not know that we are in pain, we *are* in pain (PI 246). Such insights not only challenge and alter the emphasis philosophers place on observation, introspection, knowledge, doubt, and certainty, but on the privileged use of the pronoun 'I.'

By playing with language we come to recognize and recover the difference between being a grammatical subject and an object. Wittgenstein reminds us that pain behaviour may point to a painful place, but the subject of pain is the person who gives it expression (PI 302). These exercises demonstrate that observing, identifying, or referring to one's own body or sensations is meaningful only under certain circumstances. And to observe and objectify oneself proves to be a challenging act of imagination. Wittgenstein draws out the implications of this insight:

> I want to say that realizing that the word "I" does not mean the same as "my body," i.e. that it is used differently, does not mean that a new entity besides the body, the ego [or soul], has been discovered. The argument that since *I* cannot be identified with my body, there must be something else, gives the impression of reporting a discovery. All that has been discovered is that "I" is not used the same way as "my body". (AWL 60)

In other words, speaking of selves, souls, or egos is not necessarily meaningless or nonsensical. Rather, Wittgenstein has only rejected the grammar that tries to force itself on us here: 'The paradox disappears only if we make a radical break with the idea that language always functions in one way, always serves the same purpose: to convey thoughts—which may be about houses, pains, good and evil, or anything else you please' (PI 304). This confirms the need to clarify our understanding of language and individual concepts.

When Wittgenstein first expresses the philosophical temptation to state that only his experiences are real, he also expresses embarrassment and discomfort. We do not always recognize the implications of our philosophical claims. Wittgenstein writes that those not calloused by doing philosophy will recognize something amiss (PI 348). His insights have profound implications for how we live and do philosophy, not only in relation to one another but in relation to the world around us:

> When we think about the relation of the objects surrounding us to our personal experiences of them, we are sometimes tempted to say that these personal experiences are the material of which reality consists.
>
> When we think in this way we seem to lose our firm hold on the objects surrounding us. And instead we are left with a lot of separate experiences of different individuals. These personal experiences again seem vague and seem to be in constant flux. (BB 45)

Wittgenstein's conceptual play challenges realism (a theory in which objects are independent of the mind), idealism (a theory in which objects are dependent on the mind), and solipsism (a theory in which only personal experiences are real) by offering a richer and deeper understanding of ourselves as human beings, and a more interactive and participatory way of doing philosophy.

Grammatical play also demonstrates the difference between practising philosophy as an art or a science. Bartley notes that:

> [Wittgenstein] never claimed that all specifically identifiable disciplines and activities in which people engaged were separate language-games each with its own sets of rules (or grammar); it is evident that he thought things more complicated than that. Many of his followers, however, made just this mistake, supposing that each individual activity – law, history, science, logic, ethics, politics, religion – has its own special grammar or logic; that mixing the grammar of one of these with that of another leads to philosophical error; and that it is the new job of the philosopher – his new research program under the Wittgensteinian dispensation – to describe in detail these separate logics or grammars. (Bartley 1985: 144)

Such research programs approach grammatical confusion as an empirical science, and respond with theory and argumentation. However, Wittgenstein's imaginative exercises exemplify a way of thinking that is characteristic of creative writers and critics:

> [It is] one that pursues essential kinds of collective and personal self-examination through the examination, in the full richness of its powers and potentialities, of the language in which we explain ourselves to ourselves and beyond that of the multifarious underlying practices that give sense to the terms of the language, and simultaneously form us as users and speakers of it. (Harrison 2016: 224)

The exercises included in this paper demonstrate that playing with language offers conceptual clarification by looking closely at how we actually use words; by inventing languages that differ from our ordinary language in order to draw comparisons and contrasts; and by imagining general facts of nature to be other than they are in order to see that there is no one necessary form of language, and to challenge our own philosophical pictures and ideas. Such clarity also makes particular philosophical problems and their theoretical solutions disappear because they are both expressions of conceptual confusion. Wittgenstein's grammatical play challenges metaphysical theories without engaging directly with them, because theories are but one form of response to such confusion. For Wittgenstein, theories express but do not solve philosophical problems and they ultimately prove not only unhelpful but superfluous. Language games and conceptual play enable us to think differently without theory or argumentation, and thus think differently *about* theory and argumentation.

Wittgenstein also plays with language because he wants to *show*—not merely *say*—how and when we go astray. For example, he writes that it makes sense to say

'I know that someone else is in pain, but not that I am,' and that it is more difficult to imagine someone else's pain on the model of one's own than to imagine actually feeling pain in their body. Such remarks are difficult if not impossible to understand on first reading, but we come to understand them after playing with language and participating in these imaginative investigations. Wittgenstein's remarks offer insights that we must discover or rediscover for ourselves. He writes, for example, that 'it is helpful to remember that it is a primitive reaction to tend, to treat, the part that hurts when someone else is in pain, and not merely when oneself is—and so to pay attention to other people's pain behaviour as one does *not* pay attention to one's own pain behaviour' (Z 540). Only after participating in these conceptual investigations does our philosophical understanding become an act of recognition.

We learn concepts when we learn language and such learning is playful. Thus, playing with language is itself a natural response to conceptual confusion. Although its conceptual implications are important and profound, the tone of such play is creative and fun.[2] Playing with language offers clarity where there is confusion; familiarity where there is alienation or loss; creativity and co-operation where there is criticism or competition; simplicity where there is difficulty and complexity; enjoyment where there is frustration or anxiety; surprise where there is complacency; and delight where there is boredom. Imagining fictitious scenes not only enables Wittgenstein to identify and express a variety of philosophical temptations and concerns, it also allows him to play them out. The examples in this paper only hint at the complexity of interconnections between concepts, and how Wittgenstein's investigations criss-cross a field in every direction. He writes that concepts occur in a field and their relations give us trouble. A proposition like 'This is *my* body' leads to investigations of the demonstrative pronoun 'this'; the personal pronoun 'I'; the act of pointing; the human body; perception; sensation; observation; identification; experience; introspection; knowledge; certainty; doubt; and belief. When concepts are hidden in relation to one another finding connections is illuminating. What is hidden is not something to be discovered, but something that lies open to view that we do not see (PI 129). Not only do we overlook simple and familiar uses of words, but their simplicity and familiarity can encourage ways of thinking that become philosophically entrenched. Wittgenstein remarks that it sometimes seems as though concepts are formed along the lines of comfortable thinking (RPP II 637). His exercises in thinking—his extraordinarily imaginative and unusual scenes—shock us out of conceptual and intellectual complacency with a surprise that challenges and delights. He also writes that 'philosophy simply puts everything before us, and neither explains nor deduces anything—Since everything lies open to view there is nothing to explain' (PI 126). Instead, playing with language offers conceptual insight and clarification without either dismissing or diminishing the significance and depth of our philosophical questions and concerns.

[2]For further discussion, see Savickey (2016).

References

Bartley, W. W. III. 1985. *Wittgenstein*, second edition. La Salle, Illinois: Open Court.
Harrison, B. 2016. 'Leavis and Wittgenstein,' *Philosophy and Literature* 40 (1): 206–225.
Moore, G. E. 1959. 'A Defence of Common Sense.' In *Philosophical Papers*. New York: The McMillan Company, pp. 35–59.
Rothman, J. 2018. 'As Real as It Gets: Are we already living in virtual reality?' *The New Yorker*, April 2, 2018: 30–36.
Savickey, B. 1999. *Wittgenstein's Art of Investigation*. London: Routledge.
Savickey, B. 2011. 'Wittgenstein's Use of Examples,' *The Oxford Handbook of Philosophy*. Eds. O. Kuusela and M. McGinn. Oxford: Oxford University Press, pp. 667–697.
Savickey, B. 2016. 'Wittgenstein's Slapstick,' *Performance Philosophy* 2 (1): 72–82.
Savickey, B. 2017a. 'Wittgenstein's Performance Philosophy,' *Aesthetics Today: Contemporary Approaches to the Aesthetics of Nature and of Art. Proceedings of the 39th International Ludwig Wittgenstein Symposium*, Kirchberg am Wechsel, Austria. Eds. Stefan Majetschak and Anja Weiberg. Austrian Ludwig Wittgenstein Society. Berlin: de Gruyter: 419–434.
Savickey, B. 2017b. *Wittgenstein's Investigations: Awakening the Imagination*. New York: Springer.
Wittgenstein, L. 1968. *Philosophical Investigations* [PI]. Trans. G. E. M. Anscombe. Oxford: Basil Blackwell.
Wittgenstein, L. 1969. *The Blue and Brown Books: Preliminary Studies for the Philosophical Investigations* [BB]. Oxford: Basil Blackwell.
Wittgenstein, L. 1979. *On Certainty* [OC]. Eds. G. E. M. Anscombe and G. H. von Wright. Trans. D. Paul and G. E. M. Anscombe. Oxford: Basil Blackwell.
Wittgenstein, L. 1980. *Remarks on the Philosophy of Psychology. Volume* II. [RPP II] Eds. G. H. von Wright and H. Nyman. Trans. C. G. Luckhardt and M. A. E. Aue. Oxford: Basil Blackwell.
Wittgenstein, L. 1981. *Zettel* [Z]. Second edition. Eds. G. E. M. Anscombe and G. H. von Wright. Trans. G. E. M. Anscombe. Oxford: Basil Blackwell.
Wittgenstein, L. 1982. *Wittgentein's Lectures: Cambridge* 1932–1935 [AWL]. Ed. Alice Ambrose. Oxford: Basil Blackwell.
Wittgenstein, L. 1984. *Philosophical Grammar* [PG]. Ed. R. Rhees. Trans. A. Kenny. Oxford: Basil, Blackwell.
Wittgenstein, L. 1988. *Lectures on Philosophical Psychology* 1946–47: Notes by P. T. Geach, K. J. Shah and A. C. Jackson [LPP]. Ed. P. Geach. London: Harvester Wheatsheaf.
Yin, S. 2018. 'There is Spirit: In Virtual Reality, How Much Body Do You Need?' *The New York Times*, May 21, 2018: D2.

Chapter 6
Wittgenstein's 'Grammatical Naturalism'

Jonathan Beale

> *Was wir liefern, sind eigentlich Bemerkungen zur Naturgeschichte des Menschen; aber nicht kuriose Beiträge, sondern Feststellungen, an denen niemand gezweifelt hat, und die dem Bemerktwerden nur entgehen, weil sie ständig vor unsern Augen sind.*
> *What we are supplying are really remarks on the natural history of human beings; we are not contributing curiosities however, but observations which no one has doubted, but which have escaped remark only because they are always before our eyes.*
> Wittgenstein, *Philosophical Investigations* §415

6.1 Naturalistic Observations

Interpreters of the later Wittgenstein as a naturalist often support their interpretation through reference to his remarks, in the first and second parts of the *Investigations*, on 'the natural history [*Naturgeschichte*] of human beings' (PI §§25, 415) and

Earlier versions of this chapter were presented to the London Wittgenstein Discussion Group in September 2018; at the University of Hertfordshire's Philosophy Research Seminar in November 2018; and at the University of Reading's Wittgenstein Forum in December 2018. Many thanks to the participants at these events for helpful points and discussion. This chapter draws upon my review of Kevin M. Cahill and Thomas Raleigh (eds.), *Wittgenstein and Naturalism* (New York: Routledge, 2018), published in *Notre Dame Philosophical Reviews*, July 22, 2018 (http://ndpr.nd.edu/news/wittgenstein-and-naturalism/). I am grateful to the publishers for permission to draw on that material, particularly to the late Founder and long-time Co-Editor-in-Chief of *Notre Dame Philosophical Reviews*, Gary Gutting, who invited me to review that book and sadly passed away earlier this year.

J. Beale (✉)
Researcher-in-Residence, Tony Little Centre for Innovation and Research in Learning, Eton College, Windsor, Berkshire SL4 6DW, UK
e-mail: j.beale@etoncollege.org.uk

S. Wuppuluri and N. da Costa (eds.), *WITTGENSTEINIAN (adj.)*,
The Frontiers Collection, https://doi.org/10.1007/978-3-030-27569-3_6

'general facts of nature' (*Naturtatsachen*, PI §142; PPF §§365–6).[1] Some passages where Wittgenstein refers to either of these suggest that he holds that these are often ignored in philosophical investigations (e.g., PI §§142, 415). A line argued by some interpreters, which this chapter defends, is that these function as 'naturalistic' phenomena to which Wittgenstein directs our attention as part of his method of providing surveyable representations (*übersichtliche Darstellungen*) of grammar. On this interpretation, Wittgenstein seeks to bring general facts about human nature into clear view as facts to which we should pay attention to elucidate certain concepts. Such 'observations' (*Feststellungen*, PI §415) about human nature function as 'reminders' (PI §127) of the ways we use words and how far detached from their use we can become when, for example, engaging in metaphysical theorizing. The purpose of such observations is to support the central aim of Wittgenstein's later philosophy: to expose misunderstandings about the workings of language, which generate the conceptual confusions underlying philosophical problems (PI §§89, 109, 122).

Daniel D. Hutto and Glenda Satne, for example, have recently endorsed this interpretation, arguing that Wittgenstein provides 'philosophical reminders about facts of our natural history' and 'perspicuous representations of general facts of nature' (Hutto and Satne 2018, 73). On this interpretation—which, as the authors acknowledge, is a development of Marie McGinn's influential interpretation of Wittgenstein as a liberal naturalist (McGinn 2010, 2013)—reminding us of and elucidating natural facts supports Wittgenstein's later method of providing surveyable representations of grammar.

Some passages where Wittgenstein refers to 'general facts of nature' and our 'natural history' suggest that he holds that these are often ignored in philosophical investigations concerning the concepts in question:

> What we have to mention in order to explain the significance, I mean the importance, of a concept are often extremely general facts of nature: such facts as are hardly ever mentioned because of their great generality. (PI §142)[2]
>
> What we are supplying are really remarks on the natural history of human beings; not curiosities, however, but facts [*Feststellungen*] that no one has doubted, which have escaped notice only because they are always before our eyes. (PI §415)

The ambiguity of *Feststellungen* is useful to illustrate Wittgenstein's position. G. E. M. Anscombe translates this in §415 as 'observations' whereas P. M. S. Hacker and Joachim Schulte translate it as 'facts'. (The translation of §415 quoted above is by Hacker and Schulte; the translation at the start of this chapter is by Anscombe.) The 'extremely general facts of nature' or facts concerning our 'natural history' Wittgenstein seeks to bring into clear view are, for example, the ways in which we

---

[1]For example, Glock (1996, 125–6), Hookway (2017, 753–5), Hutto and Satne (2018, 73), Macarthur (2018, 42–3), McGinn (2010, 2013), Snowdon (2018, 20).

[2]This remark was a note inserted in the manuscript of the Philosophical Investigations and was published as a note at the bottom of p. 56 in G. E. M. Anscombe's translation, on the same page as §142. I have cited it as PI §142 because it is clearly connected with that remark and is published as a note at the end of that remark in the recent fourth edition of the Investigations, edited by P. M. S. Hacker and Joachim Schulte.

learn our mother tongues and learn to follow rules. His remarks about such facts are 'remarks on the natural history of human beings' insofar as they concern human nature. Such observations are intended to remind us about how we use words, the ordinary uses of words and the roles certain concepts play within our lives, to see how far detached from these we can become when engaged in philosophical theorizing. The role of such observations is to help elucidate concepts, such as meaning and understanding.

'Naturalistic' observations emerge from the very beginning of the *Investigations*. Following the quotation from Augustine's *Confessions* which opens the book, Wittgenstein writes:

> Augustine does not speak of there being any difference between kinds of word. If you describe the learning of language in this way you are, I believe, thinking primarily of nouns like "table", "chair", "bread", and of people's names, and only secondarily of the names of certain actions and properties; and of the remaining kinds of word as something that will take care of itself. (PI §1)

Throughout the passages discussing the Augustinian picture which follow, Wittgenstein draws attention to the ways in which a child learns a language (PI §§5, 7, 9), sometimes by comparison with Augustine's recollection of how he learnt language as a child (e.g., PI §32). Wittgenstein considers how we learn the uses of words and their rules for use, and thereby learn the meanings of words (PI §35, 54, 77). Such observations draw attention to obvious facts about how we learn a language and are 'naturalistic' in the sense that they are things that human beings do in the process of learning how to engage in the world. Yet, such facts, Wittgenstein argues, are often ignored in philosophical investigations and ignorance of them can engender conceptual confusions, for example of the kind emerging from the Augustinian picture. By making clear such 'natural facts', Wittgenstein aims to elucidate concepts and thereby dissolve the conceptual confusions that underlie and generate philosophical problems. In this sense, naturalistic observations support his later method of grammatical investigation.

Certain notions in Wittgenstein's later philosophy have a 'naturalistic' character, in that they draw attention to facts about human nature on which Wittgenstein lays emphasis to elucidate concepts. One such notion is 'training' (*Abrichtung*).[3] Several interpreters of Wittgenstein as a naturalist have used this notion to support their interpretation. Training sometimes appears within the context of Wittgenstein's observations about how we learn a language and use words (e.g., PI §§5, 6, 9, 27). In German, '*Abrichtung*' has the connotation of animal training, especially that of dogs; it is not typically used to denote the 'training' of human beings. Wittgenstein uses this notion to emphasise the importance of a certain kind of learning. We teach children and in some contexts adults to follow rules in a similar way, in some respects, to how we train animals: through conditioning. This fundamental aspect of the way in which we learn is 'naturalistic' in the sense of being a feature of our natural history, one that we share in basic respects with animals. *Abrichtung*, interpreters have emphasised, draws attention to features about our natural history, such as learning our first

[3] See for example Hutto and Satne (2018, 69), Smith (2018, 249–50), Snowdon (2018, 21).

language and learning to follow rules, just like other features of our natural history such as learning to walk, eat, drink and play (*cf.* PI §25).

The connection between this concept and teaching emerges most clearly in *Zettel*:

> Any explanation has its foundation in training [*Abrichtung*]. (Educators ought to remember this.) (Z §419)

Wittgenstein's lesson for educators is that in the early stages of education in many contexts, an important way of learning is a straightforward process of repeating simple tasks to master fundamental basics, without a vision of what purpose the education is ultimately aimed towards fulfilling. For example, rote learning, such as a child learning the sequence of intervals in the major scale by repeating 'tone, tone, semitone, tone, tone, tone, semitone' without a clear vision of why exactly they are doing this, other than being told that it is a basic component of music theory. Similarly, when learning a language, a child might repeat counting from one to ten until they find it harder to get the order wrong than right. Such basic processes of learning are similar in some respects to how we train animals: a dog has no idea of why it is being repeatedly told to 'sit' or 'roll over'; following many repetitions, with rewards and sanctions, it will, typically, eventually respond to such commands in the desired way. It is this kind of learning that Wittgenstein has in mind with *Abrichtung*.

Two of the most important notions in Wittgenstein's later philosophy also have a naturalistic character: 'language games' (*Sprachspiele*) and 'forms of life' (*Lebensformen*). Hans-Johann Glock writes that our natural history is 'the history of cultural, language-using creatures' (Glock 1996, 125–6). Wittgenstein draws our attention, Glock writes, to 'anthropological phenomena' which form part of our 'natural history' (RPP I §1109). An example within the context of Wittgenstein's remarks on our natural history is the following:

> Giving orders, asking questions, telling stories, having a chat, are as much a part of our natural history as walking, eating, drinking, playing. (PI §25)

'These activities', Glock writes, 'are cultural activities, forms of social interaction'. To understand a language we have to imagine it within our normative practices. This involves understanding the place of words within the context of the language games and the non-linguistic activities—i.e., forms of life—interwoven within those games, for 'the *speaking* of language is part of an activity, or a form of life (PI §23; *cf.* §7). The emphasis upon the speaking of language as an activity motivates the use of the term 'language game': 'I shall also call the whole, consisting of language and the actions into which it is woven, the "language-game"' (PI §7). To imagine a language is to imagine a 'culture' (BB 134) and 'To imagine a language means to imagine a form of life' (PI §19).[4] Imagining a language game draws attention to aspects of human nature.

Wittgenstein's naturalism is therefore anthropological, Glock writes. The 'facts of life' to which Wittgenstein alludes 'are the specific patterns of behaviour which

[4]Wittgenstein replaces Kultur (culture) with Lebensform (form of life) in 1937; the former does not appear in the Philosophical Investigations.

together constitute *a* form of life' (Glock 1996, 125).[5] With notions such as language games and forms of life, Wittgenstein draws our attention to the place that language has within the context of our lives as a whole, which involves attention to our linguistic and non-linguistic activities. These notions are naturalistic in the sense that they concern human nature.

Before proceeding, it should be clarified that Wittgenstein did not ever use the terms 'naturalism' or 'naturalist'. However, throughout the *Investigations* he uses the German equivalent of the English adjective 'natural', '*natürlich*', to make various points. '*Natürlich*' can also be used to mean 'naturally', which has a more common usage in German than in English. Wittgenstein employs *natürlich* and its cognates in this way and related ways throughout the *Investigations*—e.g., 'it comes naturally to us', or 'it is most natural to…' (PI §§16, 185, 256, 308, 318, 595–6, 642, 647, 691); but given its common usage in German, this in itself does not reveal anything of significance. Wittgenstein also writes of the 'feeling' (*Gefühl*) of 'naturalness' (*Natürlichkeit*) and 'unnaturalness' (*Unnatürlichkeit*) (PI §596); he uses phrases such as 'what would be more natural in this language-game' (*natürlicher*, PI §48); he talks about natural laws (*Naturgesetze*, PI §§54, 325) and natural science (*Naturwissenschaft*, PI §§81, 89, 392); the phrase 'natural language', appears in the quote from Augustine's *Confessions* in the opening passage of the *Investigations* (PI §1; '*verbis naturalibus*'); and he speaks of 'natural expressions' (*natürlichen Äußerungen*, PI §§209, 256, 647).

In addition to *Naturgeschichte* and *Naturtatsachen*, the final of the above—'natural expressions' (*natürlichen Äußerungen*)—is of most relevance to Wittgenstein's naturalism, because it bears a connection with the conception of meaning as use, a notion that is arguably connected with the kind of 'naturalistic' enterprise in which Wittgenstein is engaged. David Macarthur, for example, in a recent article arguing that the later Wittgenstein is a liberal naturalist, connects *Naturgeschichte* and *Naturtatsachen* with meaning as use, writing that 'Wittgenstein calls our attention to … the "natural history of the human"' when he reminds 'us of the ordinary use of words as they function in the everyday lives of human beings considered as natural beings in a natural world' (Macarthur 2018, 42). The naturalistic aspect emerges primarily from Wittgenstein's emphasis on descriptions of phenomena as we describe them in everyday life. This is a play on meaning as use, a point also recently discussed by Paul F. Snowdon and by Hutto and Satne in articles on Wittgenstein's naturalism (Snowdon 2018, 20–1; Hutto and Satne 2018, 63).

[5]This supports the so-called 'naturalistic' interpretation (as opposed to the 'transcendental' interpretation) of Wittgenstein's suggestion that forms of life provide the foundations of language, in the sense of constituting 'the given' on the basis of which words gain meaning: 'What has to be accepted, the given, is—so one could say—*forms of life*' (PPF §226). On the naturalistic and transcendental interpretations, see Glock 1996, 125–6.

6.2 Naturalism as Anti-metaphysics

There is another sense in which Wittgenstein's later philosophy can be described as 'naturalistic': his opposition to metaphysics. Wittgenstein held that metaphysics is the main cause of the conceptual confusion targeted in his later work. This is because he holds that metaphysics is the area of philosophy that most blurs the distinction between conceptual and empirical investigations, a point he makes clear in *Zettel*:

> Philosophical investigations: conceptual investigations. The essential thing about metaphysics: it obliterates the distinction between factual and conceptual investigations. (Z §452)

The distinction between these two types of investigation is of paramount importance in Wittgenstein's later philosophy. Wittgenstein holds that if this distinction is blurred, we risk falling into methodological confusion where we approach philosophical questions in a scientific manner—e.g., thinking we need to offer theories to solve philosophical problems and consequently engaging in a process of quasi-theory construction. This alleged source of confusion can be traced to a tendency he sought to expose: an 'irresistible temptation' to 'ask and answer questions in the way science does':

> Our craving for generality has another main source: our preoccupation with the method of science. … Philosophers constantly see the method of science before their eyes, and are irresistibly tempted to ask and answer questions in the way science does. This tendency is the real source of metaphysics, and leads the philosopher into complete darkness. I want to say here that it can never be our job to reduce anything to anything, or to explain anything. Philosophy really is "purely descriptive". (BB 18)

This 'irresistible temptation' is something with which Wittgenstein is concerned in PI §109, where he writes of our 'urge to misunderstand' the workings of language—an urge that generates the kind of confusion that underlies philosophical problems. Metaphysics, according to Wittgenstein, finds its source in this urge.

Macarthur uses Wittgenstein's opposition to metaphysics to support his argument that the later Wittgenstein was a liberal naturalist. Macarthur argues that Wittgenstein endorses a non-doctrinal, 'non-explanatory liberal naturalism', aimed at helping us overcome 'our tendency to metaphysicalize' (Macarthur 2018, 45–6). Macarthur connects this tendency with Wittgenstein's views on human nature. Macarthur interprets Wittgenstein's aim of 'bring[ing] back words from their metaphysical to their everyday use' as 'bringing ourselves back from unnaturalness to naturalness' (Macarthur 2018, 43). This is inspired by Stanley Cavell's account of Wittgenstein's naturalism. Cavell reads Wittgenstein's diagnosis of the human tendency to fall into metaphysical confusion as an aspect of Wittgenstein's account of human nature: 'we might … wish to say that what is natural to the human is precisely the unnatural' (Cavell 2004, 276–7; Macarthur 2018, 43). While Macarthur's account is convincing, this chapter will argue that although Wittgenstein's opposition to metaphysics and his observations concerning natural facts constitute what can plausibly be called 'naturalistic' aspects of his philosophical method, we should endorse those points but stop short of using these to motivate the ascription of liberal naturalism to Wittgenstein, given the objections facing that move.

There is another way in which we can identify a connection between Wittgenstein's opposition to metaphysics and naturalism. The later Wittgenstein is opposed to argumentative moves that make appeal to supernatural phenomena to shed light on philosophical problems. An important feature of his later work is his emphasis on our ordinary, everyday experience as the subject matter for dissolving philosophical problems. Snowdon has recently illustrated this by appealing to the absence in Wittgenstein's discussions of meaning and understanding of 'the ontology of thoughts (and propositions), and their components, plus the hypothesized psychological relation to them' as 'things which are not part of the natural realm' (Snowdon 2018, 26). Snowdon offers the example of Frege's conception of the nature of thoughts, where Frege postulates a 'third realm' which exists apart from the mental and physical world.[6] Snowdon argues that 'Wittgenstein's complete avoidance' of such ontologies of phenomena 'is deliberate' and 'should be counted as an element in his naturalism' (Snowdon 2018, 26). He suggests that we can understand Wittgenstein's discussions of notions with a normative character, such as understanding and rule-following, 'as attacks by him on what ... "non-naturalistic" models [of such notions] involve' (Snowdon 2018, 26). On this reading, Wittgenstein's accounts of normative notions are partly constructed around an attack on non-naturalistic explanations of such notions—e.g., explanations of the kind offered by Frege.[7]

So there are two senses in which the later Wittgenstein's philosophy can be considered 'naturalistic'. First, observations about our 'natural history' and 'general facts of nature', such as the ways in which we learn a language, are employed to support Wittgenstein's method of providing surveyable representations of grammar. Second, Wittgenstein's opposition to metaphysics, where this is either made explicit or where there is an absence of recourse to anything metaphysical in his philosophical investigations.

6.3 Liberal Naturalism

The later Wittgenstein has been interpreted as a naturalist of various kinds. The most common interpretation is that he was a 'liberal naturalist', of varying sorts.[8] A number of possible naturalistic aspects of Wittgenstein's later philosophy have been outlined in the existing literature; the two above are perhaps the most saliently 'naturalistic' ones and have been used by several interpreters to support their interpretations that Wittgenstein was a naturalist. Readings of the later Wittgenstein as a liberal naturalist have increased in recent years, as has interest in his relation-

[6]See Frege, 'The Thought' ['*Der Gedanke*'] (1918).

[7]For an excellent historical analysis of Wittgenstein's change in thought from his early to later work in terms of a movement back towards the Fregean idea that the representational contents of thought cannot be reduced to particulars, see Charles Travis's recent article on Wittgenstein's naturalism, 'The End of an Affair' (Travis 2018).

[8]See, for example, Crary (2007), Hutto and Satne (2018), Livingston (2015); Macarthur (2009, 2015, 2018), McGinn (2010, 2013), Putnam (2008, 108), Smith (2017, 2018).

ship with naturalism, indicated, for example, in the publication last year of the first volume on his relationship with naturalism, *Wittgenstein and Naturalism*.[9] Several other forms of naturalism are ascribed to him in that volume. For example, Snowdon argues that Wittgenstein is committed to 'naturalismR' (Snowdon 2018, 16) and Eugen Fischer argues that 'meta-philosophical naturalism' is 'consistent with broadly Wittgensteinian aims and strictures' (Fischer 2018, 260–1). Several contributors, such as Macarthur and Benedict Smith, endorse the more orthodox position that Wittgenstein is a liberal naturalist (Macarthur 2018; Smith 2018). Hutto and Satne describe Wittgenstein as a 'liberating naturalist', in the sense that he is concerned with liberating us from 'certain mystifying tendencies of thought', such as the tendency to engage in metaphysics. Instead, Wittgenstein offers accounts of notions such as meaning through attention to our everyday practices, such as the ways we use words (Hutto and Satne 2018, 62–3). Wittgenstein is also a major influence for the most well-known forms of liberal naturalism, advanced by John McDowell and Hilary Putnam respectively. Several philosophers who read Wittgenstein as a liberal naturalist or a naturalist endorsing a related position, such as 'liberating naturalism', endorse the two points outlined in the previous sections as part of their argument, that Wittgenstein was a naturalist.

Broadly speaking, naturalism can be defined as the view that all that exists is the natural world and it should be studied as such (Williamson 2007, 18). Naturalism thus involves ontological and methodological claims. If we accept the ontological claim that all that exists is the natural world, the best candidate for finding out about the natural world is the natural sciences. So naturalism could be defined a little less crudely as the view that all that exists is the natural world and the best way to find out about it is through the methods of the natural sciences (see Williamson 2011).

Sometimes the methodological and ontological theses are separated, yielding a distinction between two forms of naturalism, *ontological* and *methodological naturalism*. Ontological naturalism is the first conjunct in the crude definition offered above: the natural world is all there is. Methodological naturalism is the attempt to find complete explanations for all phenomena without appeal to anything beyond the natural world. Methodological naturalism does not entail an ontological claim about what exists; it goes by a principle of parsimony to impose a methodological limit on the ontological *explananda*. Methodological naturalism is, therefore the weaker of the two positions.[10]

Scientific naturalism is the view that the sciences can, in principle, provide a complete ontology, and the discoveries and explanations offered by the sciences determine what we can know. Like all major types of naturalism, scientific naturalism takes a variety of forms; some of these vary in the degree to which they are reductionist. More reductive forms claim that we only need the resources of the natural sciences, although an important place is often given to mathematics or mathematical logic—

[9] Cahill and Thomas (2018). For a critical overview, see Beale (2018).

[10] For a useful summary of ontological and methodological naturalism and of the role of naturalism in contemporary philosophy, see Cottingham (2005, 109–113).

e.g., in Quinean naturalism. Less reductive forms include other sciences within their conceptual resources, such as the social sciences.[11]

'Liberal naturalism' has been a term of art since the publication in 2004 both of the volume *Naturalism in Question* and Gregg Rosenberg's *A Place for Consciousness*.[12] Forms of what have been retrospectively dubbed 'liberal naturalism' were propounded earlier than those outlined in *Naturalism in Question* (e.g. the naturalism of McDowell); but, to my knowledge, no philosopher self-identified as a liberal naturalist until the publication of the aforementioned works. In the case of *Naturalism in Question*, which contains papers by the likes of McDowell, Putnam, Donald Davidson, and Richard Rorty, one of the volume's central concerns is to adumbrate as 'an alternative to scientific naturalism … a new nonreductive form of naturalism' which is 'philosophically more liberal'. The more liberal aspect is its 'more inclusive conception of nature' (De Caro and Macarthur 2004, 1).[13] This is what fundamentally distinguishes scientific from liberal naturalism: the latter allows for a plurality of ways of understanding what counts as 'natural'—i.e. more than just those recognised by the sciences.

The central tenets of liberal naturalism, which it shares with scientific naturalism and which earn it the title of 'naturalism', are, first, its rejection of all forms of supernaturalism and, second, its concern with taking the sciences seriously (De Caro and Macarthur in Putnam 2012, 16–17). Liberal naturalism is therefore less reductive than scientific naturalism. The only reductive aim of liberal naturalism might be the first tenet: to reject all forms of supernaturalism. The primary *desideratum* motivating liberal naturalism is to conceive of a more inclusive ontology than that offered by scientific naturalism, yet not one overly inclusive—an aim some argue it does not always fulfil.[14]

All forms of liberal and scientific naturalism share a commitment to the ontological tenet: all that exists is the natural world. But they disagree about how we should define 'natural', with liberal naturalism taking a broader view about how this should be understood. Consequently, liberal naturalists disagree with scientific naturalists about what the methodological tenet amounts to. Some liberal naturalists argue that we require a plurality of ways of understanding the natural world.

[11]See Macarthur (2018, 34) for a clear overview of scientific naturalism in contrast to the liberal naturalism that he argues Wittgenstein endorses.

[12]De Caro and Macarthur (2004); Rosenberg (2004).

[13]Rosenberg's use of the term is different but fundamentally consistent with this. Rosenberg 'proposes a place for consciousness in nature' by developing a framework that 'ties experience into a theory of the categorical foundations of causation'; he uses liberal naturalism as the 'name for views of this type' (Rosenberg 2004, ix).

[14]E.g. Hutto and Satne (2018, 61–2).

6.4 Reasons Wittgenstein Can Be Considered a Liberal Naturalist

There are at least two reasons Wittgenstein can be considered a liberal naturalist. First, he endorses a plurality of ways of understanding reality. Wittgenstein does not hold that the methods of the natural sciences exhaust our understanding of reality. This is so whether 'science' covers the natural sciences or the sciences more widely—e.g., a less strict form of scientific naturalism. Even construing science as *Wissenschaft*—a far wider concept than that denoted by 'science'—would not be acceptable as a rubric for the discipline that exhausts our understanding of reality. Wittgenstein endorses the view other disciplines and areas of life also provide ways of understanding reality, such as the arts.

Consider the following passage from 1939–40, where Wittgenstein expresses his opposition to a widespread view of the role of the arts in Western culture:

> People nowadays think that scientists are there to instruct them, poets, musicians, etc. to entertain them. *That the latter have something to teach them* never occurs to them. (CV 42)

Wittgenstein is opposed to the view that the arts are not didactic, but are merely forms of entertainment. His opposition to this is related to the disdainful attitude he held towards his time and Western culture, one strongly influenced by Oswald Spengler's *The Decline of the West* (*Der Untergang des Abendlandes*, 1918). Therein Spengler offers an organic conception of the life of a culture, arguing that Western culture in the twentieth century had descended into its uncreative final stage, 'civilization' (CV 73; *cf.* 8). A result of this, Spengler argues, is a loss of appreciation for the arts, which is signified by their perception as mere sources of entertainment, rather than sources of understanding. In the stage of civilization, Spengler argues that a veneration of science occurs such that its products and accomplishments are regarded as the true expression of what has most value or significance. For example, theoretical knowledge is given pride of place over the value of the arts (see Spengler 1918, 177–8 and 424).

The wider significance of Wittgenstein's view that the arts provide forms of understanding can be illustrated by comparing this with the kind of understanding he thinks philosophy provides. Philosophy brings about the 'kind of understanding that consists in "seeing connections"' (PI §122). This is, Ray Monk writes, 'non-theoretical': it aims at dissolving conceptual confusion, which does not require theory construction (Monk 1999, 66). On this conception of philosophy, the aim of philosophy is to provide clarity about concepts and language in order to dissolve philosophical confusion; we are not aiming to acquire new information about the world.

Monk describes 'non-theoretical understanding' as the kind of understanding 'we have when we say we understand a poem, a piece of music, a person or even a sentence' (Monk 1999, 66). These are non-theoretical because we do not need a theory to understand such things. Theoretical understanding involves gaining new knowledge about the world; it is generated by testing hypotheses and constructing theories which are always subject to revision through discoveries. According to Wittgenstein, dissolving philosophical problems does not require any such theoretical endeavours. We need, rather, to gain clarity about the nature of such problems and their underlying conceptual confusions; gaining such clarity would make them disappear. Philosophy, on this view, does not provide knowledge, but provides *understanding*, in the form of conceptual clarity. If we wish to call that a form of knowledge, it is a form of ability knowledge: knowledge of how to dissolve philosophical problems and avoid the confusions that generate them (Glock 1996, 283; Beale 2017, 62).[15]

The second reason Wittgenstein can be considered a liberal naturalist is that his philosophy is at odds with scientific naturalism. It does not follow from this that he was a liberal naturalist; but since liberal naturalism is marked out primarily in opposition to scientific naturalism, and since there is at least one other reason for describing Wittgenstein as a liberal naturalist in addition to the other ways in which his later philosophy can be considered 'naturalistic', his opposition to scientific naturalism supports the view that he was a liberal naturalist.

There are at least four reasons why Wittgenstein cannot be considered a scientific naturalist:

1. Scientific naturalists often conceive of the relation between philosophy and science in a way anathema to the later Wittgenstein's conception of this relation.
2. Scientific naturalists often theorize about philosophical issues. This involves, *inter alia*, engaging in theory construction in philosophy, a methodological idea to which the later Wittgenstein was strongly opposed.
3. Wittgenstein is concerned with 'natural facts' and our 'natural history' in a way markedly different from scientific naturalists.
4. Scientific naturalism is often accused of being scientistic; and it is clear that Wittgenstein was strongly opposed to scientism.

Concerning (1), scientific naturalists conceive of philosophy and science as continuous with one another. For example, the most influential scientific naturalist, W. V. Quine, conceived of philosophy as 'continuous with the rest of science'.[16] Some define naturalism in general as a position holding that philosophy and science are continuous with one another. For example, in a paper on Quine on naturalism, Sander Verhaegh writes: 'Naturalism is the metaphilosophical thesis that philosophy ought to be in some sense continuous with science' (Verhaegh 2017, 96). The later Wittgenstein, by contrast, drew a firm distinction between philosophy and science.[17] He

[15]For further discussion of Wittgenstein's disdain for time in which he lived, its connection with scientism and Spengler's influence upon Wittgenstein's worldview, see Beale (2017, 66–9).

[16]Quine (1986, 430–1); see Macarthur (2018, 36). The conception of philosophy as continuous with science has a long history—consider for example Locke's conception of philosophy as science's 'underlabourer' (Locke 1689, 9–10). Quine's naturalistic conception of philosophy has been particularly influential in recent decades. The influence of naturalistic approaches influenced by Quine is sometimes cited as one of the main causes for Wittgenstein's waning influence (see, for example, Schroeder 2006, 238–241).

[17]Sometimes Quine is contrasted with Wittgenstein in terms of Quine being scientistic and Wittgenstein being opposed to scientism (e.g. Hacker 1996, 33; Winblad 1996, 97; Putnam 2011, 482).

conceived of philosophy as an entirely a priori discipline, concerned only with dissolving philosophical problems.[18]

One might question (2) by arguing that if Wittgenstein is committed to any form of naturalism, it follows that he endorses a theory. Even liberal naturalism can be construed as a theory, since it offers an ontology—i.e., a wider conception of the natural order that can accommodate phenomena that resist description within the conceptual framework of the natural sciences, such as normative phenomena. Most forms of liberal naturalism share this ontological aim.

This objection has greater traction if one describes Wittgenstein as a liberal naturalist whose position involves claims beyond the two naturalistic aspects of his philosophical method outlined earlier. For example, if one claims that Wittgenstein conceives of a wider ontology of what counts as natural than that offered by natural science, to accommodate entities resistant to accommodation within the conceptual framework of the natural sciences. But it would be highly controversial to claim that Wittgenstein offers any kind of ontology or engages in theory construction. If, however, we read Wittgenstein's naturalism as purely methodological, the objection carries less weight. His 'naturalism', on the interpretation defended here, supports his aim of dissolving philosophical problems. This makes it negative.[19] He is not offering a theory of what counts as natural or, for example, of the place of normativity within the natural order. He is, rather, making observations which might be described as 'naturalistic' as part of his method of elucidating concepts. Wittgenstein's 'naturalism', on this interpretation, is entirely methodological and negative.

Macarthur and Snowdon have each recently offered responses to this objection. Macarthur argues that Wittgenstein escapes this objection because his liberal naturalism is non-doctrinal and non-explanatory (Macarthur 2018, 33–4). Snowdon argues that Wittgenstein's later philosophy involves theorizing, so there is nothing wrong with claiming that Wittgenstein was a naturalist, even if doing so implies that Wittgenstein endorses a theory (Snowdon 2018, 19–22). The form of naturalism Snowdon ascribes to the later Wittgenstein is 'naturalismR', the theory that the truths of some domain are grounded in or constituted by some 'natural range specified in a general way' (Snowdon 2018, 16). Snowdon argues that Wittgenstein thereby offers a 'naturalistic … theory, of language' (Snowdon 2018, 21).

Both Macarthur and Snowdon are open to objections. Macarthur has to address the tension between holding that Wittgenstein endorses a form of naturalism while defending the claim that he does not propose any theories. Snowdon has to defend the controversial position that Wittgenstein proposes theories. While Macarthur offers a convincing argument for his position, this chapter will argue that given the obstacles facing the description of Wittgenstein as a liberal naturalist, a less objectionable position is to endorse the observations Macarthur and some other interpreters have

[18]For further discussion of this point in connection with Wittgenstein's anti-scientism, see Beale 2017, 60–65.

[19]For a lucid account of Wittgenstein's later philosophical method as entirely negative in this way, see Schroeder (2006).

made about some of the naturalistic aspects of Wittgenstein's philosophy, but to stop short of using these to motivate the ascription of liberal naturalism to Wittgenstein.

Concerning (3), Wittgenstein's methodological concern with 'natural facts' is markedly different from the ordinary sense of this notion—i.e., the way it is employed within the natural sciences and by scientific naturalists. Wittgenstein considers 'natural facts' in his philosophical investigations to be general observations about human nature; and he draws attention to these not as a source of scientific investigation, but to elucidate concepts to help expose and dissolve philosophical confusion (Hutto and Satne 2018, 72). These points are expressed most clearly in the second part of the *Investigations*:

> We are, indeed, also interested in the correspondence between concepts and very general facts of nature [*Naturtatsachen*]. (Such facts as mostly do not strike us because of their generality.) But our interest is not thereby thrown back on to these possible causes of concept formation; we are not doing natural science; nor yet natural history – since we can also invent a fictitious natural history for our purposes. (PPF §365)

A 'fictitious natural history' is, for example, the builders' language game. Wittgenstein invites us to conceive of a 'complete primitive language' that fits with the Augustinian picture of words primarily functioning as nouns representing objects, employed as a system of communication (PI §2). The aim of this fictitious language is to expose conceptual confusion underpinned by such a picture of meaning:

> If one looks at the example in §1, one can perhaps get an idea of how much the general concept of the meaning of a word surrounds the working of language with a haze which makes clear vision impossible. – It disperses the fog if we study the phenomena of language in primitive kinds of use in which one can clearly survey the purpose and functioning of the words. (PI §5)

The general methodological point Wittgenstein makes in these early sections of the *Investigations* is that we can make clear where confusions about the nature of meaning lie by describing cases to provide surveyable representations of the concepts in question—for example, a primitive language 'for which the description given by Augustine is right' (PI §2). This primitive language is part of a fictitious natural history which Wittgenstein uses as part of his attempt to try to expose conceptual confusions about the nature of meaning.

Concerning (4), Wittgenstein's opposition to scientism is an important feature of his later thought.[20] 'Scientism' is difficult to define and takes many forms; we can generally define it as excessive belief in the power or value of science. This could be manifested, for example, in a dogmatic assumption that the only genuine kind of knowledge is scientific knowledge or that the methods of natural science can be applied in all areas of inquiry; or in the claim that natural science is far more valuable than all other areas of inquiry or culture.[21] Wittgenstein would be vehemently opposed to such manifestations of scientism as well as others. Scientific

[20]On Wittgenstein's anti-scientism, see Beale and Kidd (2017).

[21]The final of these is the definition of scientism offered by Tom Sorell in his seminal 1991 book on the theme (Sorell 1991, x, 1, 9).

naturalism is much more open to accusations of scientism than liberal naturalism, given the emphasis the former lays on the natural sciences for giving us an account of what exists and what we can know, and given that scientism is typically concerned with the natural sciences.[22] It is much more difficult to level accusations of scientism against liberal naturalism, given its wider conception of what counts as 'natural'.

If scientific naturalism can be shown to be scientistic, this gives us greater reason to distance Wittgenstein from it. Trying to show that scientific naturalism is scientistic would take us beyond the scope of this chapter. Let it just be said that, the fact that scientific naturalism is much more open to accusations of scientism than liberal naturalism is another reason Wittgenstein would not endorse scientific naturalism.

6.5 Reasons Wittgenstein Cannot Be Considered a Liberal Naturalist

So there are at least two reasons for describing the later Wittgenstein as a liberal naturalist. But we have seen that there are at least two obstacles standing in the way of his description as such. On an orthodox interpretation of the later Wittgenstein, the following two claims are uncontroversial:

i. Wittgenstein is not concerned with offering an ontology.
ii. Wittgenstein does not engage in theory construction.

We have seen that some philosophers have offered ways of overcoming these two obstacles in order to ascribe liberal naturalism to Wittgenstein. Macarthur, for example, argues that Wittgenstein's liberal naturalism is non-doctrinal and non-explanatory; if Macarthur's account is correct, the two claims above cease to be obstacles standing in the way of interpreting Wittgenstein as a naturalist. But even if we accept Macarthur's account, there are two further obstacles. Recall that liberal naturalism takes the sciences seriously and rejects all forms of supernaturalism. However:

iii. Wittgenstein sometimes appears to be hostile towards science.
iv. Wittgenstein does not reject all forms of supernaturalism.

Concerning (iii), some of Wittgenstein's remarks that suggest an anti-scientific stance are not indicative of hostility towards science, but rather of his anti-scientism or his dislike for popular science, the latter of which he thought gave people the false impression of understanding complex scientific ideas on the basis of superficial understanding. Some of Wittgenstein's anti-scientistic remarks reflect his disdainful attitude towards his time—a time in which he felt scientism was pervasive. Wittgenstein's attitude towards his time was vehemently negative. G. H. von Wright recalls that Wittgenstein's attitude was 'one of censure and even disgust' (Wright 1982,

[22]For an overview of common definitions of scientism and the connection between scientism and the natural sciences, see Beale (2019).

110). The most important place this finds expression is in his rebuking of 'the darkness of this time' in the preface to the *Investigations* (PI, p. 4). His negative appraisal of his time, influenced by Spengler, is partly driven by the scientism he felt to be present in Western culture in the twentieth century.

We saw such worries expressed in Wittgenstein's remark where he criticizes the widespread attitude that the sciences teach us new knowledge whereas the arts merely entertain us (CV 42). Following Spengler, among the ways of thinking Wittgenstein thought were being increasingly 'elbowed aside' in his time were those of the arts (CV 69). For Spengler and Wittgenstein, scientism pushes out other aspects of culture, by dominating our ways of life or assimilating other areas. Wittgenstein's hyperbolic remarks on science such as, 'Man has to awaken to wonder … Science is a way of sending him to sleep again' (CV 5), are indicative of anti-scientism and its relation to his cultural pessimism rather than hostility towards science (Child 2017, 84).

Notwithstanding this caveat, there are nonetheless remarks that suggest that Wittgenstein was hostile towards science. In a recent article on Wittgenstein's anti-scientism, William Child offers a convincing interpretation of the ways in which some of Wittgenstein's remarks on science suggest such hostility.[23] Child draws attention to Wittgenstein's hyperbolic remarks on science, such as:

> It isn't absurd, e.g., to believe that the age of science and technology is the beginning of the end for humanity … [or] that there is nothing good or desirable about scientific knowledge and that mankind, in seeking it, is falling into a trap. (CV 56)

> [T]he [atom] bomb offers a prospect of the end, the destruction, of an evil – our disgusting soapy water science. And certainly that's not an unpleasant thought. (CV 49)

These polemic remarks reflect Wittgenstein's anti-scientism. His claims that 'it isn't absurd to believe that the age of science and technology is the beginning of the end for humanity' and that it is 'not unpleasant' that an end might be on the horizon for 'our disgusting soapy water science' are indicative of his cultural pessimism, which was partly rooted in what he perceived as the hubristic place of science in his time. As Child writes, 'part of the hostility in these comments is directed at the social institution of science … rather than at scientific method or scientific enquiry itself' (Child 2017, 85). However, these remarks, particularly the first, also suggest that Wittgenstein endorses an 'outright hostility to science as such' (Child 2017, 85). To be sure, Child notes, Wittgenstein 'does not commit himself to this … extreme view'; rather, he says only that it 'isn't absurd … that there is nothing good or desirable about scientific knowledge'. But Wittgenstein 'plainly regards it as a real possibility that there is not' anything good or desirable about scientific knowledge. This suggests 'hostility to scientific enquiry and scientific knowledge as such, not simply hostility to the overgeneralization of scientific thinking, to the social institution of science, or to the spirit in which science is currently pursued' (Child 2017, 85–6).

To be sure, there is no clear sense in which Wittgenstein's philosophy is anti-scientific. But if Child's interpretation is correct, remarks such as the two cited above suggest that, at best, Wittgenstein was not completely anti-scientific, at worst,

[23] William Child, 'Wittgenstein, Scientism, and Anti-scientism in the Philosophy of Mind', in Beale and Kidd (2017).

he was hostile towards science. Given the way in which liberal naturalism is typically defined, it would, therefore, be misleading to state that Wittgenstein takes the sciences seriously *enough* to justify being labelled a liberal naturalist.

Concerning (iv), one of the few criteria that unites almost all forms of naturalism is the rejection of supernatural entities within our ontology.[24] We have seen that it is controversial to claim that Wittgenstein offers any kind of ontology. It is also misleading to claim that he outright rejects supernaturalism.

Wittgenstein's opposition to metaphysics illustrates a clear sense in which he rejects supernaturalism. There is, however, a sense in which he does not reject supernaturalism. In his conception of religious belief, which focuses on a particular kind of Christianity which he revered, Wittgenstein does not outright deny the supernatural character of concepts such as God or the Last Judgement, or the idea that Christ was a supernatural being.[25] This would not create an obstacle towards describing Wittgenstein as a naturalist if he regarded Christian belief as incoherent. While there were undoubtedly forms of Christian belief that Wittgenstein saw as incoherent, he was extremely—arguably excessively—sympathetic towards religious belief, or at least to an idiosyncratic kind of Christianity to which he was attracted: that of the 'honest religious thinker' (CV 84).

To be sure, a leitmotif of Wittgenstein's remarks on religious belief is that it should not be assumed to involve metaphysical claims, nor should religious propositions simply be taken to be reducible to a set of metaphysical doctrines. He held that much confusion about the nature of religious belief is generated by seeing it in such ways. That is a central theme both of his 'Lectures on Belief' (1938)[26] and his *Remarks on Frazer's* Golden Bough (1931).[27] Moreover, central to Wittgenstein's conception of religious belief is that it is a way of living and seeing one's life—a point that has motivated several interpreters to read Wittgenstein as an 'expressivist' about religious belief.[28] Yet, Wittgenstein does not deny that religious belief, even the form he held in reverence and much desired, involves taking certain supernatural claims to be true.

Wittgenstein's view that religious belief involves each of these components—first, living and seeing one's life in a particular way, second, endorsing metaphysical

[24]Maximilian de Gaynesford offers an account of naturalism that illustrates how it is consistent to endorse the existence of supernatural phenomena while remaining a naturalist: 'It is theoretically possible to be a pluralist or non-reductionist naturalist, claiming that the natural order and natural explanation exist as one among many. But unless the natural is emphasized within the overall scheme proposed, there is no good reason to emphasize it by referring to the position as a whole as 'naturalist'. (So, for example, it would be very odd but not contradictory to describe as naturalists those who believe in the co-existence of natural and supernatural beings, or of types of behaviour whose explanation requires appeal to both natural and supernatural forces.)' (de Gaynesford 2004, 50–51). Since this would, as de Gaynesford points out, 'be very odd' and since the orthodox definition of liberal naturalism includes a rejection of supernaturalism, I will not consider this a problem for my arguments.

[25]I am grateful to Severin Schroeder for drawing my attention to this point.

[26]Published in LC, 53–72 (Wittgenstein's 1966).

[27]Wittgenstein (1993).

[28]For objections to expressivist interpretations of Wittgenstein on religious belief, see Schroeder (2007, 443–445).

beliefs—is suggested in the following remark from 1947. This has been the subject of much attention in work on Wittgenstein's philosophy of religion and has often been used in support of an expressivist interpretation:

> It appears to me as though a religious belief could only be (something like) passionately committing oneself to a system of coordinates // a system of reference. Hence, although it is *belief*, it is a way of living, or a way of judging life. (CV 73)

Here we see both components: Wittgenstein claims that religious belief involves 'belief' (*Glaube*)[29]—i.e., taking certain doctrines to be true—while it is also a way of seeing and living one's life. Both are essential to religious belief as he conceived of it.

This point has been convincingly argued by Severin Schroeder, in an article offering an account of the form of religious belief to which Wittgenstein was attracted.[30] Following an analysis of the above remark, Schroeder provides an overview of several passages in which Wittgenstein expresses the view that religious belief involves taking various supernatural claims to be true (Schroeder 2007, 445–446). Schroeder draws attention to the following passage, which suggests that Wittgenstein holds that belief in the doctrine of the resurrection is essential to Christian belief:

> If he did not rise from the dead, then he decomposed in the grave like every human being … & can no longer *help*; & we are once more orphaned & alone. (CV 38)

Schroeder also draws attention to the notes taken from Wittgenstein's lectures on belief in 1938, where he appears to reject the suggestion that belief in life after death is reducible to an expression of 'a certain attitude', devoid of belief in anything supernatural:

> Suppose someone, before going to China, when he might never see me again, said to me: "We might see one another after death" – would I necessarily say that I don't understand him? …
>
> *Lewy* [Casmir Lewy, one of the students attending Wittgenstein's lectures]: "In this case, you might only mean that he expressed a certain attitude."
>
> I would say "No …" … It says what it says. Why should you be able to substitute anything else? (LC 71)

Schroeder also draws attention to remarks from 1937, where Wittgenstein writes about Christianity involving belief in 'redemption through Christ's death' (D 193) and belief in 'eternal life & eternal punishment' (D 188). Wittgenstein states that he shares neither belief (D 190).

Several other passages where Wittgenstein talks about Christian belief involving assent to metaphysical doctrines are accompanied by his refusal to endorse such doctrines. Another example cited by Schroeder is:

---

[29]In German, '*Glaube*' denotes both faith and belief. A clear distinction between these cannot, therefore, always be drawn when investigating Wittgenstein's views on religious belief, especially when his remarks were originally written or spoken in German. For the purposes of this discussion I treat them interchangeably.

[30]Severin Schroeder, 'The Tightrope Walker' (Schroeder 2007).

> I cannot call [Jesus] *Lord*; because that says absolutely nothing to me. … I cannot utter the word "Lord" meaningfully. *Because I do not believe* that he will come to judge me. (CV 38)

Schroeder also draws attention to a remark from 1944 where Wittgenstein discusses miracles, describing a miracle as 'a gesture which God makes'. Of this, Wittgenstein writes, 'do I believe that this happens? I don't' (CV 51).

If Schroeder's account is correct, religious belief, for Wittgenstein, involves commitment to a particular way of living and seeing one's life, *and* supernatural beliefs. It seems to have been the latter that prevented Wittgenstein from becoming religious, despite desperately longing to be so. Wittgenstein lived what would be seen in many respects as the lifestyle of a devout Christian and revered the form of faith he took to be exemplified by the 'honest religious thinker'. He desperately longed for this form of faith and felt impoverished without it; but he could never bring himself to endorse it, since, he remarked, although he yearned to kneel to pray, his knees were too stiff to bend (CV 63). From remarks such as those above, it seems that the main obstacle standing in the way of his becoming an honest religious thinker was that doing so required assenting to supernatural claims which he could never bring himself to believe.[31]

One might respond that Wittgenstein's writings on religion are relatively minimal compared with his writings on themes more important to his thought, such as language and the nature of philosophy; so, even if Wittgenstein does not rule out supernaturalism in his conception of religious belief, this does not threaten his description as a liberal naturalist. To be sure, the only primary materials we have on Wittgenstein's views on religion are his 'Lectures on Religious Belief'—a twenty-page collection of notes taken by a few students from three lectures he gave in 1938 on 'Belief'; his *Remarks on Frazer's* Golden Bough (1931); various scattered remarks, most of which are published in *Culture and Value*; and his 'Lecture on Ethics' of 1929, in which he discusses religious belief. However, some remarks suggest that religion plays a far more significant role in his philosophy than is typically acknowledged. Take, for example, the following remark, to his close friend and former student Maurice Drury:

> I am not a religious man but I cannot help seeing every problem from a religious point of view. (Drury 1981, 94)

In an earlier publication, Drury recollects Wittgenstein's remark slightly differently: 'I am not a religious man, but I cannot help seeing every *thing* [*sic*] from a religious point of view' (Drury 1973, xiv, my emphasis added). Whichever Wittgenstein said, it led Drury to wonder 'whether there are not dimensions in Wittgenstein's thought that are still being largely ignored' (Drury 1981, 94) and fascinated Norman Malcolm so much that he spent his final book—*Wittgenstein: A Religious Point of View?* (1993)—investigating what Wittgenstein might have meant by it.[32]

---

[31] I discuss these points and the importance of religious matters to Wittgenstein's thought in general in Beale (2011, 107–108).

[32] For further discussion, see Beale, ibid.

Another reply might be to try to distinguish between Wittgenstein the *philosopher* and Wittgenstein the *person*. This response would argue that although religion occupies an important place in Wittgenstein's personal life, it does not constitute an obstacle towards defining Wittgenstein as a naturalist, since in his later philosophy, there is no place for anything metaphysical.[33] So, by the lights of, for example, the *Investigations*, it is uncontroversial to state that Wittgenstein was opposed to supernaturalism.

There are several reasons for rejecting this response. First, it is sometimes untenable to draw a firm distinction between the philosophical and personal views of a philosopher. Second, we have seen that it is no exaggeration to say that religious matters dominated Wittgenstein's thought, which makes it more difficult to draw a dichotomy between his 'personal' views on religion and his philosophical views. Third, we have seen that we have reason to think that Wittgenstein's views on religion may play a far more significant role in his thought than is often acknowledged. Therefore, although supernaturalism does not occupy a place in Wittgenstein's main philosophical writings, this does not give adequate support to the claim that Wittgenstein outright rejects supernaturalism.

The later Wittgenstein therefore resists description as a naturalist, given (i)–(iv). We have seen that some philosophers, such as Macarthur, have offered accounts of Wittgenstein's liberal naturalism that aim to overcome (i) and (ii). However, (iii) and (iv) are more difficult obstacles to overcome in order to ascribe liberal naturalism to Wittgenstein, given the way in which that position is typically described.

6.6 Grammatical Naturalism

Even if we cannot read the later Wittgenstein as a naturalist, we have seen that there are fundamental aspects of his philosophy that can plausibly be described as 'naturalistic'. If we are to describe the later Wittgenstein as a naturalist, what should we call his position?

It would be misleading to describe Wittgenstein's position as a kind of 'methodological naturalism', since this is a term of art used to denote forms of naturalism that state that the methods of the natural sciences are the best way of finding out about the natural world. Since 'liberal naturalism' is a misleading label for Wittgenstein

[33]Some interpreters argue that Wittgenstein's hostility to metaphysics is the most important strand of his later philosophy. For example, in the introduction to the recently published *Wittgenstein and Naturalism*, the editors write that 'Wittgenstein saw philosophy as either metaphysics or critique of metaphysics' (Cahill and Raleigh 2018, 10). This, however, goes too far. Wittgenstein's later philosophy is primarily concerned with dissolving philosophical problems and he saw metaphysics as the most common cause of the conceptual confusion generating such problems. It is therefore of primary concern to the later Wittgenstein, but it is misleading to claim that it is the sole source of philosophical confusion. For example, Wittgenstein's attempts to expose misleading pictures which generate conception confusion do not always target metaphysics (on this point, see the final section of Beale 2018).

for the reasons described, his naturalism is perhaps best captured as 'grammatical naturalism': a form of naturalism that functions as part of his method of grammatical investigation. What is unique to, or at least original about, Wittgenstein's naturalism is that it is geared towards dissolving philosophical confusion, rather than offering any kind of ontology.[34] In this way, it is a purely methodological and negative form of naturalism, supporting Wittgenstein's later philosophical method of providing surveyable representations of grammar in order to elucidate the workings of language, to help expose conceptual confusions and dissolve philosophical problems.

Grammatical naturalism is consistent with (i)–(iv). It is not concerned with offering an ontology (i). It is not a theory (ii). It makes no claim about the respectability of science (iii). And we can allow that Wittgenstein might be sympathetic towards the existence of supernatural entities without this threatening the claim that he endorsed a form of naturalism (iv).

One might question whether grammatical naturalism should be considered a form of naturalism at all, since it neither shares the aims nor fulfils the function of most forms of naturalism; i.e., it neither offers an account of what counts as 'natural' nor a method for how we should go about finding out about the natural world. In response, first, it has been argued that there are at least two fundamental aspects of Wittgenstein's later philosophy that merit the description of 'naturalistic'. Second, the term 'naturalism' is employed so broadly in contemporary philosophy that Wittgenstein's position arguably merits this label at least as much as many other so-called forms of 'naturalism'. Third, my argument is not that we *should* describe Wittgenstein as a naturalist; indeed, given the obstacles raised, it would be best not to. Rather, the circumspect conclusion to be drawn is that if we are to describe Wittgenstein's position as a form of naturalism, the most plausible position to ascribe to him is grammatical naturalism. But he could endorse the methodological aspects 'grammatical naturalism' picks out without needing to call it a form of naturalism or labelling Wittgenstein a naturalist.

6.7 Coda: Why Describe Wittgenstein as a Naturalist?

Before closing, it is worth considering why we should attempt to describe Wittgenstein as a naturalist at all. It is of course useful to categorize philosophers in terms of the various 'isms' they endorse or which can be applied to them: it can help to understand a philosopher's views and their place in intellectual history; and it can help to assess whether a philosopher is open to certain objections. So the question of whether Wittgenstein was a naturalist and, if so, of what kind is interesting and important. But we should be careful to place Wittgenstein into this broad and multifaceted category, for at least two reasons.

[34]Similarly, I have argued elsewhere that what is unique about Wittgenstein's anti-scientism is that it is geared towards dissolving philosophical confusion. See Beale (2017, 71).

First, naturalism is the orthodox stance in contemporary philosophy and to be a 'naturalist' has become something of a badge of honour. Or, rather, to *not* be a naturalist has come to be widely regarded as highly objectionable, such that if a philosopher is regarded as such, their views are, at best, less attractive, at worst, untenable ab initio. The following hyperbolic excerpt from Putnam's 'The Content and Appeal of "Naturalism"' (Putnam 2004) perhaps sums up a prevalent attitude in contemporary philosophy:

> Today … philosophers – perhaps even a majority of all the philosophers writing about issues in metaphysics, epistemology, philosophy of mind, and philosophy of language – announce in one or another conspicuous place … that they are "naturalists" …; this announcement, in its placing and emphasis, resembles the placing of the announcement in articles written in Stalin's Soviet Union that a view was in agreement with Comrade Stalin's; as in the case of the latter announcement, it is supposed to be clear that any view that is not "naturalist" (not in agreement with Comrade Stalin's) is anathema, and could not possibly be correct. (Putnam 2004, 59)

Although overstating the case, Putnam—himself a liberal naturalist—draws attention to what has perhaps become a widespread tendency in contemporary philosophy: to make it clear that one is a naturalist or that the respective philosopher being defended is a naturalist, otherwise the views of the author or philosopher in question immediately risk looking suspect.

To be sure, there are popular non-naturalist positions in contemporary philosophy. While card-carrying substance dualists are rare nowadays and the divine command theory remains highly controversial, influential forms of non-naturalism are endorsed widely in several areas of philosophy. In the philosophy of mathematics, for example, mathematical Platonism—some forms of which are non-naturalist—is the orthodox position about the ontological status of mathematical entities. In moral philosophy, while Moore's non-naturalism was widely criticized by prominent anti-realists during much of the first half of the twentieth century, non-naturalism has been a relatively orthodox position in contemporary meta-ethics for at least the past five years. Several leading contemporary ethicists endorse non-naturalist positions: an aggressive, robust form of non-naturalism is endorsed by ethicists such as Terence Cuneo, David Enoch, Michael Huemer, and Russ Shafer-Landau; and a less aggressive form—which has changed the topography of debates concerning naturalism in ethics—is endorsed by ethicists such as Derek Parfit and Tim Scanlon.[35]

Notwithstanding these influential non-naturalist positions, naturalism is the orthodoxy in contemporary philosophy and an attitude of the kind Putnam describes is, arguably, commonplace. The suggestion that the later Wittgenstein is not a naturalist might therefore be met with hostility by some, since, in the light of contemporary philosophical trends, this suggestion would make his philosophy seem even more controversial than it is generally taken to be.

Second, placing the later Wittgenstein neatly in an 'ism' box is always difficult. While numerous philosophers have offered convincing arguments for the naturalistic

[35]I am grateful to Doug Kremm and Richard Rowland for helpful discussions on the place of non-naturalism in contemporary meta-ethics.

positions they ascribe to Wittgenstein, the obstacles outlined above prevent us from interpreting him as a liberal naturalist and perhaps, for that matter, a naturalist of any kind. When it comes to the question of whether Wittgenstein should go into one of the many boxes on the side of naturalism rather than into one of the more controversial set of boxes on the side of non-naturalism, if what has been argued above is convincing, he might be better placed somewhere in the latter. Or, this chapter proposes, we should not place Wittgenstein into any of these boxes and stop at the claim that there are several important 'naturalistic' aspects to his later work, rather than going further and claiming that he was a liberal naturalist. If we insist upon a box for that position, we could label it 'grammatical naturalism'. But that box does not sit neatly on either side of the divide between naturalism and non-naturalism. 'The difficulty here', to borrow a quote from *Zettel*, is 'to stop' (Z §314).

6.8 Concluding Remarks

The interpretation that the later Wittgenstein was a liberal naturalist is untenable. This is for four reasons: first, liberal naturalism offers an ontology; second, liberal naturalism can be construed as a theory; third, Wittgenstein sometimes appears to be hostile towards science; fourth, Wittgenstein does not reject all forms of supernaturalism. While convincing arguments have been offered to overcome the first and second obstacles—notably by David Macarthur—the third and fourth are more difficult for this interpretation to overcome.

Wittgenstein's later philosophy can nonetheless be described as 'naturalistic' for at least two reasons. First, observations concerning natural facts occupy an important role within his philosophical method of providing surveyable representations of grammar. Second, he is strongly opposed to metaphysics and there is an absence of any metaphysical explanations in his philosophical investigations.

The label 'naturalism' for this position is misleading given the four points raised. If we insist on calling this a form of naturalism, 'grammatical naturalism' is a fitting label: a form of naturalism that functions as part of his method of grammatical investigation. This methodological naturalistic approach does not face the objections faced by interpreting the later Wittgenstein as a liberal naturalist.

References

Beale, Jonathan, 2011. 'The Possibility of Authenticity: On Schönbaumsfeld's Wittgenstein'. *Ratio* XXIV, 107–115.

Beale, Jonathan, 2017. 'Wittgenstein's Anti-scientistic Worldview', in Beale and Kidd (eds.), 2017.

Beale, Jonathan, 2018. 'Review of Cahill, Kevin M. and Thomas Raleigh (eds.), *Wittgenstein and Naturalism*' (New York: Routledge, 2018).' *Notre Dame Philosophical Reviews* (22 July 2018): http://ndpr.nd.edu/news/wittgenstein-and-naturalism/.

Beale, Jonathan, 2019. 'Scientism and Scientific Imperialism'. *International Journal of Philosophical Studies*, 27: 1, 73–102, https://doi.org/10.1080/09672559.2019.1565316.
Beale, Jonathan and Ian James Kidd (eds.), 2017. *Wittgenstein and Scientism*. London: Routledge.
Cahill, Kevin M. and Thomas Raleigh (eds.), 2018. *Wittgenstein and Naturalism*. New York: Routledge.
Cavell, Stanley, 2004. 'Postscript (2002) to 'The Investigations' Everyday Aesthetics of Itself" in Mario De Caro and David Macarthur (eds.), *Naturalism in Question* (Cambridge, MA: Harvard University Press, 2004, 275–9).
Child, William, 2017. 'Wittgenstein, Scientism, and Anti-scientism in the Philosophy of Mind' in Beale and Kidd 2017.
Cottingham, John, 2005. *The Spiritual Dimension: Religion, Philosophy and Human Value*. Cambridge: Cambridge University Press.
Crary, Alice, 2007. 'Wittgenstein and Ethical Naturalism', in Guy Kahane, Edward Kanterian, and Oskari Kuusela (eds.) *Wittgenstein and His Interpreters: Essays in Memory of Gordon Baker* (Oxford: Blackwell, 2007).
De Caro, Mario and David Macarthur (eds.), 2004. *Naturalism in Question*. Cambridge, MA: Harvard University Press.
de Gaynesford, Maximilian, 2004. *John McDowell*. Cambridge: Polity.
Drury, Maurice O'C., 1973. *The Danger of Words* (ed. R. F. Holland). London: Routledge and Kegan Paul.
Drury, Maurice O'C., 1981. 'Some Notes on Conversations with Wittgenstein' in Rush Rhees (ed.), *Ludwig Wittgenstein: Personal Recollections* (Oxford: Blackwell, 1981).
Fischer, Eugen, 2018. 'Wittgensteinian "Therapy", Experimental Philosophy, and Metaphilosophical Naturalism' in Cahill and Raleigh 2018.
Frege, Gottlob, 1918. '*Der Gedanke*', *Beiträge zur Philosophie des deutschen Idealismus*, v. 1, n. 2, 58–77.
Glock, Hans-Johann, 1996. *A Wittgenstein Dictionary*. Oxford: Blackwell.
Hacker, P. M. S., 1996. 'Wittgenstein and Quine: Proximity at Great Distance' in Robert L. Arrington and Hans-Johann Glock (eds.), *Wittgenstein and Quine* (London: Routledge, 1996).
Hookway, Christopher, 2017. 'Wittgenstein and Naturalism' in Hans-Johann Glock and John Hyman (eds.), *A Companion to Wittgenstein* (West Sussex: Wiley-Blackwell, 2017), 746–56.
Hutto, Daniel D. and Glenda Satne, 2018. 'Naturalism in the Goldilocks Zone: Wittgenstein's Delicate Balancing Act', in Cahill and Raleigh (eds.), 2018.
Livingston, Paul, 2015. 'Naturalism, Conventionalism and Forms of Life: Wittgenstein and the "Cratylus"'. *Nordic Wittgenstein Review* 4 (no 2), 7–38.
Locke, John, 1975 [1689]. *An Essay Concerning Human Understanding* (ed. Peter H. Nidditch). Oxford: Clarendon Press.
Macarthur, David, 2009. 'Review of *Understanding Naturalism* by Jack Ritchie' (Acumen 2008). *Notre Dame Philosophical Reviews*, 2009.11.10. http://ndpr.nd.edu/news/24219-understanding-naturalism/.
Macarthur, David, 2015. 'Liberal Naturalism and Second Personal Space: A Neo-Pragmatist Response to "The Natural Origins of Content"'. *Philosophia*, Vol. 43 no. 3, 565–578.
Macarthur, David, 2018. 'Wittgenstein's Liberal Naturalism of Human Nature' in Cahill and Raleigh (eds.), 2018.
McGinn, Marie, 2010. 'Wittgenstein and Naturalism' in Mario De Caro and David Macarthur (eds.), *Naturalism and Normativity* (New York: Columbia University Press).
McGinn, Marie, 2013. 'Liberal Naturalism: Wittgenstein and McDowell', in Matthew C. Haug (ed.), *Philosophical Methodology: The Armchair or the Laboratory?* (New York: Routledge, 2013).
Monk, Ray, 1999. 'Wittgenstein and the Two Cultures'. *Prospect Magazine*, July 1999, 66–7. Available online under the title 'Wittgenstein's Forgotten Lesson': http://www.prospectmagazine.co.uk/magazine/ray-monkwittgenstein/#.UrGuo42zccs.
Putnam, Hilary, 2004. 'The Content and Appeal of "Naturalism" ', reprinted in Putnam 2012, 109–25.

Putnam, Hilary, 2008. '12 Philosophers: And Their Influence on Me'. *Proceedings and Addresses of the American Philosophical Association*, Vol. 82, No. 2 (November 2008), 101–115.

Putnam, Hilary, 2011. 'Wittgenstein: A Reappraisal' in Putnam 2012, 482–92.

Putnam, Hilary, 2012. *Philosophy in an Age of Science: Physics, Mathematics, and Skepticism* (ed. Mario De Caro and David Macarthur). Cambridge, MA: Harvard University Press.

Quine, W. V., 1986. 'Reply to Putnam' in *The Philosophy of W. V. Quine* (ed. L. E. Hahn and P. A. Schillp). La Salle: Open Court Press.

Rosenberg, Gregg, 2004. *A Place for Consciousness: Probing the Deep Structure of the Natural World*. Oxford: Oxford University Press.

Schroeder, Severin, 2006. *Wittgenstein: The Way Out of the Fly Bottle.* Cambridge: Polity.

Schroeder, Severin, 2007. 'The Tightrope Walker'. *Ratio*, XX.

Snowdon, Paul, 2018. 'Wittgenstein and Naturalism' in Cahill and Raleigh 2018.

Smith, Benedict, 2017. 'Wittgenstein, Naturalism, and Scientism' in Beale and Kidd 2017.

Smith, Benedict, 2018. 'Wittgenstein, Hume, and Naturalism in Cahill and Raleigh 2018.

Sorell, Tom, 1991. *Scientism: Philosophy and the Infatuation with Science*. London: Routledge.

Spengler, Oswald, 1926 [1918]. *The Decline of the West: Form and Actuality* [*Der Untergang des Abendlandes: Gestalt und Wirklichkeit*] (tr. Charles Francis Atkinson). New York: Alfred A. Knopf.

Travis, Charles, 2018. 'The End of an Affair' in Cahill and Raleigh (eds.), 2018.

Verhaegh, Sander, 2017. 'Quine on the Nature of Naturalism'. *The Southern Journal of Philosophy* 55:1 (March 2017), 96–115.

von Wright, G. H., 1982. 'Wittgenstein in Relation to his Times' in *Wittgenstein and his Times* (ed. Brian McGuinness) (Oxford: Blackwell, 1982).

Williamson, Timothy, 2007. *The Philosophy of Philosophy*. Oxford: Blackwell.

Williamson, Timothy, 2011. 'What Is Naturalism?' *The New York Times* (September 4th). http://opinionator.blogs.nytimes.com/2011/09/04/what-is-naturalism/.

Winblad, Douglas G., 1996. 'Scepticism, Science, Quine and Wittgenstein' in Robert L. Arrington and Hans-Johann Glock (eds.), *Wittgenstein and Quine* (London: Routledge, 1996).

Wittgenstein, Ludwig, 1958 [1953]. *Philosophical Investigations* [*Philosophische Untersuchungen*] (tr. G. E. M. Anscombe). Oxford: Blackwell.

Wittgenstein, Ludwig, 1965. 'A Lecture on Ethics'. *The Philosophical Review*, Vol. 74, No. 1 (January 1965), 3–12.

Wittgenstein, Ludwig, 1966. *Lectures and Conversations on Aesthetics, Psychology and Religious Belief* (ed. Cyril Barrett). Oxford: Blackwell. (Cited as 'LC'.).

Wittgenstein, Ludwig, 1972 [1958]. *The Blue and Brown Books: Preliminary Studies for the Philosophical Investigations* (Second Ed.). Oxford: Basil Blackwell. (Cited as 'BB'.).

Wittgenstein, Ludwig, 1981 [1967]. *Zettel* (Second Ed.) (ed. G. E. M. Anscombe and G.H. von Wright; tr. G.E.M. Anscombe). Oxford: Blackwell. (Cited as 'Z'.).

Wittgenstein, Ludwig, 1990. *Last Writings on the Philosophy of Psychology*, Vol 1 (ed. & tr. By C. Grant Luckhardt and Maximilian A. E. Aue). Chicago: University of Chicago Press. (Cited as 'RPP I'.).

Wittgenstein, Ludwig, 1993 [1967]. *Remarks on Frazer's* Golden Bough, in James C. Klagge and Alfred Normann (eds.), *Philosophical Occasions 1912–1951*. Indianapolis, IN: Hackett Publishing Co..

Wittgenstein, Ludwig, 1997. *Denkbewegungen. Tagebücher 1930–1932/1936–1937 (MS 183)* (edited by I. Somavilla). Innsbruck: Haymon. (Quoted by MS page numbers; cited as 'D'.)

Wittgenstein, Ludwig, 1998 [1977]. *Culture and Value* [*Vermischte Bemerkungen*] (Revised Ed.) (ed. G.H. von Wright & Heikki Nyman; Revised Ed. by Alois Pichler; tr. Peter Winch). Oxford: Blackwell. (Cited as 'CV'.).

Wittgenstein, Ludwig, 2009 [1953]. *Philosophical Investigations* [*Philosophische Untersuchungen*] (tr. G.E.M. Anscombe, P.M.S. Hacker and Joachim Schulte). West Sussex: Wiley-Blackwell. (Cited as 'PI'; 'Part Two' cited as 'PPF', to denote Hacker and Schulte's titling of Part Two as *Philosophy of Psychology: A Fragment*.).

Chapter 7
Wittgenstein's *On Certainty* as Pyrrhonism in Action

Duncan Pritchard

7.1 Exegetical Issues

Wittgenstein's (1969) final notebooks, published as *On Certainty*, make for fascinating reading. These four notebooks constitute a sustained treatment of questions concerned with knowledge and certainty, and with the status of so-called Moorean certainties in particular. As notebooks, the remarks are impressionistic, and sometimes contradictory. As such, the notebooks raise all kinds of exegetical issues. For example, there is the fact that the first notebook seems to be more concerned with G. E. Moore's remarks on idealism rather than the Moorean certainties that are the primary concern of the other three notebooks.[1] There is also the issue of the proximal intellectual impetus for these notebooks. In their introduction, the editors of this work, Anscombe and von Wright, claim that Wittgenstein's concern was G. E. Moore's remarks on everyday certainties, at least as those remarks were interpreted and responded to by Norman Malcolm, who Wittgenstein stayed with around this time. This is certainly the most plausible impetus for these notebooks, but I think that it is at least contentious. In particular, I would suggest that there is an alternative way of thinking about these notebooks available that takes the work of John Henry Newman as a more primary source.[2]

Even granted these exegetical issues, the consensus regarding *On Certainty* tends to cluster around two kinds of proposal. On the one hand, there are those who

[1]The contrast here is especially clear when one compares the focus of Moore (1925) and Moore (1939). The exegetical point in play is one emphasised by Williams (2004, 2005) and Pritchard (2015a, part two; *forthcomingb*).

[2]The text of particular interest in this regard is Newman (1979 [1870]). See especially Pritchard (2015b). See also Kienzler (2006) and Pritchard (2011b, 2017a, 2018b).

D. Pritchard (✉)
University of California, Irvine, USA
e-mail: dhpritch@uci.edu

University of Edinburgh, 3 Charles Street, Edinburgh EH8 9AD, Scotland, UK

S. Wuppuluri and N. da Costa (eds.), *WITTGENSTEINIAN (adj.)*,
The Frontiers Collection, https://doi.org/10.1007/978-3-030-27569-3_7

think that there is a canonical reading of what Wittgenstein had in mind in these remarks, whereby Wittgenstein is advancing a concrete philosophical thesis (or theses).[3] Accordingly, proponents of this view are obliged to claim that some of the remarks should not be taken at face-value, in order to preserve the overall consistency of the notebooks. Given that we are dealing with notebooks which were unedited by the man himself, and never intended for publication, this is an entirely plausible line to take. Indeed, one would find it surprising if a set of notebooks didn't contain apparent inconsistencies. Moreover, there is also the point that it is undeniable that Wittgenstein often explores ideas in his notebooks without ever explicitly advertising this fact. Accordingly, where remarks don't fit with the wider interpretation, there is always the option to maintain that Wittgenstein isn't advancing the claims that he seems to be putting forward in the problematic passages.

Still, while one can plausibly derive a consistent proposal from the notebooks that make up *On Certainty*, since they are unedited notebooks it is hardly surprising that there are many commentators who eschew the task of offering an overarching interpretation and content themselves instead with the much weaker goal of presenting a position that is merely inspired by Wittgenstein's remarks. This leaves the further question of whether Wittgenstein himself would endorse the position so described as at least not settled.[4] Indeed, in some cases, commentators who fall into this camp go so far as to suggest that their Wittgensteinian proposal is not one that they think Wittgenstein himself would endorse.[5]

I want to explore a different way of thinking about these notebooks, which does not fall into either camp. This is the idea that Wittgenstein might have been undertaking a kind of intellectual exercise that is broadly Pyrrhonian in spirit. In particular, the thought is that he isn't trying to advance philosophical theses in these works at all, but rather working through opposing arguments as a means of cultivating within himself a kind of sceptical attitude. Recall that Pyrrhonism was an ethical stance, in the sense that it was offering a radical conception of the life of flourishing, one on which the sceptical attitude was required for *eudaimonia*.[6] This was achieved via the sceptical modes. These were designed to create an opposition (*isosethenia*)

---

[3]The list of commentators who fall into this camp (at least to some significant degree anyway) includes Morawetz (1978), Strawson (1985), McGinn (1989), Stroll (1994), Moyal-Sharrock (2004), and Schönbaumsfeld (2016).

[4]For some of the key Wittgenstein-inspired proposals, see Williams (1991), Wright (2004), Coliva (2010, 2015), and Sosa (2013). I offer my own Wittgenstein-inspired stance—which I call the *non-belief reading* (since a key part of the view is the claim that the propositional attitudes involved in our hinge commitments are not to be understood as beliefs, at least in the knowledge-apt sense of that term)—in Pritchard (Pritchard 2015a, part two). See also Pritchard (2012, 2014, 2018a). (Note that in earlier work I advanced a very different Wittgenstein-inspired proposal, one which treated our hinge commitments as knowledge, at least in an externalist, non-rationally grounded, sense—see Pritchard (2001; 2005)) See also endnote 5.

[5]This is the line I have taken in previous work myself, as it happens. See, especially, Pritchard (2015a, part two). For a recent survey of interpretations of *On Certainty*, see Pritchard (2017b). See also Pritchard (2011c).

[6]I describe Pyrrhonism as a 'stance' in order to distance it from a philosophical position or proposal. Clearly the Pyrrhonians couldn't consistently advance any kind of philosophical thesis.

that engenders a neutral attitude (*epoche*) which, in turn—at least if the process is repeated enough times anyway—eventually leads to a tranquil and untroubled state of mind (*ataraxia*). Sceptical doubt is thus in the service of the good life. Could Wittgenstein be attempting something similar with his notebooks?

This way of thinking about *On Certainty* would certainly be very different from those proposals that maintain that Wittgenstein was advancing a particular philosophical thesis. After all, on this reading, Wittgenstein is not meant to be understood as advancing any kind of philosophical thesis. He is, instead, like the Pyrrhonians before him, simply manifesting a certain kind of approach to philosophical issues, without endorsing any particular claim. Relatedly, if one took this line about *On Certainty*, then it would be a mistake to set aside the general exegetical question of what Wittgenstein is up to in these notebooks and instead focus on a particular train of thought that one could plausibly take to be inspired by them. That's not to say that the Wittgenstein-inspired view would be lacking in philosophical import. Rather, the point is that one would be missing something very significant about *On Certainty* by not recognizing the distinctive philosophical approach that Wittgenstein is manifesting in this text.

There are some points that count in favour of this way of thinking about *On Certainty*. We know that Wittgenstein was deeply interested in Pyrrhonian scepticism, and was very influenced by the work of the Austro-Hungarian (near) contemporary Fritz Mauthner who was attempting to develop an updated version of the Pyrrhonian stance.[7] Indeed, it is arguable that Pyrrhonian scepticism informs all of Wittgenstein's work (going right back to the *Tractatus*), though I will focus my attentions to its influence on *On Certainty*. In addition, reading *On Certainty* this way means that we don't need to explain away the apparent contradictions in these notebooks. Instead, the contradictions could be treated as being instead evidence in favour of this interpretation; as Wittgenstein generating the kind of opposition needed to fuel his Pyrrhonian stance. Indeed, there are also some interesting overlaps between what Wittgenstein seems to be attempting in *On Certainty* (on this reading anyway) and the practices of Pyrrhonism, at least on a plausible interpretation of the latter. Moreover, reading *On Certainty* this way enables us to better understand the specific kind of philosophical quietism that Wittgenstein seems to be embracing in his philosophical method, one that is also in a certain important sense *dis*quieting. I will be explaining this feature of the view in terms of the phenomenon of *epistemic vertigo*. Finally, and relatedly, although this Pyrrhonian way of reading *On Certainty* diverges from the two interpretative camps noted above, it does have some affinities with the influential 'therapeutic' way of thinking about Wittgenstein's work in general, whereby we are

[7]See Sluga (2004) for further discussion of how Mauthner's brand of Pyrrhonian scepticism influenced Wittgenstein. Interestingly, Weiler (1958) argues that Wittgenstein's use in the *Tractatus* of the metaphor of the ladder that one should discard once employed, which will be familiar to anyone who has read Sextus Empiricus, was gained from reading Mauthner's work (though this metaphor also appears in other work that Wittgenstein is likely to have read, such as Schopenhauer (1909, 256)). See Wittgenstein (1922, §6.54), and Sextus Empiricus (2005, 183, II 480–81).

not to understand what Wittgenstein is doing (e.g., in the *Tractatus*) as defending philosophical theses.[8]

Before we jump into considering these issues, however, I first want to register the spirit in which this enterprise is being undertaken. My claim is only that this is a fruitful way of thinking about *On Certainty*, and no more than that. That's a weak claim, from an exegetical point of view, to be sure. But it also has a hidden strength. For if this claim is true, then the import of thinking about *On Certainty* in this fashion would survive even the unearthing of definitive evidence that this is not what Wittgenstein intended with these notebooks (e.g., a letter from him, written at the time, that explicitly disavows this possibility). *On Certainty* is an important philosophical work, and what such works have in common is that they can inspire interesting philosophical ideas that diverge from what the author intended. In any case, in what follows my primary motivation is to persuade the reader that this way of responding to *On Certainty* is philosophically enlightening, regardless of whether it turns out to be an accurate representation of Wittgenstein's motives behind this work.[9]

7.2 Pyrrhonism as Perpetual Inquiry

There is a common way of thinking about Pyrrhonism according to which it involves a rejection of the demands of rationality. Indeed, such a reading can seem obvious given that this is a form of scepticism, albeit one that is cast as an ethical stance. As Michael Williams (1988, 561–2) once provocatively described the view, one can think of Pyrrhonism as a kind of inversion of Stoicism. While the Stoics (very roughly) thought that the good life, the life of flourishing, involved suppressing the passions and living by reason, the Pyrrhonians were effectively advocating that one should suppress reason so that one can live by the passions (though they obviously would never put it in this way, since it is part of the Pyrrhonian stance that one wouldn't advocate a theoretical claim of this kind). It is thus reason, or at least the desire to present oneself as rational, that is problematic in terms of living the good life.[10] The sceptical modes are thus a means to an end (bearing in mind that we are taking an external perspective on Pyrrhonism here, in that they themselves advance no ends, and thus no means to ends either), in that they are designed to provoke a response, over time, such that one relinquishes rational inquiry altogether and acquiesces in non-voluntary responses to stimuli.

[8] I have in mind here especially Cavell (e.g., 1979). For something like the therapeutic line applied to *On Certainty*, see Conant (1998), Crary (2005), and Maddy (2017). See also the closely related 'resolute reading' of the *Tractatus* as offered by Diamond (1991) and Conant (1989).

[9] I have previously explored the relationship between Wittgenstein's *On Certainty* and Pyrrhonism in Pritchard (2011a; *forthcomingc*).

[10] There are very interesting commonalities here between Pyrrhonism (on this specific reading anyway) and the Madhyamaka Buddhist school of thought. For further elaboration, see Brons (2018).

While this is certainly a plausible way of thinking about Pyrrhonian scepticism—and note that the exegetical issues surrounding Wittgenstein's relatively recent, historically speaking, notebooks are magnified several times when it comes to the relevant ancient texts—I don't think it is really true to how Sextus Empiricus (2000), who is our key source when it comes to Pyrrhonian scepticism, describes the stance. Right at the very beginning of *Outlines of Pyrrhonism* (SE 3), for example, Sextus emphasises a crucial difference between, on the one hand, the Dogmatists and the Academic sceptics—who, respectively, "have claimed to have discovered the truth" and "have asserted that it cannot be apprehended"—and, on the other hand, the Pyrrhonian sceptics who simply "go on inquiring." If we take this description of Pyrrhonism seriously, then we are led to a conception of this stance whereby it isn't rejecting reason altogether, but rather rejecting a particular conception of rationality that embraces closed inquiries, whether in the positive sense of generating putative knowledge or the negative sense of generating (settled) doubt. The thought is that either outcome is contrary to the good life of human flourishing. In contrast, the Pyrrhonian skeptical attitude is meant to promote a genuinely open inquiry, one that would thereby be perpetually ongoing. In this way we have a model for thinking about Pyrrhonian scepticism whereby it embodies a rational perspective of sorts, albeit one that is principally opposed to reaching dialectically closed conclusions.

It is also significant that the passage just cited is fairly representative of how Sextus describes Pyrrhonian scepticism. For example, after describing the ethical goal of Pyrrhonian scepticism as being "quietude in respect of matters of opinion and moderate feeling in respect of things unavoidable" (SE I 19), he goes on to contrast the "quietude" gained by living the sceptical life with the "disquietude" of those who do not suspend judgement:

> "For the man who opines that anything is by nature good or bad is for ever being disquieted; when he is without the things which he deems good he believes himself to be tormented by things naturally bad and he pursues after the things which are, he thinks, good; which when he has obtained he keeps falling into still more perturbations because of his irrational and immoderate elation, and in his dread of a change of fortune he uses every endeavour to avoid losing the things which he deems good. On the other hand, the man who determines nothing as to what is naturally good or bad neither shuns nor pursues anything eagerly; and, in consequence, he is unperturbed." (SE I 19)

Again, then, we find that the sceptical good life of quietude is not fuelled by doubt as such (i.e., a settled conviction that one lacks knowledge), much less the suspension of all rational inquiry, but rather results from keeping the flames of inquiry alight, where this means ensuring that nothing is 'determined', whether positively (in the manner of Dogmatism) or negatively (in the manner of Academic scepticism).[11]

Indeed, it is arguable that by treating Pyrrhonism as a form of scepticism as we now understand that term, and hence as directed at bringing about doubt, we are implicitly appealing to a conception of scepticism that would be alien to the Pyrrhonians themselves (but which, for example, is very present in later forms of

[11]For a recent defence of this reading of Pyrrhonian scepticism, see Perin (2012). For an earlier, and influential, account of this reading, see Barnes (1982).

scepticism, such as Cartesian scepticism and its 'method of doubt'). *Skepsis*, after all, means inquiry rather than doubt. Moreover, doubt is too much like a settled conclusion for Pyrrhonian tastes. What they regard continual inquiry as generating is not the settled negative conclusion of doubt (e.g., regarding the possibility of knowledge in the target domain), which would be one, negative, way of concluding a rational inquiry, but rather the more unsettled *suspension of judgement*, something which could in turn generate further inquiry, and so on ad infinitum. On this reading, then, Pyrrhonism is to be thought of as a way of living, whereby the employment of the sceptical modes will keep one's judgements suspended and hence ensure that inquiry is forever on-going.[12]

This way of thinking about Pyrrhonian scepticism also dovetails with another exegetical question about the stance, which is what the sceptical modes were aimed at. If the scope is any and all of one's beliefs, then that would suggest a radical form of scepticism. With the scope so broad, the traditional problem for Pyrrhonism of explaining how it offers a psychologically possible sceptical strategy would be acute.[13] But this isn't a very credible way of understanding the stance. This is because Pyrrhonism doesn't seem to be concerned with ordinary spontaneous belief at all, but rather with the kind of (putatively) rationally grounded theoretical claims that are characteristically not like ordinary, unreflective, spontaneous beliefs. Indeed, in the most straightforward case, these rationally grounded theoretical claims are the claims made by the philosopher. Here is Robert Fogelin on this point:

> "[...] the attacks of the Pyrrhonian skeptic are directed against the dogmas of 'Professors'—not the beliefs of common people [...] the Pyrrhonian skeptic leaves common beliefs, unpretentiously held, alone." (Fogelin 2004, 163)

This point is sometimes expressed in terms of two styles of assent: one that is theoretical, putatively rational, non-spontaneous, and dogmatic (and so subject to sceptical doubt); and one that is ordinary, unreflective, spontaneous, and non-dogmatic (and hence avoids sceptical doubt). Here, for example, is how Williams expresses the idea of a style of assent that is immune to Pyrrhonian doubt:

> "[...] the Pyrrhonian has a distinctive style of assent: spontaneous, involuntary submission to his unrationalised impulses. Assent is a pathos, something that comes over one. [...] Ordinary life, as Sextus sees it, is much more a matter of impulse and habit than of judgement properly so called."
>
> (Williams 1988, 561–2)

Again, we see why our ordinary unreflective judgements should be thought of as not the target of sceptical challenge.[14]

This way of characterizing the scope of Pyrrhonian scepticism accords with our earlier presentation of it in terms of perpetual inquiry. Our ordinary, spontaneous,

---

[12]For an overview of Pyrrhonian scepticism, roughly so conceived, see Vogt (2015, §31.1; 2017, 91–92).

[13]For some useful discussions of the 'liveability' of Pyrrhonian scepticism, see Naess (1969, ch. 2), Burnyeat (1980), Stough (1984), and Ribeiro (2002).

[14]For some other key discussions in this regard, see Burnyeat (1980; 1984) and Frede (1984).

unreflective beliefs are perfectly in order as they are, in the sense that they do not generate the disquiet that the Pyrrhonian sceptical modes are designed to treat. It is only once we begin making judgements of the problematic kind that we are led into disquiet, and away from *eudaimonia*. The sceptical modes are then needed to tackle this form of disquiet and bring about a certain kind of peace. But notice that the quietude that results from the employment of these sceptical techniques is not to be thought of in terms of a settled stance of its own. Indeed, this is the sense in which there can be no peace once one takes up inquiry, since the path to flourishing is now continual inquiry. Put another way, while there is quietude of sorts, it is of a completely different kind to the 'quietude' available to those who unreflectively live their lives, in that it is active rather than passive (for there is something that needs to be quiet*ed*).

This interpretation of Pyrrhonian scepticism suggests that the successful application of the sceptical modes does not return one to a state of intellectual innocence, of a kind available to the unreflective. This point is crucial, since it entails that the Pyrrhonian strategy is not to be understood as *undoing* anything. Indeed, one could claim that the *eudaimonia* that result from employing this strategy is of a higher order, given its active nature, though this would bring difficulties of its own. (For example, how are we to mark this greater value, given that there is no philosophical thesis in play here that we can appeal to?)[15] In any case, for our purposes what is important is just that the Pyrrhonian sceptical modes are targeted at a particular kind of theoretical claim, that they lead to perpetual inquiry (i.e., rather than a settled state, including one of doubt), and that the *eudaimonia* that results is thereby by its nature also in one sense unsettling/disquieting, as distinct from the kind of *eudaimonia* available to those who are essentially unreflective.

7.3 Wittgensteinian Quietism

The conception of Pyrrhonian scepticism just presented aligns closely with the specific brand of philosophical quietism that Wittgenstein expresses, and which is most apparent in his later works, like the *Philosophical Investigations* and (I would say, *especially*) *On Certainty*.[16] In its most austere form, philosophical quietism simply holds that there is nothing to be gained from engaging in philosophical reflection, and hence that it should be abandoned. This is clearly not Wittgenstein's view, as he practiced philosophy right until the end of his life. But he evidently does regard

[15]Since drafting this paper I have come across some very interesting unpublished work by Gutschmidt (2018a, b). He argues for the related, though stronger, thesis that the employment of the Pyrrhonian sceptical modes is *transformative*, in the technical sense offered by Paul (2016). Like me, he also draws a connection between the Pyrrhonian sceptical strategy, so construed, and the kind of Wittgensteinian quietism that I describe below (indeed, he also makes appeal to my notion of epistemic vertigo in this respect, which we will discuss below).

[16]Though it was also present in the *Tractatus*. See, especially, the 'resolute reading' of this work as offered by Diamond (1991) and Conant (1989).

philosophical problems as being in a certain sense illusory. Indeed, he maintains that philosophy is the source of philosophical problems. Nonetheless, he makes clear that philosophy—at least of the kind that he practices—is also the way to resolve them. ("To show the fly the way out of the fly bottle." (Wittgenstein 1953, §309)) Wittgensteinian philosophical quietism thus sees philosophy as playing an important role in resolving philosophical difficulties, in contrast to a more austere form of philosophical quietism whereby philosophy could serve no useful purpose.[17]

In order to understand Wittgenstein's philosophical quietism, however, it is important to recognise that Wittgenstein regards philosophy as a certain kind of *activity*, rather than involving the advancement of philosophical theses.[18] In particular, Wittgenstein doesn't resolve philosophical problems by putting forward philosophical theses of his own (which would obviously be in tension with his philosophical quietism). Rather, his philosophical approach involves showing how the philosophical puzzle in hand is illusory by constant engagement with the phenomena in hand, by focussing on the details of cases and so forth.

We can bring this point out by considering one of the core lines of thought that is found in *On Certainty*. In the notebooks that make up this work Wittgenstein is clearly showing the illusory nature of a philosophical puzzle regarding knowledge (Cartesian scepticism, in short), by teasing out how it rests on the implicit import of a philosophical conception of the structure of rational evaluation that should be abandoned.[19] This is in contrast to the actual structure of rational evaluation found in our everyday epistemic practices, misdescribed by the philosopher, which are shown (through adducing many examples) to be entirely in order as they are. In a nutshell, what Wittgenstein demonstrates in *On Certainty* is that the traditional epistemological project of defending knowledge from sceptical attack presupposes, like the Cartesian form of scepticism that it is responding to, that there can be such a thing as universal rational evaluations. In contrast, Wittgenstein highlights how our ordinary practices of rational evaluation tacitly incorporate what he calls *hinge* commitments, which are essentially arational commitments that enable rational evaluations and hence are not themselves subject to rational evaluation.[20] Moreover, he makes clear

---

[17]See McDowell (2009) for a similar reading. See also Lear (1982). For further discussion of different varieties of quietism, see Virvidakis (2008).

[18]He makes this point explicit in the following famous passage from the *Tractatus*:

"The object of philosophy is the logical clarification of thoughts.

Philosophy is not a theory but an activity.

A philosophical work consists essentially of elucidations.

The result of philosophy is not a number of "philosophical propositions", but to make propositions clear.

Philosophy should make clear and delimit sharply the thoughts which otherwise are, as it were, opaque and blurred." (Wittgenstein 1922, §4.112).

[19]Elsewhere I have argued that the anti-sceptical import of Wittgenstein's remarks only applies to one aspect of the Cartesian sceptical problematic. See Pritchard (2015a) for the details.

[20]He actually refers to them as hinge *propositions* (OC, §§341–3), though as I explain in Pritchard (Pritchard 2015a, part two) it is more helpful to focus on the nature of the propositional commitment involved rather than on the particular propositional content that the subject is committed to (as the latter can lead to confusion).

that it is in the nature of rational evaluations that they involve hinge commitments in this way, and hence that this is not an incidental feature of our system of rational evaluations.

It follows that the philosophical problem of radical scepticism (of showing that knowledge is impossible), along with the traditional philosophical response to radical scepticism (of showing that knowledge is possible), is based on a faulty picture of the nature of our rational evaluations, one that illicitly imports a problematic conception of these rational evaluations as being, potentially anyway, harmlessly universal. In contrast, Wittgenstein's philosophical reflections help us to understand that our everyday epistemic practices, which have hinge commitments at their heart, are entirely in order as they are. He is thus offering what I have elsewhere called an *undercutting* response to the philosophical paradox involved in Cartesian scepticism, whereby what looks like a genuine philosophical paradox is shown to rest on illicit philosophical assumptions that should be abandoned. Our ordinary epistemic practices are fine as they are, with the apparent philosophical puzzle arising not out of these practices at all, but rather from a problematic philosophical account of those practices, one that we need to abandon. Undercutting responses to philosophical paradoxes are to be contrasted with overriding responses which effectively concede the paradoxical nature of the puzzle in hand (and thus there is something genuinely amiss with our everyday practices in the relevant respects), but nonetheless maintain that we can revise our everyday practices in certain respects in order to deal with the paradox.[21]

Notice, however, that it is crucial to Wittgenstein's approach that he is *showing* how we should deal with philosophical puzzles rather than advancing an opposing philosophical thesis. This is a natural consequence of his stance, for recall that he is trying to make clear how our everyday epistemic practices are fine as they are, and thus that the philosophical puzzlement arising from those practices is in fact the result of philosophical theses being smuggled into the description of them (a 'false philosophical picture'). The relentless focus on examples helps us to see that what we thought was a genuine philosophical puzzle about our everyday epistemic practices was in fact nothing of the kind. Wittgenstein shows that the puzzle is illusory not by arguing directly that this is the case, but rather by focussing on the nature of our everyday epistemic practices in such a fashion that we can come to understand how those practices can be in order as they are.

This is why there is nowhere in the notebooks that make up *On Certainty* a clear statement of a view (even though we can easily extract such a statement on his behalf, as indeed we just did). This is not an incidental lack, but part-and-parcel of the philosophical enterprise that Wittgenstein is engaging in, which does not involve the advancement of philosophical theses, but rather concerns a kind of philosophical therapy. There is thus a world of difference between, on the one hand, showing how a

---

[21] I discuss the contrast between undercutting and overriding strategies for dealing with (putative) philosophical paradoxes in more detail in Pritchard (2015a, part one). Similar distinctions regarding types of responses to philosophical paradoxes are advocated by Williams (1991, ch. 1) and Cassam (2007, ch, 1).

philosophical inquiry can undercut a putative philosophical paradox and, on the other hand, advancing a concrete philosophical thesis. The former can lead to quietude for those perplexed by the philosophical puzzle in play (I claim it can lead to *dis*quietude too, as I explain below). But if philosophy, in the sense of advancing philosophical theses, is the source of philosophical puzzles, then it is important to ensure that in offering the former one is not tempted into thereby offering the latter too, and hence being sucked back into the very faulty philosophical enterprise that was generating the fake conundra in the first place.

7.4 Epistemic Vertigo

That Wittgenstein isn't advancing philosophical theses in *On Certainty* removes one key potential inconsistency between his philosophical stance in this work and that offered by Pyrrhonian scepticism (which clearly does bring the sceptical modes to bear on the advancement of philosophical theses). But there is a second element of Wittgenstein's philosophical approach that is centrally important to this way of understanding *On Certainty*, and more generally to understanding the particular brand of philosophical quietism that he (and, arguably, Pyrrhonism) represents. This is that it is part of the very style of philosophical quietism that Wittgenstein offers that it is in a certain sense also deeply disquieting. Elsewhere I have characterised this point via the notion of *epistemic vertigo*.[22]

One thing that we should remember about Wittgenstein's approach to philosophical therapy is that it is explicitly designed to help those who are already subject to philosophical difficulties. In short, it is directed at treating the afflicted. Those who have never succumbed to philosophical concerns will find no use in these reflections, which will simply seem strange and affected. Moreover, since there is nothing philosophically amiss with our ordinary beliefs on the Wittgensteinian conception, hence there is nothing for philosophical theorizing to contribute to their lives anyway. Note that we saw earlier that something very similar is true of the Pyrrhonian sceptical modes, at least on one way of understanding that form of scepticism, such that its target was not ordinary unreflective beliefs at all, but rather the kind of theoretical claims characteristically made by the philosopher.

It is a crucial feature of Wittgenstein's account of hinge commitments that we are ordinarily completely unaware of these commitments and the special role that they play in our everyday epistemic practices. This is something that Wittgenstein repeatedly emphasises with his examples. One is not taught that one has two hands, for example (much less that one knows this), but rather that one can do things with one hands, the existence of one's hands being part of the accepted backdrop of certainty against which one evaluates other specific claims that are raised. (OC, §153) If one does not ordinarily encounter hinge commitments *qua* hinge commitments, and the nature of their epistemic status is raised only in a philosophical context, then it follows

[22]See especially Pritchard (2015a, part four). See also Pritchard (*forthcominga*; *forthcomingc*).

that the epistemic status of hinge commitments is not something that is at issue in ordinary contexts. As Wittgenstein puts it at one point, the hinge commitments "lie apart from the route travelled by inquiry." (OC, §88)

What Wittgenstein demonstrates with his examples is that there is nothing at all epistemically amiss in our everyday use of hinge commitments, and that it only seems that there is because we have bought into a faulty philosophical picture (one that is not licensed by our everyday practices at all) according to which there can be universal rational evaluations. Our everyday epistemic practices are thus revealed to be entirely in order as they are. But notice an important difference between philosophers and the rest of the folk once this point is made. Those who are unselfconsciously engaged in these everyday epistemic practices, and hence are (rightly) unaware of the role that hinge commitments play in those practices, will also be accordingly unaware of why this challenge to the legitimacy of these practices is unsound. But our situation, as philosophers, is very different. We have adopted (or at least attempted to adopt, anyway) the disengaged philosophical stance and have accordingly become aware of our hinge commitments and the role that they play. This discovery cannot be unlearnt. It means that even after coming to recognise that there is nothing amiss in our having hinge commitments, we cannot return to the state of epistemic innocence prior to their recognition.

Even in the grip of an adequate philosophical understanding of our epistemic position, then—one that in the most important sense "leaves everything as it is." (Wittgenstein 1953, §I.124), since it demonstrates that the everyday practices in question are unproblematic—one is nonetheless destined to take an 'unnatural' perspective on it. This is where the epistemic vertigo comes in. In becoming disengaged from our everyday epistemic practices, and thereby taking a detached perspective on them such that we can recognise our hinge commitments for what they are, one is naturally made intellectually uneasy, even if one is persuaded by the fact that these practices, and thus the arationality of our hinge commitments, is perfectly in order. We are now fated to be aware, as Stanley Cavell (1988) memorably put it, of the 'uncanniness of the ordinary.'

That vertigo is a phobic reaction is important here: epistemic vertigo is not the same as the *epistemic angst* that the radical sceptical problem gives rise to, as the latter concerns an anxiety about the legitimacy of our ordinary epistemic practices. For once the sceptical problem has been dissolved, and these practices are shown to be in order as they are, then there is no longer any impetus for such *angst*. But just as one can recognise, while high up, that one is perfectly safe and yet have a phobic reaction to heights nonetheless, so one can recognise, from a detached intellectual perspective, that there is nothing epistemically amiss with our hinge commitments being arational and yet feel a deep sense of epistemic insecurity regardless.[23]

The phenomenon of epistemic vertigo generalises. If philosophical problems more generally rest on illicit philosophical claims, of a kind that are produced by adopting a detached perspective that hides salient features of our actual practices, then

[23]The pedantic reader should, of course, replace the 'vertigo' metaphor with (less catchy) *acrophobia*.

one will expect a corresponding vertigo effect to occur as regards these other philosophical puzzles also. Once the disengaged stance is adopted, and one comes to see how features of our practices function in ways that were hitherto hidden, then even though one might further come to understand how these features of our practices are unproblematic, one can nonetheless no longer return to a philosophically innocent state where these features of our practices are once again tacit.[24]

In any case, what is relevant for our purposes is that a Wittgensteinian quietism, while in a sense leaving everything as it is, is also fundamentally disquieting. The positive effect of the quietism is the recognition that the philosophical challenge in play is illusory. But there is also an important negative effect too, in that one cannot simply return to one's old practices as before because one has now come to recognise, from the detached perspective of philosophical inquiry, those features of one's everyday practices that are ordinarily left hidden. There is thus no return to a state of philosophical innocence, prior to engagement with this problem. One is instead left in a kind of intellectual exile. Wittgensteinian quietism treats the afflicted, in the sense that it shows them how our everyday practices are not subject to the philosophical challenges that we thought they were. But it does not offer a way back from the intellectual exile prompted by their attempt to adopt the disengaged philosophical perspective, as there is no cure for this affliction.

7.5 Wittgensteinian Pyrrhonism

Let's review where we are with regard to understanding what a specifically Wittgensteinian brand of Pyrrhonism would look like. One immediate worry, relating to how Wittgenstein seems to put forward philosophical theses in response to the problem of radical scepticism has been dealt with, since properly understood he is doing nothing of the kind. Wittgenstein's approach to philosophical problems is not to advance philosophical theses, but to manifest a certain kind of activity. We have also seen how Wittgensteinian quietism, while in a sense leaving our everyday practices exactly as they are, also has an intellectually unsettling effect on the practitioner. On the one hand, the quietism leads the practitioner away from the philosophical *angst* that the problem in hand was generating. But on the other hand, however, this doesn't lead the practitioner back to a state of intellectual innocence, prior to her engagement with the philosophical problem in hand. Rather it generates a new kind of unsettling state—philosophical vertigo—whereby the practitioner is aware of the fact that our everyday practices are not subject to the philosophical challenge posed for them while also now being unavoidably cognisant of intellectually unsettling features of those practices that are ordinarily hidden from view. The result is that one is now in a certain sense alienated from this practice.

---

[24]I explore the more general philosophical consequences of epistemic vertigo in Pritchard (*forthcominga*). I take the general idea in play here to be one that is explored quite extensively, albeit often under a very different guise, in the work of Cavell (e.g., 1979, 1988, 2003).

Wittgensteinian quietism, so construed, may appear to be a rather dispiriting stance to take. But once we view this form of quietism through the lens of Pyrrhonian scepticism, then the outlook changes considerably. Recall that the favoured interpretation of the latter is as generating a permanently open-ended inquiry, one that is no more inclined towards doubt than it is towards claims to knowledge, where this is serving the ethical goal of generating quietude, and thereby *eudaimonia*. The sceptical modes are thus designed not to promote doubt, but to keep inquiry ongoing. The modes, recall, are not aimed at ordinary unreflective belief but at the very different kinds of reflective claims that depart from the ordinary, with philosophical claims being the clearest example of these. It is the latter that are contrary to *eudaimonia*, and must be opposed. The former, in contrast, are not disquieting in the relevant sense, and so are no barrier to *eudaimonia*. The proper application of the modes, however, does not return one back to ordinary unreflective belief, but rather enable one, as a reflective subject, to gain quietude through engagement in perpetual sceptical inquiry.

The parallels with Wittgensteinian quietism should be obvious. Just as the Pyrrhonian sceptical modes are not aimed at our everyday beliefs, so those who have not engaged with philosophical problems, and so are not excised by them, have no need of Wittgenstein's philosophical therapy, and hence no need for the undercutting anti-scepticism that he presents. Since Wittgenstein's philosophical quietism does not consist in the advocacy of philosophical theses—but is rather, like Pyrrhonian scepticism itself, a kind of philosophical activity—it is not in tension with the Pyrrhonian sceptical modes (which would be naturally applied to the defence of philosophical claims). Crucially, however, neither Wittgensteinian quietism nor Pyrrhonian scepticism are geared towards leading the philosophically engaged subject back to a state of innocence prior to their entanglement with philosophical concerns—this is simply not the kind of quietude that they offer.

For the Pyrrhonian, the road to *eudaimonia* for such a subject involves perpetual inquiry, and hence essentially involves a kind of intellectual disquietude, of a kind that can drive inquiry. One can now regard the Wittgensteinian quietist, in a Pyrrhonian spirit, as utilising philosophical vertigo to disquiet the philosophical subject and thereby keep the flames of inquiry alight. This is very clear in the case of the specifically epistemic vertigo at issue in *On Certainty*. Epistemic vertigo is the result of being aware of the (ordinarily tacit) role of arational hinge commitments in our epistemic practises, even while recognising that the philosophical challenges to those practices is based on faulty philosophy. The awareness of the hinge commitments prevents one from participating in those everyday practices as one did prior to engaging in philosophy, even now that one recognises that they are entirely in order as they are. It also means that one is aware of the essentially localised nature of rational evaluations, and that acts as a guard against the impetus towards philosophical dogmatism (indeed, dogmatism generally).

Epistemic vertigo—indeed, philosophical vertigo more generally—is thus a means towards generating the kind of *isosethenia* that would lead to *epoche*. In this way, Wittgensteinian Pyrrhonism, and the particular brand of quietism that goes along with it, is neither about advancing philosophical claims nor about eschewing philosophical inquiry altogether, but rather involves embracing an undogmatic, open-ended philosophical inquiry.

References

Barnes, J. (1982). 'The Beliefs of a Pyrrhonist', *Proceedings of the Cambridge Philological Society* 208, 1–29.

Brons, R. (2018). 'Life Without Belief: A Madhyamaka Defence of the Liveability of Pyrrhonism', *Philosophy East and West* 68:2.

Burnyeat, M. (1980). 'Can the Skeptic Live his Skepticism?', *Doubt and Dogmatism: Studies in Hellenistic Epistemology*, (eds.) J. Barnes, M. Burnyeat & M. Schofield, ch. 3, Oxford: Clarendon Press.

Burnyeat, M. (1984). 'The Sceptic in His Place and Time', *Philosophy in History: Essays on the Historiography of Philosophy*, (eds.) R. Rorty, J. B. Schneewind & Q. Skinner, ch. 10, Cambridge: Cambridge University Press.

Cassam, Q. (2007). *The Possibility of Knowledge*, Oxford: Oxford University Press.

Cavell, S. (1979). The *Claim of Reason: Wittgenstein, Skepticism, Morality, and Tragedy*, Cambridge, MA: Harvard University Press.

Cavell, S. (1988). *In Quest of the Ordinary: Lines of Skepticism and Romanticism*, Chicago, IL: University of Chicago Press.

Cavell, S. (2003). *Disowning Knowledge in Seven Plays of Shakespeare*, Cambridge: Cambridge University Press.

Coliva, A. (2010). *Moore and Wittgenstein: Scepticism, Certainty, and Common Sense*, London: Palgrave Macmillan.

Coliva, A. (2015). *Extended Rationality: A Hinge Epistemology*, London: Palgrave Macmillan.

Conant, J. (1989). 'Throwing Away the Top of the Ladder', *The Yale Review* 79, 328–64.

Conant, J. (1998). 'Wittgenstein on Meaning and Use', *Philosophical Investigations* 21, 222–50.

Crary, A. (2005). 'Wittgenstein and Ethics: A Discussion with Reference to *On Certainty*', *Readings of Wittgenstein's On Certainty*, (eds.) D. Moyal-Sharrock & W. H. Brenner, 275–301, London: Palgrave.

Diamond, C. (1991). 'Throwing Away the Ladder: How to Read the *Tractatus*', in her *The Realistic Spirit*, Cambridge, MA: MIT Press.

Fogelin, R. J. (2004). 'The Skeptics are Coming! The Skeptics are Coming!', *Pyrrhonian Skepticism*, (ed.) W. Sinnott-Armstrong, 161–73, Oxford: Oxford University Press.

Frede, M. (1984). 'The Sceptic's Two Kinds of Assent and the Question of the Possibility of Knowledge', *Philosophy in History: Essays on the Historiography of Philosophy*, (eds.) R. Rorty, J. B. Schneewind & Q. Skinner, ch. 11, Cambridge: Cambridge University Press.

Gutschmidt, R. (2018*a*). 'Beyond Quietism: Transformative Experience in Pyrrhonism and Wittgenstein', *unpublished manuscript*.

Gutschmidt, R. (2018*b*). 'Transformative Experience in Skepticism: The External Standpoint and the Finitude of the Human Condition', *unpublished manuscript*.

Kienzler, W. (2006). 'Wittgenstein and John Henry Newman On Certainty', *Grazer Philosophische Studien* 71, 117–38.

Lear, J. (1982). 'Leaving the World Alone', *Journal of Philosophy* 79, 382–403.

Maddy, P. (2017). *What Do Philosophers Do?*, Oxford: Oxford University Press.

McDowell, J. (2009). 'Wittgenstein's 'Quietism'', *Common Knowledge* 15, 365–72.

McGinn, M. (1989). *Sense and Certainty: A Dissolution of Scepticism*, Oxford: Blackwell.

Moore, G. E. (1925). 'A Defence of Common Sense', *Contemporary British Philosophy* (2nd series), J. (ed.) H. Muirhead, London: Allen & Unwin.

Moore, G. E. (1939). 'Proof of an External World', *Proceedings of the British Academy* 25, 273–300.

Morawetz, T. (1978). *Wittgenstein and Knowledge. The Importance of On Certainty*, Atlantic Highlands, NJ: Humanities Press.

Moyal-Sharrock, D. (2004). *Understanding Wittgenstein's On Certainty*, Palgrave Macmillan, London.

Naess, A. (1969). *Scepticism*, London: Routledge & Kegan Paul.

Newman, J. H. (1979 [1870]). *An Essay in Aid of a Grammar of Assent*, Notre Dame, IN: University of Notre Dame Press.

Paul, L. A. (2016). *Transformative Experience*, Oxford: Oxford University Press.

Perin, C. (2012). *The Demands of Reason: An Essay on Pyrrhonian Scepticism*, Oxford: Oxford University Press.

Pritchard, D. H. (2001). 'Radical Scepticism, Epistemological Externalism, and "Hinge" Propositions', *Wittgenstein-Jahrbuch 2001/2002*, (ed.) D. Salehi, 97–122, Berlin: Peter Lang.

Pritchard, D. H. (2005). 'Wittgenstein's *On Certainty* and Contemporary Anti-Scepticism', *Investigating On Certainty: Essays on Wittgenstein's Last Work*, (eds.) D. Moyal-Sharrock & W. H. Brenner, 189–224, London: Palgrave Macmillan.

Pritchard, D. H. (2011*a*). 'Wittgensteinian Pyrrhonism', *Pyrrhonism in Ancient, Modern, and Contemporary Philosophy*, (ed.) D. Machuca, 193–202, Dordrecht, Holland: Springer.

Pritchard, D. H. (2011*b*). 'Wittgensteinian Quasi-Fideism', *Oxford Studies in the Philosophy of Religion* 4, 145–59.

Pritchard, D. H. (2011*c*). 'Wittgenstein on Scepticism', *The Oxford Handbook on Wittgenstein*, (eds.) O. Kuusela & M. McGinn, 521–47, Oxford: Oxford University Press.

Pritchard, D. H. (2012). 'Wittgenstein and the Groundlessness of Our Believing', *Synthese* 189, 255–72.

Pritchard, D. H. (2014). 'Entitlement and the Groundlessness of Our Believing', *Contemporary Perspectives on Scepticism and Perceptual Justification*, (eds.) D. Dodd & E. Zardini, 190–213, Oxford: Oxford University Press.

Pritchard, D. H. (2015*a*). *Epistemic Angst: Radical Skepticism and the Groundlessness of Our Believing,* Princeton, NJ, Princeton University Press.

Pritchard, D. H. (2015*b*). 'Wittgenstein on Faith and Reason: The Influence of Newman', *God, Truth and Other Enigmas*, (ed.) M. Szatkowski, 141–64, Berlin: Walter de Gruyter.

Pritchard, D. H. (2017*a*). 'Faith and Reason', *Philosophy* 81, 101–18.

Pritchard, D. H. (2017*b*). 'Wittgenstein on Hinge Commitments and Radical Scepticism in *On Certainty*', *Blackwell Companion to Wittgenstein*, (eds.) H.-J. Glock & J. Hyman, 563–75, Oxford: Blackwell.

Pritchard, D. H. (2018*a*). 'Epistemic *Angst'*, *Philosophy and Phenomenological Research* 96, 70–90.

Pritchard, D. H. (2018*b*). 'Quasi-Fideism and Religious Conviction', *European Journal for Philosophy of Religion* 10, 51–66.

Pritchard, D. H. (*Forthcominga*). 'Epistemic Vertigo', *Ambivalence*, (eds.) B. Brogaard & D. Gatzia, London: Routledge.

Pritchard, D. H. (*Forthcomingb*). 'Scepticism and Certainty: Moore and Wittgenstein on Commonsense', *Cambridge Companion to Common Sense Philosophy*, (eds.) R. Peels & R. van Woudenberg, Cambridge: Cambridge University Press.

Pritchard, D. H. (*Forthcomingc*). 'Wittgensteinian Epistemology, Epistemic Vertigo, and Pyrrhonian Scepticism', *Epistemology After Sextus Empiricus*, J. Vlasits & K. M. Vogt, Oxford: Oxford University Press.

Ribeiro, B. (2002). 'Is Pyrrhonism Psychologically Possible?', *Ancient Philosophy* 22, 319–31.

Schönbaumsfeld, G. (2016). The Illusion of Doubt, Oxford: Oxford University Press.

Schopenhauer, A. (1909). The World as Will and Idea Vol. II, London: Kegan Paul.

Sextus Empiricus (2000 [= SE]). *Outline of Pyrrhonism*, (eds.) J. Annas & J. Barnes, Cambridge: Cambridge University Press.

Sextus Empiricus (2005). *Against the Logicians*, (ed.) R. Bett, Cambridge: Cambridge University Press.

Sluga, H. (2004). 'Wittgenstein and Pyrrhonism', *Pyrrhonian Skepticism*, (ed.) W. Sinnott-Armstrong, 99–117, Oxford: Oxford University Press.
Sosa, E. (2013). 'Intuitions and Foundations: The Relevance of Moore and Wittgenstein', *The A Priori in Philosophy*, (eds.) A. Casullo & J. C. Thurow, 186–204, Oxford: Oxford University Press.
Stough, C. (1984). 'Sextus Empiricus on Non-Assertion', *Phronesis* 29, 137–64.
Strawson, P. (1985). *Naturalism and Skepticism. Some Varieties*, London: Methuen.
Stroll, A. (1994). *Moore and Wittgenstein on Certainty*, New York: Oxford University Press.
Virvidakis, S. (2008). 'Varieties of Quietism', *Philosophical Inquiry* 30, 157–75.
Vogt, K. (2015). 'Why Ancient Skeptics Don't Doubt the Existence of the External World: Augustine and the Beginnings of Modern Skepticism', *Roman Reflections: Studies in Latin Philosophy*, (eds.,) G. D. Williams & K. Volk, ch. 13, Oxford: Oxford University Press.
Vogt, K. (2017). 'Ancient Skepticism', *Cambridge History of Moral Philosophy*, (eds.) S. Golob & J. Timmerman, 88–99, Cambridge: Cambridge University Press.
Weiler, G. (1958). 'On Fritz Mauthner's Critique of Language', *Mind* 67, 80–87.
Williams, M. (1988). 'Scepticism Without Theory', *Review of Metaphysics* 41, 547–88.
Williams, M. (1991). *Unnatural Doubts: Epistemological Realism and the Basis of Scepticism*, Oxford: Blackwell.
Williams, M. (2004). 'Wittgenstein's Refutation of Idealism', *Wittgenstein and Scepticism*, (ed.) D. MacManus, 76–96, London: Routledge.
Williams, M. (2005). 'Why Wittgenstein Isn't a Foundationalist', *Readings of Wittgenstein's On Certainty*, (eds.) W. Brenner & D. Moyal-Sharrock, London: Palgrave Macmillan.
Wittgenstein, L. (1922). *Tractatus Logico-Philosophicus*, (tr.) C. K. Ogden & F. Ramsey, London: Kegan Paul.
Wittgenstein, L. (1953). *Philosophical Investigations*, (eds.) G. E. M. Anscombe & R. Rhees, (tr.) G. E. M. Anscombe, Oxford: Blackwell.
Wittgenstein, L. (1969 [= OC]). *On Certainty*, (eds.) G. E. M. Anscombe & G. H. von Wright, (tr.) D. Paul & G. E. M. Anscombe, Oxford: Blackwell.
Wright, C. J. G. (2004). 'Warrant for Nothing (and Foundations for Free)?', *Proceedings of the Aristotelian Society* (supp. vol.) 78, 167–212.

Chapter 8
In Quest of a Wittgensteinian Hinge Epistemology

Annalisa Coliva

8.1 Introduction

Hinge epistemology is a family of theories about justification (and knowledge) which give centre-stage to Wittgenstein's notion of a "hinge". In the following, I will first put forward some basic methodological considerations regarding the relationship between merely exegetical work on, in particular, Wittgenstein's texts, and more theoretically committed work, which aims at developing suggestions that can be found in the texts, even though they are not clearly attributable as such to their author (§8.2). I will then summarize the main tenets of what, to date, is still the most widespread reading of *On Certainty* (1969) (§8.3)—the so-called "framework reading". In light of the initial methodological considerations and of this exegesis of *On Certainty*, I will then review some contemporary attempts at developing Wittgenstein's ideas in an anti-skeptical direction, such as Crispin Wright's, Michael Williams' and Duncan Pritchard's (§8.4). I will argue that, their intrinsic merits notwithstanding, they fail to take proper measure of Wittgenstein's own position. I will then close by sketching my own version of hinge epistemology and by highlighting points of contact and disagreement with Wittgenstein's own views as portrayed by the framework reading (§8.5).

8.2 Some Basic Methodological Considerations

Let me start with some platitudes. When engaged in doing *history* of philosophy—analytic or otherwise—as opposed to doing philosophy as such, we are first and foremost involved in a process of interpretation of more or less remote texts. Such

---

A. Coliva (✉)
Department of Philosophy, UC Irvine, Irvine, USA
e-mail: a.coliva@uci.edu

S. Wuppuluri and N. da Costa (eds.), *WITTGENSTEINIAN (adj.)*,
The Frontiers Collection, https://doi.org/10.1007/978-3-030-27569-3_8

interpretations cannot be assessed based merely on their own specific philosophical interest and merits, if they are meant as *exegeses*. Rather, they will have to be evaluated with respect to the text(s) of which they are purported interpretations, as well as to other historical evidence pertaining to what we may generically call the "context" in which the work was created.[1] I believe there is a strong tendency, especially in analytic philosophy, to forget such a basic methodological requirement and thus to end up presenting more how a text *struck* one, as it were, than what that text itself tried to say, in light of its author's intentions and context. However, to put it baldly, rational reconstructions without textual and contextual evidence, or else supported by very little textual and contextual evidence, seem to me to be "empty"—to use a piece of Kantian terminology. That is to say, they are devoid of historical content and therefore interest, as exegeses.

As an analytic philosopher, however, I am sensitive to what I think are, upon close examination, just rational reconstructions. Take for instance Saul Kripke's work on Wittgenstein's remarks on rule following and the so-called "private language argument". Its exegetical accuracy is extremely debatable—indeed, as a historian of analytic philosophy, I would rank myself among those who think that "Kripkenstein" is clearly not Wittgenstein. Yet, there is no denying that that reading is intellectually engaging and philosophically fruitful. This creates a tension between, as it were, rational reconstructions that are historically empty and historical reconstructions that might end up being philosophically "blind"—to pursue our Kantian comparison—by being exegetically accurate but not philosophically very illuminating or fruitful.

How can one then resolve the pressure between, on the one hand, offering interpretations that may be accurate from a textual point of view, yet not philosophically germane to future investigation and, on the other hand, presenting rational reconstructions that may end up having little to do with the text they aim to be exegeses of, while spurring—in the best case scenario—a lot of philosophical interest?

I think the tension can be resolved by being very explicit about the nature of the claims one is putting forward. If they are exegetical ones, textual and contextual evidence should be provided. If, in contrast, they are possible developments of ideas that have occurred to one while reading the text, they should be advertised as such. There is a strong tendency, especially when Wittgenstein's texts are concerned, to be inexplicit about the nature of the claims one is advancing. Here I will not try to venture an explanation of such an attitude. I think, however, that being explicit about the nature of the claims one is putting forward will have the methodological advantage

[1]For instance, *On Certainty* cannot be read without reference to G. E. Moore's epistemological papers, especially "A defence of common sense" (1925) and "Proof of an external world" (1939). It will also be important to consider that Wittgenstein got interested in it through discussions with Norman Malcolm, who, at the time, was writing on the topic and was criticizing Moore for his use of "I know" in connection with his truisms and premises of his proof. In effect, however, Malcolm was applying to those propositions considerations about the use of "I know" that Wittgenstein himself had developed in the 1930's in connection with psychological avowals. I discuss this issue at length in Coliva 2010, Chap. 2. In addition, the influence on Wittgenstein of Cardinal Newman's *An Essay in Aid of the Grammar of Assent* should be taken into proper consideration. On the latter topic, see Kienzler (2006) and Pritchard (2019).

of determining the criteria to be used in order to evaluate them. If one's claims are exegetical in nature, they will have to be evaluated based on textual and contextual evidence in their support. If, in contrast, they are meant as personal developments of ideas, which can be found more or less vaguely expressed in the text, they will have to be assessed on philosophical merit. Furthermore, along this dimension, there will be more or less *X*-ian, i.e. Wittgensteinian, Kantian, Aristotelian, etc., philosophical proposals, depending on how close they are to the texts by which they have been inspired. In the second half of this paper, I will evaluate three specimens of hinge epistemology along this dimension, leaving for other occasions their assessment on purely philosophical merit.

8.3 The Framework Reading of *On Certainty*

What, to date, remains the most widespread reading of *On Certainty* is the so-called "framework reading".[2] This, however, is not a unified view, but a family of interpretations, which fluctuate along various dimensions. Yet, central to all of them is the idea that "hinges" play a peculiar, rule-like role (OC 95) which ultimately makes them immune to skeptical assaults while also divesting them of the status of knowledge. Accordingly, propositions like "Here's my hand" (in Moore-like situations), "Nobody has ever been on the Moon" (around 1949), "The Earth has existed for a very long time", "My name is AC", "There are physical objects", are rules of "evidential significance" (Wright 1985), or "norms of description" (OC 167, 321). For instance, they have to stay put for us to be in a position to take perceptual evidence to bear on the truth of specific propositions about material objects; or for us to be able to take geological evidence to bear on the determination of the precise age of our planet or of a fossil that we have found on it; or to take our personal memories to bear on the truth of propositions regarding our own past, etc. Since justification and knowledge, to the extent that knowledge involves justification, depend on holding them fast, hinges, for Wittgenstein, are neither justified nor unjustified, neither known nor unknown (OC 91, 110, 137, 151, 466, 559, 613–614). In other words, justification and knowledge, as well as doubt, are possible only within the limits set by taking hinges for granted. It is worth stressing that even though, for Wittgenstein, the origin of hinges lies in our shared practices, hinges do play a regulative and normative role in guiding us in the practice of acquiring and assessing evidence for or against ordinary (mostly) empirical propositions, thus rendering the latter justified, or unjustified, known or doubtful. Thus, as we might put it, they are constitutive elements of epistemic rationality.

Furthermore, hinges are not bipolar—that is, liable to truth and falsity (OC 94, 197, 199, 200, 205)—for Wittgenstein. At most, they may be taken to be "unipolar" (OC 206, 514–515), insofar as truth may be predicated of them only in a minimalist

[2] See, for instance, Wright (1985), McGinn (1989), Stroll (1994), Moyal-Sharrock (2005), Coliva (2010), Schönbaumsfeld (2017).

or deflationary sense. Yet, they are propositions (OC 95–99, 309, 318–21 cf. PI 65).[3] In particular, while playing a rule-like role, they retain some descriptive content for they constitute a *Weltbild*, i.e. a world-*picture* (OC 93–95, 233, 262). Just as those pictures that sometimes guide us in performing certain activities, like assembling a chest of drawers, so they guide us in our treatment of any kind of evidence, let it be perceptual, memory-based, geological, etc.

While they do not normally bear saying, since their function is not descriptive, but prescriptive and enabling with respect to our epistemic practices, we sometimes use hinges in perfectly legitimate language games, like teaching children about them, or reminding deluded subjects, or philosophers, of their role (OC 106, 165, 233, 281, 350, 355, 374, 406–423, 450, 467, 674).

Concerning philosophical skepticism, *On Certainty* presents a two-pronged strategy. On the one hand, skeptical doubts are said to flout the conditions that govern ordinary ones. In particular, they are not sustained by reasons (OC 231, 342, 452–4, 458) as there cannot be reasons to doubt those hinges that allow us to acquire reasons for or against ordinary empirical propositions in the first place. Thus, skeptical doubts are not rational, since they are unsupported by reasons (OC 56, 151, 318, 342). On the other hand, since only ordinary doubts that are supported by reasons are meaningful, for Wittgenstein, skeptical doubts are not meaningful either (OC 347–55, 383, 676). It may seem otherwise, at first glance, but only because we project sense onto the words used to express them from other contexts of their use in which those very words do have a meaning.[4]

Finally, the title given to Wittgenstein's collection of notes by the curators—i.e. *On Certainty*—is entirely appropriate. For hinges are certain in a grammatical (or "logical") sense as opposed to a psychological (or subjective) one (OC 21, 26, 43, 51, 53, 57–60, 82, 98, 110, 136, 155, 194, 308, 319, 358–361, 375, 447, 454, 501, 618, 628), due to their role of basic rules of our (linguistic and) epistemic practices. Furthermore, our attitude with respect to them is one of acceptance (or trust), which shows no doubt and displays itself mostly in action: we behave in accord with them, accept reproach for not doing so and pass them on to children (OC 150, 275, 337, 344, 360, 509, 599–600, 603, 672).

8.4 Hinge Epistemologies: Wright, Williams and Pritchard

8.4.1 Wright and Williams

Crispin Wright has repeatedly advertised his version of hinge epistemology—as I would call it—as Wittgensteinian. Central to it, it is the idea that there are enti-

[3] This point has been vigorously denied by Moyal-Sharrock (2005), but it is fair to say that her extreme view, according to which hinges are not propositions, since they aren't bipolar, has not been met by consensus, not even among supporters of a framework reading of *On Certainty*.

[4] See Conant (1998).

tlements—i.e. non-evidential warrants—for basic, general presuppositions, such as "There is an external world", "My sense organs work mostly reliably", "I am not the victim of a lucid and sustained dream", etc.

Michael Williams too, in recent years, has proposed an epistemic reading of *On Certainty.*[5] Accordingly, we may have entitlements, i.e. non-evidential justifications, and perhaps even knowledge, of contextually determined hinges, which could be regarded as minimally truth-apt propositions. As he writes:

> Holding some particular propositions fast need not be a matter of credulity: to hold them fast is reasonable. So while *they* are (in one sense) ungrounded, *we* are justified in cleaving to them. In this way, they can be the objects of beliefs that are true (in a deflationary sense of 'true') and justified (though not derived from evidence). Thinking of 'justified true belief' in this second way, even basic certainties can amount to knowledge". (2004, p. 280)

With respect to their favourite hinges, both Wright and Williams would concur that skeptical doubts would be unmotivated and based on the mistaken idea that all justification (and knowledge) should be evidential in nature and, in particular, based on sensory experience. Hence, (at least part of) the skeptic's mistake would depend on too narrow a notion of justification.

Before dwelling further on Williams' proposal, let us ask how Wittgensteinian would this kind of hinge epistemology be. No doubt, it is easy to imagine for instance a Crispin Wright who might want to protest that even if entitlements are not Wittgenstein's creatures, they are in his spirit. It is equally easy to imagine a Michael Williams who, besides defending the idea of entitlements in this way, may actually say that an externalist conception of knowledge is in keeping with Wittgenstein's own pronouncements in *On Certainty*.[6]

In response to such a kind of defense, I would only stress that nothing seems to me firmer than a negation like the one we find in a passage, which, ironically, is often appealed to by supporters of the epistemic reading, which is worth citing in full and in its proper context:

> The *questions* that we raise and our *doubts* depend on the fact that some propositions are exempt from doubt, are, as it were hinges [*die Angeln*] on which those turn. (OC 341)

> That is to say, it belongs to the *logic* (my emphasis) of our scientific investigations that certain things are *in deed [in der Tat]* not doubted. (OC342)

> But it *isn't* (my emphasis) that the situation is like this: we just *can't* investigate everything, and for that reason we are forced to rest content with assumption. If I want the door to turn, the hinges *must* (my emphasis) stay put (OC 343).

---

[5] See Williams (2004). Williams, in conversation, has resisted being considered a hinge epistemologist, for reasons that largely evade me. In fact, to my knowledge, he has been one of the first epistemologists who have developed a Wittgenstein-inspired theory of knowledge, with a characteristic contextualist flavor to it. The fact that it differs from other ones is unimportant, once hinge epistemology is considered a family of views that give center-stage to Wittgenstein's hinges and to their peculiar role in the structure of justification and knowledge. Another contemporary supporter of an epistemic reading of *On Certainty* is Kusch (2016).

[6] An earlier supporter of such a reading of Wittgenstein is Morawetz (1978).

Here Wittgenstein is introducing the idea of hinges and is making clear that it belongs to the logic—i.e. to the norms that regulate our investigations—that certain propositions are in fact not doubted. This, however, *isn't* due to the fact that since we cannot—pragmatically—investigate everything, if we ever want to start investigate anything, we have to content ourselves with assumptions, as Wittgenstein explicitly acknowledges in OC 343.

By contrast, on Wright's view, it is precisely the fact that certain cognitive projects are valuable or even indispensable to us that forces us, in order to get them started and to pursue them, to rest content with their characteristic assumptions. Here is a telling passage:

> If a cognitive project is indispensable, or anyway sufficiently valuable to us (…) and if the attempt to vindicate (some of) its presuppositions would raise presuppositions of its own of no more secure an antecedent status, and so on ad infinitum, then we are entitled to (…) the original presuppositions without specific evidence in their favour. (2004, p. 192)

Yet, for Wittgenstein, saying that it belongs to the *logic* of our investigations that certain propositions aren't doubted simply means that it is a constitutive element of our empirical inquiries that, whenever evidence is collected for or against any empirical proposition, certain other propositions have to be kept fixed. Hence, it is constitutive of empirical inquiries, and not just a fact about our nature or pertaining to pragmatics, that certain propositions are taken for granted.

However, saying this does not corroborate the idea that hinges are therefore *reasonable* and reasonable because they would be somewhat epistemically justified, albeit non-evidentially, as they would be required in order to collect evidence for or against ordinary empirical propositions. Indeed Wittgenstein repeatedly points out that they are neither reasonable nor unreasonable. Yet he does say that they are required in order to collect evidence for or against ordinary empirical propositions. What that means, I take it, is that they are rationally mandated. Yet a rational mandate is not a *sui generis* kind of justification. It is a requirement for reasons and reasoning to be possible in the first place, not a product of reason, which, like any proper epistemic good, makes certain propositions somehow justified and therefore reasonable. Hence, hinges must stay put, and thus behave like rules, for us to be in a position to acquire evidence that may add up to justification and sometimes to knowledge, or that can motivate rational doubts and eventually the dismissal of a previously held belief. Thus, I think there is nothing in the key passages usually appealed to by supporters of entitlements to corroborate their view.

Nor is there anything which suggests that knowledge, for Wittgenstein, is to be understood somewhat in an externalist spirit and that it can be attained even if one is in no position to produce reasons for one's own true beliefs. In fact, the idea that Wittgenstein would allow for the possibility of knowledge even when no reasons can be produced in favor of one's claim is based on disregarding the distinction between empirical (ordinary) and grammatical uses of "I know" he draws in *On Certainty*. As Wittgenstein clearly writes in OC 58–59:

> If "I know etc." is conceived as a grammatical proposition, of course the "I" cannot be important. And it properly means "There is no such thing as a doubt in this case" or "The

> expression 'I do not know' makes no sense in this case". And of course it follows from this that "I know" makes no sense either.
>
> "I know" is here a logical insight. Only realism can't be proved by means of it.

It should be kept in mind that, for Wittgenstein, only grammatical uses of "I know" are not backed by reasons. Furthermore, and as weird as that may be, they actually don't express, for him, an epistemic relation between a subject and a proposition. Rather, they express a kind of certainty, which would be more perspicuously expressed by saying "Here a doubt is (logically) impossible". Thus, we do not have anything in *On Certainty* resembling the externalist idea that, even if we cannot produce reasons in favor of knowledge claims regarding hinges, we would know them nonetheless. Rather, we have the suggestion that the use of "I know" in their connection is a misleading expression of the kind of certainty, which characterizes them. Namely, a certainty that has nothing subjective or psychological about it, but which depends on the role hinges play in the relevant epistemic practices.

That is, all we get from these and other key passages in *On Certainty* is the idea that justification and knowledge do not take place in a vacuum. They always depend on there being certain hinges, which, as such, cannot themselves be justified or known, yet allow us to acquire evidence for or against ordinary empirical propositions. They are therefore constitutive of the practice that in turn determines what being epistemically rational amounts to. Some of them can change in time—"Nobody has ever been on the Moon" is no longer a hinge for us—and according to context—sometimes "Here is my hand" is an empirical proposition we subject to verification and control. Yet, there are several of them we simply cannot revise, e.g. "The Earth has existed for a very long time", "There are physical objects", etc. on pain of giving up all our system of beliefs. This is not to say that they are metaphysically necessary but only that they play such a fundamental role in our *Weltbild* that, from *within it*—as we in fact are—we can't actually find any reason to doubt them.

At this stage, one may ask whether entitlements are any good, *as such*. In response, it is worth pointing out that Wright's entitlements are said to be epistemic warrants, yet they are also said not to speak to the likely truth of the propositions they warrant. This gives them a strange status, which runs the risk of conflating pragmatic and epistemic warrants. He writes:

> The unified strategy (…) concedes the basic point of the skeptical arguments to which it reacts, namely that we indeed have no claim to know, in any sense involving possession of evidence for their likely truth, that certain cornerstones of what we take to be procedures yielding knowledge and justified belief hold good (2004, p. 206).

Pragmatic warrants are such that they do not speak to the truth of the proposition they warrant. They merely give us practical reasons to believe/endorse/act based on a given proposition. So, even if the kind of project involved is one of complying with epistemic rationality, pointing out that, to that end, we should trust in certain assumptions, does not make those assumptions more likely true than not. Yet, it may be a perfectly good reason—from a practical point of view—to go on as if they were true.

In Wright (2014, p. 214) entitlements are introduced as "warrant(s) to accept a proposition as true". Yet no argument is given to that effect, save one concerned with the kind of attitude involved. Namely, an attitude of trust, which entails trusting in the truth of what is being relied upon. This obviously is no argument for the truth of the proposition trusted on, though. For it merely makes explicit the meaning of "trust"; or, if you will, it makes explicit the norm of trust. That is, what makes trust correct, when complied with, or incorrect otherwise. Yet, this is not a good response to the challenge of providing epistemic warrants—albeit non-evidential ones—for the relevant hinges. For epistemic warrants are such that the proposition thereby warranted is, as such, more likely to be true than not. Merely noticing that we trust hinges and thus accept them as true, does not make their actual truth any more likely, though. Thus, the take home message is that entitlements are not genuinely epistemic warrants; for they do not speak to the likely truth of the proposition they are supposed to warrant.

Let us take stock. I have argued that Wright's and Williams' readings of *On Certainty* do not find adequate support in the text. I have also raised some doubts about the soundness of the notion of entitlement as proposed by Wright.[7] To the extent that Williams' entitlements resemble Wright's—something on which there is no clear textual evidence as far as I can tell—the worries raised against the latter apply to the former. If, in contrast, as one may surmise, Williams' entitlements are in the vein of Burge's externalist notion of justification, and, together with a deflationary conception of truth, were used to bolster the idea that hinges are known—albeit in an externalist sense—we have seen how remote and at odds with Wittgenstein's own writings that proposal is.

8.4.2 *Pritchard*

Let us now turn to Pritchard's (2016) recent rational reconstruction of *On Certainty*, which is at the service of a "bioscopic" strategy against radical skepticism. Pritchard is explicit that his aim is not so much to engage in exegesis, but rather to develop Wittgensteinian ideas in a more systematic way. Thus, he is not guilty, by my lights, of attributing to Wittgenstein his own epistemological views. Still, one may raise the issue of how Wittgensteinian Pritchard's position is, and of how successful it is. I will not address the latter question directly here.[8] Yet, the following remarks, while speaking more to the former issue, will also cast some doubts about the prospects of success of Pritchard's approach, in my opinion.

There are three key elements to Pritchard's strategy. First, his idea that we have "hinge commitments", as he calls them, and that acknowledging their proper role will help one resist Closure-based arguments for radical skepticism. Second, that there is nothing in *On Certainty* that speaks to Underdetermination-based skeptical

[7]For further criticism, see Coliva (2019).

[8]For a fuller discussion of Pritchard (2016), see Coliva (2016) and (Coliva 2018b).

arguments. Third, that this, in its turn, motivates supplementing Wittgenstein's own insights in *On Certainty* with epistemological disjunctivism. Hence, first appearances notwithstanding, hinge epistemology and disjunctivism are mutually supportive. For, on the one hand, the former shows the locality of epistemic evaluation and therefore sets the boundaries within which the factive rational support for our beliefs disjunctivists hold we have can in fact obtain. On the other hand, disjunctivism supplements hinge epistemology by providing it with the means to deal with Underdetermination-based skepticism.

A hinge commitment, for Pritchard, who refuses the non-propositional reading of hinges and does not consider them as rules but as propositions, is a peculiar kind of attitude, different from belief, or acceptance, which is a-rational and has visceral and animal elements to it. It is helpful, in connection with Closure-based arguments for skepticism,[9] according to Pritchard, because it shows how Closure is not applicable when hinges are concerned. For, hinges are not the object of belief, belief is necessary for knowledge, and Closure-based arguments for skepticism concern the possibility of extending knowledge from known premises to the conclusion of valid arguments through known entailments. In particular, according to Pritchard, we do not acquire a belief in, let alone knowledge of, a hinge, like "there is an external world" or "I am not a BIV", by reasoning from "Here is a tree", "If there is a tree here, I'm not a BIV/there is an external world", to "I'm not a BIV/there is an external world". Thus, Closure cannot be applied to entailments figuring hinges as consequents and cannot be used, by contraposition, to claim that since we do not have knowledge of the conclusion of the previous inference, we do not have knowledge of its premises either, and, in particular, of ordinary empirical propositions such as "Here is a tree".

In other writings,[10] I have raised some doubts about this strategy, which seems to me to conflate Closure with Transmission-failure, and which does not seem to me to be able to save the unconditional validity of Closure, contrary to Pritchard's aim. Indeed, it can be re-described as showing that Closure holds only restrictedly. That is, for the subset of all propositions, including hinges, which contains ordinary empirical propositions only. In fact, the non-applicability of Closure to hinges would follow more naturally if one denied, like Moyal-Sharrock's Wittgenstein, that hinges are propositions, or if one denied they are so in the sense of the term usually assumed by supporters of Closure, which is such to exclude the possibility for propositions believed or known to be rules.

What is also striking, from a Wittgensteinian perspective, is Pritchard's insistence on the visceral elements of our attitude towards hinges together with the denial that it is a kind of acceptance or trust, as Wittgenstein repeatedly calls it in *On Certainty*, that mostly displays itself in action (see §8.3). As a result, the notion of commitment, in relation to hinges, Pritchard places so much emphasis on seems to veer towards

---

[9]Closure-based arguments for skepticism have the following form: you don't know that *p* (where *p* is a ordinary empirical proposition) unless you know you are not a BIV (or that there is an external world). Since you cannot know the latter, you don't know that *p* either.

[10]See, in particular, Coliva (2018b).

elusiveness, in a way in which, I think, Wittgenstein's own remarks, while certainly open to interpretation, do not.

Turning to Underdetermination-based skepticism, the issue is this. If you hold that it is possible for a subject to have the same kind of evidence irrespective of whether she is actually perceiving a tree in front of her, or whether she is merely hallucinating one, it follows that, whatever is rationally available to a subject actually falls short of providing her with knowledge of (or justified belief that) "Here is a tree". According to Pritchard, Wittgenstein says nothing that could speak to this worry—whence the need to supplement his position with disjunctivism, which allows us to have factive reasons for perception-based beliefs about physical objects in our surroundings. For, when you see a tree in front of you, you take in that very fact, while, when you are hallucinating one, you merely have an experience or an appearance as of a tree.

I have to confess that here too I think that the strategy seems to be based on failing to appreciate key elements of *On Certainty*. For Wittgenstein's own insistence on the locality of reasons is matched by the idea that knowledge and justification *do* take place. Yet, they are not direct or immediate, since they take place within a system of hinges. The latter are not themselves known or justified, yet stand fast for all of us and actually allow us to form justified beliefs and to gain knowledge of (among others) ordinary empirical propositions about physical objects in our surroundings. That is, by means of hinges such as "There is an external world" and "I'm not a BIV", by having a tree-like experience, absent defeaters, one would thereby possess a justification for "Here is a tree". That is, thanks to those hinges, one would be entitled to take one's perceptual experience at face value as favoring "Here is a tree" (as opposed to its skeptical counterpart, e.g. "I am a BIV hallucinating seeing a tree"). Thus, thanks to hinges, our perceptual experiences provide us with decisive, albeit fallible and defeasible, reasons in favor of the corresponding ordinary empirical propositions.

Furthermore, on this picture one can actually offer one's perceptions as reasons in support of one's claim that, say, there is a tree where one is looking at. For, given one's available evidence and the suitable hinges, absent defeaters, one can trust one's senses, and justifiably believe that there is a tree in virtue of one's current perceptions. Hence, I do not see why Pritchard thinks we should turn to disjunctivism to solve the Underdetermination-based skeptical paradox. Probably it is the quest for factive, indefeasible reasons that motivates him. Yet one would need arguments to follow Pritchard on that quest.

Let me close this section by considering the compatibility between Wittgenstein's views regarding (ordinary) knowledge and disjunctivism. There is a striking passage in *On Certainty*, in which he seems to be discussing *ante litteram* disjunctivism. It is worth quoting it in full:

> "I know" has a primitive meaning similar to and related to "I see" ("wissen", "videre"). And "I knew he was in the room, but he wasn't in the room" is like "I saw him in the room, but he wasn't there". "I know" is supposed to express a relation, not between me and the sense of a proposition (like "I believe") but between me and a fact. So that the fact is taken into my consciousness. (Here is the reason why one wants to say that nothing that goes on in the outer world is really known, but only what happens in the domain of what are called

> sense-data.) This would give us a picture of knowing as the perception of an outer event through visual rays which project it as it is into the eye and the consciousness. Only then the question at once arises whether one can be certain of this projection. And this picture does indeed show how our imagination presents knowledge, but not what lies at the bottom of this presentation. (OC 90)

Wittgenstein was struck by the fact that in German the verb for "to know" is "wissen", which has the same root as the Latin "videre" that obviously means "to see". He also recognized the factivity of "to know". For instance, in OC 178, he writes

> the wrong use made by Moore of the proposition "I know…" lies in his regarding it as an utterance as little subject to doubt as "I am in pain". And since from "I know it is so" there follows "It is so", then the latter can't be doubted either.

Taking knowledge to be akin to seeing, together with its factivity, produces a captivating "picture", as he calls it. Namely, a picture according to which knowledge would consist in taking outer facts into one's consciousness. Yet this is just a picture, and a bad one at that, for him. We do not take outer facts into consciousness; nor do we have knowledge only of inner facts, since the latter, being mental in nature, might be thought of being taken in consciousness directly. Rather, we only have perceptual experiences, testimonies, and reasoning procedures, which can lend support to our beliefs thanks to pertinent hinges. Our knowledge, as fallible as is, just consists in having true beliefs supported by defeasible reasons. Indeed, it is of the essence of knowledge, for Wittgenstein, that it be defeasible and this can be so only if knowledge does not depend on or entail having factive reasons for the propositions known. As he writes in OC 12,

> 'I know' seems to describe a state of affairs which guarantees what is known, guarantees it as a fact. One always forgets the expression 'I thought I knew'.

Disjunctivism, with its insistence on factive reasons, is forgetful of this grammatical fact, as Wittgenstein would consider it. For, saying that one does know that *p*, iff one has factive reasons for *p*—viz. because one has taken in the very fact that *p*—excludes the possibility of being wrong—that one's reasons may not be conclusive—and that *p* isn't the case. Yet Wittgenstein insists that it is only when that possibility remains open that we do have, and are allowed to claim, knowledge. When that possibility is not open, then we are actually stumbling on something categorially different. Namely, a hinge. To put it otherwise, for Wittgenstein it is of the essence of knowledge that it be fallible. Since knowing that *p* entails *p*, the only room for fallibility there is has to do with the kind of reasons one has for believing that *p*. If those, in turn, are factive, then no room is left for knowledge of *p* to be fallible. Hence, by contraposition, reasons for knowledge of *p* cannot be factive.

Thus, to sum up, I think Wittgenstein is more explicit than Pritchard about our attitude towards hinges and that he would have had stronger arguments than Pritchard's to hold that Closure is simply not applicable to entailments involving hinges: either because ultimately they are not propositions, or because their rule-like role would prevent them from being objects of belief and knowledge. Furthermore, I have maintained that there is enough in *On Certainty* to counter Underdetermination-based forms of skepticism without endorsing disjunctivism. Finally, I have claimed that disjunctivism would be at odds with Wittgenstein's own views about knowledge.

8.5 A More Wittgensteinian Hinge Epistemology

Let me close by presenting a different hinge epistemology—something Wittgenstein wasn't interested in developing and would have considered a misguided project—but which *I* think would be of interest to contemporary practitioners of the discipline. Moreover, let me suggest how such a version of hinge epistemology would remain more faithful than its previously examined kin to the core ideas presented in *On Certainty*.

First, contrary to Wittgenstein's own pronouncements, it had better divorce epistemic considerations from semantic ones and focus on the former rather than the latter. For considering skeptical doubts meaningless would require endorsing a view of linguistic meaning that is very remote from mainstream philosophy nowadays and for good reasons. By so doing, it would immediately foreclose on a therapeutic response to skepticism of the kind propounded by Conant (1998), and more recently by Penelope Maddy (2017), but it would also foreclose on those developments of Wittgenstein's views which capitalize on the idea that skeptical doubts are nonsensical, or illusory, like Schönbaumsfeld's (2017). Second, it should explicitly embrace the idea that hinges are propositions, which are at least minimally truth-apt.[11] That is to say, they have semantic content, which can be meaningfully negated, or embedded in conditional statements or become the object of a propositional attitude of acceptance. Third, it should insist on the idea that justification and knowledge do not take place in the void. Rather, they always depend on there being certain hinges, which, as such, cannot themselves be justified (evidentially or non-evidentially), or known, yet allow us to acquire evidence and justification for or against ordinary empirical propositions. Finally, I think it should take a firm stance on the kind of propositions that can play such a hinge-role. In my view, at least when dealing with skepticism,[12] hinges should be confined to general propositions, like "There is an external world" or "My sense organs are generally reliable", or even "I am not the victim of sustained and lucid dreams", which are necessary in order for us legitimately to take perceptual experiences to bear onto beliefs about mid-size physical objects. These propositions, that is, allow us to transcend our "cognitive locality".[13] In this case, by allowing us to take mind-dependent evidence to bear onto beliefs about mind-independent objects.

It is worth pointing out that, on such a view, certain epistemic principles would fail. For instance, there wouldn't be transmission of justification from a set of evi-

[11]For further considerations regarding the kind of truth enjoyed by hinges, see Coliva (2015), Chap. 4 and Coliva (2018a).

[12]It remains an open issue what kind of hinges would be needed to acquire testimonial justification, for instance; or if there could be more local ones specific to particular kinds of inquiry.

[13]I am borrowing Wright's (2004) expression here.

dentially justified premises for a given argument to its conclusion, even if the latter were entailed by them, when hinges figure as conclusions. Consider the following argument:

MOORE

(I) Here is a hand
(II) If there is a hand here, there is an external world
(III) There is an external world.

If (III) has to be presupposed in order to have perceptual justification for (I), it can't itself be justified by an argument such as MOORE. Yet the transmission of justification wouldn't fail because a *justification* for (III) would be needed already in order to have justification for (I). Rather, it would fail simply because we cannot acquire justification to believe the conclusion of an argument, which—as unjustified and unjustifiable as it is—has to be taken for granted in order for its premises to be justified in the first place. Hence, as I have argued elsewhere,[14] we would have a subtly different kind of transmission failure from the one originally advertised in Wright's (1985) writings on the topic.

Moreover, we would have failure of Closure for epistemic operators such as justification and knowledge. For even if we do possess justification or even knowledge for propositions like (I), we would have none for (III), which is entailed by them and that we do know it is entailed by them. I think this result is acceptable, for reasons that I have explored in more detail elsewhere.[15] To put it briefly, because Closure would hold for arguments containing known entailments in which the conclusion is not a general proposition of the kind I would allow to play a hinge-role. Such arguments are the ones we are usually interested in in our everyday epistemic inquiries, when general presuppositions like "There is an external world" are taken for granted. Furthermore, Closure would fail in a limited amount of cases—i.e. when such hinges figure as the entailed conclusions—for principled reasons. That is to say, because of a conception of the structure of perceptual justification, which is independently plausible.[16] Such a conception could in fact represent a desirable mid-way position in between so-called "liberal" and "conservative" accounts of perceptual justification, such as the ones respectively put forward by Pryor (2004) and Wright (1985, 2004), that impose too lax and too stringent requirements upon empirical justification, respectively. I have dubbed such a view the "moderate" conception of the architecture of perceptual justification. Central to it is the idea that perceptual justification is possible only thanks to taking for granted the relevant hinges, absent defeaters, while having the relevant course of experience.

Furthermore, we would get an answer to at least some form of skepticism, of Humean descent, along the following lines.[17] Regarding ordinary empirical propo-

---

[14]Coliva (2012), (2014), (2015).

[15]Cf. Coliva (2012).

[16]I have presented it in Coliva (2012b), (2014) and have defended it in more detail in Coliva (2015).

[17]As to Cartesian skepticism, by denying Closure, we would block the consequence that since we do not know that we are not BIVs we do not know there is a hand here. Moreover, if "I am not a BIV" were a hinge, then it too would turn out to be mandated by the lights of epistemic rationality.

sitions, we can have justification and even knowledge of them, whenever we have an appropriate course of experience, absent defeaters, thanks to the relevant hinges, even if the latter are not justified or known. True, hinges themselves are beyond epistemic appraisal, but, actually, we are mandated by epistemic rationality itself to hold on to them. For they are constitutive of the practice of gathering evidence for or against ordinary empirical propositions on the basis of certain assumptions, which is itself constitutive of epistemic rationality. On a Wittgenstein-inspired epistemology, notions such as the one of epistemic rationality do not hang in the air. Rather, they depend on human practice and, in particular, on the practice of forming, assessing and withdrawing from beliefs about physical objects based on one's perceptual evidence. Hence, either a skeptic is a skeptic about epistemic rationality itself, or else, if he is merely a skeptic about a class of targeted propositions, like hinges, he will not be allowed to conclude that they fall beyond epistemic rationality. For, though unwarrantable, they are part of epistemic rationality as they are its constitutive elements. To exclude them from epistemic rationality would then be a categorial mistake just like the mistake of excluding rules from a game simply because they are not moves *within* the game. The correct notion of epistemic rationality could then be defined thus:

> **Epistemic Rationality$_{\mathbf{Ext}}$**: a proposition is epistemically rationally believed or assumed if and only if either there is a (evidential) justification for it, or, while being unjustifiable, its assumption makes the acquisition of epistemic justifications possible.

The skeptical mistake would thus reside in holding on to too narrow a notion of epistemic rationality, which confines it to evidentially warranted propositions only. Moreover, it would depend on a misconception of the structure of justifications and knowledge, which requires them to spring only from more basic, yet similarly justified or even known propositions. This view can be defined as follows:

> **Epistemic Rationality$_{\mathbf{sk}}$**: a proposition is epistemically rationally believed if and only if there is a (evidential) justification for it.

Finally, the animal element in this Wittgenstein-inspired response would also be preserved, to some extent, in that it is acknowledged that hinges are what they are in virtue of a human practice that has not developed out of ratiocination. However, it is the normative element, which on my reading of Wittgenstein is the core aspect of *On Certainty*, that is given centre-stage insofar as certain unwarrantable propositions are allowed to play a constitutive role with respect to evidential justification and knowledge and, therefore, with respect to epistemic rationality itself.[18]

---

[18] I have presented such an extended conception of epistemic rationality in Coliva (2012b, 2014) and further developed it in Coliva 2015.

References

Coliva, A. 2010a *Moore and Wittgenstein: Scepticism, Certainty and Common Sense*, Palgrave, Macmillan.

Coliva, A. 2012a 'Varieties of failure (of warrant transmission—what else?!), *Synthese* 189/2, pp. 235–254.

Coliva, A. 2012b 'Liberals and conservatives. Is there a (Wittgensteinian) third way?', in A. Coliva (ed.) *Mind, Meaning and Knowledge. Themes from the Philosophy of Crispin Wright*, OUP, pp. 323–351.

Coliva, A. 2014 'Moderatism, transmission-failures, closure and Humean scepticism', in D. Dodd and E. Zardini (eds.) *Scepticism and Perceptual Justification*, OUP, pp. 248–272.

Coliva, A. 2015 *Extended Rationality. A Hinge Epistemology*, London, Palgrave Macmillan.

Coliva, A. 2016 "Review of Duncan Pritchard *Epistemic Angst. Radical Skepticism and the Groundlessness of Our Believing*", *Notre Dame Philosophical Reviews*. URL: https://ndpr.nd.edu/news/epistemic-angst-radical-skepticism-and-the-groundlessness-of-our-believing/.

Coliva, A. 2018a "What anti-realism about hinges could possibly be", in C. Kyriacou and R. McKenna (eds.) *Metaepistemology. Realism and Anti-Realism*, London, Palgrave Macmillan, pp. 267–288.

Coliva, A. 2018b "Strange bedfellows. On Pritchard's disjunctivist hinge epistemology", *Synthese*. https://doi.org/10.1007/s11229-018-02046-z.

Coliva, A. 2019 "Against neo-Wittgensteinian entitlements", in P. Graham and N. Pedersend (eds.) *New Essays on Entitlements*, Oxford, Oxford University Press, forthcoming.

Conant, J. 1998 'Wittgenstein on meaning and use', *Philosophical Investigations* 21, pp. 222–50.

Kienzler, W. 2006 'Wittgenstein and John Henry Newman on certainty', *Grazer Philosophische Studien* 71, pp. 117–138.

Kusch, M. 2016 "Wittgenstein on mathematics and certainty", in A. Coliva & D. Moyal-Sharrock (eds.) *Hinge Epistemology*, Leiden, Brill, pp. 48–71.

Maddy, P. 2017 *What Do Philosophers Do? Skepticism and the Practice of Philosophy*, Oxford, Oxford University Press.

McGinn, M. 1989 *Sense and Certainty: A Dissolution of Scepticism*, Oxford, Blackwell.

Morawetz, T. 1978 *Wittgenstein and Knowledge. The Importance of* On Certainty, Atlantic Highlands N. J., Humanities Press.

Moyal-Sharrock, D. 2005 *Understanding Wittgenstein's* On Certainty, London, Palgrave Macmillan.

Pritchard, D. 2016 *Epistemic Angst. Radical Skepticism and the Groundlessness of Our Believing*, Princeton, Princeton University Press.

Pritchard, D. 2019 'Wittgenstein on faith and reason: the influence of Newman', in M. Szatkowski (ed.) *God, Truth and Other Enigmas*, Berlin, De Gruyter, forthcoming.

Pryor, J. 2004 "What's wrong with Moore's argument?", *Philosophical Studies* 14, pp. 349–378.

Schönbaumsfeld, G. 2017 *The Illusion of Doubt*, Oxford, Oxford University Press.

Stroll, A. 1994 *Moore and Wittgenstein on Certainty*, Oxford, Oxford University Press.

Williams, M. 2004 'Wittgenstein, truth and certainty', in M. Kölbel & B. Weiss (eds.) *Wittgenstein's Lasting Significance*, Routlede, pp. 249–84.

Wittgenstein, L. 1969 *On Certainty*, Blackwell.

Wright, C. 1985 'Facts and certainty', *Proceedings of the British Academy* 71, pp. 429–472.

Wright, C. 2004 'Warrant for nothing (and foundations for free?), *Aristotelian Society Supplementary Vol.* 78/1, pp. 167–212.

Wright, C. 2014 'On epistemic entitlement (II). Welfare state epistemology', in D. Dodd and E. Zardini (eds.) *Scepticism and Perceptual Justification*, OUP, pp. 213–247.

Chapter 9
Paradigms and Self-reference: What Is the Point of Asserting Paradoxical Sentences?

Jakub Mácha

9.1 Paradoxes

Let me begin with the notion of a paradox. A paradox is usually defined as a self-contradictory (or at least unacceptable) conclusion that is derived from true premises by sound reasoning. For example, Russell's paradox is derived from the axioms of naïve set theory, and this derivation is sound. A self-contradictory conclusion indicates that there is something wrong with the original premises (or with the rules of reasoning). The original Greek meaning of 'paradox' is, however, different. *παράδοξος* (parádoxos) is something contrary to common belief (doxa). A paradox is, thus, something unexpected, strange, something surprising.[1]

Wittgenstein's understanding of paradox goes back to this original Greek meaning. We can find passages in his writings where the word 'paradox' is used as an alternative for 'surprise': '[such a] representation of it makes a situation surprising, or astonishing, even paradoxical' (RFM I, appdx. II, §1). Here is Wittgenstein's most prominent characterization of a paradox: 'Something surprising, a paradox, is a paradox only in a particular, as it were defective, surrounding. One needs to complete this surrounding in such a way that what looked like a paradox no longer seems one' (RFM VII, §43). Or, in a similar manner: 'A sentence [Satz] is a paradox only if we abstract from [absehen] its use' (RPP I, §65). Hence, paradoxes arise when we take sentences out of their context of use. A sentence appears paradoxical, i.e. surprising, only if we abstract away from its context of use—only if its context of use is missing. One way of dealing with paradoxical sentences is just to imagine the missing context of use, i.e. the missing language-game.

There is a long history of paradoxes of self-reference, going back to the Epimenides paradox. I will begin, however, with Frege's concept 'horse' paradox. For

[1] Cf. Kutschera (1989) for an overview.

J. Mácha (✉)
Masaryk University, Brno, Czech Republic
e-mail: macha@mail.muni.cz

S. Wuppuluri and N. da Costa (eds.), *WITTGENSTEINIAN (adj.)*,
The Frontiers Collection, https://doi.org/10.1007/978-3-030-27569-3_9

Frege (1892), concepts are functions, unsaturated entities. Functions are saturated by objects, which are saturated entities. This distinction between concepts and objects is derived from a grammatical analysis of the use of definite and indefinite articles. A noun with the definite article refers to an object ['auf einen Gegenstand hinweist'], whereas a noun with the indefinite article refers to a concept. Frege was aware of several exceptions to this rule (due to the logical imperfections of ordinary language). This intuitive distinction leads, as Bruno Kerry observed, to a paradox if we consider the expression 'the concept "horse"'. In this expression, 'concept()' is a concept and 'horse' is an object. This concept is then saturated in the expression 'the concept "horse"' and becomes an object. Thus, the concept 'horse' is an object. Or, in other words, the concept 'horse' is not a concept. This is an absurd consequence, which Frege was not able to deal with. It was, for him, 'a certain inappropriateness of linguistic expression', that is, a distortion of natural language that could be overcome in a logically perfect language.

For Frege, an object cannot be a concept and *vice versa*. There is an unbridgeable gap between these two classes of entities. Kerry's suggestion, however, is that there is an entity that is both a concept and an object at the same time. 'Horse' is an object if it is an argument of the function 'concept()' and it is a concept, at the same time, if the application of the same function is supposed to yield the truth.

This paradox of the concept 'horse' is circumvented in the *Tractatus*. For Wittgenstein, the problem lies in the idea of assimilating philosophical (as well as logical and linguistic) vocabulary into language itself.[2] This philosophical vocabulary includes concepts such as 'concept', 'object' and 'name'. 'Concept' is a formal concept (or pseudo-concept) which—unlike proper concepts—is not supposed to be represented by a function or by a class, but by a propositional variable (4.126). 'The concept "horse"' is, thus, *not* an application of the function concept() to the name 'horse', and therefore 'the concept "horse"' is not a name, but a variable. It makes no sense to consider the application of a formal concept to something, because formal concepts are, quite simply, not something that can be applied. To be a concept is a formal characteristic (or feature) of the name 'horse'. This feature cannot be expressed in a proposition, but can only be shown. The introduction of formal concepts thus effectively avoids the 'horse' paradox. This circumvention is incorporated into the conceptual framework of the *Tractatus*. Names refer to objects. If there were a name X of a name Y, then Y would be an object. Objects, however, are simple; they cannot mean anything. Hence, there is no name of Y.

The gap between concepts and objects is not closed in the *Tractatus*. It is instead transformed into the gap between what can be said and what can be shown: 'What can be shown, *cannot* be said' (4.1212). Let us now move on to Wittgenstein's later philosophy, where, I will argue, this gap is bridged.

---

[2]Cf. Diamond (1995, p. 184): 'The predicate "concept" cannot be predicated of any concept: we think we want it in order to say things about concepts, but to think that it will enable us to do so is confusion, confusion about what a concept is. Wittgenstein thought that the *whole* philosophical vocabulary reflected confusion.'

9.2 Paradigms, Rules and Private Language

Let a *paradigm* of X be a *material object* together with the *praxis* of applying this object in a given situation. I call the object of a paradigm a *paradigmatic sample*. The term 'object' is understood in the broadest way possible. Paradigmatic samples are real material things, ranging from clearly defined objects such as the metre stick or a colour plate to intricate structures such as formalizations of mathematical proofs or works of art. A paradigmatic sample is a model of a situation (what Wittgenstein called a 'model' in his early *Notebooks*, and an 'image' [Bild] in the *Tractatus*). A paradigm is, thus, a model along with a method for its projection or, rather, a method of comparison.[3] A paradigmatic sample is an object of comparison (PI, §130).[4]

Wittgenstein considers many instances of giving a definition of a certain expression by means of a paradigm. The most familiar ones are the definition of 'one metre' and 'sepia' by reference to the standard metre rod (or another yardstick [Maßstab]) and the standard sepia sample. Is this technique in any way general? Are we qualified to say that every expression must be defined by means of a paradigm? The later Wittgenstein's method involves comparing language to (or looking at language as) a rule-governed activity.[5] There are two points to bear in mind. Firstly, Wittgenstein often compares using language and measuring with a yardstick [Maßstab].[6] This is a general analogy. Note that Wittgenstein conceives of paradigms and yardsticks at two distinct levels: at the general level of this analogy and at the local level of defining a particular expression by means of a paradigm. Second, Wittgenstein at times uses the expressions 'rule' and 'paradigm' (or 'yardstick', 'model', 'object of

---

[3]Cf. Wittgenstein's remark that a paradigmatic sample (yardstick) alone is a dead object: 'Lay the yardstick alongside a body; it doesn't say that the body is such-and-such a length. Rather, I want to say, in and of itself it is dead, and it achieves nothing of what thought achieves' (BT, pp. 70e–71e). On the other hand, the materiality of a paradigmatic sample must not be taken as its imperfection: 'In the case of the measuring stick you can't say: "Yes, the measuring stick measures length in spite of its corporality; but a measuring stick that had only length would be the ideal, would be the *pure* measuring stick". No, if a body has length, then there can't be any length without a body—and even if I do understand that in a certain sense only the length of the measuring stick does the measuring, yet what I put into my pocket is still the measuring stick—the body and not the length.' (BT, p. 352e).

[4]Every rule and concept has a pictorial (iconic, natural) component and a conventional (habitual, customary) component (the praxis).

[5]Though there are other objects of comparison that do not involve rules; cf. Kuusela (2019, §§5.5 and 6.1) for details.

[6]Cf. the following very early remark from 1914: 'Proposition and situation are related to one another like the yardstick and the length to be measured' (NB, p. 32). This idea is developed in the *Tractatus* (2.1512) and further in Wittgenstein's later thought. For instance, 'der Satz wäre wie ein Maßstab an die Wirklichkeit angelegt' (Ms-109, 272), or in the *Big Typescript*: 'When I compared a proposition to a yardstick, then strictly speaking all I did was to take a proposition that relies on a yardstick to state a length and use it as an example of a proposition./If one conceives of propositions as instructions for making models, their picture-like quality becomes even clearer' (BT, p. 67e). In Ms-109, 176, Wittgenstein says: '"In der gleichen Sprache ausdrücken" heißt mit dem gleichen Maßstab messen.' Or simply: 'Vergleiche die Grammatik von "Satz", "Maßstab" und "Gedanke".' (TS-219, 3).

comparison') interchangeably or for textual variation. So, for instance, in the *Big Typescript* we read that 'the steps in the chain of equations have been taken in accordance with *these* particular rules (or paradigms)' (BT, p. 460e).[7] It thus seems that comparing language to a rule-governed activity is close to comparing it to measuring with a yardstick or paradigm. These two objects of comparison are very close to each other. In fact, I would like to claim that, beyond these exegetical considerations, they amount to the same thing.

I shall now argue that paradigms involve rules and rules involve paradigms. The former claim is fairly obvious. Defining an expression using a paradigm certainly involves some rules; namely the rules of how the comparison with (or the measurement by) the paradigmatic sample is to be carried out. In PI, §53, Wittgenstein restates paradigms in the form of a chart [Tabelle] and further explains that 'we call such a chart the expression of a rule of the language-game' (cf. also PI, §§73, 74, 162, 265). The latter claim merits closer consideration. Can any rule be restated in terms of a paradigm? In other words, does every rule involve some comparison with a paradigmatic sample? Let us consider public and private rules separately. Can there be a public rule that does not involve any paradigm? The negative answer to this question is guaranteed by a certain flexibility of the concepts of paradigm and rule. The notion of a paradigmatic sample is defined very broadly. A paradigm can be thought of as any rule-based technique that, at some points, takes or depends on external objects as its inputs. In that case, the concepts of a public rule and of a paradigm are effectively equal.

What about private rules? The considerations discussed in the previous paragraph would entail that a rule of this sort that does not make use of any paradigmatic sample would be a private rule. The argument could be made by pointing to the result of the private language argument, namely that private rules are impossible. I think an even stronger conclusion is available if the private language argument is examined more closely. We have to realize that the aim of an ostensive definition of a general term (as opposed to a proper name) is to introduce a sign by pointing at a paradigmatic sample and stipulating in which respect entities of that kind are similar to the sample (i.e. stipulating the method of comparison or measurement).[8]

There have been many interpretations of the private language argument. I can focus only on those that are relevant to the overall topic of this essay, and shall leave others aside without commenting on them. Several commentators (McGinn 1997; Stroud 2001; Ahmed 2017) have restated the problem with private rules as a failure

[7]Here is another formulation from the early 1930 s period: talking about the equation $a + (b + c) = (a + b) + c$, Wittgenstein says: 'For if the proposition is a *rule*, a *paradigm*, which every calculation has to follow, then it makes no more sense to talk of working out the equation, than to talk of working out a definition' (PG, p. 395, my italics).

[8]This precise idea is expressed in the *Big Typescript*: 'in an ostensive definition I do not state anything about the paradigm (sample), but only use it to make statements; that it belongs to the symbolism and is not one of the objects to which I apply the symbolism' (BT, p. 408e). Or we can refer to §1 of the *Investigations*: the shopkeeper possesses paradigmatic samples for 'apple', 'red' and numbers, and uses them as described in the passage. 'It is in this and similar ways that one operates with words.'

of a *private* ostensive definition. In PI, §258, Wittgenstein considers establishing the meaning of the sign 'S' by concentrating one's attention on a sensation. This is effectively a kind of ostensive definition. This private ostensive definition is, however, futile; it fails to determine the subsequent use of 'S'. But why is this so? McGinn and Stroud argue that the problem with a private ostensive definition is that the private diarist fails to specify the respect in which a future episode will resemble the original sensation. The private diarist fails to specify that 'S' is 'a concept of a certain kind, i.e. [has] a particular use or grammar' (McGinn 1997, p. 131). Ahmed provides another reason for the failure of a private ostensive definition. A private ostensive definition is underdetermined between the following cases: '"S" refers to the *original* felt quality of the initial sample', and 'at any later time *t*, "S" refers to his *memory at time t* of the initial sample' (Ahmed 2017, p. 47). That is to say, 'S' refers to the initial sensation or to the diarist's memory of it. On all these interpretations, the problem with a private language or private rules thus boils down to defining a sign by means of a private paradigm (McGinn and Stroud effectively say that what is problematic is the method of comparison, whereas Ahmed says that the problem lies in the indeterminacy of the paradigmatic sample).

Before moving on, let us pause for a moment and consider the very idea of a private sample. In the early 1930s, Wittgenstein contrasts a yardstick [Maßstab] or a measuring instrument [Meßinstrument] with an *Augenmaß*, which, in this context, can be translated as 'private sample' or 'private yardstick'.[9,10] He maintains that such a private yardstick is not an inaccurate yardstick, but something completely different from a physical yardstick (Ms-109, 244[2]). A private table [Tabelle] is of no use: 'Looking up a table in the imagination is no more looking up a table than the image of the result of an imagined experiment is the result of an experiment' (PI, §265). Here, the idea of a private paradigm is not wholly inconsistent. It is clear, however, that we can use private samples only if there is a way of using a public paradigm in the same situation (or we must at least be able to imagine an actual measurement of this sort).

Let me conclude this section by saying that, for Wittgenstein, the grammar of the words 'paradigm', 'yardstick' and 'measure' is closely related to the grammar of the word 'rule' (cf. PI, §150).[11]

---

[9]Cf. BT, p. 43; Ms-107, 264[3]; Ms-109, 244[2]; Ms-110, 280[2].

[10]There is an analogous idiom in English: to *eyeball* something means to judge it by eye. Here, however, the intimate connection to a measurement is lost.

[11]See Diamond (2001, pp. 125–132) for another elaboration of this parallel between the understanding of measurement and of rules.

9.3 Paradigm Paradox

We can find a restatement of the concept 'horse' paradox in terms of paradigmatic samples in Wittgenstein writings: 'one sentence can never describe the paradigm in another, unless it ceases to be a paradigm' (PG, p. 346).[12] Let me call this the *paradigm paradox*.

This way of putting things is striking for its generality, which is unusual for Wittgenstein even in the 1930s. There are several instances scattered throughout Wittgenstein's writings: 'one cannot say of a group of strokes serving as a paradigm of 3, that it consists of 3 strokes' (PG, p. 346)—or of the standard metre rod that it is or is not one metre long, or of the standard sepia sample that it is this colour or it is not (PI, §50). Why can we not say this? To be one metre long means being the same length as the standard metre when compared—and there must be a standard method of comparison.[13] To say that the standard metre is one metre long requires us to compare the standard metre with itself. For this peculiar kind of comparison to make sense, we have to specify how it works. We can say that the length of the standard metre is always identical with itself. That is, the standard metre is always one metre long, whatever happens. But this is an *ad hoc* stipulation that, moreover, does not explain its negation, i.e. saying that the standard metre is not one metre long.

What we can do, however, is describe a paradigm by using another paradigm. Then, of course, the original paradigm ceases to be a paradigm in this usage. We can say, for example, of the standard metre that it is 1.0936133 yards long. Then, however, the paradigm of length is the standard yard, not the standard metre.

Remember that, according to Wittgenstein, paradoxes arise when we take sentences out of their context of use. When someone says this rod is one metre long, there is always an implicit commitment to the act of comparing the rod with the standard metre. This praxis is part of the meaning of 'one metre long'. In everyday situations, we do not do this, of course (no real shopkeeper uses paradigmatic samples as the shopkeeper from §1 of the *Investigations* does). But this praxis must at least be possible. A proposition may be a paradox, if we abstract from this praxis, from this context of use, and take 'one metre long' to be a function ranging over all material objects. Then the situation may occur that we apply this function to an object that appears in its definition, that is, to the standard metre rod.

[12]'Und ein Satz kann das Paradigma im andern nie beschreiben, sonst ist es nicht das Paradigma.' From the German wording it is clear that the pronoun in the second clause 'es' refers to the paradigm from the first clause.

[13]This is, of course, a crude simplification of the actual praxis, which, however, does not impact on the overall argument.

9.4 Russell's Paradox

The (intuitive) belief that we can express in language any property whatsoever has been captured by various *principles of unrestricted abstraction* or comprehension (e.g. Frege's Basic Law V). The principle states that for any property nameable in language, there is a concept expressing this very property. This principle implies that we can express in language properties such as 'being one metre long', 'being sepia-coloured' or 'having three elements'. We can, of course, do this intuitively. Our intuition says (or tempts us to accept) that these concepts (whose existence is postulated by the principle) can be truthfully applied to *any* objects having this or that property. The principle of abstraction is *unrestricted*; this means that nothing can be excluded from the range of the universal quantifier. For instance, the concept 'being one metre long' can be truthfully applied to *any* object with a length of one metre. The standard metre cannot be excluded. This captures our everyday intuition that the standard metre is one metre long.

This intuition is wrong, however. It is well known that the principle of unrestricted comprehension leads to a contradiction. *Russell's paradox* follows from a set-theoretical version of the principle:

> For any formula $\phi(x)$ containing x as a free variable, there will exist the set $\{x: \phi(x)\}$ whose members are exactly those objects that satisfy $\phi(x)$.

And again, this principle is very intuitive. If $\phi(x)$ stands for 'x is one metre long', then $\{x: \phi(x)\}$ will be the set containing all objects that are one metre long, including the standard metre. Russell's paradox follows if we let $\phi(x)$ stand for $\neg(x \in x)$. Russell's initial solution or, rather, response to his paradox is to adopt the *vicious circle principle*. This principle states that 'whatever involves all of a collection must not be one of the collection'; that is, a collection must not contain members that are presupposed by the whole collection.[14]

Let us return to paradigms. We can regard paradigms as characteristic functions of collections. A paradigm generates a collection of objects that are in accord with this paradigm. Accordingly, a paradigmatic sample is precisely an object that is presupposed by the collection of objects that are defined by this paradigm. The question of whether we can describe a paradigmatic sample using the very same paradigm is structurally equivalent to the question of whether a set contains itself as a member. From this structural equivalence it follows that the vicious circle principle effectively counteracts the paradigm paradox as well. Or, in other words, Wittgenstein's claim that 'one proposition can never describe the paradigm in another' is a version of the vicious circle principle.

Russell's vicious circle principle is a prohibition on possible members of a collection. The principle states something we *must not* do.[15] Wittgenstein's own principle, however, continues: 'unless it ceases to be a paradigm'. This means that if we attempt to apply a paradigm to its own paradigmatic sample, it ceases to be a paradigm in this

[14]This account follows Irvine and Deutsch's (2016) entry on Russell's paradox.

[15]Russell's theory of types is a more sophisticated version of this prohibition.

usage. This is the point that goes beyond Russell's vicious circle principle. There is no such absolute prohibition in Wittgenstein's version of the principle. Propositions describing the paradigm are not prohibited. They are, however, paradoxical. But remember, again, that a sentence will only appear paradoxical if we are not aware of its context of use. Our task is to identify a language-game that involves such sentences, and the paradox will then vanish.[16] What I shall now attempt to do is work out what a use of paradoxical sentences of this kind might be.

9.5 The Self-reference of a Paradigm, the Self-membership of a Set

We shall look for sentences that assert the self-reference of a paradigm or the self-membership of a set. Let us consider Wittgenstein's own examples first. §50 of the *Philosophical Investigations* is well known. In this passage, we read: 'There is *one* thing of which one can say neither that it is one metre long, nor that it is not one metre long, and that is the standard metre in Paris.' But this prohibition is restricted to 'the language-game of measuring with a metre-rule' (ibid.) I take this to mean that we cannot ascribe the property of being one metre long to the standard metre by using the sentence 'The standard metre is one metre long'. This does not, however, exclude the possibility that there is another use of the sentences 'The standard metre is one metre long' and 'The standard metre is not one metre long'. Let us move on to a noteworthy passage from the final part of the *Remarks on the Foundations of Mathematics*, which opens as follows:

> What *sort* of proposition is: 'The class of lions is not a lion, but the class of classes is a class'? How is it verified? How could it be *used*?—So far as I can see, only as a grammatical proposition. To draw someone's attention to the fact that the word 'lion' is used in a fundamentally different way from the name of a lion; whereas the class word 'class' is used like the designation of one of the classes, say the class *lion*. (RFM VII, §36)

This remark says that paradoxical sentences like 'The class of lions is not a lion, but the class of classes is a class' can be used as a grammatical reminder.[17] Or—as Wittgenstein also suggests—we can employ the Tractarian method of analysis and say that 'lion' is used in two different ways or with two different meanings here (e.g. 'Lion' as the name of a particular lion or "Cat" as the name of a cat). Another suggestion is that we can 'ascribe a sense out of politeness' to these sentences. I take this sense to be whatever sense is ascribed by an accidental stipulation—analogous to the stipulation that the standard metre is (always) one metre long.

These are, however, ways to analyse the paradox away by getting rid of the self-referential structure. Wittgenstein wants to take the sentences in the right way [richtig auffaßen]—not to analyse them away but to find language-games for them.

---

[16]This could be seen as a general characterization of the later Wittgenstein's method.

[17]Cf. PI, §127: 'The work of the philosopher consists in assembling reminders for a particular purpose.'

Animal fables [Tierfabeln] are the first context of language-games that Wittgenstein considers:

> The fable says: 'The lion went for a walk with the fox', not a lion with a fox; nor yet the lion so-and-so with the fox so-and-so. And here it actually is as if the species lion came to be seen as a lion.

Wittgenstein undertakes a grammatical analysis based on the use of the definite and indefinite articles. The very same analysis is the basis of Frege's distinction between concepts and objects, as we saw above. The definite article refers to an object (for Frege) or to an instance of a noun that is identifiable from the context. 'An instance of a noun' is, in this context, analogous to 'an element of a set'. There are, however, many exceptions to this rule.[18] We can analyse the definite article as referring to every member of a set (e.g. in 'The fox is a predatory animal'), i.e. as the universal quantifier. Yet this is clearly not the analysis that fits the context of the fable. Wittgenstein considers and opposes another analysis that was suggested by Lessing:

> (It isn't as Lessing says, as if a particular lion were put in the place of some lion or other. 'Reynard the Fox' ['Grimmbart der Dachs'] does not mean: a fox of the name 'Reynard'.) (RFM VII, §36)

Lessing says, in his *Essays on Fable*, that the fox represents a distinctive feature of its species (e.g. cunningness), and continues: 'It would be a childish abuse of words to say that the particular is *similar* to the general, the individual to the species, and the species to the genus' (Lessing 1825, p. 72). Lessing's view is, thus, that in a fable a particular animal *exemplifies* a certain feature of its species. Accordingly, Reynard the Fox represents the characteristics of being thoughtful and calm, for instance. However, exempla are not mere instances; by exemplifying a certain feature, they involve every member of its species (of a set, of a collection), and in this respect they are analogous to paradigms. Lessing's view is not as simple as putting a particular animal (lion, fox, etc.) in the place of some animal in a fable—as Wittgenstein seems to read him. Despite this inaccurate interpretation of Lessing's views, Wittgenstein's point is that the definite article must not be analysed away by interpreting it in terms of the indefinite article and that it is not childish to say that the particular is similar to the general—as we shall see in the next section.

9.6 Asserting Self-referential Sentences

Finally, let us move on to Wittgenstein's positive examples of this kind of self-reference. The passage quoted above continues:

[18]This rule is rather complex. There are differences between particular languages, such as English and German. Moreover, there are languages that do not use articles, such as Latin and Slavic languages.

> Imagine a language in which the class of lions is called 'the lion of all lions', the class of trees 'the tree of all trees', etc.—Because people imagine all lions as forming *one* big lion. (We say: 'God created man [Gott hat den Menschen geschaffen]'.) (RFM VII, §36; Ms-124, 125–126)

In this passage, which curiously enough has escaped the attention of commentators, Wittgenstein discusses language-games that deal with self-membered classes. In 'God created man' (in German with the definite article 'den Menschen', accusative singular), 'man' refers to all men and at the same time to a particular man Adam. The quotation from Genesis 1:27 continues with 'in his own image, in the image of God created he him'. The first man and man, the class of men, is created as an image of God. The first man is himself an image of man; he is, so to speak, an archetype of man—a paradigmatic sample. In the same way, the lion of all lions can refer to the paradigm of a lion, or the song of songs to a paradigmatic song.[19]

I take from these considerations that there is a structural analogy between a noun being employed as a self-membered set and a paradigmatic sample being included in or excluded from the set it generates. If there is a use for employing self-membered sets, there might be a use for describing paradigmatic samples using the very same paradigm. Wittgenstein's vicious circle principle says that if we attempt to use such a self-referential description, the paradigm ceases to be a paradigm. What, then, would be the use of the sentence 'the standard metre is *not* one metre long'? To point out two different uses of 'metre'? This would be too confusing, and besides, there is a better way of putting this point. It cannot mean that the standard metre has changed its length either, for this must be put in terms of another paradigm, e.g. by saying that the standard metre is not 3.28 feet long any more. The only employment of this sentence I can imagine is to indicate a change in the praxis with the standard metre or, more precisely, a change in the method of comparison (e.g. with respect to the accuracy of the comparison, the material the standard metre is made of, the storing of the standard metre in a certain environment, or its replacement by a beam of light). Let us imagine that we can use the standard metre to measure another object using the old method, and subsequently we can use this object to measure the standard metre using the new method. Then we might say that the standard metre is *not* one metre long.

Is the sentence 'The standard metre is not one metre long' paradoxical? It may be if 'paradox' means *surprise* and if we are not envisioning its context of use, i.e. an attempt to compare a stick with itself. However, even if we are aware of the context, we can nevertheless be surprised, because then this sentence indicates that the method of comparison (or the praxis in general) has changed. Moreover, there is another surprising element: the way in which the praxis has changed is completely arbitrary,

---

[19]The construction 'X of Xs' ('song of songs', 'holy of holies', 'king of kings', etc.) is common in the Old Testament. It expresses a *superlative*. Although it is in the singular, it does not necessarily express a single element (cf. Keel 1994, pp. 38–39). Today, there is a similar idiomatic construction: 'the mother of all somethings', e.g. Douglas Engelbart's 'The Mother of All Demos'. This refers to something's being superlative not in the sense of having the greatest degree of a certain characteristic, but rather in having the greatest number of characteristics of X. There is a single thing that has the greatest number of characteristics of X, the most X, the paradigmatic sample of X.

if we have no further context than this sentence. The old paradigm has ceased to exist and no new paradigm has been established by asserting the sentence.[20]

It is often said that a paradox can cause confusion if it goes undetected. If 'paradox' means *surprise*, there are no undetected paradoxes. We cannot be surprised without knowing it. However, one can cause confusion by describing paradigmatic samples using the very same paradigm *without realizing it*. This situation arises if we derive a measuring instrument (e.g. a ruler) from the standard metre and subsequently use this instrument to measure the standard metre without realizing that the measured object is the standard metre. That leads to the conclusion that one can confuse a paradigmatic sample with an object generated by this paradigm. This is analogous to confusing a set with its member (which, as we saw above, is also possible) and, of course, to confusing a concept with an object.

This series of (possible) confusions leads to the following question: can one confuse the concept of X with an X itself? Given the discussion above, the answer must be affirmative. But how is this possible? How can one possibly confuse the word 'lion' with a lion? Neither the graphic (written) form of 'lion', nor the acoustic (spoken) form of 'lion' resembles a real lion. Such a resemblance occurs only in rare onomatopoeic expressions such as 'bark', 'cry', 'whisper' or 'beep', or shape words (phenomimes) such as 'U-turn'. Or we can imagine a fully pictorial language where every sign is a picture of the thing it refers to. However, this is not the confusion in question here. The quotation marks refer to the graphic form of an expression, not to a concept; they refer to a sign, not to a symbol (in the terminology[21] of the *Tractatus*). The question, however, is: how can one confuse the symbol of X, i.e. 'X' plus the sense of X, with an X? 'X' refers to an X in a given context. This amounts to saying that the reference of 'X' is an X. If 'X' is actually referring, then all we get is an X, and then the confusion ensues. In order to avoid the confusion, what we need is to capture (or, better, freeze) the sense of X when it is not actually referring.[22] The sense of X can be taken as a paradigmatic sample of X together with the *praxis* of applying this paradigmatic sample. To take the sense of X when it is not referring is to disregard the praxis. What remains then is the paradigmatic sample alone. To confuse

[20]Cf. Antonio Gramsci's famous remark: 'the old is dying and the new cannot be born; in this interregnum a great variety of morbid symptoms appear.' (Gramsci 1971, pp. 275–276).

[21]'The sign is the part of the symbol perceptible by the senses.' (TLP 3.32).

[22]This idea is captured by Agamben (1993, p. 72): 'Actually, since every term refers by definition to every and any member of its extension, and can, furthermore, refer to itself, one can say that all (or almost all) words can be presented as classes that, according to the formulation of the paradox, both are and are not members themselves.

It is not worth objecting against this that one never mistakes the term "shoe" for a shoe. Here an insufficient conception of self-reference blocks us from grasping the crux of the problem: What is in question is not the word "shoe" in its acoustic or graphic form (the *suppositio materialis* of medieval logicians), but the word "shoe" precisely in its signifying the shoe (or, *a parte objecti*, the shoe in its being signified by the term "shoe"). Even if we can completely distinguish a shoe from the term "shoe," it is still much more difficult to distinguish a shoe from its being-called-(shoe), from its *being-in-language*. Being-called or being-in-language is the non-predicative property par excellence that belongs to each member of a class and at the same time makes its belonging an aporia.'

a paradigmatic sample of X with an X itself is something we can now imagine very clearly.

It seems, then, that self-membership is much more common than one would expect. Every predicate is prone to cause the confusion between its sense and its reference, and thus generate a paradox. We can, however, intentionally force people to run up against this paradox in order to indicate a change in the praxis with this predicate, i.e. to generate a new sense.[23]

References

Agamben, G. (1993) *The Coming Community: Theory of Bounds*, University of Minnesota Press.

Ahmed, A. (2017) 'Triangulation and the Private Language Argument', *Belgrade Philosophical Annual* 30, 35–52.

Diamond, C. (1995) *The Realistic Spirit*, MIT Press.

Diamond, C. (2001) 'How Long Is the Standard Meter in Paris?', in: T. McCarthy and S. C. Stidd (eds) *Wittgenstein in America*, Oxford University Press, 104–139.

Frege, G. (1892) 'Über Begriff und Gegenstand', *Vierteljahresschrift für wissenschaftliche Philosophie* 16, 192–205.

Gramsci, A. (1971) 'Wave of Materialism' and 'Crisis of Authority', in: Q. Hoare and G. Nowell-Smith (eds) *Selections from the Prison Notebooks*, International Publishers.

Irvine, A. D., and Deutsch, H. (2016) 'Russell's Paradox', *The Stanford Encyclopedia of Philosophy* (Winter 2016 Edition), Edward N. Zalta (ed.), URL = <http://plato.stanford.edu/archives/win2016/entries/russell-paradox/>.

Keel, O. (1994) *The Song of Songs: A Continental Commentary*, Fortress Press.

Kutschera, F. von (1989) 'Paradox', in: K. Gründer (ed.) *Historisches Wörterbuch der Philosophie*, vol. 7, Schwabe Verlag, 81–97.

Kuusela, O. (2019) *Wittgenstein on Logic as the Method of Philosophy*, Oxford University Press.

Lessing, E. G. (1759/1825) *Fables and Epigrams; with Essays on Fable and Epigram*, John and H. L. Hunt.

McGinn, M. (1997) *Wittgenstein and the* Philosophical Investigations, Routledge.

Stroud, B. (2001) 'Private objects, physical objects and ostension', in: D. Charles and W. Child (eds) *Wittgensteinian Themes: Essays in Honour of David Pears*, Oxford University Press, 143–162.

[23] I have benefited from discussions with Wolfgang Kienzler, Andrew Napthine, Rupert Read and Oskari Kuusela. This work has been supported by the Czech Science Foundation, project no. GA19-16680S.

Chapter 10
"Life is Very Complicated": Remarks on a Recurring Adjective

David R. Cerbone

10.1 A Not so Simple Adjective

In the very first remark of *Philosophy of Psychology: A Fragment*, which accompanies *Philosophical Investigations* (and was previously referred to as the second part of that work), Wittgenstein writes regarding "the manifestations of hope" that they "are modifications of this complicated form of life."[1] The question I want to explore in this paper concerns the adjective Wittgenstein here uses to qualify "form of life."

[1] *Philosophy of Psychology—A Fragment*, edited by P. M. S. Hacker and Joachim Schulte; translated by G. E. M. Anscombe, P. M. S. Hacker, and Joachim Schulte. Chichester: Wiley-Blackwell, 2009, § 1. Further citations to this work will be made parenthetically using the abbreviation PPF. Citations from *Philosophical Investigations*, which appears in the same volume, will be made parenthetically using the abbreviation PI.

Here I quote from one of Wittgenstein's later manuscripts, the second volume of *Remarks on the Philosophy of Psychology*, edited by G. H. von Wright and Heikki Nyman, translated by C. G. Luckhardt and M. A. E. Aue. Chicago: University of Chicago Press, 1980. The full remark reads:

> If you consider the reasons someone might have for stifling pain, or simulating it, you will come up with countless ones. Now why is there this multiplicity? Life is very complicated. There are a great many possibilities.
> But couldn't other men disregard many of these *possibilities*, shrug them off, as it were? (§ 639)

The stretch of remarks in which this one appears will be of particular concern throughout this paper. Further citations from this work will be made parenthetically using the abbreviation RPP II.

D. R. Cerbone (✉)
Department of Philosophy, West Virginia University, Morgantown, WV 26506, USA
e-mail: cerbone@wvu.edu

S. Wuppuluri and N. da Costa (eds.), *WITTGENSTEINIAN (adj.)*,
The Frontiers Collection, https://doi.org/10.1007/978-3-030-27569-3_10

Put most succinctly, what does Wittgenstein mean here by "complicated"?[2] And, as seems fitting in a discussion of Wittgenstein, in asking after that meaning, my question concerns the *use* to which he is putting "complicated" here. What work is being done in describing something as complicated? And complicated, we might well ask, as opposed to what? The most obvious answer to this last question is simply, "As opposed to being *simple*." But this obvious answer itself turns out to be, well, complicated.

To explain what I mean with this last remark, it will help to consider how pervasive Wittgenstein's use of this one adjective—*complicated*—is, spanning, as it does, his early and late work. The adjective appears more than once in one of the lengthier remarks in the *Tractatus*—4.002—where Wittgenstein turns his attention to "everyday language," which he there characterizes as "part of the human organism" and as "no less complicated than it."[3] The remark ends with another use of the adjective that concerns us: "The tacit conventions on which the understanding of everyday language depends are enormously complicated."[4] The meaning of "complicated" is here bound up with a particular kind of difficulty, which Wittgenstein notes immediately after linking everyday language to the human organism, thereby binding the complexity of the former to the latter; he writes that "it is not humanly possible to gather immediately from it what the logic of language is." And in the

[2]Newton Garver is the one prior commentator I've come across who asks specifically about the adjective here rather than the whole phrase. In his collection of essays, published as *This Complicated Form of Life* (Chicago: Open Court, 1994), he writes:

> It is interesting not only that there is a single form of life determined by mastery of the use of a language, but also that it is complicated. Why "complicated"? It is true that there are many subtle nuances of hope, but they would seem to be a rather insubstantial basis for calling the form of life complicated. What complicates this form of life are all the other possibilities besides just the phenomena of hope. (p. 253).

[3]*Tractatus Logico-Philosophicus*, translated by D. F. Pears and B. McGuinness. London: Routledge and Kegan Paul, 1981.

[4]Here is the passage in full (other portions of it will be attended to in due course):

> Man possesses the ability to construct languages capable of expressing every sense, without having any idea how each word has meaning or what its meaning is – just as people speak without knowing how the individual sounds are produced.
>
> Everyday language is part of the human organism and is no less complicated than it.
>
> It is not humanly possible to gather immediately from it what the logic of language is.
>
> Language disguises thought. So much so, that from the outward form of the clothing it is impossible to infer the form of the thought beneath it, because the outward form is not designed to reveal the form of the body, but for entirely different purposes.
>
> The tacit conventions on which the understanding of everyday language depends are enormously complicated.

Chon Tejedor's "Tractarian Form as the Precursor to Forms of Life" (*Nordic Wittgenstein Review*, October 2015 <https://www.nordicwittgensteinreview.com/article/view/3358>) provides an illuminating discussion of this remark in relation not just to the *Tractatus*' overarching concern with *form*, but to the evolution of that concern in Wittgenstein's later philosophy.

subsequent paragraph, although still within the same numbered remark, he notes that "it is impossible to infer the form of the thought" that lies beneath the "outward form of the clothing" of everyday language. This outward form "disguises thought." The difficulty described here—the impossibility of inferring the form of the thought from the outward form—redoubles the prior claim concerning the (human rather than absolute) impossibility of gathering "immediately" what the logic of language is. Rather than any kind of immediate or direct inference, the thought expressed via everyday language is reached only via the "tacit conventions" that are themselves "enormously complicated."

Before probing further the sense of "complicated" in this passage from the *Tractatus*, first consider the following passage, from the second volume of *Remarks on the Philosophy of Psychology*, written roughly three decades after the *Tractatus* remark:

> We judge an action according to its background within human life, and this background is not monochrome, but we might picture it as a very complicated filigree pattern, which, to be sure, we can't copy, but which we can recognize from the general impression it makes. (RPP II, § 624)

Despite the decades separating them, there is a striking resonance between this passage and the one from the *Tractatus* we have been considering. Most immediately, of course, there is the appeal to something as *complicated*. Here what is complicated is a "background" against which we "judge an action," with the suggestion being that we pick out—make sense of—an action as the kind of action it is by seeing it against, or in relation to, this background. We see someone raise her arm: what action[5] has she performed? Set that action against the background of an urban milieu replete with cars for hire, and we might judge her as *hailing a taxi*. Set the person and her movements against the background of a lecture hall, as a member of an audience being addressed by a speaker, and we might judge her as *trying to ask a question*. There is nothing especially esoteric in these judgments: most of us are able most of the time to make them and many others like them. But their ubiquity and ease serve to mask the complexity of the "pattern" we are sensitive to in making them. Judging someone as hailing a taxi, for example, encompasses a tremendous range of knowledge: of cars, streets, and cities, of hailing, hiring, and paying, of getting from one place to another and how that might be done, and so on.[6] In order to see someone

---

[5]Even seeing the movement as an *action*—any action—already involves seeing it against a background "within human life," which is complicated in a way that resists spelling out. Consider Wittgenstein's question in the *Investigations*: "What gives as *so much as the idea* that beings, things, can feel?" (PI, § 283) Replace "feel" in this question with "act" and we are apt to feel an analogous sense of puzzlement as the one he is trying to induce at this juncture of the *Investigations*.

[6]Notice what Wittgenstein says very early in the *Investigations*: "One has already to know (be able to do) something in order to be capable of asking a thing's name. But what does one have to know?" (PI, § 30) Analogously, we might ask what someone "has already to know" in order to judge an action in the sense we've been considering. That such judgments register, or enlist a sensitivity to, a background tempers the expectation of an exhaustive answer here. I take it that achieving this same kind of tempering of expectations is one of Wittgenstein's motives in raising the question about naming in the *Investigations* passage. In other words, what is most important about Wittgenstein's raising that question at the end of § 30 is his leaving it *unanswered*.

as hailing a taxi—and to differentiate that action from the action of trying to ask a question—requires situating what is a more or less indistinguishable physical motion in relation to that range of knowledge. But the knowledge is *background* insofar as it is not anything we bring explicitly to bear in so judging the action, nor is it for the most part something that we *can* bring to bear in an explicit way. If I try to spell out what I, or what one, must know in order to be able to judge that movement as *hailing a taxi*, it is not entirely clear where to begin, nor is it clear where it is that I should end. Urban environments with their myriad possibilities make a distinctive "general impression," just as classrooms and lecture halls do, but without my being able to unpack that impression in a way that is exhaustive or definite.

The "background" Wittgenstein appeals to here has much in common with the "tacit conventions" noted in the *Tractatus* passage. In both cases there is something—something *complicated*—that we are attuned to or pick up on in order to understand or make sense of something, an utterance in everyday language or an action in everyday life. And in both cases, one dimension of that complexity is precisely its "tacit" character: we are in both cases able to do something without really knowing—and certainly without having to know—how we do it. *Tractatus* 4.002 begins: "Man possesses the ability to construct languages capable of expressing every sense, without having any idea how each word has meaning or what its meaning is—just as people speak without knowing how the individual sounds are produced." So I might know that someone is hailing a taxi by raising her arm—rather than trying to ask a question or greet someone—or I might know that someone is asking me how I'm doing rather than for a description of what might be at a higher altitude than my head in uttering, "What's up?" without really knowing *how* I know that and without my being able to say all the things I have to register in order really to know how. Hence the sense of something as "not humanly possible to gather immediately," which finds an echo in the "filigree pattern" that we "cannot copy."

In both the *Tractatus* and the later passages there is a curious dialectic between the complicated and the simple: uttering a meaningful sentence is (usually) easy, just as judging someone to be hailing a taxi does not require any kind of effort or deliberation. In both cases, there is something *simple* in terms of how it *feels* to do these things.[7] And yet that simplicity rests, as it were, on something more complicated. Just consider how much is going on in terms of my lungs, throat, and mouth when I utter a sentence. It is surely incredibly complicated, but I am mostly oblivious to that complexity as I chatter on, uttering sentence after sentence. Indeed, I am so oblivious that I can barely begin to tell you what is going on with my lungs, throat, and mouth beyond the obvious reports of breathing, opening and closing my mouth, and changing the shape of my lips and the position of my tongue in a variety of ways (although the details of that variety elude me). Only on those occasions where something is awry, such as when I am out of breath or when I've returned

[7]I am indebted here to J. F. M. Hunter's "'Forms of Life' in Wittgenstein's 'Philosophical Investigations'," American Philosophical Quarterly, Vol. 5, No. 4 (October, 1968), pp. 233–243, which contrasts the simple and the complicated in this manner with particular reference to 4.002 of the *Tractatus*. See especially p. 238.

from the dentist with my tongue and lips still numb from novocaine, do I notice the workings of the "machinery" of my speech and how fine-grained those movements have to be. Wittgenstein's suggestion in the first sentence of 4.002 is that there is an analogous obliviousness when it comes to the *meanings* of my words. It is not that I do not know what I am saying—what I mean by saying what I do—but I am oblivious to the *how* of that meaning, to how each word contributes to the meaning of the sentence I utter. The analogy prepares us for the next paragraph of the remark and its likening of the complexity of everyday language to the complexity of the "human organism." In the later passage, the "filigree pattern" that serves as "background" figures into my judgment of the individual action. I am in some way sensitive to it, just as I am sensitive to word-order and the movements of my tongue, but there is no more than what Wittgenstein refers to as a "general impression" that affords a kind of recognition but nothing that would facilitate copying it out explicitly.

In the *Tractatus* passage, there is a further step in the dialectical interplay between the simple and the complicated that serves to reverse their positions in terms of which one is surface and which one is depth. The complexity of everyday language serves to obscure what Wittgenstein refers to as the "logic of language." We cannot "gather immediately" the underlying logic or "infer" the underlying "thought" because of the ways in which the "clothing" of our ordinary language hides the logical "body" in its drapes and folds. Hence the need for an equally—indeed enormously—complicated range of conventions to connect the surface complexity of everyday language with the underlying logic of the thoughts that are expressed in that language. The contrasts in play here—between *surface* and *depth*, *immediate* and *mediated*, *direct* and *indirect*—carry with them the implication that the complexity Wittgenstein appeals to here can be regarded as, at the very least, *lamentable*, and ultimately perhaps *dispensable*. If we return to our question of "Complicated as opposed to what?" the obvious answer, "As opposed to simple," relies here in the *Tractatus* on an idea of a language that does not "disguise" thought, such that the route from language to thought is immediate or direct. (I will return to this idea later in my discussion.)

So we have here two contrasting ways in which the simple and the complicated relate to one another: there is a surface simplicity in our uttering sentences and judging actions, surrounded and underwritten by a scarcely-registered complexity. Here, the sense of "simple" accords with something's being *easy to do* or *done without difficulty*, in contrast to the myriad things we are in some way sensitive to without being able to articulate explicitly. But in the reversal effected in the *Tractatus* passage, there is an underlying simplicity, hidden or disguised by the surface complexity of everyday language and its tacit conventions. Here the sense of "simple" moves in another direction, as what is simple in *this* sense may be difficult to discern. This second kind of interplay between what is simple and what is complicated persists into at least some of the later work. For example, in *The Big Typescript*, Wittgenstein writes that "philosophy unravels knots in our thinking; hence its results must be simple, but its activity is as complicated as the knots it unravels."[8] Here what is simple

[8] *The Big Typescript: TS 213*, edited and translated by C. G. Luckhardt and M. A. E. Aue. Oxford: Blackwell Publishing, 2005. Citation from p. 311e.

lies at the end of a difficult and often frustrating activity. Wittgenstein sometimes likens this simplicity to a kind of *clarity* (see PI, § 133). What makes the end result something simple is the *absence* of vexing problems, whose complexities are a result of confusions that arise from our overlooking or obscuring something that is itself ultimately simple: in the *Investigations*, Wittgenstein writes of the "aspects of things that are most important for us" as being "hidden because of their simplicity and familiarity." (PI, § 129) There also continues into the later work vestiges of the theme of something's being obscured or hidden by the surface trappings of language. This theme is announced as early as § 11 of the *Investigations*, where Wittgenstein writes that "what confuses us is the uniform appearance of words when we hear them in speech or see them written or in print," and toward the very end, at § 664, where he invokes a contrast between "surface" and "depth" grammar:

> In the use of words, one might distinguish 'surface grammar' from 'depth grammar'. What immediately impresses itself upon us about the use of the word is the way it is used in the sentence structure, the part of its use – one might say – that can be taken in by the ear. – And now compare the depth grammar, say of the verb "to mean", with what its surface grammar would lead us to presume. No wonder one finds it difficult to know one's way about.

The depth that's registered here differs dramatically from the kind invoked in the *Tractatus*. What removes the presumption we are led to by the "surface grammar" of "to mean" is not a matter of drilling down, as it were, to the disguised thought, as much as it is a matter of casting a wider net in terms of recollecting the uses of "to mean" and comparing them to the uses of other verbs. This way of proceeding is already indicated in the continuation of the passage from § 11 just cited; the uniformity of the appearances of words is counteracted by attending to their *use*: "For their *use* is not that obvious. Especially when we are doing philosophy!"

10.2 From Simple to Simples: Language "on Holiday"

In my first pass at explicating Wittgenstein's invocation of a "background" as a "filigree pattern," I started by treating the notion of a background in a fairly literal sense as more or less the *background scenery* in which an action is situated. But as we saw straight away, that scenery has to not just be there; it has to figure into our *judging* of the action, so that in seeing the action as the action it is, we must see it in relation to that scenery. So we must be sensitive to that background. That sensitivity too is a kind of background. But the background, as encompassing these different senses, extends beyond just what is literally there as surroundings in the moment. The background is a "background within human life," which extends temporally and practically beyond that moment of judgment. Consider a remark from early in the *Investigations*, toward the end of Wittgenstein's first examination of the notions of ostensive teaching and ostensive definitions. Here, he wants us to think further about *pointing to something's color* versus *pointing to something's shape*. He writes, starting in the second paragraph of the remark:

> Suppose someone points to a vase and says, "Look at that marvelous blue – forget about the shape". – Or: "Look at the marvelous shape – the colour doesn't matter." No doubt you'll do something *different* in each case, when you do what he asks you. But do you always do the same thing when you direct your attention to the colour? Imagine various different cases! To indicate a few:
>
> "Is this blue the same as the blue over there? Do you see any difference?" –
>
> You are mixing paints and you say, "It's hard to get the blue of this sky".
>
> "It's turning fine, you can already see blue sky again."
>
> "Not how different these two blues look."
>
> "Do you see the blue book over there? Bring it here."
>
> "This blue light means …"
>
> "What's this blue called? – Is it 'indigo'?"
>
> One attends to the colour sometimes by blocking the contour from view with one's hand, or by not focusing on the contour of the thing, or by staring at the object and trying to remember where one saw that colour before. (PI, § 33)

Wittgenstein's exhortation to "imagine various cases" is meant to lead us away from fixating on there being some one thing that happens or one thing that we do that *by itself determines* that we are pointing to something's color. The short list that Wittgenstein provides situates attending to something's color within a variety of different activities, where the color is attended to in different ways and for different reasons. Given its brevity, the list is by no means exhaustive—Wittgenstein presents them as just a "few examples"—but even in these seven sketches we can begin to see how attending to something's color enlists a variety of skills: matching colors, mixing colors, noting the appearance of a color, contrasting colors, identifying an object by means of its color, using color to indicate or signal something, naming different shades of the same color. Moreover, we can see these skills as being enlisted in a variety of activities; here are some that incorporate and track Wittgenstein's own examples: deciding between two shirts, mixing paint, tracking and noting the changing weather, comparing two paintings, fetching something for someone, reviewing an evacuation plan, trying to recall the name of a particular shade. The list could be continued and developed in various ways. Attending to something's color is something we do within those activities in conjunction with a wide variety of skills. These variegated activities and skills make up the *circumstances* in which we do that and the variety of those circumstances cuts against the idea that managing to point to or attend to something's color can be condensed into a single kind of gesture or single kind of experience, which by itself makes it the case that one has pointed or attended to the color. To be sure, the gesture in each case is often quite simple, as are the words one might utter or the thought that might run through one's mind. There isn't anything that *feels* complicated about aiming one's finger and saying, "Notice how much darker the red is here," but that simplicity hides the way doing *that* is situated in something far more complicated. In the final paragraph of the remark, Wittgenstein elaborates on the role of these *circumstances*:

> One attends to the shape, sometimes by tracing it, sometimes by screwing up one's eyes so as not to see the colour clearly, and so forth. I want to say: this and similar things are what one does *while* one 'directs one's attention to this or that'. But it isn't only these things that make us say that someone is attending to the shape, the colour, etc. Just as making a move in chess doesn't consist only in pushing a piece from here to there on the board – nor yet in the thoughts and feelings that accompany the move: but in the circumstances that we call "playing a game of chess", "solving a chess problem", and the like. (PI, § 33)

Wittgenstein's suggestion here that we compare attending to a color or a shape with making a move in chess further cuts against the idea of finding something to fill the role of *consists-only-in*. In thinking that there should be something to fill that role, we are misled by the ease and simplicity with which we point to colors or move chess pieces, thereby thinking that what gives those gestures their meaning must be something there in the moment of pointing or moving alone. The gesture of course singles out a color right there and then, just as I move the knight right at the moment when my hand slides it along the board, but only if these momentary movements are situated in diffuse and complicated circumstances.

The immediate aftermath of this remark in the *Investigations* consists of an extended critique of what Wittgenstein casts as an attractive line of resistance to this appeal to "circumstances" and their attendant complicatedness (its attractiveness is attested to by the extent to which Wittgenstein himself was at one time drawn to it). This resistance reveals another sense of *simple*, invoked to affirm a more direct connection between language and reality, as that connection seems too diffuse and tenuous in Wittgenstein's appeal to circumstances and surroundings. The sense of "simple" at issue here concerns *simples* in an *absolute* sense, as fundamental elements of meaning and, correspondingly, fundamental elements of reality. (Note here that "simple" here does not contrast with "complicated" or with "difficult" or "not easy," but instead with *composite* or *compound*.) Wittgenstein's assault on this sense of "simple" exemplifies his more general concern with "a tendency to sublimate the logic of our language." (PI, § 38) Against this tendency, Wittgenstein notes in § 38 that "we call very different things 'names', the word 'name' serves to characterize many different, variously related, kinds of use of a word." Here Wittgenstein builds upon § 23's insistence that there are "countless kinds" of sentences and the deliberately jumbled list of language-games that follows.[9] This insistence on variety informs § 24 ("Remember how many different kinds of things are called 'description'") and § 27 ("As if there were only one thing called 'talking about things'. Whereas in fact we do the most various things with our sentences"). Ironically, then, sublimation turns out to be a kind of condensation. By overlooking or effacing the vast variety in our uses of language, our uses of words, the tendency to "sublime the logic of our language" is thereby encouraged, prompting the search for *the* basic form of the proposition, *the* relation between a name and what it names, and so on. We again see a kind

---

[9]In the continuation of the passage from Garver cited in Note 3 above, he likewise links the appeal to what is complicated to Wittgenstein's insistence in § 23 of *Investigations* that there are "countless kinds of sentences." So "complicated" in "this complicated form of life" indicates not just multiplicity and variety, but also refers to something about that multiplicity and variety, as being, for example, countless. I return to this sense of indefiniteness at the conclusion of the paper.

of reversal in the relations between the simple and the complicated: the fixation on simples is counteracted by returning us back to the wide—and complicated—variety in our uses of names.

Wittgenstein's charge in § 38 is that a raft of words—"name," "this," "simple"—has gone "on holiday," where that means something like thinking about language—and using language to express that thinking—without really attending to language at work, to the "countless kinds" of sentences, the "different kinds of things [that] are called 'description'." A principal example for Wittgenstein at this point in the *Investigations* is Russell's insistence that "this" is the only genuine proper name. What kind of "work" does "this" really do? How could "this" take over—do the job of—all the kinds of names we find in our actual day-to-day language use, where we name people and pets, characters and constellations, businesses and teams, chronic conditions and car models, and on and on? Somehow replacing—or analyzing away—all names with the admittedly useful but at best only name-like "this" would prove intractable in terms of our really being able to talk to one another about anything. (Small wonder, then, that Russell admitted that a "logically perfect language" would at the same time be "very largely private to one speaker.")[10] How, for example, could I re-write—ahem—*this* paragraph, where I've named two philosophers, one philosophical work, a passage from that work, as well as a number of examples, using only "this"? How could "this" do the work that the words in this paragraph do?

While, like a souvenir shirt that proves unwearable upon returning home, "this" as the only genuine proper name loses its allure when the work of language resumes, there are deeper considerations that make such an idea appear attractive in the first place (just as the special combination of sun, sand, and sangria, perhaps, made that shirt look good while on vacation). The remarks immediately following § 38 ferret out these deeper motivations, which turn on the idea, voiced in § 39 with particular emphasis, that a "name ought really to signify a simple." The idea here is roughly this: if the meaning of a name is bound up with the name's bearer, then names without bearers face the threat of meaninglessness. Ordinary names, whose putative bearers come and go (or never existed in the first place, as is the case with names for fictional characters), face this kind of threat. To preserve the meaningfulness of proper names, they must be in some way constructed from, and so analyzable into, other names (the genuine names) that do not, and indeed cannot, face such a threat of lacking or losing a bearer. These names must track something about reality not subject to change: they cannot have parts, they can be neither created nor destroyed, and so they must be absolutely simple.

Wittgenstein develops a series of worries about this line of reasoning over the next several remarks that put pressure on the identification of the meaning of a name and the bearer of the name (§ 40), explore different, equally intelligible ways of handling or responding to bearer-less names via an extension of his builders scenario (§§ 41–

[10]Russell, "The Philosophy of Logical Atomism," in *Logic and Knowledge*, edited by R. C. Marsh. London: Unwin Hyman, 1956. The citation is from the second lecture, p. 198. The privacy asserted here is bound up with the way "this" is attached to a simple sensory element with which the speaker is *acquainted* in Russell's special sense, where no two speakers are ever acquainted with the same elements.

42), and return to the case of "this" (§ 46). At § 46, Wittgenstein recapitulates the reasoning behind the postulation of simples, this time invoking Plato's *Theaetetus* and again calling out his earlier views (as the author of the *Tractatus*), as well as Russell's atomism. At this juncture—§ 47—Wittgenstein returns to the themes of § 38, in particular, the idea of language "on holiday," here glossed as speaking "outside a particular language game."

Here's one, admittedly tempting, way of unpacking § 47: the aim of the remark, as well as the stretch of remarks in which it is embedded, is to show that there is no such thing as an absolute simple. Read this way, Wittgenstein uses the appeal to ordinary language—to language games and the idea of speaking "outside" them—as a way of reaching a conclusion about the nature of reality, metaphysics via reflection on language alone. The problem with this way of construing Wittgenstein's criticisms is that it assumes precisely the thing he is out to challenge, namely, that we know what we are talking about when we take ourselves to be in search of "absolute simples." If, as we might put it, "absolute," like "metaphysical," is a kind of *holidaymaker*, then we have not got as far making a fully intelligible *claim* about reality whose merits can then be adjudicated. "'Simple' means: not composite. And here the point is: in what sense 'composite'? It makes no sense at all to speak absolutely of the 'simple parts of a chair'." And if it "makes no sense at all," then the denial or the rejection of "absolute simples" is no less problematic than "their" endorsement. As the passage proceeds, Wittgenstein takes the same kind of "back to work" approach as he had in the preceding remarks about names and descriptions, and again the emphasis is on the *variety* of ways of drawing a distinction between simple and composite: "We use the word 'composite' (and therefore the word 'simple') in an enormous number of different and differently related ways." We draw such a distinction for different reasons and in different ways depending on what we are talking about and why we want to make the distinction. In different circumstances, different criteria of simplicity and compositeness will be in play. This need not entail that there is anything "vague" or "ambiguous" about the words "simple" and "composite," or the contrast between them. What counts as simple and what counts as composite can be perfectly clear in all manner of circumstances in the sense that almost no one will be unable to follow what's being distinguished and why, but it does not follow that the *same* distinction is being made in the *same way* across all these different circumstances. An insistence on sameness here is a first step on the way to sublimation.

10.3 De-Sublimating the Complicated

Wittgenstein's critique of "simple" and "composite," which questions whether we are clear about what we are after in trying to determine what is simple "in an absolute sense," allows us to consider the juxtaposition of our opening two passages—from the *Tractatus* and the much later psychology manuscripts—in a different light. While there is a striking consonance between the two, despite the many decades separating them, there is an important difference that emerges when we consider further Wittgenstein's critique of "simple." Recall Wittgenstein's imagery in the *Tractatus* likening everyday language to clothing: language "disguises" the (underlying) thought in much the same manner as clothing hides the body. That analogy can be further developed by considering how the revelation of the body is effected by the *removal* of the clothing that obscured it. What goes proxy for the removal of clothing on the language-thought side of the analogy is *analysis*. And just as the entirety of one's clothing can be removed—leaving only the body, unadorned and naked—so too can there be a *complete* analysis, leaving only a concatenation of names for simples. Those concatenations would be expressions of *the thoughts themselves*, whose sense would be determinate and independent of any contingencies.[11] That Wittgenstein has such a model of analysis in his sights is evident from the continuation of his critique of simples in the *Investigations*. This is especially evident in the comedy[12] of § 60's treatment of the mundane declaration, "My broom is in the corner." Just as the broom consists of a broomstick and brush—and so can be separated into those two components—one might then be led to think that the *sense* of "broom" should likewise be decomposable into the senses of "broomstick" and "brush." These senses and their relation to one another are "hidden in the sense of the first sentence" and "*articulated* in the analysed sentence." And just as in *grasping* the broom we inevitably grasp the broomstick-plus-brush combination, so too, this line of thought continues, in grasping the sense of "broom," we grasp the senses of "broomstick" and "brush." Finally, since decomposing the broom into a broomstick and a brush hardly marks the end of such a process—the broomstick and brush can be further decomposed into bits of wood, bristles, and so forth—decomposing the sense of "broom" will not stop at the senses of "broomstick" and "brush."

The seamless-seeming slide from the broom to the sense of "broom," such that the decomposition of the latter parallels the decomposition of the former, is interrupted by considering further the crucial claim that we grasp the sense of "broom" by grasping the senses of terms that refer to the broom's constituent parts. Does "someone who says that the broom is in the corner really mean: the broomstick is there, and so is the brush, and the broomstick is fixed in the brush?" It is hard to know what the criterion of "really means" might be. Wittgenstein's first pass is to consider a likely response to this question. To the extent that someone is not simply baffled by the question, her response is likely to be that she "had not specially thought of either the broomstick or the brush." There is thus nothing to appeal to at the level of what we think as or when we speak to motivate the claim that an utterance or thought about a broom consists of an utterance or thought about the broom's constituent parts. If we switch from a psychological to a practical perspective, there seems to be little motivation on that front either: telling someone to fetch a broomstick attached to a

[11]Consider, for example, *Tractatus* 3.23: "The requirement that simple signs be possible is the requirement that sense be determinate," and shortly thereafter, at 3.25: "A proposition has one and only one complete analysis."

[12]Here I draw on Richard Rorty's characterization of Wittgenstein's later philosophy as "pure satire." See his "Keeping Philosophy Pure: An Essay on Wittgenstein" in his *Consequences of Pragmatism*. Minneapolis: University of Minnesota Press, 1982; pp. 19–36. See especially p. 34.

brush is not a more clearly articulated utterance than the request to fetch the broom. If anything, in most contexts the elongated request is apt to elicit puzzlement and so is, for the most part, far *less* clear: "Isn't the answer: 'Do you want the broom? Why do you put it so oddly?'" These considerations are hardly decisive, but they do begin to challenge a line of thinking that conceives of meaning or sense in terms of overt surface expressions and hidden underlying thoughts, where that contrast in turn fuels a complicated process (analysis) leading to something simple (names for ultimate constituents). This is not to say that what we mean when we say something cannot be clarified by replacing what we say with a different utterance. In some cases, that might even be a matter of replacing a term for something with terms for its parts. If someone is puzzled at my request for a particular wrench, I *might* clarify what I asked for by specifying that I want the one with the serrated jaws and the red grip. But in other instances, I may clarify what I mean by replacing terms referring to parts with a term for the whole: "Not the one with the red handle. I need the *crescent wrench.*" The point is that there is no one *direction* for that activity of clarification to go.

If we see Wittgenstein's criticisms here as dissuading us from picturing the understanding of meaning in terms of a grasp of underlying constituents whose exact contours are revealed through analysis, there nonetheless remains an important aspect of the overall Tractarian image that sustains a kind of continuity with the later passage. I don't mean here the image of clothing and the body, but rather the idea of there being "enormously complicated" and "tacit" conventions that are involved in the understanding of "everyday language." While there is reason to be suspect of understanding as involving the grasp of *hidden*—and even unknown—elements, that still leaves in place the idea of our grasping of the sense that is *there on the surface* as involving some kind of registering of, or sensitivity to, those complicated "conventions." While *tacit*, such conventions need not be thought of as *hidden* in the same way that the elements revealed through analysis are conceived of as being. They can instead be conceived of—to use the language of the later passage—as making a "general impression" without our being able to articulate or explicate just what that impression *is*. So while the later perspective encourages us to jettison as fantasy simple names corresponding to simple elements, it leaves in place the idea of there still being something "enormously complicated" involved in even our most simple-feeling utterances and actions.

At the same time, the critique of simples and analysis in the early stages of the *Investigations* is instructive for delving further into what Wittgenstein is up to in characterizing the "background of human life" as "complicated," as its warnings about language going "on holiday" serve to caution us against trying to construe that background as being complicated in an "absolute" or "metaphysical" sense. Wittgenstein's critique of the desire for such absolute senses should be seen as a kind of cautionary tale in order to avoid tempting, but ultimately misleading, ways of understanding this appeal to what is complicated. One such way is to see this appeal to a background as "complicated" as carrying some kind of special *explanatory* force.[13] Another, which might be understood as the result of a kind of recoil from this first temptation, is to *mystify* this notion of background, as something that eludes our comprehension or grasp.[14] That the background is understood as mysterious is not entirely separate from the idea of the background as explanatory, since what is mystifying here *would be* explanatory if only we were able to comprehend it. In this way, the two temptations may be regarded as two sides of the same coin.

In a 1931 passage that appears among the remarks collected in *Culture and Value*, Wittgenstein writes:

> Perhaps what is inexpressible (what I find mysterious and am not able to express) is the background against which whatever I could express has its meaning.[15]

What accounts for this sense of there being something "mysterious" and "inexpressible" in the notion of a "background"? Consider again the example of someone's making a move in chess, as well as the importance of what Wittgenstein refers to in the *Investigations* as simply "circumstances." The import of those circumstances lies in their contribution to the *sense* of that gesture. When we judge the sliding of the chess pieces as a *move*, we see it as happening within those circumstances that constitute a kind of background for the particular movement. In shifting our attention from the particular movement to the circumstances in which the movement is made, we are inclined to see those circumstances as playing some kind of *explanatory* role in terms of how that movement manages to *be* just the movement that it is. And there is something correct about this insofar as *this* (the piece's being slid from one square to another) is only a move in chess because it is takes place *within* a game of chess. Had my hand only brushed the board while I was chatting with a friend or had my dog just happened to nudge a piece from one square to another: we might joke about these as brilliant chess moves, but they would not really be moves at all.

While there is nothing wrong with seeing this "because" as having *some* explanatory force, there is apt to be something distinctly unsatisfying about the way such explanations are framed. If we acknowledge that "making a move in chess doesn't consist only in pushing a piece from here to there on the board … but in the circumstances that we call 'playing a game of chess', 'solving a chess problem', and the like," what is doing the explanatory work in terms of telling us what a move in chess consists in risks a kind of triviality, since we are explaining what makes something a move in chess by appeal to the game of chess (where any particular game itself consists of moves in chess). Since *chess* appears twice over, in what is being explained

[13]Denis McManus has in more than one place drawn attention to the difficulties of appeals to a background as playing a fundamental explanatory role. See his "Rules, Regression, and the 'Background': Dreyfus, Heidegger and McDowell." *European Journal of Philosophy* 16, no. 3 (December, 2008): 432–458, as well as *Heidegger and the Measure of Truth*. Oxford: Oxford University Press, 2016 (see especially Chap. 4).

[14]I examine this tendency toward mystification at greater length in my "Ground, Background, and Rough Ground: Dreyfus, Wittgenstein, and Phenomenology" in *Normativity, Meaning, and the Promise of Phenomenology*, edited by. M. Burch, I. McMullin, and J. Marsh. London: Routledge, 2019.

[15]*Culture and Value*, translated by P. Winch. Chicago: University of Chicago Press, 1984, p. 16e.

and what is doing the explaining, there is a sense that some further explanation is needed. That is, the appeal to circumstances—as involving just more in the way of chess—would appear to invite the same kind of consists-in question all over again: if the question of what makes this movement a move in chess is answered by appealing to its being "located" in a game of chess, then the same sort of question arises about the answer: what does its *being a game of chess* consist in? And just as answering the first question led to our looking at how that movement is located in something broader, so too are we led to look for some kind of larger set of circumstances for the whole game. The "circumstances that we call 'playing a game of chess', 'solving a chess problem', and the like," which serve as a kind of background to the making of a move in chess—a background against which we judge an action as a move in chess—themselves need to be seen against a background so as to be judged as just those sorts of circumstances. To borrow Wittgenstein's phrasing from § 201 of the *Investigations*, which is often read as the pivot-point of his rule-following remarks, we place one background "behind another, as if each one contented us at least for a moment, until we thought of yet another lying behind it."

That there is this kind of continually dissolving contentment indicates that something is amiss in this way of understanding the background as explaining the sense of what lies in the foreground. In order *really* to be explanatory, the thought goes, what lies in the background must be understood as complicated in a *special*—perhaps even a *mysterious*—way, so as to provide a kind of final or master explanation of how "whatever I could express has its meaning." But while the circumstances are indeed "a very complicated filigree pattern" that "we can't copy," and while such circumstances are indeed determinative for making sense of any particular action or utterance, it turns out that any bit of such a pattern, when foregrounded so to speak, turns out to be more of the kinds of things we were seeking to explain in the first place. The practice of playing chess has, no doubt, a long and complicated history and is itself but a fragment of a much longer history of human beings playing games with one another. That long and complicated history plays a role in making sense of something someone does as a move in chess, but there is nothing especially mysterious in that, but just a whole lot of chess history and a whole lot of chess practices.

After describing the background as a complicated "filigree pattern," Wittgenstein continues:

> The background is the bustle of life. And our concept points to something within *this* bustle. (RPP II, § 625)

As a bustle (the German here is *Getriebe*), the background is something complicated, but that is not to suggest that what lies in the background is any different in kind from what stands in the foreground. Consider what Wittgenstein writes four remarks later:

> How could human behavior be described? Surely only by showing the actions of a variety of humans, as they are all mixed up together. Not what *one* man is doing *now*, but the whole hurly-burly, is the background against which we see an action, and it determines our judgment, our concepts, and our reactions. (RPP II, § 629)

So here we have a background that "determines our judgment, our concepts, and our reactions," but that background—the "whole hurly-burly"—just consists of "the *actions of a variety of humans*, as they are all mixed up together." The upshot of Wittgenstein's idea that we "judge an action according to its background within human life" is that we judge an action by seeing it in relation to *other actions* that, as "all mixed up together," form a variety of patterns which the present action can be judged as continuing (or deviating from). There is thus not some thing—a background—that is complicated in itself or in some kind of absolute sense. What is complicated instead is the way all the various actions—those actions "of a variety of humans"—are "all mixed up together." There is no danger here of language going "on holiday" in the manner of the quest for what is absolutely simple.

These last remarks suggest a further sense of "complicated." Here "complicated"—as referring to "the whole hurly-burly" and what is "all mixed up"—conveys a sense of being messy, indeterminate, or indefinite. Indeed, this sense has been in play from the start of the *Investigations*, when Wittgenstein talks about there being "countless kinds" of sentences and there being "different, variously related, kinds of use of a word." But in the psychology manuscripts Wittgenstein is more explicit about this sense of indeterminacy. After invoking the notion of a *bustle*, he writes:

> And it is the very concept 'bustle' that brings about this indefiniteness. For a bustle comes about only through constant repetition. And there is no definite starting point for 'constant repetition'. (RPP II, § 626)

The difficulty of what is complicated is not here a matter of resolving it into, or replacing it with, what is simple, but of interrogating the desire for such resolution or replacement.[16] The difficulty, in other words, lies in determining what stands in the way of *accepting*[17] that our "human form of life" is indeed complicated.[18]

---

[16]Compare Wittgenstein's remark on the desire for sharp boundaries:

> If someone were to draw a sharp boundary, I couldn't acknowledge it as the one I too always wanted to draw, or had drawn in my mind. For I didn't want to draw one at all. It can then be said: his concept is not the same as mine, but akin to it. The affinity is that of two pictures, one of which consists of colour patches with blurred boundaries and the other of patches similarly shaped and distributed by with sharp boundaries. The affinity is just as undeniable as the difference (PI, § 76).

[17]Recall: "What has to be accepted, the given, is—one might say—*forms of life*" (PPF, § 345).

[18]Thanks to Edmund Dain, Duncan Richter, and especially Denis McManus for their helpful suggestions when discussing some of the ideas in this paper.

Chapter 11
Wittgenstein on Truth

Paul Horwich

11.1 Introduction

This paper will address four related questions.—What is the account of truth that Wittgenstein gives in the *Tractatus Logico-Philosophicus*?[1] To which view of that concept does he turn in his *Philosophical Investigations*?[2] Is this a move in the right direction? And how does it relate to other important differences between his early and late philosophy. Is it a cause of those changes of mind, a mere effect of them, or fairly independent of them?

11.2 The *Tractatus* View of Truth

What seems to jump out of the initial pages of the *Tractatus* is something like a *correspondence* theory of truth. But I say "something like", because a couple of qualifications must be made. In order to explain them, a few preliminaries are needed.

First: keep in mind that although we can speak of *sentences* (such as "snow is white") as "true" or "false", we more often apply those terms to *the things expressed by sentences (*such as the *proposition* that snow is white)—the objects of belief and assertion. Philosophers dispute which of these two ways of speaking is the more fundamental one; but most of us would agree that they are not the same: so distinct accounts of what we mean by them are called for.

[1] Wittgenstein (1922).

[2] Wittgenstein (1953).

P. Horwich (✉)
Department of Philosophy, New York University, 5 Washington Place, New York, NY 10003, USA
e-mail: ph42@nyu.edu

S. Wuppuluri and N. da Costa (eds.), *WITTGENSTEINIAN (adj.)*,
The Frontiers Collection, https://doi.org/10.1007/978-3-030-27569-3_11

Second: it's far from obvious how to bring this distinction to bear on Wittgenstein's remarks about truth in the *Tractatus*. For his terminology sometimes diverges considerably from what's typically employed nowadays. In particular:

- He *uses the* term, "propositional sign" for what *we* might call a "purely physical sentence"—that is, a "sentence conceived of as a sequence of mere noises or inscriptions".
- He uses "proposition" (translated from the German, "Satz") for what we might say is a "significant sentence" or, in other words, a "physical sentence together with an interpretation of it, conceived of as an assignment of referents to its components noises or inscriptions".
- And he uses "the *sense* of a proposition" for what we (and he) call "the *possible fact* that's represented by a significant sentence", or perhaps "the *Russellian proposition* (composed of objects and properties) that's expressed by a significant sentence".[3]

Translating his terminology into ours, it's fairly clear from the text that his overt theory of truth concerns truth for significant *sentences*, and not for what they express, or represent as being the case. His view (as we would put it) is that

A sentence is trueiff

i. it represents a certain possible fact; and

ii. that possible fact is *actual*

which is tantamount to

$$\text{S is a true sentence} = \text{S represents a } fact$$
$$(= \text{an } actualized\ possible\ fact)$$

To see why this might aptly be termed a *correspondence* theory of truth we must look at Wittgenstein's distinctive account of *representation*.

His basic idea is that we should answer the relatively hard question of how a *sentence* (—a string of significant signs—) is able represent something (—*that such-and-such is the case*—) by beginning with the relatively easy question of how a realistic *picture*, or a map, or an architect's model, represents what it does, and then proceeding to show that, initial appearances to the contrary, sentences represent in exactly the same way. Sentences *are* pictures!

More specifically, his view is that (a) a pictorial representation consists of elements arranged with respect to one another in a certain way; (b) each such element has a referent; and (c) the *actual* fact that certain pictorial elements are arranged as they

[3]Wittgenstein insists (3.13) that a *proposition* (in his usage of the term) does not contain its *sense* (in his usage). So the material that must be added to the mere propositional *sign* in order to arrive at a proposition is *not* the possible fact it represents. Rather (and as just suggested in the text) it's an assignment of referents to the component words. Granted, this assignment will *determine* which possible fact is represented. But that possible fact will not be *contained* in the proposition.

are represents the *possible* fact that the referents of those elements are also arranged in that way.

Consider, for example, a map of Italy that has shaped dots, "•", "◇", and "■", whose referents are (respectively) Pisa, Rome, and Naples. The actual fact that, on the map, "◇" is between "•" and "■", *depicts* the possible fact that Rome is between Pisa and Naples. Here, the common arrangement of pictorial elements and their referents is ***that x is between y and z***.

Now, as I've just mentioned, Wittgenstein holds that we can and should regard a sentence (e.g. "John loves Mary") as a kind of pictorial fact (—the fact that "John" is just to the left of "loves" which is just to the left of "Mary"). In this example we are to suppose that:

- The pictorial fact has three referring elements – namely, the words, "John" and "Mary", and the relation, *x is just to the left of "loves" which is just to the left of y*.[4]
- The pictorial *arrangement* of these elements is not *spatial*, but *logical*: it's the abstract logical structure, ***#(x, y)***. This form is what the representing fact and the represented possible fact have in common.
- The referents of the three elements are, respectively, John, Mary, and the relation, *x loves y*.
- Therefore—in light of principle (c) above—the depicted (represented) possible fact is just what we pre-theoretically know it to be—namely, the possibility *that John loves Mary*.

This is Wittgenstein's famous 'picture theory of meaning'.[5] Clearly, *depicting* a possible fact is being treated as a way of *corresponding* to it. So it would seem that we *can* aptly say that Wittgenstein is proposing a "correspondence theory of truth".

But now let me explain the pair of reservations, to which I alluded at the outset, about the applicability of that label.

One of them is that, arguably, the above-sketched picture-theory of sentential representation is supposed by Wittgenstein to apply *only* to sentences that don't contain any *logical* vocabulary.[6] And if this is right then Wittgenstein's correspondence the-

---

[4]Note Wittgenstein's 3.1432:

We must not say, "The complex sign '*aRb*' says '*a* stands in relation *R* to *b*'"; but we must say "*That* '*a*' stands in a certain relation to '*b*' says *that aRb*".

Clearly, the "certain relation" is: —*x is just to the left of "R" which is just to the left of y.*

[5]For elaboration of this reading of Wittgenstein's picture theory of representation, see Horwich (2012)

[6]Why this restriction? According to Wittgenstein, "My fundamental thought is that the logical constants don't represent" (4.312).—That is to say, the words, "and", "not", "or", and so on, don't stand for bits of reality. And, in that case, how could his picture theory of representation conceivably work for sentences containing those words? Consider, for example, "It's raining or snowing". If his theory were to explain how this logically complex sentence represents what it does, the word, "or", would either have to be part of a representing component of the sentence, or else it would have to be part of the pictorial arrangement of those components. But Wittgenstein's "fundamental thought" appears to preclude the first of these options. And his requirement that the represented entities

ory of truth must also be restricted to such sentences. Further principles will have to be added in order to extend that limited account into one that can cover logically complex sentences. And such principles are indeed supplied in the *Tractatus*: they are the rules implicit in Wittgenstein's *truth-tables*—rules which specify how the truth or falsity of negations, disjunctions, conjunctions, and so on are determined by the truth and falsity of the sentences to which the logical terms have been applied. Thus, what Wittgenstein really proposes is a *two*-stage theory of sentential truth, only the first of which invokes *correspondence*.

A second misgiving one might well have about calling Wittgenstein's view a "correspondence" theory of truth emerges from reflection on what he has to say about the other brand of truth I mentioned at the start: truth, not for *sentences,* but for what they express—that is, for what we nowadays call *propositions*.

One conclusion we *might* reach is that Wittgenstein does not have, and cannot have, *any* theory of truth of that sort. That's because the only kind of thing he countenances that resembles what are now called "propositions" are *possible facts*, and possible facts aren't the sorts of things that it makes sense to speak of as "true or false".—Rather, such things can only be "actual or non-actual" (that is, "actual or *merely* possible").

Alternatively, we might be inclined to think that, for us, "The proposition *that k is f* is true" and "The possible fact *that k is f* is actual" are just two ways of saying the same thing. In other words, we might suppose that Wittgensteinian *possible facts* should be identified with what we call "propositions", so that they *can* perfectly well be spoken of either as "actual or non-actual" or as "true or false". In which case we'd conclude that Wittgenstein is, after all, *implicitly* committed to a view of the notion that we call "propositional truth"—namely, the view of it that coincides with what he explicitly says about *possible-fact actuality*.

I myself am not entirely sure which of these answers is best (although I'm inclined towards the second). For, on the one hand, I really don't see how, for example, "Massimo thinks *that Mars is green*" could be *ambiguous*: in one sense relating Massimo to a *Russellian proposition* and in another sense to a *possible fact*. On the other hand it's indeed hard to accept that what's said to be "true" could with equal propriety be described as "actual", and vice versa.

But whichever choice is made here, we can see that the *core* of Wittgenstein's proposal about *sentential* truth is the idea that only when the things meant (or *expressed*, or *represented*) have a certain special quality (—either *being actual,* or *being factual,* or *obtaining, or existing,* or *agreeing with reality*, ...—) can the sentences with those meanings be true. Note:

2.202 The picture represents a possible state of affairs in logical space.

---

exhibit exactly the *same* arrangement with respect to one another as the representing components appears to preclude the second. (Since surely the *noise*, "or", doesn't feature in the possible fact that it's raining or snowing!).

Admittedly, Wittgenstein's formulations often seem to allow that his picture theory applies across the board. But in some passages (e.g. 4.0311) the restriction to *logically simple* sentences is explicit. And we have just seen why that's called for.

2.21 The picture agrees with reality or not; it is right or wrong, true or false.
2.22 The picture represents what it represents, independently of its truth or falsehood, through the form of representation.
2.221 What the picture represents is its sense.
2.222 In the agreement or disagreement of its sense with reality, its truth or falsity consists.

Thus one might well suppose that discerning the character of this special quality will be crucial in judging whether his overall view deserves to be called "a *correspondence* theory of truth".

And there appears to be no role for *correspondence* at this fundamental stage. The only point at which that notion enters Wittgenstein's story is in his theory of *representation*—in the relationship between sentences and possible facts. We get to sentential *truth* only by relying on the concept of '*agreeing with reality*' (or '*obtaining*', or ...), which Wittgenstein does not explain.[7] So we might reasonably conclude that what we are given is *not* really a correspondence theory of *truth* but rather a limited correspondence theory of *representation* plus a primitivist *non*-theory of when the represented entities are facts.[8]

11.3 The *Investigations* View of Truth

Wittgenstein's remarks on truth in his *Philosophical Investigations* suggest a position that's very different from the one to be found in the *Tractatus*. Pretty clearly he is pivoting to a perspective that these days would be classified as "deflationary"—the term applied to accounts emphasizing that:

- Truth has *no* traditional explicit definition or reductive analysis (e.g. in terms of *correspondence*, or *coherence*, or *verifiability*, or *utility*, or *consensus*).
- Instead, the nature of the concept is implicitly fixed by the way that each statement specifies its own condition for being true—e.g. the statement *that lying is wrong* is true if and only if lying is wrong.

---

[7]To be discussed further in Sect. 4.

[8]Hans-Johann Glock puts the point nicely as follows: "The *Tractatus* marries a correspondence theory of depiction to an obtainment theory of truth". ("Wittgenstein on Truth", in Löffler and Weingartner (2004).

He goes on to allow that it is still appropriate to call Wittgenstein's view "a correspondence theory" since the similar view of truth that's proposed by both Moore and Russell around 1912 is standardly taken to be *paradigmatic* of such theories.

But it's worth noting a relevant *dis*-similarity. As we've seen, the foundation of Wittgenstein's account is his distinction between those possible facts that are *actual* and those that are not; and *actuality* is not a *correspondence* notion. But there's no analogous distinction in the account offered by Russell and Moore. They hold, for reasons we'll be examining in Sect. 4, that there can be no such things as *false propositions*. And this reasoning would lead them to the same skeptical conclusion about *merely possible facts*.

- It's an extremely *superficial* concept. There are hardly any concepts that are defined in terms of TRUTH, or whose possession requires prior possession of the concept, TRUTH.
- It's merely a useful *expressive* device, enabling certain generalizations to be formulated – for example, "All propositions of the form, <p or not-p>, are true", and "A belief is correct if and only if it's true".[9]

There's a lot of evidence in favor of attributing some such perspective to the later Wittgenstein.

First: we have the *Investigations,* Sect. 136:—"**p** is true = p". He's claiming here that ascribing truth to a proposition is equivalent to asserting the proposition itself.

Second: he was aware of, and influenced by, Frank Ramsey's advocacy of precisely that view.[10] They were together at Cambridge University in the early 1930s: and in the Preface of the *Investigations* he credits Ramsey with having been an enormous influence on his thinking.

Third: in accord with the deflationary definition and expressive raison d'etre of TRUTH, this concept is given no important role in the *Investigations.*

And fourth: going hand-in-hand with deflationism about truth is the idea that our notions of predicative satisfaction and nominal reference are fixed by the schemata

$$\text{Predicative concept } \mathbf{f}() \text{ is true of } x = f(x)$$
$$\text{Nominal concept } \mathbf{n} \text{ refers to } x = n \text{ is identical to } x$$

specifying what a given concept applies to, *but only given prior possession of that concept* (which is deployed on the right hand side of the relevant equation)—a possession that, on pain of circularity, cannot derive from knowing the concept's reference but must instead consist in mastery of its *use*. Thus Wittgenstein's move from the Tractarian *referential* conception of word-meanings to the Investigations *use* conception permits him to adopt a deflationary view of *reference*—and that's exactly what one would expect from a deflationist about *truth*.

Assuming it's right that the *Investigations* view of truth is deflationary, one might wonder which one of the various brands of deflationism on the market these days Wittgenstein favored, or would have favored. Would it be *disquotationalism*, according to which sentences (rather than propositions) are the bearers of truth, and the schema, "p" is true $\leftrightarrow$ p, is the core of what implicitly defines the truth predicate? Or *pro-sententialism*, which denies that "true" is a genuine (logical) predicate, and which stresses instead the analogy between a pronoun and the sentence-type, "That is true", insofar as both inherit their content from another expression—one that's contextually salient? Or the *redundancy theory*, whereby "The proposition *that p* is true" means exactly the same as just "p"? Or the *sentence-variable analysis*, which analyzes truth-talk in terms of quantification into sentence positions :—"x is true" is taken to mean "$(\exists p)(x = \langle p \rangle \; \& \; p)$"? Or *Tarski's theory*, which explains the

[9]For amplification of these points see Horwich (2010)

[10]Ramsey (1927).

truth condition of each sentence of a language in terms of the referential properties of its component words (characterized disquotationally) and the logical structure in which the words are embedded? Or *minimalism*, which resembles disquotationalism, except that it takes propositions (rather than sentences) to be the fundamental bearers of truth, and according to which our possession of the concept TRUTH is said to consist in our underived inclination to accept instances of the schema, "The proposition *that p* is true ↔ p"?

We don't have enough evidence to decide. But my guess is that Wittgenstein's aversion to philosophical theorizing would push him away from those versions of deflationism that come with a substantial amount of theoretical baggage, making them not *fully* deflationary.

So the redundancy theory would be disliked for its artificial and implausible conception of propositional identity (whereby, x can be the same proposition as y even though x involves the concepts of TRUTH and of PROPOSITION, and y does not.

Disquotationalism would be disliked for its misguided naturalistic presupposition that *propositions* are too weird to exist, and for its resulting revisionist disrespect for our *actual* use of "true"—our more normal application of it to *the things we believe and disbelieve*, rather than to our *expressions* of such attitudes.

Tarski's theory would be disliked even more. For not only does it involves the same mis-motivated abhorrence of propositions and resulting focus on the truth of linguistic expressions—but, in addition, it repeats the Tractarian assumption (which Wittgenstein came to regard as deeply mistaken) that any meaningful sentences is either an elementary sentence, or else is equivalent to the result of logical operations on elementary sentences.

The sentence-variable analysis would be disliked for its different form of revisionism. It defines a concept of 'truth' in terms of unfamiliar kinds of variable and quantifier. So the concept of truth that it defines cannot possibly be ours.

And the pro-sentential theory would be disliked for its scientistic overstretching of the analogy between our use of "That's true" to avoid repeating someone's recent assertion, and our use of "she" to avoid repeating some recent use of a name or description (e.g. "the discoverer of radium"). The most valuable of our deployments of "true" (as in "Some truths are unverifiable" and "Goldbach's conjecture is true") cannot be assimilated to that paradigm without absurd contortions.

Which leaves me thinking he'd go for minimalism (What a surprise!).

11.4 Which of Wittgenstein's Two Views of Truth Is Better?

Well my opinion about that is already obvious. But let me supply some justification for it.

As indicated above, Wittgenstein's account in the *Tractatus* explains truth in terms of an unexplained (but equally puzzling) notion of *fact*. We aren't told what it means to say "It's *fact* that p", although this idea is obviously no less mysterious (and no less in need of explanation) than "It's *true* that p", which is what it's supposed to explain. This is a considerable defect. And, of course it won't help to define a "fact" as a possibility that "is actualized" or "obtains" or "exists" or "agrees with reality" unless we have decent explanations of at least one of these—which, as will become clear, we don't have.

A second difficulty with Wittgenstein's early position is that it's hard to see how it could respect the evident difference between *actual* facts and merely *possible* facts. After all, whether it's an actual fact *that Mars is green*, or merely a possible fact that *Mars is green*, we're going to have the same constituents—namely, the object Mars and the property of being green—embedded in the same logical structure! Or to put the point in a slightly different way: the fact is a certain logical arrangement of certain entities; but the merely possible fact is exactly that arrangement of exactly those entities. So how could there exist any *merely possible facts*? Why wouldn't all *possible* facts have to be *actual*?

It's extremely surprising that Wittgenstein didn't address this second problem, since his teacher and mentor, Bertrand Russell, made such a fuss about it. In his 1912 book-chapter, "Truth and Falsehood", and in subsequent writings, Russell argued in just this way that there could be no such things as *false* propositions.[11] He concluded that one cannot regard beliefs as relations between people and propositions (since many beliefs are mistaken). And so he devised a novel account of belief in which, for example, Massimo's belief *that Mars is green* does not consist in a two-place relation between Massimo and the proposition (or possible fact) *that Mars is green*, but rather in a three-place relation between Massimo, Mars, and greenness. Russell showed this work to Wittgenstein, who dismissed the new theory of belief (on the grounds that it "didn't prohibit believing nonsense"). And Russell was demoralized, confiding to his then girlfriend that although he didn't really understand the objection, his respect for Wittgenstein's insight made him feel that it must be right.[12] Still—and this is my main point—it's surprising—given that Wittgenstein had never complained about Russell's argument for the non-existence of false propositions—that didn't appreciate how this argument would count equally against his own commitment to non-actual possible facts.[13]

These two defects in the *Tractatus* account of truth are related to one another as follows. In response to the first one it might be protested that Wittgenstein *does* address what it is for a possible atomic fact to be *actual* (or to *exist*, or to *obtain*, ...). He says that this will be so when the constituents of the possible fact are *combined*. And this might seem to provide an illuminating analysis. But that's an illusion. The objection to it is *not* that *nothing* is said about what it is for objects to be *combined*. For Wittgenstein tells us that "In the atomic fact the objects hang one in another, like

[11]Chapter 13 of The Problems of Philosophy (1912)

[12]This biographical material is taken from pp. 80–82 of Monk (1990).

[13]I don't see how we can follow Goldfarb (1997) and Zalabardo (2019) in maintaining that the existence of merely possible facts is not endorsed by Wittgenstein. For this would go against his 2.202 and 2.221 (quoted above) in which the *sense* of a proposition is explicitly identified with *what it represents*, which is in turn explicitly identified with a possible fact.

the links of a chain" (2.03). Nor is the objection that this mere metaphor is woefully insufficient. For he makes it clear that the mode of combination is literally *logical*: it's for example, that Mars and being red (pretending for a second that these are basic entities) are embedded in the structure, '#(x)', or 'x exemplifies #-ness'. The real objection with analyzing a fact as a combination of constituents, is the *second* of the just-sketched defects in the *Tractatus* account of truth – which is that this analysis fails to do what it was mainly supposed to do – namely, to distinguish the *actual* atomic facts that from those that are merely *possible*. For, in both cases the same objects are embedded in the same structure. Thus we are left with no clue as to what it is for a possible fact to obtain, hence what it is for the sentence representing that possible fact to be true.[14]

And a third objectionable feature of Wittgenstein's early theory is that the picture theory of representation, on which his view of sentential truth depends, deploys a unexplained relation of *reference* between words and things—a notion that's in no less need of definition than the notion of truth it's being used to define.

All of these related defects in the *Tractatus* account of truth—its reliance on mysterious notions of *actuality* and *reference*, and its failure to explain how *merely* possible facts could exist—can be removed in one fell deflationary swoop.

To begin with the problem of 'merely possible facts'—let's call it "Russell's problem". Wittgenstein can solve his own version of it by giving up the idea that there must be some *intrinsic* quality of a possible fact that makes it actual, and instead specifying conditions of *actuality* by means of the schema:

The possible fact *that p* is actual *iff* p

And, of course, the parallel reply to Russell's argument against false propositions is that we can identify *the facts* with *the propositions that are true,* and then supply the conditions for propositions to be true by means of the schema:

---

[14]Zalabardo—in support of his above-mentioned view that Wittgenstein wasn't ontologically committed to *possible* facts.—argues (op.cit. 2019) that the picture theory of representation enables Wittgenstein to define sentential truth *directly*, that is, without bringing in *possible facts* (or any other represented items) as intermediaries. For an elementary proposition consists in *referring elements, e1, e2, ..., ek being arranged in a certain way, A*. And Wittgenstein's definition (according to Zalabardo) is *that such a proposition* is true if and only if the referents of the elements are also arranged in that way. In other words: *the propositional fact* <u>*that A[e1, e2, ..., ek)]*</u> *is true if and only if A*[*Ref*(*e1*), *Ref*(*e2*), ..., *Ref*(*ek*)].

But Wittgenstein doesn't proceed in this way. Instead, he begins by explaining, via his picture theory, how the *sense* of a proposition—a possible fact—is determined; and he then goes on to tell us (in 2.222, quoted above) that that the proposition is true or false depending on whether that sense agrees or disagrees with reality. Thus senses (qua possible facts) are invoked, as is the idea that some of them agree with reality and others do not.

No explanations are given of how the arrangements that constitute merely possible facts would differ from those that constitute actual facts, or of what it is for a sense to *agree with reality*. And, granted, these problems would have been avoided had Wittgenstein defined sentential truth in the way that Zalabardo thinks he does. However, that 'better' account amounts to the deflationary idea (i) that the possible fact <u>*that p*</u> is actual ↔ p, and (ii) that s represents <u>*that p*</u> → (s is true <u>*iff*</u> p). (See the text below). And Wittgenstein didn't appreciate the merits (and radical implications) of this approach until well after his *Tractatus* had been published.

<p> is true *iff* p[15]

Again, the mistake was to think that there must be something in the *intrinsic nature* of a fact that distinguishes it from a mere proposition.

This stone also kills the second bird. What I mean is that in resolving the Russellian objection to Wittgenstein's early theory we're also addressing the objection that Wittgenstein's attempts to demystify TRUTH by explaining it in terms of FACT, which is equally mysterious. For, as we have seen, the solution again is to say:

It's a fact *that p iff* p

Thirdly, regarding his reliance on an unexplained relation of reference: Wittgenstein's coming to appreciate that the meaning of a word is *not* constituted by its reference, but rather by how the word is used, allows him to explicate and demystify reference via a pair of schemata along the lines of

"n" refers to x ≡ n is identical to x (for names)

and

"f" is true of x ≡ f(x) (for predicates)

That's because our understanding of their right hand sides will *not* already presuppose knowledge of the facts of reference that the left hand sides are supposed to specify.

The overall moral is that the *Tractatus* theory of sentential truth needs an injection of deflationism if it is to be saved.

11.5 The Role Played by Wittgenstein's Accounts of Truth Within His Earlier and Later Philosophies

However we might well wonder whether it's really possible to coherently inject the deflationary component of Wittgenstein's *Investigations* philosophy into his *Tractatus* philosophy.

I would suggest that the answer is no. That's basically because, as we've seen, deflationism about truth goes hand-in-hand with deflationism about reference; and deflationism about reference goes hand-in-hand with a use-theoretic view of meaning. But Wittgenstein's later commitment to 'meaning as use' is the principal basis for

---

[15]This is not the so-called *Identity Theory of Truth*, which is advocated by Hornsby (1997). Yes, she too maintains that true propositions are facts. But her idea is that this equation will define "true" in terms of "fact"—where the latter, roughly in the spirit of the *Tractatus*, is taken either to be a primitive or to be defined as a "combination of objects and properties". But my deflationary suggestion goes in the opposite explanatory direction—relying on the trivial Equivalence Schema to fix the concept of TRUTH, and then proceeding to define "fact' as "true proposition".

his recognizing many of what he regards as his earlier "grave mistakes".[16] These included his failure to see:

- The fundamentally *instrumental* purpose of language
- The limitless variability of the functions of the different words in a language, and of the kinds of rules that must govern the uses of different words in order for them to perform this variety of functions
- The pervasiveness of vagueness and other forms of indeterminacy
- The absence of a small set of primitive words in terms of which all others can be defined
- The philosophical irrelevance of logic: in particular, that it does not provide the structure of language and thought
- The impossibility of drawing a line in advance around everything a language can be used to say
- The *real* source of philosophical confusion and pseudo problems—which is *not* the enormous conceptual distance between ordinary language expressions and their fundamental analyses (but rather our scientistic inclination to overstretch linguistic analogies)
- The contradiction between, on the one hand, the *theoretical* nature of his commitments (—to primitive terms, to the ideal of determinacy, to the fundamentality of logic, and to the exclusive role of *final conceptual analysis* in dissolving philosophical problems—) and, on the other hand, his bottom-line view that *philosophy cannot be theoretical.*[17]

In short, once we let the genie out of the bottle—that is, the deflationary and use-theoretic cat out of its bag (the *Investigations*)—then the fundamental assumptions of his early philosophical system become evidently untenable.

So my conclusion is that Wittgenstein's magnum opus, *Philosophical Investigations,* is a great advance on the *Tractatus*, his brash, brilliant, *initial* attempt at a radically anti-theoretical philosophy. And this progress is almost entirely due to the way in which his new view of truth improves on the old one.[18]

---

[16]See the Preface to *Philosophical Investigations.*

[17]As far as I know, Wittgenstein never explicitly renounced his picture theory of elementary sentential representation and its truth-functional extension to logically complex sentences. But there can be little doubt that he came to recognize that no such account is needed. For it's a fundamental triviality that—no matter which meaningful categorical sentence we substitute for "p"—the sentence, "p", means <u>*that p*</u>.

[18]The present publication is a slightly revised version of an article of mine that appeared in *Argumenta*, Issue 3, December 2016—available online at http://wwww.tandfonline.com/ [Article DOI]. I am most grateful to the editors of *Argumenta* for permitting me to reprint that paper.

My thanks also to Hans-Johann Glock for the stimulus of his excellent above-cited essay on this topic, to Charles Djordjevic for an enjoyable and illuminating conversation about these matters, to Massimo Dell'Utri and Alex Miller for their astute comments on earlier versions of the present paper, to José Zalabardo for the stimulating challenges presented by his alternative interpretation of the *Tractatus*, and to all those who helped me by raising tough questions during my discussions of these ideas at the University of Zurich (June 2016), the University of Sassari, Sardinia (September 2016), and the University of Minnesota (November 2016).

References

Frank Ramsey, "Facts and Propositions" (1927), in D. H. Mellor (ed.) *Foundations: Essays in Philosophy, Logic, Mathematics and Economics*, Routledge and Kegan Paul, London, 1978.

Horwich P "What is truth?", Chapter 1 of my Truth-Meaning-Reality, Oxford University Press, 2010.

Horwich P Wittgenstein's Metaphilosophy, Oxford: Oxford University Press, 2012, pp. 80–83.

Jennifer Hornsby "Truth: The Identity Theory", *Proceedings of the Aristotelian Society*, 97: 1–24, 1997.

Jose Zalabardo "The Tractatus on Unity", *Australian Philosophical Review*, 2019.

Ray Monk Ludwig Wittgenstein: The Duty of Genius, The Free Press, New York, 1990

The Problems of Philosophy, New York: H. Holt and Company, 1912.

Warren Goldfarb "Metaphysics and Nonsense: On Cora Diamond's *The Realistic Spirit*", *Journal of Philosophical Research* 22:57–74, 1997.

W. Löffler and P. Weingartner (eds.), *Knowledge and Belief* , Hölder-Pichler-Tempsky, Vienna 2004, 328–346.

Wittgenstein, L. *Tractatus Logico-Philosophicus,* London, Kegan Paul, 1922. Translated by C. K. Ogden.

Wittgenstein, L. *Philosophical Investigations,* Oxford, Blackwell, 1953. Translated by G. E. M. Anscombe and G. H. von Wright.

Part II
Theoretical Sciences

Chapter 12
Wittgenstein's Philosophy: Reorientating Science

A Sketch by Michael Nedo

Michael Nedo

Much of what once was praised as scientific and technological progress has shown itself to be in conflict with the world, the world man has inherited, that we hold in trust.

As scientists are not malicious and their aims on the whole are not necessarily wrong there must be something wrong in the way scientists view, describe and explain the world, which needs to be corrected. It seems that scientists have to re-orientate themselves as Wittgenstein argues in remark 108 of his *Philosophical Investigations*:

> Die Betrachtung muß gedreht werden, aber um unser eigentliches Bedürfnis als Angelpunkt.[1]
>
> We must rotate the observation, but about the pivot of our actual need.

12.1 The Scientific Concept of Nature

Science as we know it today began in the 17th century with Copernicus, Kepler, Galileo and Newton. Although their views of the world were still very much those of the Middle Ages: that the world was created by God, only to be understood as a whole and only through God's word. But their interests had shifted, as Galileo put it during his trial by the Inquisition: 'from God's word to God's deeds'. For Newton (about one hundred years after Galileo) the world could be understood in parts and independent of God, but he admitted:

M. Nedo (✉)
Wittgenstein Archive, Cambridge, UK
e-mail: mn15@cam.ac.uk

S. Wuppuluri and N. da Costa (eds.), *WITTGENSTEINIAN (adj.)*,
The Frontiers Collection, https://doi.org/10.1007/978-3-030-27569-3_12

> This most beautiful system of the sun, planets and comets, could only proceed from the counsel and dominion of an intelligent and powerful Being. [···] This Being Governs all things, not as the soul of the world, but as Lord over all: And on account of his dominion he is wont to be called Lord God [···][2]

Classical physics, which had evolved from Newton's mechanics, has been, up to the beginning of the 19th century, successfully applied to all fields of natural science and to industry. Based on the mathematical presentation of scientific findings, laws could be formulated through which sciences could develop instruments and methods to see ever deeper into the micro and macro cosmos and technologies were developed utilising the forces of nature to be employed by industry.

These laws, describing for instance the falling of an object (or Newton's apple), were subsequently applied not only to understand the movement of the moon around the earth but to the whole of the universe. Those methods and technologies have been changing our notion and our understanding of nature, by enabling scientists to see ever further and deeper into the macro- and micro-cosmos, into our world and into the universe than we had previously been able through our naked senses.

The term 'nature' as the object of natural sciences included now all fields into which man could penetrate by the means of science and technology and all these descriptions and explanations where based on the understanding of atoms, of elementary particles, their mass, movement and positions in space and time.

In the second half of the 19th century this materialistic, object based description of the world had to be extended as the phenomena of electricity and magnetism and their interrelation could only be described and explained by energy fields, structures without material substance, without objects like elementary particles.

With the discovery of quantum physics at the beginning of the 20th century classical mechanics was finally replaced: the objective reality of elementary particles was replaced by mathematics representing the relation between object and observer. The scientist, as Niels Bohr put it, had himself become an actor and part in the drama of life.

12.2 Sciences, Technology and Industry

During the past 200 years, technology has been both a consequence and a precondition of natural sciences by further developing the methods and instruments by which scientists were studying and describing the world.

In the 18th and early 19th century technologies were developed which were mainly simulating the hand of man thus improving craftsmanship. A decisive change came with the understanding of the relation between electricity and magnetism by Michal Faraday and its further development by Maxwell and Hertz, the development of the electro technology in the second half of the 19th century, a technology which was no longer related to craftsmanship but one that led to the establishment of a new forms of industry, vastly extending the power of man.

While these technologies originally evolved from science, it was now industry that determined more and more their further development, which, employed by scientists, determined their research, a phenomenon vastly accelerated during the two world wars. These new technologies were no longer the result of studying and understanding nature but using nature in the interest of industry. It is therefore not surprising that their application is creating ever more conflicts with and within our world, a development which, if not halted, might destroy this world, that is, make it uninhabitable to men.

12.3 Wittgenstein and Science

Wittgenstein's philosophy does not represent an alternative concept of science, a new form of science but in his attention to what is unser eigentliches Bedürfnis[3], what are our actual needs, his philosophy presents means and methods, which will enable us to overcome the problems of today's sciences. His philosophical method enables a scientist to distinguish between the real and the fantastic, a distinction, which characterises Wittgenstein's work up to the very last sentence he wrote. It is the loss of this distinction which leads to the kind of specialisation that characterises todays sciences: for instance to development of the atomic bomb, which neither man, beast or plant needs or indeed anything on the surface this whole world.

With a selection of remarks that I have chosen from Wittgenstein's writings and from other writers, I will show how Wittgenstein's philosophy and his philosophical method could change the way we see and understand this world, our relation to and within this world; it could lead to a reorientation of science.

To separate Wittgenstein's remarks more clearly from quotes not by Wittgenstein, as well as from my introductions into and between his remarks, I show Wittgenstein's remarks red in their original German and blue in their translations into English.

A surprising and remarkable aspect of Wittgenstein's philosophy is its poetic dimension, the poetic quality of his philosophy, of his philosophical writing, including his writing and thinking about science and in particular of mathematics. It shows how closely for Wittgenstein science, mathematics and philosophy are related to each other, that he sees them all rooted in poetry and it further shows how closely Wittgenstein's thinking and writing is related to that of the German Romantic, in particular to Novalis and his credo to poeticise science, his ambition of the 'Poetisierung der Wissenschaften'. In 1946 Wittgenstein described how his philosophy was related to poetry:

> O, warum ist mir zumute, als schriebe ich ein Gedicht, wenn ich Philosophie schreibe? Es ist hier, wie wenn hier ein Kleines wäre, was eine herrliche Bedeutung hat. Wie ein Blatt oder, eine Blume.[4]
> Why is it, when I write philosophy, it is as if I was writing a poem? It is as if there was something tiny, something with a wonderful meaning. Like a leaf or a flower.

Six years earlier he had written about a common and still widespread misunderstanding of the importance of sciences and mathematics versus that of poetry and the arts in general:

> Die Menschen heute glauben, die Wissenschaftler seien da, sie zu belehren, die Dichter und Musiker etc, sie zu erfreuen. *Daß diese sie etwas zu lehren haben*; kommt ihnen nicht in den Sinn.[5]
> People today believe scientists are there to teach them; poets, musicians etc., to please them. *That they have something to teach*, doesn't occur to them.

Wittgenstein's criticism of science and mathematics of his time, which is often seen as polemic, resembles that of other poetic thinkers like Novalis, Nietzsche and Valéry. He shares their view that the route of creativity, including of sciences and mathematics and of philosophy is not sophistication but instinct, what Novalis called 'die unbestimmten Empfindungen', the undefined emotions. In the context of a long discussion about mathematics Wittgenstein referred to this idea on 7 November 1929 relating Dedekind's concept of infinity in which truth is only on the side of probability to Austrian literature and music where truth is never on the side of probability:

Ich glaube, das gute Österreichische (Grillparzer, Lenau, Bruckner, Labor) ist besonders schwer zu verstehen. Es ist in gewissem Sinne *subtiler* als alles andere, und seine Wahrheit ist nie auf Seiten der Wahrscheinlichkeit.[6]
I think the good Austrian (Grillparzer, Lenau, Bruckner, Labor) is particularly difficult to understand. It is, in a sense, more subtle than anything else, and its truth is never on the side of probability.

And in 1949 he contrasts, with a stark picture, his own poetic view to the sophisticated one by the scientists and mathematicians of his time:

In den Tälern der Dummheit wächst für den Philosophen noch immer mehr Gras, als auf den kahlen Höhen der Gescheitheit.[7]
For the philosopher there grows always more grass in the valleys of stupidity then on the barren heights of cleverness.

In the Anglo-Saxon world Wittgenstein's philosophy, his philosophical thinking, is often seen as idiosyncratic, unique and unrelated to other thinkers. The following quotes from Lessing show how deeply Wittgenstein was rooted in the European, in particular in the German speaking culture. What Wittgenstein is expecting of his work in the context of science and mathematics, is quite similar to what Lessing is arguing in his pamphlet *Die Erziehung des Menschengeschlechts*, *Educating Mankind* in the context of religion:

Ein beßrer Pädagog muß kommen, und dem Kinde das erschöpfte Elementarbuch aus den Händen reißen. –[8]
A better teacher must come and tear the primer from the child that has outgrown it.

The primer, the Elementarbuch that Lessing is referring to is the Old Testament, a book of rules for children up to a certain age:

Das ihm entwachsene Kind länger, als die Meinung gewesen, dabei zu verweilen, ist schädlich. Denn um dieses auf nur einigermaßen nützliche Art tun zu können, muß man mehr hineinlegen, als darin liegt; mehr hineintragen, als es fassen kann. Man muß der Anspielungen und Fingerzeige zu viel suchen und machen, die Allegorien zu genau ausschütteln, die Beispiele zu umständlich deuten, die Worte zu stark pressen. Das gibt dem Kinde einen kleinlichen, schiefen, spitzfindigen Verstand; das macht es geheimnisreich, abergläubisch, voll Verachtung gegen alles Faßliche und Leichte.[9]

To persist in using a primer longer than intended, when a child has outgrown it, is harmful. For in order to do this in a somewhat useful kind of way, one must read more into it than it can contain. One must look for and make too many allusions and hints; squeeze too much out of the allegories; interpret examples too circumstantially; squeeze the words too hard. This gives the child a narrow, skewed, hair-splitting understanding; it makes it secretive, superstitious, full of contempt for all that is comprehensible and easy to understand.

Wittgenstein's argument for a reorientation of sciences and in particular of mathematics is quite similar to what Lessing is arguing in respect to religion: The primer(s) of science and mathematics have to be removed, the self-righteousness of rule-following has to be replaced by moral responsibility of each individual scientist and/or mathematician as in the parable of 'The Pharisee and the Tax Collector', in the New Testament, Luke 18:9–14. Such a re-orientation of sciences and mathematics is one, if not the main interest in Wittgenstein's work. His philosophy and his philosophical method could be a guide to scientists and mathematicians to help them re-oriented themselves in this world and in their work, but by 1947 Wittgenstein had given up this hope as he writes resignedly:

Nichts kommt mir weniger wahrscheinlich vor, als daß ein Wissenschaftler, oder Mathematiker, der mich liest, dadurch in seiner Arbeitsweise ernstlich beeinfußt werden sollte. (In sofern sind meine Betrachtungen/Warnungen/ wie die Plakate an den Kartenschaltern der englischen Bahnhöfe "Is your journey really necessary?" Als ob Einer, der das liest, sagen würde "Wenn ich's recht bedenke nein/On second thoughts, no" /.)[10]

Nothing seems to me more unlikely than that a scientist or mathematician who reads me should be seriously influenced thereby in the way he works. (In that respect my observations/warnings/ are like the posters on ticket offices at English railway stations "Is your journey really necessary?" As if anyone reading that would say to himself "On second thoughts, no".)

12.4 Wittgenstein's Philosophy

When Wittgenstein was asked to describe his work for the 1944 edition of Chambers's Encyclopaedia, he wrote—as Rush Rhees told me—just one sentence:

> He was primarily concerned with questions about the foundations of mathematics.

Wittgenstein's philosophy is often misunderstood as if he, like Russell and others at the time, was working on an intellectual edifice, a self-contained 'Denkgebäude'. But Wittgenstein's aim was the opposite: He was not interested in laying down new foundations nor was he interested in building new structures thereupon, his interest was, as he wrote on 6 November 1930, to investigate the foundations of existing buildings, for instance those of the foundations of mathematics:

> Es interessiert mich nicht ein Gebäude aufzuführen sondern die Grundlagen der möglichen Gebäude durchsichtig vor mir zu haben.[11]
>
> I am not interested in erecting a building but in having the foundations of possible buildings transparently before me.

And in May 1932 he was contrasting his work in mathematics to that of the mathematicians of his time:

> Den Mathematiker muß es bei meinen mathematischen Ausführungen grausen, denn seine Schulung hat ihn immer davon abgelenkt sich Gedanken und Zweifeln, wie ich sie aufrolle, hinzugeben. Er hat sie als etwas verächtliches ansehen lernen und hat um eine Analogie aus der Psychoanalyse[···] zu gebrauchen einen Ekel vor diesen Dingen erhalten, wie vor etwas Infantilem.[12]
>
> A mathematician must shudder at my mathematical elaborations, as his training has kept him from thoughts and doubts as the ones I am concerned with. He has learned to disregard them as something contemptible and, to use an analogy from psychoanalysis, [...] has perceived a kind of squalor in these things, as something infantile.

And before, on 16 February 1932 he said what he expected of philosophy, of his philosophy, for the future of mathematics:

> Der Philosoph spürt Wechsel im Stil einer Ableitung, an denen der Mathematiker von heute, mit seinem stumpfen Gesicht ruhig vorübergeht. – Eine höhere Sensibilität ist es eigentlich, was den Mathematiker der Zukunft von dem heutigen unterscheiden wird; und *die* wird die Mathematik - gleichsam - stutzen; weil man dann mehr auf die absolute Klarheit als auf ein/das/ Erfinden neuer Spiele bedacht sein wird.[13]
> A philosopher feels changes in the style of a derivation which a contemporary mathematician passes over calmly with a blank face. What will actually distinguish the mathematician of the future from those of today will be a greater sensitivity, and *that* will - as it were - prune mathematics; since people will then be more intent on absolute clarity than on the invention of new games.

On 23 May 1932 Wittgenstein illustrated the effect he expected of philosophy, his philosophy, for mathematics with a picture that today requires some explaining as the potatoes we buy at the supermarket today do not develop the long and fantastic sprouts Wittgenstein refers to. In his days a year's supply of potatoes was stored in dark cellars where towards the end of the winter they would have developed the most fantastic sprouts and potatoes had became ever more soggy and tasteless. Exposed to sunlight these useless sprouts would wither, changing into roots or leaves:

> Die philosophische Klarheit wird auf das Wachstum der Mathematik den gleichen Einfluß haben, wie das Sonnenlicht auf das Wachsen der Kartoffeltriebe. (Im dunklen Keller wachsen sie meterlang.)[14]
> Philosophical clarity will have the same effect on the growth of mathematics as the sunlight has on the growth of potato sprouts. (In a dark cellar they grow metres long.)

Scientists today explain by dividing the world into disciplines assigned to specialists. They argue for instance that Wittgenstein, whom they have classified as a philosopher of 'ordinary or natural language', was mistaken in his work on the foundations of mathematics, just as they said that Goethe was mistaken by seeing himself as a naturalist. It is largely this misleading 'classification' of Wittgenstein's philosophy together with the incomplete and problematic edition of his writings, in particular of his work in mathematics, which is causing difficulties understanding his work. It is therefore not surprising that Wittgenstein's work on the foundations of mathematics is largely ignored, it is even considered 'unworthy of the author of the *Tractatus* and of the *Philosophical Investigations*':

> The laconic, fragmentary and disorganised character of the Remarks on the Foundations of Mathematics, and the apparent eccentricity of the views therein expressed, have made it easy for philosophers, including many who would profess to find much of value in the Investigations, to dismiss and, in some cases, to abuse it.[15]

This widely shared view in the literature about Wittgenstein's work in the philosophy of mathematics poses a serious problem for anyone studying his philosophy.

It took the genius of Frank Ramsey—who to this day exerts a major influence onto mathematics—to recognise the importance of Wittgenstein's work in this field, in particular in the foundations of mathematics and in mathematics in general. 1924, after Ramsey had translated Wittgenstein's *Logisch-Philosophische Abhandlung*, the so-called *Tractatus*, into English, he visited Wittgenstein in Austria from where he wrote to G. E. Moore:

> In my opinion Mr Wittgenstein is a philosophical genius of a different order from anyone else I know. [⋯] From his work more than that of any other man I hope for a solution of the difficulties that perplex me both in philosophy generally and in the foundation of mathematics in particular.[16]

And on 14 July 1929, after Wittgenstein's return to Cambridge and to writing philosophy again after a break of more then 10 years, Ramsey, in another letter to Moore, specifically referred to his initially doubts resulting from Wittgenstein's lack of formal training in mathematics:

> At first I was afraid that lack of mathematical knowledge and facility would proof a serious handicap to his working in this field. But the progress he has made has already convinced me that this is not so, and that here too he will probably do work of the first importance.[17]

But Wittgenstein was not only interested in mathematics and its the foundations, he was also investigating the foundations of our intellectual edifices in general, that what Nietzsche in *Über Wahrheit und Lüge im außermoralischen Sinne* called our 'Begriffsdom [...] das in's ungeheure aufgetürmte Fachwerk', our 'cathedral of concepts [⋯] that vast towering Fachwerk' the framework into which the whole of the empirical world was to be filled, i.e. our anthropomorphic world. These investigations are at the very heart of Wittgenstein's philosophical work. They show his powerful and on-going struggle to attain truth and above all clarity, his continuous effort to distinguish what is real from the fantastic, the metaphysical, nonsensical from the physical i.e. the meaningful, or from being true to the rules in contrast to useful.

In his first book, the *Logisch-Philosophische Abhandlung*—in its English translation the *Tractatus Logico-Philosophicus*—which he finished on his last leave from the front, in the autumn of 1918, Wittgenstein concerns himself largely with the philosophical problems and the expectations of Russell and Whitehead in *Principia Mathematica* and of similar intellectual frameworks: Fantastic problems resulting from our confused, entangled and knotted minds, which in his book—as he argues in its preface—he has resolved once and for all:

> Das Buch behandelt die philosophischen Probleme und zeigt - wie ich glaube - daß die Fragestellung dieser Probleme auf dem Mißverständnis der Logik unserer Sprache beruht.
> [···]
> Ich bin also der Meinung, die Probleme im Wesentlichen endgültig gelöst zu haben. Und wenn ich mich hierin nicht irre, so besteht nun der Wert dieser Arbeit zweitens darin, daß sie zeigt, wie wenig damit getan ist, daß diese Probleme gelöst sind.[18]

> The book deals with the problems of philosophy and shows, as I believe, that the method of formulating these problems rests on the misunderstanding of the logic of our language.
> [···]
> I am, therefore, of the opinion that the problems have in essence been finally solved. And if I am not mistaken in this, then the value of this work secondly consists in the fact that it shows how little has been done when these problems have been solved.

The importance of Wittgenstein's preface for the understanding of his book—a text all too often ignored by Wittgenstein scholars—becomes clear in a letter Wittgenstein wrote in 1919 to Ludwig Ficker whom he had approached as a potential publisher of the book:

Ich wollte einmal in das Vorwort einen Satz geben, der nun tatsächlich nicht darin steht, den ich Ihnen aber jetzt schreibe, weil er Ihnen vielleicht ein Schlüssel sein wird: Ich wollte nämlich schreiben, mein Werk bestehe aus zwei Teilen: aus dem, der hier vorliegt, und aus alledem, was ich nicht geschrieben habe. Und gerade dieser zweite Teil ist der Wichtige. Es wird nämlich das Ethische durch mein Buch gleichsam von innen her begrenzt; und ich bin überzeugt, daß es, *streng*, NUR so zu begrenzen ist. Kurz, ich glaube: Alles das, was *viele* heute *schwefeln*, habe ich in meinem Buch festgelegt, indem ich darüber schweige. Und darum wird das Buch, wenn ich mich nicht sehr irre, vieles sagen, was Sie selbst sagen wollen, aber Sie werden vielleicht nicht sehen, daß es darin gesagt ist. Ich würde Ihnen nun empfehlen das *Vorwort* und den *Schluß* zu lesen, da diese den Sinn am unmittelbarsten zum Ausdruck bringen.[19]

I once intended to put a sentence into the preface of my book which is now not included but I am writing it to you now as it might serve you as a key: I intended to write that my book consists of two parts, of what is in front of you and all that I have not written, and it is this second part which is the important one. My book is in a way delimiting ethics from within and I am convinced that *strictly* it can ONLY be delimited IN THAT WAY. In short, I am convinced that I have determined what *many* today *gossip* about by remaining quiet. The book will therefore - if I am not mistaken - tell you a lot of what you yourself would want to say, but you might not see that it is being said. I would now like to advise you to read the *preface* and the *end* as they express the meaning in the most immediate fashion”.‾

The last sentences of his book make clear what Wittgenstein meant when he said that by the propositions of his book he has once and for all solved all the problems caused by the misunderstanding of the logic of language, that with his book he has disentangled the knots in our minds, just as we would have to disentangle the knots in wool messed up by a cat if we want to continue knitting. But this necessary and complicated task loses its meaning, its Sinn, the moment we have completed the task, that is, the moment we have disentangled the mess:

Meine Sätze erläutern dadurch, daß sie der, welcher mich versteht, am Ende als unsinnig erkennt, wenn er durch sie - auf ihnen - über sie hinausgestiegen ist.
(Er muß sozusagen die Leiter wegwerfen, nachdem er auf ihr hinaufgestiegen ist.)
Er muß diese Sätze überwinden, dann sieht er die Welt richtig.[20]
My propositions are elucidatory in this way: he who understands me finally recognizes them as senseless, when he has climbed out through them, on them, over them.
(He must so to speak throw away the ladder, after he has climbed up on it.)
He must surmount these propositions; then he sees the world rightly.

With the completion of the *Logisch-Philosophische Abhandlung*, the *Tractatus*—the only philosophical text Wittgenstein published in his lifetime—he was convinced that he had exhausted his means to do philosophy, that he had lost his vocation of philosophy. He told Frank Ramsey, the translator of his book who had visited him in Austria, where Wittgenstein was working as a schoolmaster, that:

[…] no one can do more than five or ten years work at philosophy. (His book took seven.)[21]

This loss of his vocation to do philosophy, which marks a break but not a discontinuity in Wittgenstein's work, lead to the problematic misunderstanding of the 'two (or even three) Wittgenstein': the 'early Wittgenstein' of the *Tractatus* supposedly dismissed by the 'later Wittgenstein' of the *Philosophical Investigations*:

After the war—Wittgenstein had joined the Austrian army as a volunteer in August 1914—he was searching for his new vocation, his Berufung, his Beruf. Possibly inspired by teachers amongst his fellow prisoners of war in Italy and possibly influenced by Tolstoy's *Kurze Auslegung des Evangeliums*, *The Gospel in Brief*, which he had read during the war, he decided to become an elementary schoolteacher and on his return to Austria in August 1919 he began a teacher training course in Vienna.

For ten years he remained in Austria, working as a schoolmaster in small villages at the Semmering in Lower Austria and as a gardener; from 1926 to 1928 he worked as an architect, building a house for his sister Margarete Stonborough in Vienna. During that time he met—arranged by his sister Margarete—the philosopher Moritz Schlick and soon participated in the discussions of a group of people around Schlick, which later became known as the Vienna Circle. This and in particular an exchange of letters with Ramsey, mediated by Moritz Schlick in July 1927 on the definition of identity, rekindled Wittgenstein's interest in philosophy, his return to philosophy.

On 18th January 1929—after the completion of his house in the autumn of 1928—Wittgenstein returned to Cambridge where on 2nd February he resumed his philosophical writings, relating to a central topic from his first book, the *Tractatus*:

> Warum ist die Philosophie so kompliziert? Sie sollte doch *ganz* einfach sein? Die Philosophie löst Knoten in unserem Denken auf die wir unsinnigerweise hinein gemacht haben; dazu muß sie aber ebenso komplizierte Bewegungen machen wie diese Knoten sind. Obwohl also das *Resultat* der Philosophie einfach ist kann es nicht ihre Methode sein dazu zu gelangen. In der Wissenschaft ist ein Resultat so einfach oder so kompliziert wie die Methode durch die wir dazu gelangen. Die Kompliziertheit der Philosophie ist nicht die ihrer Materie sondern die unseres verknoteten Verstandes.[22]
>
> Why is philosophy so complicated? It ought to be *entirely* simple. Philosophy unties the knots in our thinking that we have, stupidly, put there. To do this it must make movements that are just as complicated as these knots. Although the *result* of philosophy it simple, its method cannot be simple if it is to succeed. In science a result is as simple or as complicated as the method by which is has been achieved. The complexity of philosophy is not a complexity of its subject matter, but that of our knotted mind.

After his return to Cambridge and to philosophy Wittgenstein's plan was to publish a second book in which he wanted to present his new thoughts after the *Tractatus*. In a series of drafts for a preface for this book he describes on 6 November 1930 the main aim of his rekindled interest in philosophy: His striving for clarity and not, what is the common aim of philosophy, to erect any new thought-buildings, but to investigate the foundations of existing ones, not only those of mathematics but in general those of the intellectual structures of his time:

> Unsere Zivilisation ist durch das Wort Fortschritt charakterisiert. Der Fortschritt ist ihre Form nicht eine ihrer Eigenschaften daß sie fortschreitet. Sie ist typisch aufbauend. Ihre Fähigkeit ist es ein immer komplizierteres Gebilde zu konstruieren. Und auch die Klarheit dient doch nur wieder diesem Zweck und ist nicht Selbstzweck.
>
> Mir dagegen ist die Klarheit die Durchsichtigkeit Selbstzweck.
>
> Es interessiert mich nicht ein Gebäude aufzuführen sondern die Grundlagen der möglichen Gebäude durchsichtig vor mir zu haben.
>
> Mein Ziel ist also ein anderes als das der Wissenschaftler und meine Denkbewegung von der ihrigen verschieden.[23]
>
> Our civilisation is characterised by the word progress. Progress is its form, not one of the properties that allow it to make progress. Typically, it builds. Its ability is to build a more and more complex structure. And clarity, again, is only a means to this end and not an end in itself.
>
> For me however, clarity and transparency are ends in themselves.
>
> I am not interested in erecting a building but in having the foundations of possible buildings transparently before me.
>
> So my aim is different from that of the scientists and my thoughts move differently from theirs.

Another important aspect of Wittgenstein's philosophy is his continuous search for truth, his distinction between what is real from the fantastic, be it in the form of belief or superstition or of 'Bildung' and 'Verbildung', that is of cultivation and corruption or the distinction between being awake and being asleep.

> Elizabeth Anscombe, Wittgenstein' s friend and pupil and one of his three heirs told me the following story of an encounter with Wittgenstein:
>
> > Wittgenstein: "Why do people say that it was natural to think that the sun went round the earth rather than that the earth turned on its axis?" I (Anscombe) replied: "I suppose, because it looked as if the sun went round the earth." "Well," he asked, "what would it have looked like if it had LOOKED as if the earth turned on its axis?"

This resembles Brecht's distinction between sehen und glotzen, between seeing, not just with your eyes, and gawping in his play *Life of Galileo*, as he is explaining the 12-year old son of his housekeeper the model of the universe by Kopernikus:

> Andrea: Aber ich sehe doch, daß die Sonne abends woanders halt als morgens. Da kann sie doch nicht stillstehen! Nie und nimmer.
> Galilei: Du siehst! Was siehst Du? Du siehst gar nichts. Du glotzt nur. Glotzen ist nicht sehen.[24]
>
> Andrea: But I can see with my own eyes that the sun goes down in a different place from where it rises. So how can it stand still? Of course it can't.
> Galileo: You can see, indeed! What can you see? Nothing at all. You just gawp. Gawping isn't seeing.[25]

About the distinction between being awake and being asleep Wittgenstein wrote in 1930:

> Zum Staunen muß der Mensch - und vielleicht Völker - aufwachen. Die Wissenschaft ist ein Mittel um ihn wieder einzuschläfern.[26]
>
> To be astounded man - and possibly peoples - has to wake up. Science is a means to put them asleep again.

This distinction between being awake or in a dream, between being conscious and being unconscious, the distinction between the real and the fantastic is a topic which reoccurs throughout Wittgenstein's work; this distinction is the theme of the last sentences he wrote on 27 April 1951, hours before he lost consciousness and died:

Wenn Einer glaubt, vor wenigen Tagen von Amerika nach England geflogen zu sein, so glaube ich, daß er sich darin nicht *irren* kann.
Ebenso, wenn Einer sagt, er sitze jetzt am Tisch und schreibe.
[…]
„Aber wenn ich mich auch in solchen Fällen nicht irren kann, – ist es nicht möglich, daß ich in der Narkose bin?" Wenn ich es bin und wenn die Narkose mir das Bewußtsein raubt, dann rede und denke ich jetzt nicht wirklich. Ich kann nicht im Ernst annehmen, ich träume jetzt. Wer träumend sagt „Ich träume", auch wenn er dabei hörbar redete, hat so wenig recht, wie wenn er im Traum sagt „Es regnet", während es tatsächlich regnet. Auch wenn sein Traum wirklich mit dem Geräusch des Regens zusammenhängt.[27]
If someone believes he flew from America to England a few days ago, then, I believe, he cannot be making a *mistake*.
Just as someone who says that he is at this moment sitting at a table, writing.
[…]
"But even if in such cases where I cannot be mistaken, isn't it possible that I am under anaesthesia?" If I am and if the anaesthesia has robbed me of my consciousness, then I am not now really talking and thinking. I cannot seriously assume that I am at this moment dreaming. Someone who, dreaming, says "I am dreaming", even if he speaks audibly in doing so, is no more right than if he said in his dream "it is raining", while it was in fact raining. Even if his dream were actually connected with the noise of the rain.

12.5 Wittgenstein's Philosophical Method

In the eighth manuscript volume, *Bemerkungen zur philosophischen Grammatik*, *Remarks on Philosophical Grammar*, that Wittgenstein had written since his return to Cambridge, he contrasts on 1 November 1931 his philosophical method to those of the intellectuals of his time, specifically to his friend Frank Ramsey, who had died aged 26 in January 1930:

Ramsey war ein bürgerlicher Denker. D.h. seine Gedanken hatten den Zweck die Dinge in einer gegebenen Gemeinde zu ordnen. Er dachte nicht über das Wesen des Staates nach - oder doch nicht gerne - sondern darüber wie man *diesen* Staat vernünftig einrichten könne. Der Gedanke daß dieser Staat nicht der einzig mögliche sei beunruhigte ihn teils, teils langweilte er ihn. Er wollte so geschwind als möglich dahin kommen über die Grundlagen – *dieses* Staates nachzudenken. Hier lag seine Fähigkeit und sein eigentliches Interesse; während die eigentlich(e) philosophische Überlegung ihn beunruhigte bis er ihr Resultat (wenn sie ein' s hatte) als trivial zur Seite schob.[28]

und sieben Bemerkungen später:

Der Philosoph ist nicht Bürger einer Denkgemeinde. Das ist, was ihn zum Philosophen macht.[29]

Ramsey was a bourgeois thinker, i.e. his thinking aimed at clearing up things in some particular community. He did not reflect on the very nature of the state - or at least he did not like doing so - but on how *this* state might reasonably be organised. The idea that this state might not be the only possible one partly unsettled and partly bored him. He wanted to get down as quickly as possible to reflecting on the foundations – of *this* state. This was what he was good at and what really interested him; whereas real philosophical reflection unsettled him until he put its result (if it had one) as trivial aside.

and seven remarks later:

The philosopher is not a citizen of any community of ideas. That is what makes him a philosopher.

At that time and in the same context Wittgenstein described what, in his view, is the central, the real problem of philosophy: Following an idea, an ideal, always aspired to but never reached like the 'Holy Grail' or in the German Romantic 'die Blaue Blume', the 'Blue Flower', instead of our actual needs unser eigentliches Bedürfnis[30]. Wittgenstein compares 'following an ideal' with an attempt to grasp an infinite extension like the ladders in Jacob's dream, which is not possibly to man, only to God. Instead, he suggests, philosophy should concentrate on soluble problems, which he compares to an infinite number of finite cross-sections of an infinite, an endless strip:

Die Unruhe in der Philosophie entsteht dadurch/kommt daher/, daß die Philosophen die Philosophie falsch ansehen, oder falsch sehen, nämlich gleichsam in (unendliche) Längsstreifen zerlegt statt in (endliche) Querstreifen. Diese Umstellung der Auffassung macht die *größte* Schwierigkeit. Sie wollen also gleichsam den Unendlichen Streifen erfassen und klagen daß es/dies/ nicht Stück für Stück möglich ist. Freilich nicht wenn man unter einem Stück einen endlosen Längsstreifen versteht. Wohl aber wenn man einen Quersteifen als Stück/ganzes definitives Stück/ sieht. – Aber dann kommen wir ja mit unserer Aufgabe/Arbeit/ nie zu Ende!
Gewiß/Freilich/ nicht, denn sie hat ja auch
keins.[31]
The disquiet in philosophy arises/comes about/ because philosophers view or see philosophy in the wrong way, as if it were dissected into (infinite) longitudinal strips rather than into (finite) cross-sections. This change of perception causes the *greatest* difficulties. They want to grab the infinite strip and then lament the fact that this cannot be done piece by piece. It is indeed impossible, if one thinks of a piece as an infinite longitudinal strip, but possible if a cross-section is seen as a piece/a complete, definitive piece/. – But then we never come to an end with our task. Quite so, since there is no end to it.

In 1930 in a draft for a lecture, which Wittgenstein wrote, like all his lecture preparations in English, he prepared his students what to expect from his lectures and for the significance of his philosophy for science, in particular mathematics. He showed them how to ask the right questions, that is how to look at philosophical problems in the right way and most importantly how to dissolve, i.e. remove the knots and confusions that we have put into our language, into our mind and into our thinking:

What I should like to get you to do is not to agree with me in particular opinions but to investigate the matter in the right way. To notice the interesting kind of things (i.e. the things which will serve as keys if you use them properly).

What different people expect to get from religion is what they expect to get from philosophy.

I don' t want to give you a Definition of Philosophy but I should like you to have a very lively idea as to the characters of philosophic problems. If you had, by the way, I could stop/start/ lecturing at once.

To tackle the philosophic problem is difficult as we are caught in the meshes of language.

"Has the universe an end/beginning/ in Time" (Einstein)

You would perhaps give up Philosophy if you knew what it is. You want explanations instead of wanting descriptions. And you are therefore looking for the wrong kind of thing.

Philosophic questions, as soon as you boil them down to . . . . change their aspect entirely. What evaporates is what the intellect cannot tackle.[32]

12.6 Wittgenstein and Science

Wittgenstein saw himself as a naturalist, a scientist; in his first book, the *Tractatus* he presents philosophy, his philosophy as a form of natural science, that is a descriptive science based on 'natural or ordinary language' related to our actual need, unser eigentliches Bedürfnis[33]:

Die richtige Methode der Philosophie wäre eigentlich die: Nichts zu sagen, als was sich sagen läßt, also Sätze der Naturwissenschaft [···][34]

The right method of philosophy would be this. To say nothing except what can be said, *i.e.* the propositions of natural science [···]

He specifies this in July 1931, while he was working on his second book-project, with a picture resembling Novalis' destinction of what he calls 'Scientologisch', descriptive science in contrast to 'Scientifisch', the explaining form of science:

[···] daß ich nämlich mein Buch mit einer Naturbeschreibung d.h. Überhaupt mit einer Beschreibung einer Situation beginnen sollte.[35]

[···] that I should begin my book with a description of nature that is a description of a situation.

In a remark immediately after his return to Cambridge in 1929 and to writing philosophy he shows that his thinking in science is not that of the scientists of his time, by comparing his fruitful conversations with Frank Ramsey on logic with the difficulties he had talking to him about science:

Ich habe sehr genußreiche Diskussionen mit Ramsey über Logic etc.
[…]
Ich gehe in der Wissenschaft nur gern allein spazieren.[36]

I have most enjoyable discussions with Ramsey on logic etc.
[…]
In science I prefer to walk on my own.

Wittgenstein's criticism of science, his disgust with the sciences of his time—which applies more so to today—becomes particularly poignant after the dropping of the atom bombs on Hiroshima and Nagasaki on 6 and 9 August 1945. His critique of a science which lead to a development like that of the atomic bomb built by specialists following the laws of nature, reminds us of Brecht, who argued in his didactic play *Leben des Galilei*, *Life of Galileo*, that Galileo had betrayed science:

G. gab den eigentlichen Fortschritt preis, als er widerrief, er ließ das Volk im Stich, die Astronomie wurde wieder ein Fach, Domäne der Gelehrten, unpolitisch, isoliert.[37]

G. abandoned real progress when he retracted, he abandoned the people, astronomy became a discipline again, a domain of the academics - apolitical, isolated.

Brecht's moral argument—exemplified in his play the *Life of Galileo*—is, that by relegating science from a means for everyone to understand the world better and with it ourselves into a subject-matter for specialists, brings sciences into conflict with men and with the world: Only in the context of specialisation a development such as the atom bomb is conceivable, the world and everything that exists on it does not need the bomb.

As an example for a consequence of specialisation resulting in a distorted perception of our world, of reality, Brecht refers to the warning Einstein wrote to the president of the United States, Harry Truman that the atom bomb, which originally he had advised to be built, must not be released to other powers, least of all to Russia:

Das brilliante Fachgehirn, eingesetzt in einen schlechten Violinspieler und ewigen Gymnasiasten mit einer Schwäche für Generalisierungen in der Politik.[38]

The brilliant brain of an expert implanted into a bad violin player and eternal grammar-school boy with a weakness for generalisations in politics.

Wittgenstein's view, although it shows at first sight some resemblances to that of Brecht, is essentially different. He argues that we only understand the world as far as we understand ourselves. On 1 August 1946 he wrote about Einstein and other scientists and scholars of his time:

> Je weniger sich Einer selbst kennt und versteht um so weniger groß ist er, wie groß auch sein Talent sein mag. Darum sind unsre Wissenschaftler nicht groß. Darum sind Freud, Spengler, Kraus, Einstein nicht groß.[39]
> The less well someone knows and understands himself the less great he is, however great his talent. For this reason our scientists are not great. For this reason Freud, Spengler, Kraus, Einstein are not great.

At the same time, on 19 August 1946, Wittgenstein made his only known remark about the atom bomb which at first appears to be rather disturbing: For Wittgenstein the bomb is the consequence of a science that has gone wrong. But he sees in the very existence of the bomb—not in its use—a kind of hope that it might bring to an end a science, which had perverted itself by its quasi religious, rule-following self-righteousness, by its specialisation and, by its ever closer association with industry:

> Die hysterische Angst, die die Offentlichkeit jetzt vor der Atom-Bombe hat, oder doch ausdrückt, ist beinahe ein Zeichen, daß hier einmal wirklich eine heilsame Erfindung gemacht worden ist. Wenigstens macht die Furcht den Eindruck der, vor einer wirklich wirksamen bitteren Medizin. Ich kann mich des Gedanken nicht erwehren: wenn hier nicht etwas Gutes vorläge, würden die *Philister* kein Geschrei anheben. Aber vielleicht ist auch das ein kindischer Gedanke. Denn alles, was ich meinen kann, ist doch nur, daß die Bombe das Ende, die Zerstörung eines gräßlichen Übels, der ekelhaften, seifenwäßrigen Wissenschaft, in Aussicht stellt. Und das ist freilich kein unangenehmer Gedanke; aber wer sagt, was auf eine solche Zerstörung *folgen* würde? Die Leute, die heute gegen die Erzeugung der Bombe reden, sind freilich der Auswurf der Intelligenz, aber auch das beweist nicht unbedingt, daß das zu preisen ist, was sie verabscheuen.[40]
> The hysterical fear of the atom bomb the public now has, or at least expresses, is almost a sign that here, for once, a really salutary discovery has been made. At least the fear gives the impression of being fear in the face of a really effective bitter medicine. I cannot rid myself of the thought that, if there were not something good here, the *philistines* would not be making an outcry. But perhaps this too is a childish idea. For all I can mean really is that the bomb creates a prospect of the end, the destruction of a ghastly evil, the disgusting dish-water science and certainly that is not an unpleasant thought; but who is to say what would *follow* such destruction? The people now making speeches against producing the bomb are undoubtedly the dregs of the intelligentsia, but even that does not prove beyond question that what they abominate is to be praised.

But less than a year later, in August 1947 he had given up this hope as he sees resignedly that the kind of science that he so detests has established itself firmly in our society, in our culture by associating itself ever closer with industry. This development, which advanced in the context of the two world wars, has not only persisted into our time but has gained speed, and education is turning more and more into a kind of drill or training to make young people fit for business, for our kind of society:

Es könnte sein, daß die Wissenschaft und Industrie, und ihr Fortschritt, das Bleibendste der heutigen Welt ist. Daß jede Mutmaßung eines Zusammenbruchs der Wissenschaft und Industrie einstweilen, und auf *lange* Zeit, ein bloßer Traum sei/ist/, und daß Wissenschaft und Industrie noch und mit unendlichem Jammer die Welt einigen werden, ich meine, sie zu *einem* Reich zusammenfassen werden, in dem/welchem/ dann freilich alles eher als der Friede wohnen wird.

Denn die Wissenschaft und Industrie entscheiden doch die Kriege, oder so scheint es.[41]

It may be that science and industry, and their progress, are the most enduring thing in the world today. That any conjecture of a collapse of science and industry for now and for a *long* time to come, is simply a dream, and that science and industry will unite the world causing infinite misery, merging it into *one* single realm, in which however anything but peace will reign.

For it is science and industry that decide wars, or so it seems.

Already in January 1947, Wittgenstein had not only expressed his doubts about the usefulness of the ever closer coalition between science and technology, but saw it as endangering the very existence of man on this planet and he argues that our trust in progress of science and industry is a dangerous fallacy:

Die apokalyptische Ansicht der Welt ist eigentlich die, daß sich die Dinge nicht wiederholen. Es ist z.B. nicht unsinnig, zu glauben, daß das wissenschaftliche und Technische Zeitalter der Anfang vom Ende der Menschheit ist; daß die Idee vom Großen Fortschritt eine Verblendung ist, wie auch von der endlichen Erkenntnis der Wahrheit; daß an der wissenschaftlichen Erkenntnis nichts Gutes oder Wünschenswertes ist und daß die Menschheit, die nach ihr strebt, in eine Falle läuft. Es ist durchaus nicht klar, daß dies nicht so ist.[42]

The truly apocalyptic view of the world is that things do not repeat themselves. It is not e.g. absurd to believe that the scientific and technological age is the beginning of the end for humanity, that the idea of the Great Progress is a bedazzlement, along with the idea that the truth will ultimately be known; that there is nothing good or desirable about scientific knowledge and that humanity, in seeking it, is falling into a trap. It is by no means clear that this is not how things are.

But today industry and politics have appeased us: They have made us believe that the atom bomb is, after all, a good thing that has granted us more than 70 years of peace; that without its peaceful variant, nuclear power, the earth would have warmed up even more quickly and large parts of the inhabited world would by now be submerged under water; that without the development of the neurotoxins—invented during the wars to kill people—into agrochemicals, so-called 'plant protection products', many people today would be starving; etc., etc., etc..

In contrast Wittgenstein argues that science has to free itself from its links with and its dependence on industry. Like Lessing in respect to rule-following religion, Wittgenstein argues that sciences must free themselves from their 'primers', from their orthodox, rule-following thinking, in which, as Lessing writes, the enforcement of rules and laws is ensured by penalties and rewards, but where questions about the rules themselves are not allowed.

Results achieved in this way are simple, they are either right or wrong; they are not to be considered under any other aspect. They are, so to speak, free of any ethical or moral 'ballast'. And if this opens up possibilities that cause conflicts with and within our world—as when Edward Teller argued that his hydrogen bomb was merely the result of the strict application of the laws of nature—society steps in with orders and interdictions. But a science that acts in a morally responsible way is in no need of an 'ethics commission'; to prevent science from getting into conflict with our world, the necessary moral and ethical criteria have to derive from within science itself and only that way will they be effective.

In remark 107 of his *Philosophical Investigations*, Wittgenstein gives an example how we can and must overcome our prejudices, the way we see and describe our world:

> Je genauer wir die tatsächliche Sprache betrachten, desto stärker wird der Widerstreit zwischen ihr und unserer Forderung. (Die Kristallreinheit der Logik hatte sich mir ja nicht *ergeben*; sondern sie war eine Forderung.)[43]
>
> The more narrowly we view actual language, the sharper becomes the conflict between it and our requirement. (For crystalline purity of logic was, of course, not a *result of investigation*: it was a requirement.)

In the following remark 108—the key remark for the understanding of his *Philosophical Investigations*—Wittgenstein shows how by changing our view, our outlook onto the world we could overcome our ever-growing, so-called ecological problems, i.e. the problems and conflicts that we are having with and within our word. He argues that science and mathematics, as well as philosophy, have to surmount their own prejudices; that they must overcome their preconceived ideals as he had to overcome the ideal from his first book, the *Tractatus*, the problem that he calls in his *Investigations* the 'Kristallreinheit der Logik'[44], 'crystelline purity of logic', by replacing it with what he calls our actual, our real needs:

Das *Vorurteil* der Kristallreinheit kann nur so beseitigt werden, daß wir unsere ganze Betrachtung drehen. (Man könnte sagen: Die Betrachtung muß gedreht werden, aber um unser eigentliches Bedürfnis als Angelpunkt.)[45]

The *prejudice*/the *preconceived idea*/ of crystalline purity can only be removed by turning our whole view round. (One might say: the view must be rotated, but on the pivot/the axis of reference/ of our actual/of our real/ need.)

He illustrates in remark 116 what he means by rotating our point of view around the pivot of our actual needs, the transformation of philosophy from the fantastic to the real, from rule-following, from following an idea, an ideal, to the meaningful, to real:

Wenn die Philosophen ein Wort gebrauchen – „Wissen", „Sein", „Gegenstand", „Ich", „Satz", „Name" – und das *Wesen* des Dings zu erfassen trachten, muß man sich immer fragen: Wird denn dieses Wort in der Sprache, in der es seine Heimat hat, je so gebraucht?

Wir führen die Wörter von ihrer metaphysischen, wieder auf ihre alltägliche Verwendung zurück.[46]

When philosophers use a word – "knowledge", "being", "object", "I", "proposition", "name" – and try to grasp the *essence* of the thing, one must always ask oneself: is the word ever actually used in this way in the language in which it has its home?

We take words back from their metaphysical to their everyday use.

By 'our actual need', the pivot, or the axis of reference around which we have to rotate our view onto the world, Wittgenstein does not mean the needs (or the greed) of stakeholders or of industry nor of specialists or politicians, but the needs of the world as a whole, of everything that exists on it, not of mankind in particular. During the Great War, in September 1916, while serving as a soldier on the Austrian front in Galicia Wittgenstein wrote about how he saw the role and the significance of man, of mankind in our world:

> Das philosophische Ich ist nicht der Mensch, nicht der Menschliche Körper oder die Menschliche Seele mit den Psychologischen Eigenschaften, sondern das Metaphysische Subjekt, die Grenze (nicht ein Teil) der Welt. Der Menschliche Körper aber, *mein* Körper insbesondere, ist ein Teil der Welt unter anderen Teilen der Welt, unter Tieren, Pflanzen, Steinen etc. etc.
>
> Wer das einsieht, wird seinem Körper oder dem Menschlichen Körper nicht eine bevorzugte Stelle in der Welt einräumen wollen.
>
> Er wird Menschen und Tiere ganz naiv als ähnliche und zusammengehörige Dinge betrachten.[47]
>
> The philosophical I is not the human being, not the human body or the human soul with its psychological properties, but the metaphysical subject, the boundary (not a part) of the world. The human body, however, *my* body in particular, is part of the world among others, among animals, plants, stones etc., etc.
>
> He who realises this will not want to assign to their own or the human body in general a privileged place in the world.
>
> He will regard humans and animals quite naïvely as similar things that belong together.

In the history of mankind no generation has left more and greater debts to this world and to future generations than we—and I don't mean financial debts, but debts that we have piled up against our world, a world which we are holding in trust and of which we are only one part amongst a multitude of others.—

Is this the result of a misunderstanding? of our self-understanding? Of our Jewish-Christian history? a consequence from the primer the 'Old Testament', Genesis chapter one, verse 28:

> "Be fruitful and multiply and fill the earth and subdue it, and have dominion over the fish of the sea and over the birds of the heavens and over every living thing that moves on the earth." ?

Is it that today's science—despite of Darwin—is still based on this primer? Is it that science is still suffering the effect of such thinking? or both?: Aristotle (in todays perception)—Thomas of Aquinas—John Locke—…?

The ideal of 'the great progress' along with the ideal that 'truth will ultimately be known' presupposes that we can see the world from outside as in Plato's simile of the cave, but this is a point of view not given to man. But only from this perspective rules and regulations created/dictated by God could be derived and these would indeed be rules and regulations which are not to be discussed by man,—the very nature and character of a primer as described by Lessing.

In contrast, Wittgenstein is describing the world not from outside but from within; his view resembles that of *Aletheia*, the ancient Greek word for truth, that what Heidegger calls 'das in's Licht getretene' the 'unclosedness or unconcealedness'. Wittgenstein is not explaining the world, but he is describing and delimiting the world from within as he put it in his *Logisch-Philosophische Abhandlung*, *Tractatus*:

> *Die Grenzen meiner Sprache* bedeuten die Grenzen meiner Welt.[48]
>
> *The limits of my language* mean the limits of my world.

Notes

1. Wittgenstein *Philosophische Untersuchungen*, Frankfurt 1963, Bemerkung 108
2. Newton *General Scholium* an essay appended to the second edition of *Pricipia*
3. Wittgenstein *Philosophische Untersuchungen*, Frankfurt 1963, Bemerkung 108
4. Michael Nedo *Wittgenstein – Ein biographisches Album*, München 2014, p. 379
5. Wittgenstein *Culture and Value*, Oxford 1998, p. 42
6. Wittgenstein *Wiener Ausgabe*, Wien – New York 1993 ff., WA2.107.4
7. Michael Nedo *Wittgenstein – Ein biographisches Album*, München 2014, p. 407
8. Gotthold Ephraim Lessing *Die Erziehung des Menschengeschlechts*, Werke, Bd 7, *Theologiekritische Schriften* I und II, München 1970 ff., § 53, p. 488
9. ibid. § 51, p. 488
10. Wittgenstein *Culture and Value*, Oxford 1998, p. 70
11. Wittgenstein *Wiener Ausgabe*, Wien – New York 1993 ff., WA3.112.2
12. Michael Nedo *Wittgenstein – Ein biographisches Album*, München, p. 277
13. Wittgenstein *Wiener Ausgabe*, Wien – New York 1993 ff., WA5.23.1
14. ibid. WA5.157.3
15. Crispin Wright *Wittgenstein on the Foundations of Mathematics*, London 1980, p.
16. G.E. Moore *Philosophical Papers*, London 1951, p. 254
17. Michael Nedo *Wittgenstein – Ein biographisches Album*, München p. 262
18. Wittgenstein *Logisch-Philosophische Abhandlung*, London 1922, p. 28
19. Wittgenstein *Briefe an Ludwig von Ficker*, Salzburg, p. 97
20. Wittgenstein *Logisch Philosophische Abhandlung* 6.54, London 1922
21. Michael Nedo *Wittgenstein – Ein biographisches Album*, München, p. 222
22. Wittgenstein *Wiener Ausgabe*, Wien – New York 1993 ff., WA1.157.6
23. ibid. WA3.112.2
24. Bertold Brecht, *Leben des Gallei* (1955/56), Stücke 5, Frankfurt 1988, p. 342
25. Bertold Brecht, *Life of Galileo*, translated by John Willett
26. Wittgenstein *Wiener Ausgabe*, Wien – New York 1993 ff., WA3.109.1
27. Wittgenstein *Über Gewißheit*, Frankfurt 1984, Paragraph 675 f.
28. Wittgenstein *Wiener Ausgabe*, Wien – New York 1993 ff., WA4.172.7
29. ibid. WA4.173.7
30. Wittgenstein *Philosophische Untersuchungen*, Frankfurt 1963, Bemerkung 108
31. Wittgenstein *Wiener Ausgabe*, Wien – New York 1993 ff., WA4.173.6
32. ibid. WA3.VII

33. Wittgenstein *Philosophische Untersuchungen*, Frankfurt 1963, Bemerkung 108
34. Wittgenstein Logisch Philosophische Abhandlung 6.53, London 1922
35. Wittgenstein *Wiener Ausgabe*, Wien – New York 1993 ff., WA3.298.8
36. ibid. WA1.3.4
37. Bertold Brecht, *Leben des Gallei* (1955/56), *Stücke* 5, Frankfurt 1988, p. 342
38. ibid. p. 346
39. Wittgenstein *Culture and Value*, Oxford 1998, p. 53
40. ibid. p. 55–56
41. ibid. p. 72
42. ibid. p. 64
43. Wittgenstein *Philosophische Untersuchungen*, Frankfurt 1963, Bemerkung 107
44. ibid. Bemerkung 108
45. ibid. Bemerkung 108
46. ibid. Bemerkung 116
47. Wittgenstein Tagebücher, Frankfurt 1984, p. 177
48. Wittgenstein Logisch Philosophische Abhandlung 5.6, London 1922

Chapter 13
Showing in Wittgenstein's *ab*-Notation

Gregory Landini

13.1 Introduction

Perhaps it is not overly pedantic to say that one will find Wittgenstein's *Tractatus Logico-Philosophicus* very difficult even if one first understands Russell's philosophical logic. The lion's share of it presupposes its readers are familiar with Russell's work and the research program of his scientific method in philosophy which, he thought, offers the last best hope of genuine progress toward solutions/dissolution of many traditional philosophical problems. In Russell's view, the scientific method in philosophy is partly synthetic *a priori* (relying on *Principia Mathematica* or some emulation of the new mathematical logic) and partly synthetic a *posteriori* (relying on the physical sciences) to undermine the metaphysicians who introduce muddles into science (mathematical and/or natural) by offering arguments for the indispensability of abstract particulars and specialized kinds of necessity (mathematical, physical/causal, biological, teleological, etc.). Russell held (at least in, and likely earlier still) that mathematical logic is the essence of philosophy as a genuine science whose aim is to reveal that logical necessity is the only necessity. Of course, Russell didn't put it that way because he thought all notions of "necessity" are confused. But that is the upshot nonetheless. And Wittgenstein, in fact, did not hesitate to make just that same point quite starkly as follows:

> *TLP* 6.37 The only necessity that exists is logical necessity.
>
> *TLP* 6.357 Just as the only necessity that exists is logical necessity, so too the only impossibility that exists is logical impossibility.

Such passages from the *Tractatus* (see Wittgenstein 1922) suggest a rather broad alliance and good deal of agreement between Russell and Wittgenstein.

G. Landini (✉)
Department of Philosophy, University of Iowa, 268 English Philosophy Building, Iowa City, IA 52242-1408, USA
e-mail: gregory-landini@uiowa.edu

S. Wuppuluri and N. da Costa (eds.), *WITTGENSTEINIAN (adj.)*,
The Frontiers Collection, https://doi.org/10.1007/978-3-030-27569-3_13

No one doubts that many ideas in the *Tractatus* began in dialogue with Russell, and this is well-corroborated by Wittgenstein himself in a letter of 13 March 1919 in which he describes his work as a work which has solved "our" problems. The exact quote is this:

> I have written a book called *Logisch-Philosophische Abhandlung* containing all my work of the last 6 years. I believe I have solved our problems finally. … in fact you would not understand it without a previous explanation as it's written in very short remarks. (This of course means that *nobody* will understand it; although I believe, it's all as clear as crystal. But it upsets all our theory of truth, of classes, of numbers and all the rest. [1]

When Wittgenstein speaks of "our" theory, the two would naturally think of their alliance in framing a scientific method in philosophy for eliminating ontological commitment to abstract particulars, e.g., classes, numbers, functions, propositions, meaning proposition, etc., from every science (mathematical and empirical). But to properly understand the *Tractatus* requires one to understand the nature and import of this "our" and the nature of this new "science."

Now I think there is very good evidence to identify what Wittgenstein tells us is the "fundamental idea" of the *Tractatus* at *TLP* 4.0312 (that logical expressions, including logical particles, are not relation signs) with what he told Russell in a letter of 19 August 1919 is the "main point," namely its *Doctrine of Showing*.[2] So the central question can be put as whether there the fundamental idea is taken by Wittgenstein to be a component that can be in collaboration with Russell's scientific method in philosophy or was there a split from Russell; and if a split, when precisely did it occur. One may wonder about the historical chronology of Wittgenstein's "discoveries" (if you will) that led up to his *Doctrine of Showing*.

The question as to whether Wittgenstein's *Doctrine of Showing* was intended by him to be compatible with Russell's scientific method in philosophy is not so straightforward to determine. It is unfortunate that many simply assume it is in diametrical opposition to Russell.[3] In his 1913 *Notes on Logic*, Wittgenstein seems to express complete agreement with Russell's idea of a scientific method in philosophy with logic as its essence.[4] We find (Wittgenstein 1979, p. 106).

> Philosophy consists of logic and metaphysics: logic is its basis.
>
> Philosophy is the doctrine of the logical form of scientific propositions (not only of primitive propositions).
>
> The word philosophy" ought always to designate something over or under but not beside, the natural sciences.

---

[1] See Russell (1968, p. 162).

[2] The *Doctrine of Showing* is explicit in the 1914 *Notes Dictated to Moore*, and the "fundamental idea" is found in an entry dated 25 December 1914 of Wittgenstein's Notebooks 1914–1916. The identification of these is quite illuminating In 2007, I have argued at length in favor of this identification.

[3] Based on this assumption we find the recent anthology edited by J. Beale & I. J. Kidd (2017).

[4] The clearest expression of this is in Russel's 1914 book *Our Knowledge of the External World as a Field for Scientific Method in Philosophy*. In that book, Russell explicitly references Wittgenstein as an ally.

These comments seem quite consonant with Russell's scientific method in philosophy. In the *Tractatus,* we find very similar entries:

> *TLP* 4111 Philosophy is not one of the natural sciences. … whose place is above or below the natural sciences, not beside them.
>
> *TLP* 4.112 Philosophy aims at the logical clarification of thoughts. Philosophy is not a body of doctrine but an activity. A philosophical work consists essentially of elucidations.

But if *showing* were incompatible with a science of philosophy, then Wittgenstein must have changed his mind. I fear that is a position widely assumed without much justification. Much turns on the meaning of "scientific" in the phrase "scientific method in philosophy." What is a "science"? If one simply rejects Russell's thesis that what it means to be "genuine science" requires there being a body of truths of a field, then the apparent tension with *showing* completely vanishes. I doubt that Wittgenstein *ever* held that mathematics, logic and philosophy (when properly conceived) are "bodies of truths." If I'm right, then he never changed his mind at all. That is the view I wish to advocate in this paper, for I hold that the *Doctrine of Showing* is already implicit in Wittgenstein's *ab*-Notation from his 1913 *Notes on Logic* and letters to Russell of that year.

Of course, empirical science requires there to be a body of truths of its field. But the alliance between Wittgenstein and Russell certainly does not require Wittgenstein to agree with Russell in holding that that every science is a body of truths. Once *showing* is in place, Wittgenstein concluded that a genuine science of philosophy cannot be a body of truths. Neither, in Wittgenstein's view, can mathematics or logic be a body of truths. Wittgenstein makes it clear that he thinks what can be *shown* cannot be *said* (*TLP* 4.1212; *Notebooks* 1914–1916, p. 34 an 29 Nov. 1914 entry). All the same, mathematics and logic and philosophy, can, in Wittgenstein's view, be *genuine sciences* "over or under but not beside the natural sciences." That is because they can be regarded a practice of revealing (*elucidating*, *showing*) relational structures. This elucidation includes e.g. his "discovery" of his *ab*-Notation and its application to quantification with identity, his discovery of his N-operator Notation, his discovery that arithmetic consists in calculations of outcomes of equations involving recursively defined functions, etc.). These "discoveries" (if you will) do not reveal truths, but rather they implement *showing*. They *show* what Russell (in Wittgenstein's opinion) mistook to be important truths. For example, they *show* that logical necessity is the only necessity. Wittgenstein's position can properly be expressed by saying that "logic and mathematics and philosophy are genuine and unique science(s)" as long as his word "science" does not mean "natural science" and thus does not connote it being a body of truths. It is clearly misguided to think that Russell's imagined that philosophy is a natural science (adopting only empirical methods). There is nothing about Wittgenstein's 1913 *Notes on Logic* which precludes his being in alliance with Russell's *scientific* method in philosophy.

The central question before us is then the question of when Wittgenstein adopted his *Doctrine of Showing*? Now in describing his philosophical insight, the word "showing" (so far as is known) first appears in the 1914 *Notes Dictated to Moore*. So we know that he held it then. But is the idea present in 1913 *Notes on Logic*? If so, how would we go about recognizing it? Obviously, it is not an appropriate necessary condition that he use the word "show" in articulating his idea of *showing* versus *saying*. But what would help us recognize it? One way to recognize it is to understand that *showing* is clearly present in Wittgenstein quest for his N-operator notation. That is not to say that the idea of *showing* enters into his work only when he introduced the N-operator notation itself into his writings (and at present it is not known when that happened). The N-operator notation presupposes that Wittgenstein has come to accept *showing* and that he is struggling to find a notation that realizes *showing*. Moreover. it is obviously not an appropriate necessary condition for discovery of the idea of the N-operation notation that he use the symbol "$N(\bar{\xi})$" with its distinctive *xi*-bar. To find evidence of this idea in Wittgenstein's work, one has to first understand what the N-operator notation is trying to do. Only then can one be justified in holding that one has found something in Wittgenstein work that prefigures it. Without such an understanding, the historian is limited to philology. There is thus controversy at every turn. But as we see, N-notation was intended to express all and only logical equivalents of quantificaiton theory with identity in one and the same way. Based on that concrete interpretation, I have found an entry of 1915 in Wittgenstein's notebooks that contains the key idea of his use of his *xi*-bar.

The question of where in the manuscripts we find the origins of the N-operator notations is intimately connected to the question of how early Wittgenstein had the idea of *showing*. By the same reasoning, however, notations which Wittgenstein thought to perform *precisely* the work that he later assigns to his N-operator also presuppose that he had *showing* in mind. This point helps to push the date of his *showing* idea well earlier in time. And indeed, this is what I have found. In Landini (2007), I argued that Wittgenstein *ab*-Notation was explicitly thought by Wittgenstein to do that very work—namely, express logical equivalent wffs of quantification theory with identity in one and the same way. Thus, the appearance of Wittgenstein's claim that his *ab*-Notation is not simply an alternative notation for truth-functions but that it can do the work of capturing the whole of quantification theory with identity) marks a boundary after which we can feel confident that Wittgenstein has embraced *showing*. It may occur yet earlier. This marks an upper bound, not a lower. The upper bound, as we shall see, is October 1913.

13.2 The *ab*-Notation

In his *Autobiography*, Russell tells an engaging story about Wittgenstein attempting to describe idea of his *ab*-notation one night when he was visiting the Whiteheads.

> Whitehead described to me the first time that Wittgenstein came to see him. He was shown into the drawing room during afternoon tea. He appeared scarcely aware of the presence of Mrs. Whitehead, but marched up and down the room for some time in silence, as at last said explosively: "A proposition has two poles. It is *apb*." Whitehead, in telling me, said:

> "I naturally asked what are *a* and *b*, but I found I had said the wrong thing. '*a* and *b* are indefinable,' Wittgenstein answered in voice of thunder."[5]

For many years this story remained obscure. But happily, that *ab*-Notation is now much better understood. It is my thesis that the intent of his *ab*-Notation is to find a mode of expression in which all and only logical equivalents have the same representation. That is a reliable indicator having intuited his *Doctrine of Showing*.

If we are to have any hope of understanding Wittgenstein's attempt to apply his *ab*-Notation to quantification theory, we must first see how the *ab*-Notation is supposed to achieve this result for quantifier-free cases—the result, that is, that all and only logical equivalents (of the same propositional bases) have one and the same expression. And at the center of this is the ways in which *negation* can be be expressed in *ab*-Notation. Consider the following:

Wittgenstein writes the figure on the right as "*b-a p b-a*." The wires (as it were) are simply rearranged spatially in the figure on the left. Note as well that though we can never change poles *a p b*, etc on a primitive letter, *p*, standing alone, we can change their poles in context to form negations of conjunctions and negations of disjunctions and so on. For example, consider the *ab*-Notation picture for "*p* • *q*":

We can then write the following alternatives for "~(*p* • *q*)":

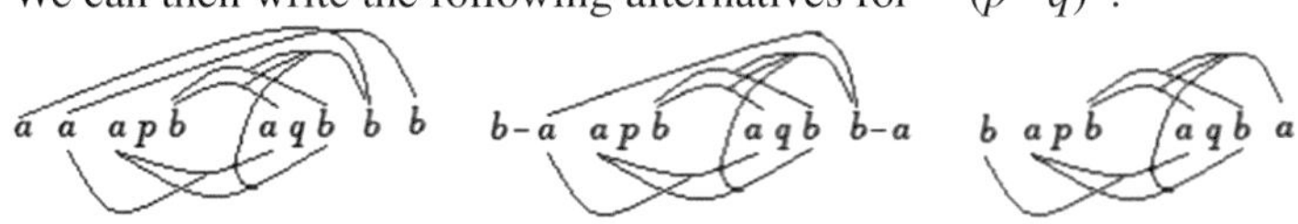

The picture on the far right simply switches the "*a*" and "*b*" poles in the picture of "*p* • *q*" to form the picture of "~(*p* • *q*)". Another switching of poles yields the following *ab*-Notation picture of "*p* • ~*q*"

A passage in the 1914 *Notes Dictated to Moore* makes the pole switching way of expressing negation clear. It comes in a discussion where Wittgenstein's raises a concern that his *ab*-Notation perhaps has a flaw and must be replaced by a *tf*-Notation. That is because nothing in using distinct poles indicates which pole, *a*-pole or *b*-pole, is the truth-pole (*t*-pole) and which is the falsehood pole (*f*-pole). He writes:

> It seems at first sight as if the *ab*-Notation must be wrong, because it seems to treat true and false as on exactly the same level. It must be possible to see from the symbols themselves that there is some essential difference between the poles, if the notion is to be right; and it seem as if in fact this was impossible.

---

[5]Russell (1968).

To illustrate this point, Wittgenstein explained as follows:

> Take $p \bullet q$ and q. When you write $p \bullet q$ in the *ab*-Notation, it is impossible to see from the symbol alone that q follows from it, for it you were to interpret the true-pole as the false, the same symbol would stand for $p \vee q$, form which q does not follow. But the moment you say which symbols are tautologies, it at once becomes possible to see from the fact that they are and the original symbol that q does follow.

Carrying out Wittgenstein's instruction, write "$p \bullet q$" as follows:

This pictures the truth-conditions of "$p \bullet q$" only if we imagine the *a*-pole to be the *t*-pole. If we uniformly[6] take the *b*-pole to be the *t*-pole, then this is the picture of "$p \vee q$." That's is, we get

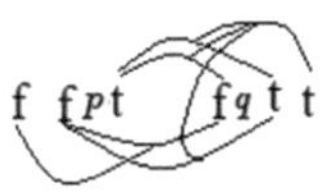

Wittgenstein's focus on negation is of central importance. Expressing negation by switching poles is a key component in the way *ab*-Notation is understood to represent logical equivalents in the *same* way.

Indeed, by pole-switching we can finally see precisely how it is that "$\sim(p \bullet q)$" is supposed to have the same *ab*-Notation as its logical equivalent "$\sim p \vee \sim q$" in accordance with a *DeMorgan* rule:

$\sim(p \bullet q)$

$\sim p \vee \sim q$

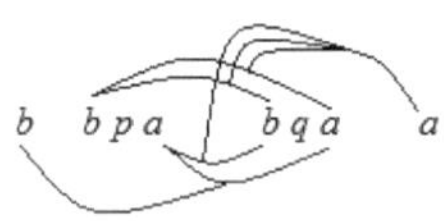

Having switched poles, the figures simply rearrange wires without altering their connectivities. Another case illustrating how switching *a*-pole and *b*-poles to express negation occurs with the *transposition* rule.

$p \supset \sim q$

$q \supset \sim p$

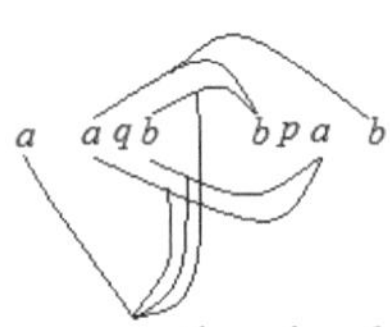

Having switched the poles to express negation, in *ab*-Notation, the *transposition* can be understood as simply a rotation of the figure. In this way, the pictures of "$q \supset \sim p$" and "$p \supset \sim q$" are the same in *ab*-Notation.

[6]The uniform impact on *apb* and *aqb* was made clear by Kevin Klement and Russell Wahl at the 2019 meeting of the Bertrand Russell Society.

In his April 1914 *Notes Dictated to Moore,* Wittgenstein explicitly connected what he is endeavoring to do with his *ab*-Notation with his notion of *showing*. The context, again, is Wittgenstein's worry that his *ab*-Notation must be replaced by a *tf*-Notation because of the problem of switching poles. He then goes on to write as follows:

That, when a certain rule is given, a symbol is tautological *shews* a logical truth.

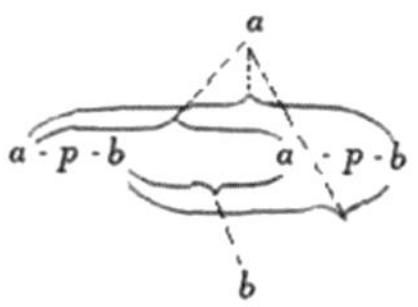

This symbol might be interpreted either as a tautology or a contradiction. In settling that it is to be interpreted as a tautology and not as a contradiction, I am not assigning a *meaning* to *a* and *b*; i.e., saying that they symbolize different things but in the same way. What I am doing is to say that the way in which the *a*-pole is connected with the whole symbol symbolizes in a *different way* from that in which it would symbolize if the symbol were interpreted as a contradiction…

The connection between *showing* and the *ab*-Notation is explicit. The purpose of the *ab*-Notation is to implement *showing*—i.e., to reveal that in *ab*-Notation tautologies are given in the notation itself. There is an oddity in the diagram, for its *a*-pole has a connection to both the *a*-pole and the *b*-pole of the proposition letter *p*. Indeed, if we interpret the *b*-pole as the *t*-pole and the *a*-pole as the *f*-pole, we get:

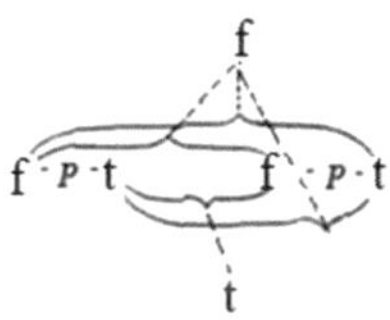

This is a jumble. What Wittgenstein was getting at is that if we were to read the *a*-pole as the *f*-pole and the *b*-pole as the *t*-pole of a tautology, then one depicts a contradiction instead of a tautology. His point—his self-imposed worry—is quite correct. There is nothing intrinsic to the *ab*-Notation that tells us what is to count as the true-pole. The *ab*-Notation is therefore replaced by the *tf*-Notation in the *Tractatus*.

I hold that the intent of the *ab*-Notation is to capture all and only logical equivalents (of the same propositional letters) in one and the same way. The *ab*-Notation is by no means unique in this feature. It is a salient feature of Venn's propositional diagrams. We can illustrate this below:

	p	q	r	$p \vee (q \bullet r)$	$(p \vee q) \bullet (p \vee r)$
1	t	t	t	t	t
2	f	t	t	t	t
3	t	f	t	t	t
4	f	f	t	f	f
5	t	t	f	t	t
6	f	t	f	f	f
7	t	f	f	t	t
8	f	f	f	f	f

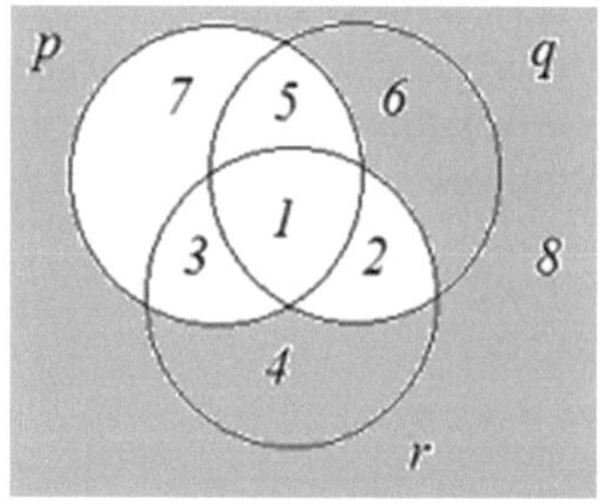

I do not know whether Wittgenstein was familiar with Venn's 1880's propositional diagrams. Venn's diagrams have precisely the feature Wittgenstein was seeking in his *ab*-Notation—namely, all and only logical equivalents (of the same propositional letters) have the same representation. It should be clear that the same thing is accomplished (in the quantifier free case) by Wittgenstein's truth-tabular notation. To picture the truth-conditions of "$p \vee (q \bullet r)$" the truth-tabular notion offers

$$(t, t, t, f, t, f, t, f)(p, q, r).$$

This is clearly the same picture of the truth-conditions for "$(p \vee q) \bullet (p \vee r)$". The truth-tabular representations picture all and only (quantifier-free) logical equivalents (of the same propositional letters) in the same way. (Of course, the truth-tabular notation requires a convention be adopted on the organization of the combinations of t and f. Various conventions may be adopted and none deserve any privileged status. Wittgenstein does not tell us which he prefers, though his examples are suggestive.)

It might at first appear, however, that the *ab*-Notation has a limitation which is not present in the Venn diagrammtical technique. It is easy to fall into the mistake of thinking that the *ab*-Notation as requiring dyadic connectivity when, e.g., we have three propositions p, q, r and so on. This is not so. The point is revealed by Wittgenstein himself in the following (admittedly difficult) passage from his 1913 *Notes on Logic* (p. 102):

> In place of every proposition "p", let us write "${}^{a}_{b}p$": Let every correlation of propositions to each other or of names to propositions be effected by a correlation of their poles "a" and "b". Let this correlation be transitive. Then accordingly "${}^{a\text{-}a}_{b\text{-}b}p$" is the same symbol as "${}^{a}_{b}p$". Let n propositions be given. I call a "class of poles" of these propositions every class of n members, of which each is a pole of one of the n propositions, so that one member corresponds to each proposition. I then correlate with each class of poles one of the two poles (a and b). The sense of the symbolizing fact thus constructed I cannot define, but I know it.

Wittgenstein is pointing out the one does not have to combine the polar connections dyadically. One can do it in a way that parallels Venn's propositional diagrams. In the dyadic case, this is clear, we have:

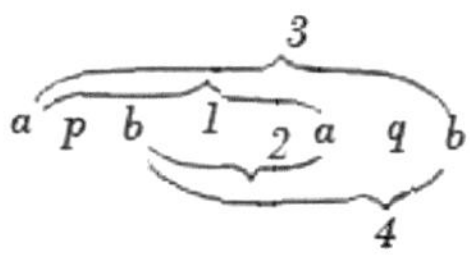

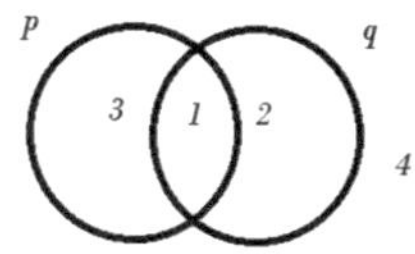

But consider, for example, the combinations of the poles of three propositions, *p*, *q*, *r*. This happens very easily in the Venn diagrams where all and only the proper combinatorial conditions are given by the areas of the overlapping circles. But it can be recovered in *ab*-Notation as we see above. This is just the basic structure of connection for all the combinations of three propositions *p*, *q*, *r* and it parallels exactly the Venn diagram for the same. In order to *say* something, one shades regions and leaves some open in a Venn diagram. In this *ab*-Notation, the analog of Venn's shading would be to simply squiggle out this or that connecting line. Alternatively, one can introduce an *a*-pole and a *b*-pole with connecting lines to what would have been the squiggled out lines.

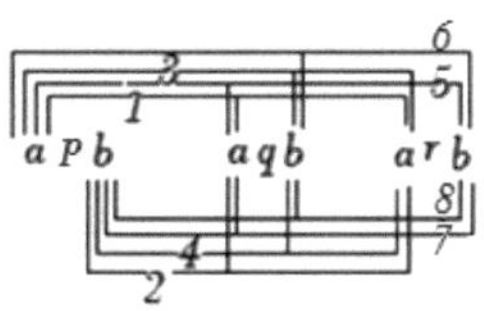

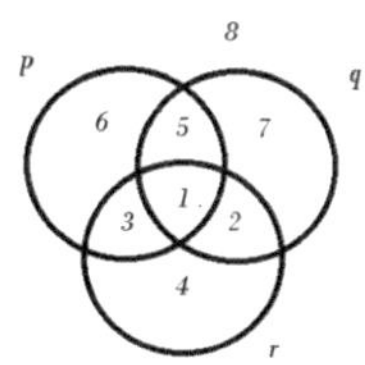

13.3 Quantification and the *ab*-Notation

It is easy to see why the truth-tabular notation is inadequate to represent quantificational *wff*s. An increase in the size of the domain of interpretation of the quantifiers will upturn the positioning of the *t*'s and *f*'s in any truth-tabular notation. There is no stability in their position as the domain's size increases (unless the language contains only monadic predicates). Wittgenstein hoped that an application of the *ab*-Notation to quantification avoids this problem entirely. It doesn't rely upon any commitments to the (non-empty) size of a domain.

In his 1913 *Notes on Logic* he clearly thought that he had found a way. Wittgenstein had written about it to Russell as follows:

> The application of the *ab*-notation to apparent variable propositions becomes clear if we consider that, for instance, the proposition "for all x, φx" is to be true when φx is true for all x's and false when φx is false for some x's. We see that *some* and *all* occur simultaneously in the proper apparent variable notation. The notation is:
>
> for $(x)\varphi x$: a-(x)-$a\varphi xb$-$(\exists x)$-b and
>
> for $(\exists x)\varphi x$: a-$(\exists x)$-$a\varphi xb$-(x)-b
>
> Old definitions now become tautologous.[7]

[7]Ludwig Wittgenstein, *Notebooks 1914–1916,* p. 96.

Interestingly, Russell brought Wittgenstein's 1913 *Notes on Logic* with him when he gave his 1914 lectures at Harvard University. Costello's notes for 11 April 1914 includes this:

> A proposition not only denotes a fact but is a fact.
>
> We want to say not-not- p is identical with p. So *not* does not occur in not p. "not" simply *reverses the sense*.
>
> $(x).\ \phi x =$ (sense = direction of meaning)
>
> $(x).\ \phi x =$ (sense = direction of meaning)
>
> $t\ (x)\ (t\ \phi x\ f)\ (\exists x)\ f$
>
> $(\exists x)\ \phi x$
>
> $t\ (\exists x)\ (t\ \phi x\ f)\ (x)\ f$[8]

Theses echo entries in Wittgenstein's *Notes on Logic*, and we see that the *ab*-Notation was changed to the *tf*-Notation. Unfortunately, there is no record at all of it being discussed in the class. Admittedly, it is hard to decipher this. Potter (2009) thinks it "… is plainly so half-baked that one suspects he had not really thought about the matter at all and simply made something up on the spot when Russell asked him about it" (p. 181). Lately Lampert (2017) has made a herculean effort, but realizes that his attempt to eliminate quantifiers in the spirt of an *ab*-Notation does not offer anything historically faithful. His rendition of what he regards as an *ab*-Notation for "$(x)(\exists y)Rxxy$" is horrifying. It turns out, however, that the matter is not "half baked" or in need of Lampert's reconstructions, and in fact it is quite representative of the "genius" that was Wittgenstein.

What has impeded the understanding of *ab*-Notation as applied to quantification is the misconception that it was to *replace* quantificational concepts. Quite to the contrary, it relies upon an independent understanding of quantification concepts. We see that the *a*-pole tells us that for *every* x we are to choose the *a*-pole of $a\ (\varphi x)\ b$. And naturally the *b*-pole of "$(x)\varphi x$" simply tells us that for *some* x we are to choose the *b*-pole of $a\ (\varphi x)\ b$. Similarly, the *a*-pole of "$(\exists x)\varphi x$" tells us that for *some* x we are to choose the *a*-pole of $a\ (\varphi x)\ b$, and the b-pole of "$(\exists x)\varphi x$" tells us that for *every* x we are to choose the

b-pole of $a\ (\varphi x)\ b$. We must not sever the quantifier from its pole designation and hence, for example, "(x)-a" or better $(x) \rightarrow a$" is a unit. Thus one might better write the following for "$(x)\varphi x$"

$$a\text{-}(x) \rightarrow a(\varphi x)b \leftarrow (\exists x)\text{-}b$$

(Hereafter, we won't use the arrows, but caution the reader to be aware.) There is no need to worry about cases such as "$(x)(y)\varphi xy$" being logically equivalent to "$(y)(x)\varphi xy$" since we can just take it as given by the *ab*-Notation itself that

$$(x)(y)\varphi xy$$

[8] See Linsky (2018).

$$a\text{-}(x)(y)\text{-}a(\varphi xy)b\text{-}(\exists y)(\exists x)\text{-}b$$

is to be regarded as the same as

$$(y)(x)\varphi xy$$
$$a\text{-}(y)(x)\text{-}a(\varphi xy)b\text{-}(\exists x)(\exists y)\text{-}b.$$

Contrary to Lambert's love's labors lost we get the following simple case:

$$(x)(\exists y)Rxxy$$
$$a\text{-}(x)(\exists y)\text{-}a(Rxxy)b\text{-}(y)(\exists x)\text{-}b.$$

The entire matter is remarkably straightforward.

What the *ab*-Notation as applied to quantifiers is designed to do is to reveal, as Wittgenstein put it, that "old definitions now become tautologous". The "old definition" he has in mind are, as Lampert himself noticed, those of *Principia Mathematica* section *9. And by their "becoming tautologous" he has in mind that all and only logical equivalents of quantification theory have one and the same *ab*-Notational tautologous representation. Interestingly, *Principia* offered a system *9 in which all subordinate quantifiers are defined in terms of quantifiers initially placed, and it demonstrated that the *9 system of quantification theory can deduce anything that can be deduced in the quantification theory set out at *Principia's* more orthodox section *10. Thus, since *10 is semantically complete (able to deduce as theorems every logical truth of quantification theory), so also is the system of *9. There is a sense in which the success of section *9 reveals that the logical truths of quantification theory (without identity) are "generalized tautologies" as Wittgenstein put it to Russell in a letter of 1913 (Wittgenstein 1979, p. 128).[9] But one must be careful. The success of *9 quite obviously does not entail that quantification theory is decidable. What it reveals is that every theorem of quantification theory can be deduced by beginning from some tautology, generalizing (universally or existentially as needed), switching the quantifiers (when permitted) and distributing them (by the definitions of *9). This does not yield a decision procedure for quantification theory. There is no recipe to determine what tautology from which to begin one's derivation.

Perhaps Wittgenstein hadn't understood this. Indeed, at first blush, it might seem that *9 does suggest a way to find a decision procedure. Consider the following:

$$(x)(\exists y)(\varphi x \supset \psi y) \supset (\exists y)(x)(\varphi x \supset \psi y).$$

The approach of *9 is to use definitions, pulling quantifiers to the front in pairs (assuming that x doesn't occur free in the *wff* φ and y does not occur free in the *wff* ψ. This yields:

[9] See Wittgenstein (1979).

$$(\exists x)(\exists y')(y)(x')(\varphi x \supset \psi y. \supset .\varphi x' \supset \psi y')$$

Next drop the quantifiers to get:

$$\varphi x \supset \psi y. \supset .\varphi x' \supset \psi y'.$$

That is not a tautology, but one can guess that the derivation of the original quantified *wff* can be accomplished by starting from the tautology:

$$\varphi x \supset \psi y. \supset .\varphi x \supset \psi y.$$

This is by no means a decision procedure. Still, Wittgenstein seems to have got the intuition that a decision procedure for quantification theory (with identity) *must* be somewhere right around the corner. In any case, he didn't let worries over decidability deter him. He was convinced that quantification theory (with identity) must be decidable and he was seeking to find an *ab*-Notation that would establish just that.

All his alternative pictures of quantifier-free truth-conditions (*ab*-Notation, truth-tabular notation, Venn diagrammatic notation) implement Wittgenstein's idea of *showing* by finding a notation in which all and only logical equivalents have the same expression and tautologies are immediately recognizable in the notation itself. But the trouble is that to make good on his doctrine of *showing*, he requires an extension to the whole of logic, so that if it is revealed by the *ab*-Notation that a *wff* is not a tautology (generalized tautology) then one can conclude that it doesn't belong to logic. In a postscript to a letter of October 1913 to Russell he wrote (Wittgenstein 1979, p. 123):

> I wrote this letter yesterday. Since then quite new ideas have come into my mind; new problems have arisen in the theory of molecular propositions and the theory of inference has received a new and very important aspect. One of the consequences of my new ideas will— I think— be that the whole of Logic follows from one Pp only!! I cannot say more about it at present.

Now clearly Wittgenstein thought that his the *ab*-Notation offers a decision procedure for propositional (quantifier-free or "elementary" statements as *Principia* calls them) tautologies. In a very important letter of November 1913, he wrote to Russell mentioning *Principia's* *9 explicitly and remarking (Wittgenstein 1979, p. 129):

> … the great question is now; How should a notation be constructed which will make every tautology recognizable as a tautology *in one and the same way*? This is the fundamental problem of logic.[10]

So it was that Wittgenstein intuited, quite mistakenly, that logic must be decidable— or as he put it, that one principle can determine which *wffs* are tautologies (i.e., belong to logic). Why should Wittgenstein have that intuition? The best answer (perhaps the only viable answer) is again that it is entailed by his *Doctrine of Showing*. Once again we can see that *showing* is driving his work This is precisely how he intuited,

[10]See Wittgenstein (1979, p. 129).

quite mistakenly, that an extension of his *ab*-Notation "must" be discoverable, or some such notation must be discoverable that applies to the whole of logic.

Admittedly, Wittgenstein also speaks as if he hadn't yet made up his mind about whether the *ab*-Notation simply offers a notional variant of the truth-functions. In this same letter he writes:

"Whether the *ab*-functions and your truth-functions are the same cannot yet be decided."[11]

Russell had insinuated that the *ab*-Notation is just a variant of truth-functional notations of *Principia*. Wittgenstein rejected that hypothesis. But what about quantification theory with identity and what about other candidates for principles of logic? How precisely are they to be represented in *ab*-Notation. In the letter, it soon becomes clear that Wittgenstein decided that his *ab*-Notation (or something in the neighborhood) must able to be extended to the entirety of logic. The best explanation of this (indeed, the only viable explanation) is that he had intuited his *Doctrine of Showing*.

Now it is illuminating to take up a few cases to see how Wittgenstein imagined that his *ab*-Notation readily justifies theorems of *Principia's* system *9 as "tautologies" (in Wittgenstein's sense of the expression). Consider the following well known theorem of quantification theory:

$$(x)\varphi x \supset (\exists x)\varphi x$$

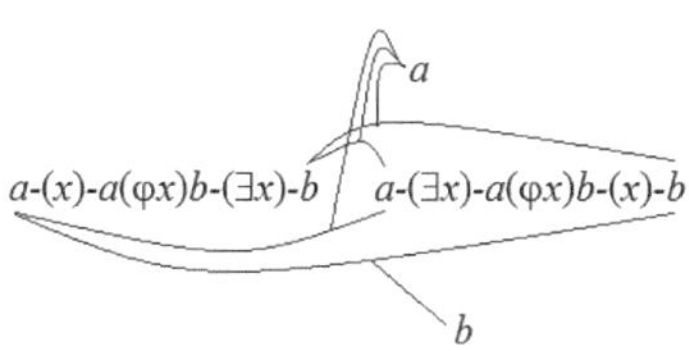

The *ab*-Notation carries the *b*-pole to the contradictory assignments (x)-b and (x)-a, thereby revealing that the expression is a tautology (in Wittgenstein's sense). That is clever indeed.

To see how the *ab*-Notation is supposed to generate such results for quantification it is best to continue to take specific cases *Principia's* *9 as illustrations. Lampert (2018) himself takes up the case of the following definition:

$$*9.01 \quad \sim(x)\varphi x = df\ (\exists x)\sim\varphi x.$$

One must reveal how the *ab*-Notational application to quantification recovers the result that "$\sim(x)\varphi x$" has the "same" *ab*-Notation as "$(\exists x)\varphi x$". The answer is certainly *not* to be found by appeal to Lampert's thesis that "…the positions of neither the quantifiers nor the intermediary poles are part of the *ab*-notation" (Lampert 2018, p. 250). The quite simple answer lies in the following. Recall that we have:

$$(x)\varphi x: \quad a\text{-}(x)\text{-}a(\varphi x)b\text{-}(\exists x)\text{-}b$$

[11] Wittgenstein (1979), p. 124.

Now we have already seen that, in context, one can form a negation in *ab*-Notation by simply switching the *a*-pole and the *b*-pole. So form the negation by switching the poles to arrive at the following

$$\sim(x)\varphi x: \quad b\text{-}(x)\text{-}a(\varphi x)b\text{-}(\exists x)\text{-}a.$$

Next just turn things around (without changing the connection of any quantifier to its pole) to form:

$$a\text{-}(\exists x)\text{-}b(\varphi x)a\text{-}(x)\text{-}b.$$

Now this is just the same as

$$(\exists x)\sim\varphi x: \quad a\text{-}(\exists x)\text{-}b\text{-}a(\varphi x)b\text{-}a\text{-}(x)\text{-}a.$$

The *ab*-Notations for "~$(x)\varphi x$" is the same as that for "$(\exists x)$~φx."

To further illustrate, let's take as well the analogous situation of showing by the *ab*-Notation that "~(x)~φx" has the same picture as "$(\exists x)\varphi x$". First, use pole-switching to get an *ab*-Notation as follows:

$$(x)\sim\varphi x: a\text{-}(x)\text{-}b(\varphi x)a\text{-}(\exists x)\text{-}b.$$

Let's next form its negation by pole-switching:

$$\sim(x)\sim\varphi x: b\text{-}(x)\text{-}b(\varphi x)a\text{-}(\exists x)\text{-}a$$

We thereby arrive at the following

$$(\exists x)\varphi x: \quad a\text{-}(\exists x)\text{-}a(\varphi x)b\text{-}(x)\text{-}b.$$

That is really all there is to it and it is all rather nicely illuminating of the connections.

We still need to explain the many other definitions of *Principia's* system *9. Perhaps the following can be taken as paradigmatic of the moves Wittgenstein intended:

$$*9.04 \quad p \vee (x)\varphi x = df\,(x)(p \vee \varphi x).$$

Here is the *ab*-Notation for $(x)(p \vee \varphi x)$:

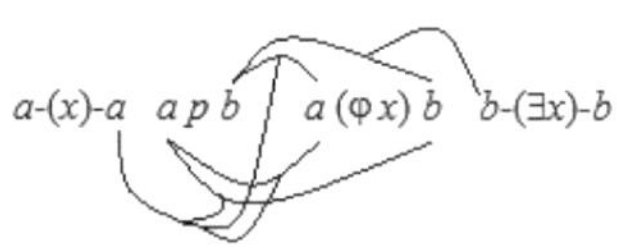

The following transformations change it into the *ab*-Notation for $p \vee (x)(\varphi x)$.

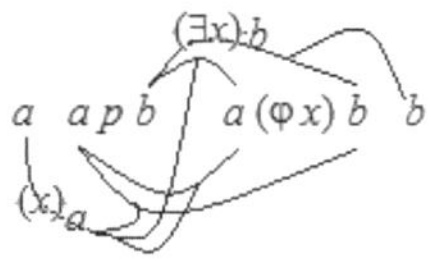

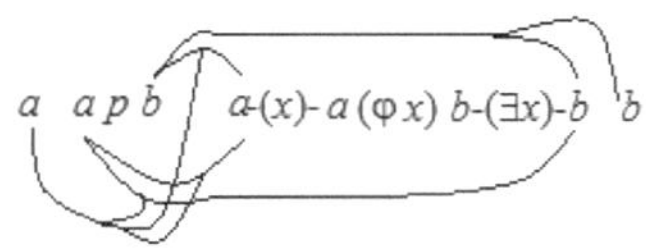

This offers no difficulty at all.

Perhaps the best case for illustrating the techniques of the *ab*-Notation to *show* logical equivalents as applied to quantifiers is the following. Start from "$(x)(\varphi x \bullet \psi x)$".

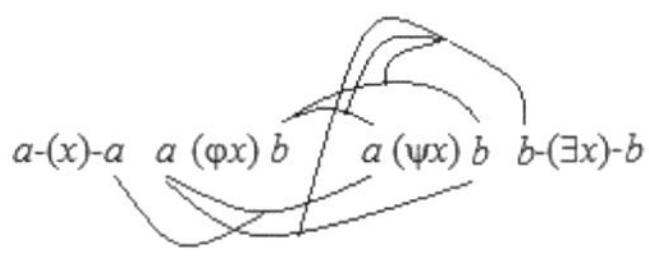

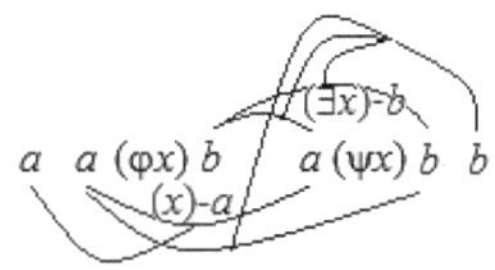

The transition pushes the quantifier expressions along their respective wires. In this case the quantifiers each double up and follow down their respective wires.

In this way the logical equivalence of "$(x)(\varphi x \bullet \psi x)$" and "$(x)\varphi x \bullet (x)\psi x$" is shown by their having "the same" *ab*-Notation.

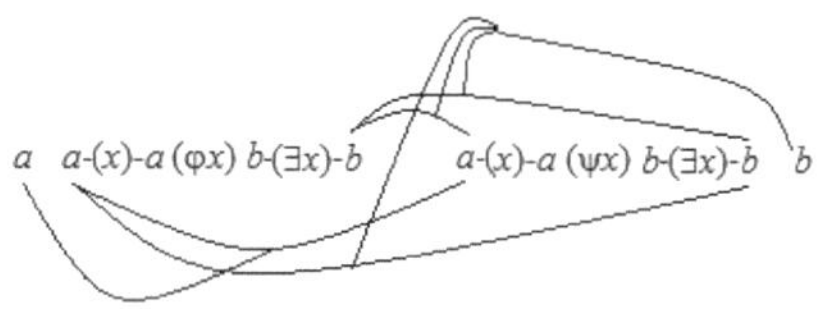

Next consider the way in which *ab*-Notation is supposed to reveal whether or not a given quantified *wff* is a "generalized tautology." Consider *Principia's*

$$*9.1 \quad \varphi x \supset (\exists y)\varphi y.$$

How shall one proceed? The problem is with the free variable. In his November 1913 letter, Wittgenstein offers the following:

> Of course, the rule I have given applies first of all only to what you called elementary propositions. But it is very easy to see that must also apply to all others. For consider your two Pps in the theory of apparent [bound] variables *9.1 and *9.11. Put then instead of φx, $(\exists y)$. $\varphi y \bullet y{=}x$ and it becomes obvious that the special cases of these two Pps like those of all the previous ones become tautologous if you apply the *ab*-Notation. The *ab*-Notation for Identity is not yet clear enough to show this clearly but it is obvious that such a Notation can be made up.

Likely, Wittgenstein was trying to deal with the free occurrence of "x" so that it can interface with the existentially bound "y." Executing his instruction in the letter, transform *Principia's* axiom schema as follows:

$$*9.1 \quad \begin{array}{l} \varphi x \supset (\exists y)\varphi y \\ (\exists y)(y = x \bullet \varphi y) \supset (\exists y)\varphi y. \end{array}$$

Next write out the *ab*-Notation as follows:

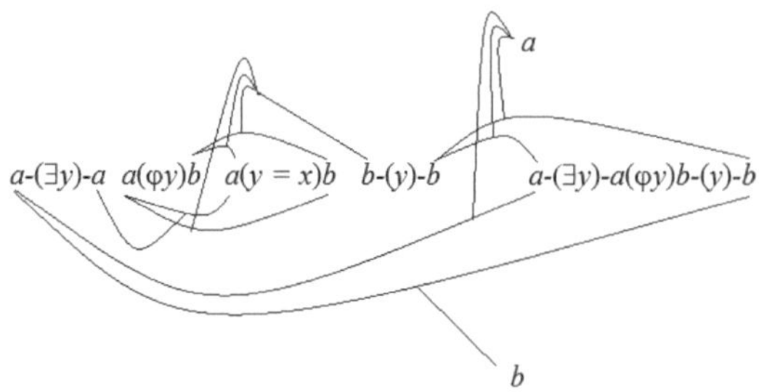

We see that the technique seems to work. We can follow the *b*-pole to contradictory assignments, both (y)-*b* and also (∃y)-*a* . But this approach is knee deep in his unresolved problem of how to treat identity. There is a better way forward in dealing with free variables that avoids identity altogether. Wittgenstein holds that there are no free variables in the genuine propositions of Logic. In his 1913 *Notes on Logic* he makes this clear, writing (Wittgenstein 1979, p. 98) "There are no propositions containing real variables." That is a help. Let us take the universal closure of *9.1 to get "$(y)(\varphi y \supset (\exists y)\varphi y)$" and write the attending *ab*-Notation as follows:

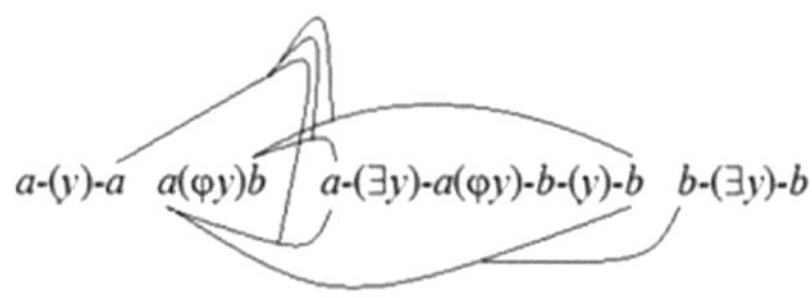

One can follow its *b*-pole to contradictory assignments *b*-(∃*y*) and the *a*-(*y*) for the *wff* φ*y*. The *ab*-Notation (or better *tf*-Notation) really quite ingenious indeed.

With these paradigm cases revealed, we can feel confident that this is how Wittgenstein came to think that his *ab*-Notation may well provide a notation in which all and only logical equivalents of quantification theory (without identity) have the same representation—a notation in which tautologies are revealed by applying one and the same *ab*-Notation rule. Of course, it does not work. Quantification theory is not decidable. The "generalized tautologies" (if you will) of quantification theory are not discoverable as such by their representation in *ab*-Notation.

Wittgenstein may well never have noticed any flaw in his *ab*-Notation until well after the *Tractatus*. But what he did notice is that he had not discovered how to apply it to include identity (where admissible). We know this because he admits in the (lately quoted) letter of November 1913 and also in an earlier letter writing to Russell on 29 October 1913 (Wittgenstein 1979, p. 123):

> Identity is the very Devil and *immensely important*; *very* much more so than I thought. It hangs—like everything else—directly together with fundamental questions, especially with the questions concerning the occurrence of the *same* argument in different places of a function. I have all sorts of ideas for a solution of the problem but could not yet arrive at anything definite. However, I don't lose courage and go on thinking.

In any case, we can see that he is completely undaunted by the problem of incorporating identity (where legitimate) into *ab*-Notation. Wittgenstein had indeed, made an important discovery. It is most clearly stated in his 1914 *Notes Dictated to Moore*.

We find that he discovered that the entirety of the theory of identity can be captured by means of just the following axiom schema[12]:

$$\text{(WI)} \quad \varphi x \equiv (\exists y)(y = x \bullet \varphi y).$$

But this went no further toward the problem of finding an approach to identity. But again Wittgenstein was undaunted and the question is: Why? The only viable answer is that he had intuited his *Doctrine of Showing*.

Observe that one of the most important features among his November 1913 letters is that it contains an unequivocal proclamation that a the *ab*-Notation has captured the nature of logic itself and that it can be used to reveal that *Principia's* *12.1.11 are do not belong to logic. On the basis of confidence in his intuition, he wrote (1979, p. 125):

> … I beg you to notice that, although I shall make use in what follows of my *ab*-Notation, the meaning of this notion is not needed, that is to say, even if this notion should turn out not to be the final correct notation what I am going to say is valid if you only admit—as I believe you must do—that it is a possible notation. Now listen!
>
> I will first talk about those logical propositions which are or might be contained in the first 8 chapters of *Principia Mathematica*. That they all follow form *one* proposition is clear because one symbolic rule is sufficient to recognize each of them as true or false.
>
> And this is the one symbolic rule: write the proposition down in *ab*-notation, trace all connections (of poles) from the outside to the inside poles: Then if the *b*-pole is connection to such a group of inside poles only as contain opposite poles of one proposition, then the whole proposition is a true, logical proposition. If on the other hand this is the case with the *a*-pole the proposition is false and logical. If finally neither is the case the proposition may be true or false, but is in no case logical.

Now the first 8 chapters consist of everything in *Principia* up until its section *12 where the so called "axioms of reducibility" (*12.1.11) occur. Wittgenstein goes on in the letter to maintain that since his *ab*-Notation rule (or some such) *must* be correct, *Principia's* *12.1 (axiom of reducibility) is not a legitimate proposition of logic! In the same letter he writes:

> … *it is not a logical proposition at all* and the same applies to the axioms of infinity and the multiplicative axiom. *If these are truth propositions they are what I shall call "accidently" true and not "essentially" true.* Whether a proposition is accidently true or essentially true can be seen by writing it down in the *ab*-notation and applying the above rule.

Of course, we now realize that all of this is misguided. It was wholly dependent on an intuition that the *ab*-Notation (or some such notation) is possible that would show whether or not a given statement is, indeed, what we nowadays call "logically true." But in fact, such a notation is impossible for it would require the decidability of quantification theory.

It is not a little ironically amusing that Wittgenstein knew fully well that he had *not* established the correctness of his *ab*-notation as capturing the essence logic

---

[12]Unaware of Wittgenstein's work on this, Quine reports its discovery to Hao Wang. See Quine (1969), p. 13.

(including quantification theory *with identity*). He offers a bit of an apology, opening his letter writing "… even if this notation should turn out not to be the final correct notation what I am going to say is valid if you would only admit—as I believe you must do—the at it is a *possible* notation." Well, it is not a possible notation and no such notation is possible. The entire letter is a massive hyperbole not a word of which seems correct. But no matter. The point I'm after is simply that the best explanation of what is motivating Wittgenstein's confidence is that he has intuited the truth of his *Doctrine of Showing*. It is this intuition driving the letter that best explains why Wittgenstein thought that there "must" be such notations.

Russell clearly must have been then resisting Wittgenstein's attack on *Principia's* *12.1.11 ("axioms of reducibility") and this explains Wittgenstein's strident tone in defending his lack of any rule or intelligible way to apply the *ab*-Notation to quantification theory in such a way that enables it to reveal *whether or not* a given *wff* is a logical truth. The axiom schemata of *Principia's* section *12 are these:

$$* \, 12.1 \quad (\exists f)(\varphi x \equiv_x f!x)$$
$$* \, 12.11 \quad (\exists f)(\varphi xy \equiv_{x,y} f!xy)$$

Wittgenstein's attempt to reveal that *12.1.11 do not belong to logic fails miserably because it relies on two false assumptions. The first false assumption is that quantification theory must be decidable (by *ab*-Notation or some variant of it) because it cannot consist of a body of truths (logical truths). The second false assumption is that logic must *not* consist of truths, and hence must not go beyond quantification theory with identity (where admissible) for that is the limit of what can be *shown.* The fact that *12.1.11 are not part of quantification theory does not preclude them from belonging to logic.

One naturally wonders where in the world is the source of all the hubris on Wittgenstein's part concerning Logic? The best explanation (the only viable explanation) is that it is coming from his having intuited his *Doctrine of Showing*. Logic *must* be decidable because its structures are *shown* in the proper grammar for empirical science and cannot be not *said* (stated as truths). Armed with his intuition of *showing*, we can understand why Wittgenstein did not give up his quest for a notation which *shows* the status of (what we call the logical truths of quantification theory with identity) as tautologies.

13.4 *N-Operator*-Notation *Instead* of *ab*-Notation

So far we have the following by way of summary. Captivated by his *Doctrine of Showing*, Wittgenstein was convinced that some notion *must* depict all and only logical equivalents of quantification theory with identity (where legitimate) in one and the same way. The *ab*-Notation was clearly adequate to quantifier free *wffs* and the remaining problem was to extend it (or find some extension or another) to quantification theory with identity (where admissible) that preserves a decision procedure. And as we have seen, in his 1913 letters to Russell, he felt that there "must" be such an extension. His feeling that the "must" be can only have been motivated by his late 1913 intuition of *showing*—though it was not explicitly stated as such until the 1914 *Notes Dictated to Moore*.

Wittgenstein must have realized his *ab*-Notation, though illuminating, was inadequate to the task of *showing* that all and only logical equivalents of quantification theory with identity (where admissible) have one and the same notation. There remained the problem of Identity. Happily, there are key passages of the *Notebooks 1914–1916* that tell us precisely how Wittgenstein came to think that he had overcome these impediments for the realization of his *Doctrine of Showing*. He came to his N-operator notation in his effort to replace his *ab*-Notation with a notation that, he thought, can apply to quantification theory with identity. At the same time, the new N-operator approach must preserve what the *ab*-Notation (*tf*-Notation) and also the truth-tabular notations all had achieved.

The way to find Wittgenstein's N-operator in his work notes is not to search for the expression "N-operator" or the expression "N($\bar{\xi}$)" but to realize that the key passages indicating the transition from the *ab*-Notation to the N-operator notation lie precisely in his discussion of "and so on" which he takes to be notion of "operation". In the *Notebooks of 1914–1916* we find the following entries concerning the ab-Notation itself (Wittgenstein, 1979, p. 54):

> [28.11. 14]
>
> Negation combines with the *ab*-functions of the elementary proposition. And the logical functions of the elementary proposition must mirror their reference, just as much as all others.
>
> [29.11. 14]
>
> The *ab*-function does not stop short of the elementary proposition but penetrates it.
>
> What can be shown cannot be said. [See also *TLP* 4.1212]
>
> I believe that it would be possible wholly to exclude the sign of identity from our notation and always to indicate identity merely by the identity of the signs… But then instead of $\varphi x \bullet \varphi y \,.\!\supset_{x,y} x = y$ one could simply write $\sim(\exists x,y)(\varphi x \bullet \varphi y)$.
>
> By means of this notation the pseudo-proposition $(x)(x = a)$ or the like would lose all appearance of justification.

Here we see that by this time Wittgenstein had settled on his approach to the elimination of the identity sign by appeal to exclusive quantifiers. All that was left was to find a replacement for the *ab*-Notation that can apply to quantification theory. Now in the *Tractatus*, Wittgenstein writes:

> *TLP* 5.502
>
> So instead of "(…..*T*)(…..)" I write "N($\bar{\xi}$)".
>
> N($\bar{\xi}$) is the negation of all the values of the variable $\bar{\xi}$.

I read the words "instead of" in this passage as intending to say that the truth-tabular representation, which owes its origins to the *ab*-Notation (*tf*-Notation) is to be *supplanted* by the N-operator representation. To understand the N-operator one must understand the use of the *xi*-bar (ie., the variable $\bar{\xi}$) and the notion of "and so on" upon which it rests. This notion—i.e., the notion of "and so on"— is quite amply discussed in the *Notebooks 1914–1916*. In an entry find the following (Wittgenstein, 1979, p. 72):

> 11 May 1916
>
> |p |(a,a)
>
> There are also operations with two bases. And the "|" –operation is of this kind.
>
> |(ξ, η)... is an arbitrary term of the series of results of an operation.
>
> $(\exists x)\varphi x$
>
> Is $(\exists x)$ etc., really an operation? But what would be its base?

And then on 9 July 1916, he writes:

> Do not forget that $(\exists x)fx$ does not mean: There is an x such that fx, but: There is a true proposition "fx".

Wittgenstein's answer to his question would eventually be: Yes. The N-operation is inspired by what was he calling the "|" operation which is Sheffer's (and which he had known in 1913 when Sheffer sent his paper to Russell), but it is quite importantly different. So we are beginning to see his idea coming together. As we noted, in order to find the earliest appearance of the idea of the N-operator, we have to understand what it is doing—i.e., how it was intended to do the work assigned to it. The work assigned to it was to find a representation within which all and only logical equivalents of quantification theory with identity have one and the same representation—so that "tautologies", i.e., logical truths of quantification theory with identity (where admissible), all have the same form. Of course, since quantification theory is not decidable there is no such form. But with his *Doctrine of Showing* driving him, he intuited that there must be such a common form. Wittgenstein is on a quest to find what he takes to be the characteristic "*form*" of a tautology (logical truth) of quantification theory?

The key notion that Wittgenstein wanted to use is the "and so on" notion. In quantification we are given a "form" and the quantifier notation together with the form is intelligible only because we understand the notion of "and so on". If we are given a form 'φx' we understand "$(x)\varphi x$" by understanding all those expressions that can be generated by applying the '*and so on*' notion to the form 'φx'. Something of this idea is already found in *Principia* which hoped to build "internal limits into the significance conditions" by semantically interpreting "$(x)\varphi x$" as saying 'φx' *always*. No doubt this is why Wittgenstein thought Russell might understand him on this matter. One we see this, we can come to realize that the following passages prefigure the N-notation *xi*-bar because it concerns the idea of capturing quantification by appeal to '*and so on*'. The key idea of the N-operator is the idea of "and so on" which would later be represented by using the *xi*-bar notations found in the *Prototractatus* and then later in the *Tractatus* itself. Here, then, is the remarkable 1915 entries on just this issue. Wittgenstein, explained as follows (1979, p. 49):

22 May 1915:

The mathematical notation" $1 + \frac{x}{1!} + \frac{x^2}{2!} + \ldots$" together with the dots is an example of that extended generality. A law is given and the terms that are written down serve as an illustration. In this way instead of $(x)fx$ one might write "$fx \bullet fy \ldots$"

The notion of an "operation" is tied by Wittgenstein to the notion of the *dots* (aka: "and so on") rendering a recipe (law) and an illustration of the general form. (This is done in the *Tractatus* by the $\bar{\xi}$ notation at 5.501.) Thus, "$(x)\varphi x$" is expressed as an expansion over an arbitrarily large n-element domain:

$$\varphi x_1 \bullet \ldots \bullet \varphi x_n,$$

where $x_1, \ldots, x_n$ are all to be interpreted *exclusively* of one another. I mark this passage, therefore, as a temporal upper limit for the "discovery" (if you will) of the N-operator. The N-operator idea was in place no later than 22 May 1915, though it may have been earler.

The superior value Wittgenstein imagined of his N-notation over the *ab*-Notation and truth-tabular notations, is that it extends the key feature that *ab*-Notation and truth-tabular notation to quantification theory with identity—and the key feature is that all and only logical equivalents have the same representation and that a tautology is recognizable by form alone. To understand how Wittgenstein thought this was supposed to work, we must understand more about the N-operator itself. So let us pause a moment to see it aright— for there have been a great many confusions over the years about it.

In 1913, Russell received Sheffer's paper on stroke and its dual (which has come to be the dagger) signs. The paper showed that each is by itself adequate to express all truth-functions of *Principia Mathematica*. Post (1920) later demonstrated the adequacy of *Principia's* primitive signs ~ and ∨ to express all truth-functions. Sheffer's stoke sign $p \mid q$ yields the following definitions:

$$\begin{aligned} \sim p &= df \quad p|p \\ p \vee q &= df \quad \sim p|\sim q \\ p \bullet q &= df \quad \sim(p|q) \\ p \supset q &= df \quad p|\sim q \end{aligned}$$

Its dual, now written with the dagger sign, yields the following:

$$\begin{aligned} \sim p &= df \quad p \downarrow p \\ p \vee q &= df \quad \sim(p \downarrow q) \\ p \bullet q &= df \quad \sim p \downarrow \sim q \\ p \supset q &= df \quad \sim(\sim p \downarrow q). \end{aligned}$$

Now the N-operator is not dyadic and thus it should not be confused with Sheffer's dyadic signs. But using the Sheffer $\downarrow (p, q)$ which is a notational alternative of $p \downarrow q$, we can easily translate anything in dagger notation into Wittgenstein's N-notation. For any *wff* in dagger notation, replace every dagger with an N. Thus, for example, consider the following:

$$\sim p\,.=.\,p \downarrow p\,.=.\downarrow(p, p)\,.=.\,\mathrm{N}(p, p)$$
$$p \supset q\,.=.\sim(\sim p \downarrow q)\,.=.\,\mathrm{NN}(\mathrm{N}p, q)$$
$$p \vee q\,.=.\sim(p \downarrow q)\,.=.\,\mathrm{NN}(p, q)$$
$$p \vee (q \vee r)\,.=.\sim(p \downarrow \sim(q \downarrow r))\,.=.\,\mathrm{NN}(p, \mathrm{NN}(q, r))\,.=.\,\mathrm{NN}(p, q, r)$$

$$p \bullet q\,.=.\sim p \downarrow \sim q\,.=.\,\mathrm{N}(\mathrm{N}p, \mathrm{N}q)$$
$$p \bullet (q \bullet r)\,.=.\sim p \downarrow \sim(\sim q \downarrow \sim r)\,.=.\,\mathrm{N}(\mathrm{N}p, \mathrm{NN}(\mathrm{N}q, Nr))\,.=.\,\mathrm{N}(\mathrm{N}p, \mathrm{N}q, \mathrm{N}r)$$

It should be noted from these examples that internal double NN's can be dropped in accordance with the fact that the N-notation is clearly not dyadic. Moreover, the following five rules of transition govern the N operator notation in the Tractatus (held over from the *ab*-Notation):

$\mathrm{N}(\ldots p \ldots q \ldots, p) \rightleftarrows \mathrm{N}(\ldots p \ldots q \ldots)$ repetitions can be dropped
$\mathrm{N}(\ldots p \ldots q \ldots) \rightleftarrows \mathrm{N}(\ldots q \ldots p \ldots)$ order doesn't matter
$\mathrm{N}(\ldots \mathrm{N}(p \ldots \mathrm{N}p) \ldots) \rightleftarrows \mathrm{N}(\ldots)$ internal $\mathrm{N}(p \ldots \mathrm{N}p)$ can be dropped
$\mathrm{N}(\ldots \mathrm{NN}(,,,) \ldots) \rightleftarrows \mathrm{N}(\ldots ,,, \ldots)$ internal double NN(, , ,) adds the content of ,,, to the ...
$\mathrm{NN}(p, \mathrm{N}(\mathrm{N}q, \mathrm{N}r) \rightleftarrows \mathrm{N}(\mathrm{N}(p, q), \mathrm{N}(p, r))$

Moreover, the same number of N's may validly be simultaneously added uniformly or subtracted uniformly from either side of these rules of calculation (though there must be at least one N). All and only tautologies have the form:

$$\mathrm{NN}(\ldots p \ldots .\mathrm{N}p \ldots).$$

To take an example, let's express "$p\ .\!\supset\!.\ q \supset p$" in N-notation. We get:

$$\sim(\sim p \downarrow \sim(\sim q \downarrow p))$$
$$\mathrm{NN}(\mathrm{N}p, \mathrm{NN}(\mathrm{N}q, p))$$

Then manipulating by the rule of N-notation that internal NN' s can be dropped, we get:

$$\mathrm{NN}(\mathrm{N}p,\ \mathrm{N}q,\ p).$$

We now see immediately in the notation that we have the form common to every tautology. Hence, all the desired features of the *ab*-Notation (when it comes to showing tautologies of quantifier-free *wffs*) are recovered by the N-notation. One can recognize a tautology from the notation alone.

Once we have the rules of calculation with N, i.e., the apparatus of elucidation (*showing*) the status of an expression as a tautology or otherwise, we can readily explain Wittgenstein's answer to Russell's concern that the Sheffer dagger has a dual—the Sheffer stroke—which is equally a candidate for doing the same work Wittgenstein does. Upon first receiving the *Tractatus*, Russell wrote to Wittgenstein asking several questions. In his remarks, Russell reveals that he thinks of the N operator as a sort of generalized Sheffer dagger. We saw that at *TLP* 6 Wittgenstein gave the general form of a truth-function (proposition). Russell asks the following question about this passage:

> General truth-function '$[p, \xi, N(\xi)]$.' Yes, this is one way. But could one not do equally well by making $N(\xi)$ mean 'at least one value of ξ is false,' just as one can do equally well with $\sim p \vee \sim q$ and with $\sim p \bullet \sim q$ as fundamental? I feel as if the duality of generality and existence persisted covertly in your system.[13]

Wittgenstein's reply was this:

> You are right that $N(\xi)$ may also be made to mean $\sim p \vee \sim q \vee \sim r \vee \ldots$ But this doesn't matter! I suppose you don't understand the notation ξ. It does not mean "for all values of ξ…". [14]

I surmise that the reason it doesn't matter is that the rules (N_1)–(N_5) are equally valid whether we regard N as akin to the Sheffer dagger or its dual the Sheffer stroke. The truth-conditions of tautologies will be pictured differently, of course. If we interpret the N-operator as picturing truth-conditions akin to the stroke, then all and only tautologies have their truth-conditions pictured by the form:

$$N(\ldots\xi, \ldots N\xi, \ldots).$$

It was because (N_1)–(N_5) apply equally on either approach that Wittgenstein felt justified in his reply.

Wittgenstein thought that the rules can be extended to quantification theory with identity. His attempt to recover this feature in *ab*-Notation extended to quantification had failed. But he mistakenly thought N-notation succeeds. A quantified statement "$(x)\varphi x$" is to be understood as an expansion given by a law indicating the form. This is the way the Tractarian N-operation $N(\bar{\xi})$ works. In N-notation "$(x)\varphi x$" is expressed by either of the following which are taken to be the same:

$$N(\varphi x)$$
$$N(N\varphi x_1, \ldots, N\varphi x_n),$$

where the variables $x_1, \ldots, x_n$ are exclusive of one another. In the first, the form φx is given as a law or recipe for expansion over an arbitrarily large n-element domain. It and instance of a law, a $\bar{\xi}$, for the case in question. It is important that the domain, though arbitrary large, be fixed at the onset by the schematic use of n. This is of utmost

[13] Bertrand Russell, *Letter of 13 August 1919*.

[14] Wittgenstein (1979, p. 131).

importance, for it avoids all the problems that otherwise would arise from questions of the scope of the quantifiers. Fogelin (1982, 1987) was perhaps first to raise this as a substantive concern.[15] How is the scope of the "x" in "NN(φx)" confined so that result is supposed to capture "$(\exists x)\varphi x$" instead of "$(x)\sim\sim\varphi x$". One might have thought that N-notation cannot possibly be expressively adequate to quantification theory without some notational apparatus of scope. Of course, Wittgenstein knew of the problem scope, for in his *Notebooks 1914–1916* we find the following entry:

> 12 October 1914
>
> If we were to try and express what we express by means of "$(x)\varphi x$" by prefixing an index to "φx", e.g., like this: "Gen . φx", it would not be adequate (we should not know what was being generalized). If we tried to show it by means of an index to the "x", e.g., like this: $\varphi(x_G)$, it would still not be adequate (in this way we should not know the scope of generality). If we thought of trying to do it by inserting a mark in the empty argument places, e.g., like this "$(G, G).\varphi(G, G)$" it would not be adequate (we could not settle the identity of the variables). All these methods of symbolizing are inadequate because they do not have the necessary logical properties. All those collections of signs lack the power to portray the requisite sense—in the proposed way [See also *TLP* 4.0411.]

So how does the N operator do it? The answer is simply that it doesn't need any scope indicator because it uses the schematic "n" and exclusive variables to fix an arbitrarily large domain at the onset. Thus:

$$\mathrm{NN}(\varphi x_1, , , \varphi x_n) \quad \text{captures} \quad (\exists x)\varphi x$$
$$\mathrm{N}(\mathrm{N}\varphi x_1, , , \mathrm{N}\varphi x_n) \quad \text{captures} \quad (x){\sim}{\sim}\varphi x.$$

In neither case is there any scope marker required. Notice that since internal NN can be dropped, the following have the same N-notations:

$$\sim(\exists x)\varphi x\ .=.\ \mathrm{NNN}(\varphi x_1, \ldots, \varphi) \rightleftarrows \mathrm{N}(\mathrm{NN}\varphi x_1, \ldots, \mathrm{NN}\varphi x_n)\ .=.\ (x){\sim}\varphi x.$$

Of course, we also have:

$$\sim(x)\varphi x\ .=.\ \mathrm{NN}(\mathrm{N}\ \varphi x_1, \ldots, \mathrm{N}\ \varphi x_n)\ .=.\ (\exists x){\sim}\varphi x.$$

Indeed, all and only logical equivalents have the "same" N-notations. Consider these:

$$\begin{aligned}
&(x)(\varphi x \bullet \psi x)\ .=.\\
&\mathrm{N}(\mathrm{NN}(\mathrm{N}\varphi x_1, \mathrm{N}\psi x_1), \ldots, \mathrm{NN}(\mathrm{N}\varphi x_n, \mathrm{N}\psi x_n)) \rightleftarrows\\
&\mathrm{N}(\mathrm{N}\varphi x_1, \mathrm{N}\psi x_1 \ldots \mathrm{N}\psi x_n, \mathrm{N}\psi x_n) \rightleftarrows\\
&\mathrm{N}(\mathrm{N}\varphi x_1 \ldots \mathrm{N}\varphi x_n, \ldots, \mathrm{N}\psi x_1 \ldots \mathrm{N}\psi x_n) \rightleftarrows\\
&\mathrm{N}(\mathrm{NN}(\mathrm{N}\varphi x_1 \ldots \mathrm{N}\varphi x_n), \ldots, \mathrm{NN}(\mathrm{N}\psi x_1 \ldots \mathrm{N}\psi x_n))\ .=.\\
&(x)\varphi x \bullet (x)\psi x.
\end{aligned}$$

[15] Geach (1981, 1982) tried to meet the concern by embellishing what he took to be N-notation with a modes theory of classes.

This was a feature of the *ab*-Notation for quantification too. But in the extension of the *ab*-Notation to quantification it was not possible to hold that all and only tautologies have the same tautologous form. There was no *ab*-Notational form for "tautologies" (i.e., what we nowadays call the logical truths of quantification theory). There is no stable truth-tabular notation of a quantified statement involving relation signs because an increase in the size of the domain shifts *t*'s and *f*'s about. But as we saw, by 22 May 1915 Wittgenstein thought the problems can be addressed by appeal to his new idea of an "operation" whose base is given by a law or recipe the can use a schematic use of "*n*" fixing a domain of arbitrary large size. This is the core of the N-operator idea. Wittgenstein had thus come to believe that he had found a common "tautologous" form all the logical truths of quantification theory with identity.

To take an example, consider "$(x)(\varphi x .\supset. q \supset \varphi x)$". The expansion over an arbitrarily large *n*-element domain is:

$$\mathrm{N}(\mathrm{N}(\mathrm{N}\varphi x_1,\ \mathrm{N}q,\ \varphi x_1), \ldots, \mathrm{N}(\mathrm{N}\varphi x_n,\ \mathrm{N}q,\ \varphi x_n)).$$

By rule of N-notation internal N(…*p*.. N*p*) may be dropped— all but the last one. So we get:

$$\mathrm{NN}(\mathrm{N}\varphi x_i,\ \mathrm{N}q,\ \varphi x_i))$$

This has the form of a tautology. I believe that Wittgenstein thought that these five "rules" of sameness for calculating the outcomes of N-operation are sufficient to realize the demands of his *Doctrine of Showing*—the demand that a notation be found which *shows* whether or not a *wff* is a tautology. Unfortunately, it fails because we need to fix the *n* in order to apply the five rules of sameness. Of course, we can apply the rules for any finite subordinate clause within an N-expression, but that is not enough to secure the result. Wittgenstein thought that expression in N-notation decides "tautologyhood" but again he was mistaken.

But perhaps we are moving a little too fast. I have marked 22 May 1915 as a date after which we can be confident that we know that Wittgenstein had the idea that the quantifiers are operations involving expansions over a fixed *n*-element domain and had as well the idea that an operation is a recipe or law determining likeness of form. But let's look at the many passages which reveal Wittgenstein's thinking. Here are some of the most important passages from the *Notebooks 1914–1916*, with dates;

> 17 December 1914
>
> The signs of the *ab*-function are not material, otherwise they could not vanish.
>
> [also *TLP* 5.44, 5.441]

> 21 January 1915
>
> How is it that *all* is a logical concept?
>
> How is it that *all* is a concept of form?
>
> How does it come about that *all* can occur in any proposition?
>
> For that is the characteristic mark of the concept of a form.

… Generality is essentially connected with elementary FORM.

23 January 1915

Negation is an operation [also *TLP* 5.2341]

24 January 1915

The logical pseudo-functions are operations.

Only operations can vanish!

7 March 1915

Just as $\sim\xi$, $\sim\xi \vee \sim\xi$ etc., are the same function, so too are $\sim\eta \vee \eta$, $\eta \supset \eta$, etc., the same—that is, the tautological—function …

8 March 1915

My difficulty is only an—enormous—difficulty of expression.

4 May 1915

The class of all propositions of the form F*x* is the proposition $(x)\varphi x$.[16]

5 May 1915

Does the general form of a proposition exist?

Yes, if by that is understood the single "logical constant" [also *TLP* 5.47].

17 August 1016

An operation is the transition from one term to the next one in a series of forms.

The operation and the series of forms are equivalents.

29 August 1916

The question is whether the usual small number of fundamental operations is adequate for the construction of all operations. It looks as if it must be so. We can also ask whether those fundamental operations enable us to pass from any expression to any related ones.

21 November 1916

We now need a clarification of the concept of the atomic function and the concept "and so on."

The concept "and so on," symbolized by "…." is one of the most important of all and like all others infinitely fundamental. For it alone justifies us constructing logic and mathematics, "so on" from the fundamental laws and primitive signs. The "and so on" makes its appearance right away at the very beginning of the old logic when it is said that after the primitive signs have been given we can develop on sign after another "and so on." Without this concept we should be stuck at the primitive signs and could not go "*on*." The concept "and so on" and the concept of the operation are equivalent. [also *TLP* 5.2523]

After the operation sign there follows the sign "…." Which signifies that the operation can in its turn be taken as the base of the operation; "and so on".

[16]It is strange that this passages has "F*x*" instead of "φx" it may be mistake in transcription.

22 November 1916

The concept of the operation is quite generally that according to which signs can be constructed according to a rule.

23 November 1916

What does the possibility of the operation depend on?

On the general concept of structural similarity.

As I conceive, e.g., the elementary propositions, there must be something common to them; otherwise I could not speak of them all collectively as the "elementary propositions" at all.

In that case, however, they must also be capable of being developed from one another as the results of operations.

24 November 1916

When the general form of operations is found we have also found the general form of the occurrences of the concept "and so on."

We can see that Wittgenstein's idea of *operation* had become central to his work, and it is of course central in the *Tractatus*. Both logic and arithmetic find their common source in the practice of calculation of outcomes of operations. Arithmetic calculates the correctness (sameness) of the outcomes of recursively defined functions appearing in its "equations", while logic calculates the correctness (sameness) of the outcomes of N-operation appearing in its "tautologies." There is an entry in the *Notebooks 1914–1916* along these lines too. We find:

24 April 1015

In logic (mathematics) process and result are equivalent. (Hence no surprises).

(See also *TLP* 6.1261).

The idea, finally worked out in the *Tractatus*, would come to be that both *arithmetic* and *logic* are deeply the same—not, as Russell thought, because the one is really just the other, but rather because both find their common source as *practices of calculating with operations.*

Of course, none of this works. But we've put that aside. We are endeavoring to uncover the earliest date of *showing* that we can find and we are looking for such a date it by looking for the manifestations in the clever notations Wittgenstein invented to try to implement *showing* in mathematical logic. It is obviously manifest in the N-operator notation whose central idea is that of a *law* or *recipe* for the bases of an operation. Thus we found date of 22 May 1915. But we were able to push the date back because Wittgenstein explicitly connected *showing* to his *ab*-Notation in his 1914 *Notes Dictated to Moore*. We found that the best (perhaps the only viable) explanation of Wittgenstein's hubris concerning an application of *ab*-Notation (or some such notation) to quantification theory and identity is that he had intuited his *Doctrine of Showing* already in 1913. We discovered that in his November 1913 letter to Russell he seemed already worried that *ab*-Notation doesn't quite work. But it was only in 1915 that he was in the process of superseding his *ab*-Notation with the

N-operator Notation. And with that, he finally rested, thinking that he had captured quantification theory with identity (where admissible).

Identity, Wittgenstein thought, is the Devil and immensely important. It turns out to have been quite important to him since he let it guide his intuition so completely that he seems willing to classify Cantor's work as being outside of mathematics. Cantor maintained the notion of *cardinal number* is a notion of one-to-one correspondence. Wittgenstein rejected this notion because the one-to-one correspondences appealed to would have to be based on a relation of identity. Now many one-to-one correspondences can be expressed by using exclusive quantifiers, but the relation of correspondence would have to be shown and not said by appeal to non-logical relationships. The fingers on my left hand are, for example, correlated by physical relations one-to-one with the fingers on my right. But mathematical logic cannot find its basis there. Wittgenstein's commitment to the elimination of the relation of 'identity' required a new non-Cantorian approach to number. Of course, he never accomplished anything close to that. Let us see how his N-operator and its exclusive free variables (all with the same scope) captured it identity—in cases where it is admissible.

Not all *wffs* involving the identity sign are admissible. He is quite explicit, for example, that there should be no analog of "$(x)(x = x)$" or its negation, and similarly he allows no analog of

"$(\exists x)(x = y)$" or its negation in his grammar. All the same, he does allow an analog of

$$(\mathrm{D}): \quad (\exists x)(\varphi x \bullet (y)(\varphi y \supset y = x)).$$

As early 1913, we find (Wittgenstein 1979, p. 129):

> The only other thing I want to say is that your Theory of Descriptions is quite *undoubtedly right*, even if the individual primitive signs in it are quite different from what you believe.

How, then, did Wittgenstein hope to do this in N-notation? In Landini (2007), I invented a notation to help the translation into N-notation of such statements. I did not attribute it to Wittgenstein, but maintained it correctly captures the spirit of what he was after with his doctrine of exclusive quantifiers. It allows no analog of "$(x)(x = x)$" etc., and builds just the right syntactic exclusiveness in to something close to the usual quantifier notations and their ordinary variable binding. In this notation, the *wff* (D) is rewritten as:

$$(\exists x)(\varphi x \bullet (\forall y^{x}){\sim}\varphi y).$$

The benefit of this is that it helps us to find the N-notation immediately. Recall that the free variables of the N-notation are exclusive of one another and that we are to use a schematic letter "n" to fix our expression at an arbitrarily large n-element domain. Thus we have:

$$\varphi x_1 \bullet (\forall y^{x_1}){\sim}\varphi y \,.\vee \ldots \vee.\, \varphi x_n \bullet (\forall y^{x_n}){\sim}\varphi y$$

$$\varphi x_1 \bullet (\sim\varphi x_2 \bullet \ldots \bullet \sim\varphi x_n) \,.\vee .$$
$$\varphi x_2 \bullet (\sim\varphi x_1 \bullet \sim\varphi x_3 \bullet \ldots \bullet \sim\varphi x_{n-1}) \,.\vee \ldots \vee .$$
$$\varphi x_n \bullet (\sim\varphi x_1 \bullet \ldots \bullet \sim\varphi x_{n-1})$$

$$\mathrm{NN}(\ \ \mathrm{N}(\mathrm{N}\varphi x_1, \varphi x_2, , , \varphi x_n), \mathrm{N}(\mathrm{N}\varphi x_2, \varphi x_1, \varphi x_3, , , \varphi x_n), , , \mathrm{N}(\mathrm{N}\varphi x_n, \varphi x_1, , , \varphi x_{n-1})\ \ ).$$

Note that in his *Notebooks 1914–1916*, Wittgenstein's comments (*op cit*, p. 34) suggest that we replace (D) with the following:

$$(\exists x)(\varphi x) \bullet \sim(\exists x)(\exists y)(\varphi x \bullet \varphi y)$$

If we translate this into N-notation it yields a logical equivalent and thus is "the same" N-notation.

A result corroborating my interpretation of N-notation comes from the way it recovers identity statements (when they are permissible). Every instance of *Principia's wff* (so called "axiom") of infinity (when shorn of the notations of its "no-classes" theory) has the form of a "tautology" in N-notation. This is a result heralded in the *Tractatus*:

TLP 5.535

This also disposes of all the problems that were connected with such pseudo-propositions. All the problems that Russell's 'axiom of infinity' brings with it can be solved at this point. What the axiom of infinity is intended to say would express itself in language through the existence of infinitely many names with different meanings.

The idea that this would be the case, however, dates at least to 1914. In the *Notebooks 1914–1916* we find not only that he had the idea of exclusive quantifiers to eliminate the identity relation, but also that it should resolve the difficulties Russell had with proving an infinity theorem:

9 October 1914

All the problems that go with the Axiom of Infinity have already to be solve in the proposition "$(\exists x)(x = x)$".

27 October 1914

"$x = y$" is not a propositional form. (Consequences.)

It is clear that "*aRa*" would have the same reference as "$aRb \bullet a = b$". So we can make the pseudo- proposition "$a = b$" disappear by means of a completely analyzed notation. The best proof of the correctness of the above remark.

28 October 1914

What the pseudo-proposition "there are *n* things" tries to express shows in language by the presence of *n* proper names with different references (Etc.)

29 November 1914

I believe that it would be possible wholly to exclude the signs of identity from our notation and always to indicate identity by the identity of the signs (and conversely). In that case, of course, $\varphi(a, a)$ would not be a special case of $(x, y)\varphi(x, y)$, and $\varphi a \bullet \varphi a$,[17] would not be a special case of $(\exists x, y)(\varphi x \bullet \varphi y)$. But then instead of $\varphi x \bullet \varphi y \;._{x,y}.\; x = y$ one could simply write $\sim(\exists x, y)(\varphi x \bullet \varphi y)$.

[See also *TLP* 5.53, *TLP* 5.533].

By means of this notation the pseudo-proposition $(x)(x = a)$ or the like would lose all appearance of justification.

Here we see clearly that Wittgenstein already has the exclusive quantifier idea and that infinity is shown by the very apparatus that eliminates the indispensability argument for a relation of 'identity. '

At its basis, the *wff* asserting infinity (as given in *Principia*) says that for any *wff* φ, if for there are exactly natural number n many entities x that are φx, then some entity x is such that $\sim\varphi x$.

For example, the following says: If there are exactly *zero*-many x that are φx, then some x is $\sim\varphi x$.

$$\sim(\exists x)\varphi x \supset (\exists x)\sim\varphi x$$
$$\mathrm{NN}(\varphi x_1, , , \varphi x_n, \mathrm{N}\varphi x_1, , , \mathrm{N}\varphi x_n).$$

This clearly has the form of a tautology in N-notation. Consider, next the following which captures the statement: If there is exactly one x that is φx, then some x is $\sim\varphi x$.

$$(\exists x)(\varphi x \bullet (y)(\varphi y \supset y = x)) \supset (\exists x)\sim\varphi x$$

Let's take it in two steps, the first of which is:

$$(\exists x)(\varphi x \bullet (\forall y^{x})\sim\varphi y) \supset (\exists x)\sim\varphi x$$
$$\mathrm{NN}(\ \mathrm{N}(\mathrm{N}\varphi x_1, \varphi x_2, , , \varphi x_n), \mathrm{N}(\mathrm{N}\varphi x_2, \varphi x_1, \varphi x_3, , , \varphi x_n), , ,$$
$$\mathrm{N}(\mathrm{N}\varphi x_n, \varphi x_1, , , \varphi x_{n-1}) \supset \mathrm{NN}(\mathrm{N}\varphi x_1, , , \mathrm{N}\varphi x_n)$$

Thus we have:

$$\begin{aligned}&\mathrm{NN}(\ \ \mathrm{N}\varphi x_1, , , \mathrm{N}\varphi x_n,\\ &\mathrm{N}(\ \ \mathrm{N}(\mathrm{N}\varphi x_1, \varphi x_2, , , \varphi x_n),\\ &\qquad \mathrm{N}(\varphi x_1, \mathrm{N}\varphi x_2, \varphi x_3, , , \varphi x_n), , ,\\ &\qquad \mathrm{N}(\varphi x_1, , , \varphi x_{n-1}, , , \mathrm{N}\varphi x_n)\ \ )\ \ )\end{aligned}$$

By N-rule of distribution, consider the internal cases :

$$\mathrm{N}(\ \ \mathrm{N}(\mathrm{N}\varphi x_1, \varphi x_2, , , \varphi x_n, \mathrm{N}\varphi x_1, , , , \mathrm{N}\varphi x_n),$$

[17] In the text this is φa. But what is intended is $\varphi a \bullet \varphi a$, which in N-notation would be φa. I have changed it here.

$$\mathrm{N}(\varphi x_1, \mathrm{N}\varphi x_2, , , \varphi x_n, \mathrm{N}\varphi x_1, , , , \mathrm{N}\varphi x_n),$$
$$\mathrm{N}(\varphi x_1, \varphi x_2, \mathrm{N}\varphi x_3, , , \varphi x_n, \mathrm{N}\varphi x_1, , , , \mathrm{N}\varphi x_n), , ,$$
$$\mathrm{N}(\varphi x_1, \varphi x_2, , , \mathrm{N}\varphi x_n, \mathrm{N}\varphi x_1, , , , \mathrm{N}\varphi x_n) \quad)$$

All but one of these internal cases will drop out, and we will arrive at a form such as this:

$$\mathrm{NN}(\varphi x_1, \varphi x_2, , , \mathrm{N}\varphi x_n, \mathrm{N}\varphi x_1, , , \mathrm{N}\varphi x_n)$$

This has the form of a tautology in N-notation. It is this result, I believe, that Wittgenstein had in mind in asserting that what *Principia* hoped to *say* with its *wff* of infinity ("axiom of infinity") is *shown* once identity has been eliminated (built into N-operator Notation).[18]

13.5 The Pièce de Résistance

Showing concerns formal "relations" (as it were) of sameness of form. Such a formal relation must be *shown* and not *said.* This doctrine naturally arises for Wittgenstein already in connection with the way the *ab*-Notation works to offer a pictorial representation of the truth-conditions of negation itself. Thus, in his struggle to understand how negation can work, *showing* was born. In light of what we have seen, there should be no further historical doubts any longer about Wittgenstein's *ab*-Notation, truth-tabular notation and N-notations. These notations are, one and all, products of his intuition of *showing*. *Showing* was precisely what both *ab*-Notation and the N-notation were devised to implement from their inception. Since the idea of the *Doctrine of Showing* is tied intimately to the *ab*-Notation we are justified in maintaining that already in the 1913 *Notes on Logic* it is an intuition of his *Doctrine of Showing* that is driving Wittgenstein ideas concerning the nature of Logic. [19] But there is yet another bit of very important evidence that ties *showing* to the 1913 *Notes on Logic*. We find the following (Wittgenstein 1979, p. 106):

> When we say "A believes *p*", this sounds, it is true, as if there we could substitute a proper name for "*p*"; but we can see that here a *sense*, not a meaning is concerned, if we say "A believes that '*p*' is true"; and in order to make the direction of *p* even more explicit, we might say "A believes that '*p*' is true and 'not-*p*' is false. Here the bi-polarity of *p* is expressed, and seems that we shall only be able to express the proposition "A believes *p*" correctly by

[18]Wehmeier and Rogers (2012) disagree. They hold that Wittgenstein's exclusive quantifiers recovers the modern view that the logical truths of quantification theory are just those that are invariant over every non-empty domain. They allow finite domains. On my interpretation, Wittgenstein's clear intent with exclusive quantifiers was to have only infinite domains. See Landini (2007). For some further criticisms see Lampert and Säbel (2016).

[19]In my book, *Wittgenstein's Apprenticeship with Russell*, I assumed that readers understood the intimate connections here made plain. See Landini (2007).

the *ab*-notation; say by making "A" have a relation to the poles "*a*" and "*b*" of *a-p-b*. The epistemological questions concerning the nature of judgment and belief cannot be solves without a correct apprehension of the form of the proposition.

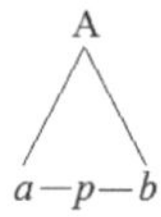

This comes as something of a shock. What bearing can the *ab*-Notation legitimately have on the analysis of the occurrence of "*p*" in a statement "A believes that *p*"? How in the world can it be that Wittgenstein could come to hold that "we shall only be able to express the proposition "A believes *p*" correctly by the *ab*-notation."

This passage is very difficult. It must be understood that Russell's worked toward offering a correspondence theory of *truth* that relied on the thesis that when a *wff* is subordinate to "... is true" it acts as a disguised definite description (an ordinary name) for a fact. That is,

$$[Rxy] \text{ is true} = df \quad \text{E}!(\iota f)(...f...)$$

Russell struggled to fill in the dots—i.e, find the correct definite description. It is often very difficult to properly describe a fact. It is often not sufficient to simply indicate constituents, since distinct facts can be composed of precisely the same constituents. (The fact of Desdemona's loving Othello is quite distinct from the fact of Othello's loving Desdemona.) But putting these complications aside, Wittgenstein seems to be suggesting that "A believes *p*" should be construed as if it were "A believes that *p* is true and not-*p* is false". In this way, he thinks we can see that the role of the occurrence of "*p*" in "A believes *p*" is not to describe (or name) a fact (i.e., as he puts it, "it is not the meaning that is concerned"). As if an oracle, he proclaims without argument in the 1913 *Notes on Logic*: "Facts cannot be *named*". He holds, *a fortiori*, that a fact cannot be given a definite descriptions. Wittgenstein is saying in this passage that the role of "*p*" in "A believes *p*" is such that it is the truth conditions (*alone*) that are concerned and this is why he thinks that only the *ab*-Notation can sort this out. Obviously, the *ab*-Notation concerns only truth-conditions since logical equivalents in *ab*-Notation are to be the same. In short, Wittgenstein is demanding that belief contexts respect *showing* already in the 1913 *Notes on Logic*. Why?

Once again there can be only one viable explanation —his *Doctrine of Showing*. It is indeed a well-known implication of *showing* that semantic relations are formal and must be *shown* and not *said* by a theory (of truth). This is precisely what we find in the 1913 *Notes on Logic* just after his shocking proclamation that only that *ab*-Notation can sort out the role of "*p*" in "A believes *p*". We writes (p. 107):

No proposition can say anything about itself, because the symbol of the proposition cannot be contained in itself; this must be the basis of the theory of logical types.

That is already a statement of the *saying* versus *showing* distinction. But if I am right, we should expect to find that Wittgenstein's intuition of *showing* lurks behind his objection to Russell's multiple-relation theory of judgment. In fact, this is precisely what we find. Compare the following two passages, the first from a 1913 letter to Russell and the second from the *Tractatus*:

> June 1913
>
> I can now express my object to your theory of judgment exactly: I believe it is obvious that, from the proposition "A judges that (say) *a* is in a relation *R* to *b*," if correctly analysed, the proposition that "*aRb* ∨ ~(*aRb*)" must follow directly *without the use of any other premiss*. This condition is not fulfilled by your theory.
>
> *TLP* 5.5422
>
> The correct explanation of the form of the proposition, 'A makes the judgement *p*', must *show* that it is impossible to judge a nonsense. (Russell's theory does not satisfy this condition.)

I have italicized the word "show" in this passage *TLP* 5.5422 from the *Tractatus* to emphasize that *showing* is the key to it. Recall again that in Wittgenstein's view, what can be *shown* cannot be *said* and that all and only logical/semantic relations pseudo-relations that are to be *shown* by the ideal grammar for empirical science. Now Griffin (1985a, b) noticed long ago that it is a contingent psychological matter that minds do not, in fact, judge nonsense—if it is indeed obviously contingently true (and Chomsky is wrong about "Green ideas sleep furiously"). How then did this contingent matter of human psychology get transformed into a matter of logical analysis? Without *showing* driving the intuition, it is not obvious at all that *logic alone* rules out nonsense judgment. The best explanation is that the "without the use of any further premise" demand is Wittgenstein's early way of expressing his view (which, by the way Russell claims he never understood) that *showing* is involved in judgment. The *Tractatus* uses the word "show" explicitly. This strongly suggests *Showing* was lurking behind Wittgenstein's objection to Russell's theory of judgment. Indeed, his objection that *logical* analysis must rule out nonsense judgment is completely unmotivated otherwise. There is no new objection to Russell's theory of judgment in the *Tractatus*—a new objection derived from the *Doctrine of Showing*. It is the same objection that dates to 1913 and the *Notes on Logic*. That is the pièce de résistance.

References

Beale, J. & I. J. Kidd. (2017)., eds., *Wittgenstein and Scientism* (London: Routledge).

Bernie Linsky, "The Near Riot Over Negative Facts," in Elkind and Landini, eds., *Logical Atomism: A Centenary Reappraisal* (Palgrave-MacMillan, 2018).

Fogelin, Robert. (1982) "Wittgenstein's Operator N," *Analysis* 42, pp. 124–128.

Fogelin, Robert. (1987) *Wittgenstein* 2nd ed., (London: Routledge & Kegan Paul).

Geach, P. T. (1981) "Wittgenstein's Operator N," *Analysis* 41, pp. 168–171.

Geach, P. T. (1982) "More on Wittgenstein's Operator N," Analysis 42, pp. 127–8.

Griffin, Nicholas. (1985a) "Russell's Multiple-Relation Theory of Judgment," *Philosophical Studies* 47, pp. 213-247.

Griffin, Nicholas. (1985b) "Wittgenstein's Criticism of Russell's Theory of Judgment," Russell 5, pp. 132–145.

Lampert, Timm and Säbel, Markus. (2016) "*Wittgenstein's Elimination of Identity for Quantifier-Free Logic*,".
Lampert, Timm. (2017) Wittgenstein's ab-Notation: An Iconic Proof Procedure," *History and Philosoph of Logic* 38 pp. 239–262.
Landini, Gregory (2007) *Wittgenstein's Apprenticeship with Russell* (Cambridge: Cam Univ. Press).
Potter, Michael. (2009) *Wittgenstein's Notes on Logic* (Oxford: Oxford University Press).
Quine, W. V. O. (1963) *Set Theory and its Logic* (Cambridge: Harvard University Press).
Russell, Bertrand. (1914a) *Our Knowledge of the External World as a Field for Scientific Method in Philosophy* (London: Open Court Publishing Company).
Russell, Bertrand. (1914b) "On Scientific Method in Philosophy," in Bertrand Russell, *Mysticism and Logic and Other Essays* (London: Longmans, Green and Co. 1921), pp. 97-124. (Delivered as the Herbert Spencer Lecture at Oxford, 1914 and originally published by Clarendon Press).
Russell, Bertrand *The Autobiography of Bertrand Russell 1914-1922* (Boston: Little Brown & Co, 1968), p. 139.
Wehmeier, Kai. & Rogers, Brian (2012) "Tractarian First-order Logic: Identity and the N-operator," *Review of Symbolic Logic* 5, pp. 538-573,
Wittgenstein, Ludwig. (1979) *Notebooks 1914–1916*, ed., by G. H. Von Wright and G. E. M. Anscombe (Chicago: University of Chicago Press).
Wittgenstein, Ludwig. (1922) *Tractatus Logico-Philosophicus*, trans. by D. F. Pears & B. F. McGuinness (London: Routledge & Kegan Paul).

Chapter 14
Thinking in Spaces: A Characteristic of Wittgensteinian Philosophy

Pascal Zambito

14.1 "Space" in Wittgenstein's Philosophy

Wittgenstein's philosophy is permeated by spatial imagery. This is true not only of the *Tractatus Logico-Philosophicus* (Wittgenstein 1922, hereafter TLP) with its prominent "logical space", but also in his so-called middle-period, especially in the time from 1929 to 1933, one of the most productive episodes of his life: in this period he created ten manuscript volumes (Wittgenstein 2015, MSS105–114) which form the backbone of much of his later philosophy. Later writings consist to a large extent in revisions of these remarks, leading to new directions the closer they get towards the *Philosophical Investigations* (Wittgenstein 2009, hereafter PI). One point where the "late" departs from the "middle" Wittgenstein is the apparent disappearance of the spatial expressions that were so conspicuous in the early 30s. What I would like to suggest in this chapter is not to neglect the methodical role of this imagery in Wittgenstein's thinking and instead to take it as a central characteristic of his philosophy. What is hinted at with the usage of "space" and "geometry" in the early and middle period survives in the later writings where it appears under different names, but with largely similar functions.

Moreover, I shall show how this "spatial" or "geometrical" mode of conceiving problems can be seen as a connection between Wittgenstein and his historical context. While he is often presented as an isolated genius, not much concerned with the work of earlier philosophers and misunderstood by his contemporaries, he himself believed that his creativity consisted mainly in taking up ideas from other thinkers

P. Zambito (✉)
University of Cambridge, Emmanuel College, Cambridge CB2 3AP, UK
e-mail: pascal.zambito@cantab.net

S. Wuppuluri and N. da Costa (eds.), *WITTGENSTEINIAN (adj.)*,
The Frontiers Collection, https://doi.org/10.1007/978-3-030-27569-3_14

for his own "clarificatory work".[1] And he also thought, as he wrote in the preface of the *Investigations*, that for several reasons his ideas "will have points of contact with what other people are writing today".[2] One of these reasons, I suggest, is the history of geometry which plays a central role in the intellectual background of his contemporaries.[3]

After briefly sketching the historical development of geometry from Kant to the 20th century, I will provide an account of how I think Wittgenstein's "logical space" can be seen as standing in that tradition; its characteristics and functions within the system of the *Tractatus* are carved out in order to establish criteria for comparison. According to these criteria I will trace the continuity of the concept of space in the middle Wittgenstein, where logical space disintegrates into a plurality of grammatical spaces which are used to account for a variety of philosophical topics, connected only by what one might call a "spatial method" which is derived from his peculiar understanding of space as a constitutive framework. Once the main features of this method have been recognized it is possible to see the same strategy at work in the later writings where Wittgenstein uses different words to describe constitutive frameworks. Central concepts are "calculus", "game", "style", "world view", "form of representation" or "form of life"—while each of them would deserve an own study to account for its idiosyncrasies, the focus here shall be on their similarity. They can all be understood as modified versions of the idea of space as a constitutive framework of possibilities which, I argue, is an integral part of Wittgenstein's grammatical method. Finally, this understanding of *a priori* frameworks in the wider sense opens up possibilities to apply Wittgensteinian thinking to other fields of knowledge.

14.2 Geometry After Kant

When I say Wittgenstein employs "space" as a metaphor this has to be taken with a pinch of salt. Although there is clearly a transfer of meaning going on, what is transferred is itself a very formal concept (not unlike "picture" in the *Tractatus*) and therefore not merely a metaphor, but a technical term used in a metaphorical way. We find such an understanding of space in geometry as this science itself seeks to abstract from objects *in* space so as to study its formal structure alone. This is how

---

[1] The most famous instance of these remarks is: "I believe there is some truth in it when I think that I am actually only reproductive in my thinking. I believe I have never *invented* a move of thought, but it has always been given to me from someone else and I have only passionately taken it up for my clarificatory work. In this way, Hertz, Boltzmann, Schopenhauer, Frege, Russell, Kraus, Loos, Weininger, Spengler, Sraffa have influenced me" (MS154, 14v).

[2] The earliest version of the preface is from 1938, the last one from 1945. There he writes that the book had occupied him "for the last sixteen years". So the word "today" refers to a larger span of time ranging from the late 20s to the 40s.

[3] Bertrand Russell's first publication on mathematical philosophy was his *Essay on the Foundations of Geometry*. Michael Friedman has pointed out the importance of the history of geometry for many members of the Vienna Circle in his *Reconsidering Logical Positivism* (1999).

Kant arrived at his view that space is, just like time, a *form* of intuition. It is a *pure* form of intuition by being abstracted from any empirical content. For Kant, geometry was an example of *synthetic a priori* knowledge. That is, it can be known prior to any experience and yet consists of true insights about the nature of space *qua* form of intuition (cf. Kant 1999, Introduction and Transcendental Aesthetic).

What became controversial about this is the claim for universality and necessity that comes with the status as *a priori* knowledge. The position Kant was known for is that Euclidean Space is the necessary and *a priori* structure of our outer intuition.[4] However, non-Euclidean geometries were developed in the 19th century which proved that other consistent geometries are at least possible. Hence Euclidean geometry cannot be *a priori* necessary. But not only spaces with positive or negative curvature were now conceivable; Bernhard Riemann developed an influential account of "manifold" which is the most abstract form of geometry one can think of: a system of differences with various dimensions and variable curvature. A concrete geometry in the narrower sense, such as Euclidean geometry, would be based on hypotheses which can be tested empirically and result in a posteriori metrics within the *a priori* structure of a Riemannian manifold. It is important to distinguish between pure geometry and physical geometry in the discussions of this time. Most thinkers who worked on non-Euclidean geometries and manifolds in the late 19th century thought of it as a purely mathematical issue; physical space was still assumed to be Euclidean. Some Kantians still sought to justify it by intuition (Hertz), others by measurement (Riemann, Helmholtz) and yet others suggested conventionalist criteria such as simplicity (Poincaré).

But the mathematical, formal, account of geometry turned out to be useful in other respects, too. Hermann von Helmholtz took up the idea of Riemannian manifolds in his theory of perceptions. Time and Space, but also colors, sounds and other sensations come in space-like structures which are *a priori* in a Kantian sense, although they do not describe Euclidean geometry. In this sense, Euclidean space is just one among many conceivable manifolds with three dimensions, time is a one-dimensional manifold, and colors or sounds come in spaces with as many extensions as are needed to account for all their possibilities of differentiation, e.g. two in the case of tones where pitch and volume can be determined (cf. Helmholtz 1903, 16f.; quoted in Hyder 2002, 26). Other physicists employed this new formal understanding of space for different purposes. Heinrich Hertz in his *Principles of Mechanics* makes it clear from the beginning that space for him is simply Euclidean space (1899, 45); but the great innovation of his book, of the scientific "image" he suggests, is a geometrical method which is based on Riemann's manifold theory. A system of points can be modeled as a *3n*-dimensional space in which various things can be said *a priori* (what Hertz does in Book I of the *Principles*). In these geometrical systems configurations of objects can be viewed as thinkable locations determined by the coordinates

[4]Kant's views on this issue changed during his career (cf. Caygill 1995, 367–373), but the *a priori* necessity of Euclidean Space became the background against which geometers and philosophers argued in the 19th century.

of these objects.[5] Another Neo-Kantian physicist, Ludwig Boltzmann, used higher-dimensional structures in order to efficiently model probabilities of distributions, thereby anticipating what today is called "phase-space", a *6n*-dimensional manifold in which not only the three-dimensional locations of particles are considered, but also their momenta.

Hertz and Boltzmann are both on Wittgenstein's list of influences for his "clarificatory work". Consequently, there is a lot of literature on their relation, alternately focusing on similarities or differences; currently the pendulum is more on the differences side. To be sure, a direct influence is unlikely for biographical reasons as well as for obvious differences between physical systems of objects and Wittgenstein's logical interest in facts and propositions (cf. Pilch 2017; Preston 2017). However, the similarity I want to highlight lies on a more abstract level: it is the creative use of geometrical methods which originate in Kant and which were further abstracted to a pure formal instrument; the use of multidimensional space-like structures as a framework of possibilities which can be known *a priori*.

14.3 Logical Space: A Universal Manifold

Jesper Lützen calls Hertz's *Principles of Mechanics* "the geometrization of mechanics" (2005, 159). In a way, Wittgenstein's *Tractatus* is the geometrization of logic. In the eyes of philosophers such as Russell and Carnap its most important insight is that logical propositions have their unique necessary status not by being the most general kind of true statements, but thanks to their propositional emptiness.[6] They are not even propositions with a sense, but "say nothing" and thereby resemble the axioms of modern formal geometry. Although Wittgenstein himself regarded the ethical implications of his book more important than the logical ones, it was clearly a central motivation to understand the capability of language to represent the world, the capability of propositions to represent possible facts. One part of the solution is: propositions are pictures and share with the pictured reality the possibility that names in the proposition are combined in the same way as objects are combined in a fact. This possibility is called "logical form" which is explicitly said to be more general than other more specific forms, such as the spatial or the temporal form which only account for possible spatial or temporal relations among objects.

---

[5]On Hertz's geometrization of mechanics see especially Lützen (2005).

[6]I quote them as leading representatives of the Cambridge and Vienna intellectuals with whom Wittgenstein was in contact: "I am convinced you are right in your main contention, that logical props are tautologies, which are not true in the sense that substantial props are true" (Wittgenstein 2004, Russell to Wittgenstein, 13 August 1919); "The most important insight I gained from his work was the conception that the truth of logical statements is based only on their logical structure and on the meaning of the terms. Logical statements are true under all conceivable circumstances; thus their truth is independent of the contingent facts of the world. On the other hand, it follows that these statements do not say anything about the world and thus have no factual content" (Carnap 1963, 24).

Logical form shares with the Kantian forms of intuition its role as a framework of possibilities and is indeed conceived in terms of spatial analogies: "The geometrical and the logical place agree in that each is the possibility of an existence" (TLP 3.411). However, Wittgenstein's logical form is more general as it can model any kind of multiplicity: space and time which were different forms of intuition for Kant are just two sub-forms of the universal logical form. Time itself can be considered a kind of space, a one-dimensional manifold as Helmholtz had said, and with which Wittgenstein implicitly agrees: "Just as we cannot think of spatial objects at all apart from space, or temporal objects apart from time, so we cannot think of *any* object apart from the possibility of its connexion with other things"—the logic of the *Tractatus* aims to account for "*any* object" which has to be conceived in a space of possibilities. "Every thing is, as it were, in a space of possible atomic facts. I can think of this space as empty, but not of the thing without the space" (TLP 2.0121; 2.013). Wittgenstein goes on to mention *specific* objects in *specific* spaces, such as spatial objects in infinite space, colored objects in color space etc. (TLP 2.0131). But they are also conceived as special cases of a more general logical space which figures prominently in the *Tractatus* and which stands in the same relation to these specific spaces as the "logical picture" stands to the "colored" or "spatial picture" (cf. TLP 2.171; 2.182).

Wittgenstein's logical space is what Kant's Euclidean Space was supposed to be, but actually is not: a truly *a priori* structure which entails all possibilities of meaningful propositions. By being a multi-dimensional manifold for every conceivable proposition—not just "geometrically conceivable" locations as in Hertz (1899, 48)—it generalizes the Kantian idea of *a priori* structure to the most abstract level. The world: a contingent and empirical totality of facts whose possibilities are modeled in a universal logical manifold. Since logic studies the conditions of these possibilities it is, echoing Kantian terminology, "transcendental" (TLP 6.13).

Precisely for this reason, however, logical propositions, i.e. tautologies, are not true in the same sense as other propositions. For a fact *in* logical space can exist or not exist, but the *possibility* that one of these is true does not tell me anything about reality (If I know that "it is raining or it is not raining" I don't know anything about the weather). The tautology "*p* or ~*p*" is necessary because it exhausts the possibilities each of which is contingent, but one of which must obtain. It determines, as it were, the candidates for truth, but is not itself a candidate for truth. The necessity of logical propositions is for Wittgenstein tied to the notion of objects. Their configurational possibilities constitute the possibilities of atomic facts; therefore it makes no sense to speak of objects' existence, for their "existence" consists precisely in the possibility to occur in a fact. We can speak of existence or non-existence with regards to facts which are by definition contingent, but not with regards to objects which constitute their possibilities and therefore have to "exist" in any case. Objects are conceived as locations within a manifold which we can know *a priori*. Without making any claim about the geometry of physical space, Wittgenstein made creative use of formal geometry, of its account of points and of multi-dimensional spaces.[7] He used these methods in order to construct his logical system "with great simplicity and brevity", as Hertz put it in his preface to justify his geometrical "image" (1899, 31). With these words Wittgenstein would surely have agreed.

Logical Space has recently received more attention as a central structure of the *Tractatus*. Martin Pilch distinguishes three central functions (cf. 2017, 18–20). The first is to provide a framework for the logical possibilities as sketched above. The second is to account for negation. How so? A proposition corresponds to a specific region of logical space. Whether it is true or not, is not a matter of logic, just like the existence of a physical object at a certain point in space is not a matter of pure geometry; the region merely indicates "the possibility of an existence". But one of the central claims of the *Tractatus* is that propositions are inherently binary, they are either true or false. Wittgenstein's account of propositions as pictures illustrates the problem, but does not solve it: a picture shows how things stand if it is true; but if it is not true it still shows the same combination of things, it has the same sense. In order to illustrate this, the space analogy is better suited than the picture theory: a proposition *p* corresponds to a certain area in logical space and its negation *~p* corresponds to the rest of logical space, its complementary area. Thereby two opposed propositions can indeed be said to refer to the same place, as it were, to the same boundary; but the negation still points to a different place by saying that it lies *outside* the negated area (cf. TLP 4.0641). While the picture theory can only account for the negative proposition by means of the positive picture, the space analogy by implying an element of direction is capable of illustrating the internal relation between a proposition and its negation more adequately.

The third function of logical space is its completeness. A proposition is indeed internally connected to its negation (by being its complementary part) as well as to other propositions that follow from it (by falling within its region in logical space). Since space was meant to capture the possible configurations of objects it must contain all transformations of the proposition, i.e. all possible combinations with other propositions in logical space. In a toy-space of two elementary propositions this would result in 16 combinations, that is, 16 distributions of true and false values as listed in TLP 5.101. These are *all* combinations that are possible in this space. "The logical scaffolding round the picture determines the logical space. The proposition reaches through the whole logical space" (TLP 3.42).

Moreover, Wittgenstein believed that logical space is necessarily universal. Subspaces are considered (cf. TLP 2.0131) but they are all representable within the *one* logical space just like "every picture is *also* a logical one" (TLP 2.182). Retrospectively, in 1937, Wittgenstein explains this with his requirement of apriority, the "crystalline purity", of logic. Since logic must be free of all experience, there could not be different *kinds* of logical spaces; for then it would be possible to discover a new kind with a different logic. Thus, the early Wittgenstein had been led to his assumption that there must be a (unique) general form of the proposition (cf.

[7] Wittgenstein was probably familiar with these developments from his studies in theoretical physics or from Russell's writings.

MS157a, 45–50). While the three functions remain important in Wittgenstein's later use of the space metaphor, it is this claim for universality and generality which turns out to be one of the "grave mistakes" in the *Tractatus* (PI, preface).

14.4 From Logical Space to Grammatical Spaces

That "spaces" are important in Wittgenstein's middle period is evident to the reader of his manuscripts from that time. Russell comments about a pre-stage of the *Philosophical Remarks* in 1930:

> He uses the words 'space' and 'grammar' in peculiar senses, which are more or less connected with each other. […] These 'spaces' are apparently given *a priori* in a Kantian sense, or at least not perhaps exactly that, but something not so very different. Mistakes of grammar result from confusing 'spaces' […] One might define a space, as he uses the word, as a complete set of possibilities of a given type. If you can say 'This is blue', there are a number of other things you can say significantly, namely all the other colours. (Russell 1968, 198)
>
> There are various relations among colours which constitute the geometry of that 'space'. All this is, in one sense, independent of experience: that is to say we need the kind of experience through which we know what 'green' is, but not the kind of through which we know that a certain patch of wall is green. Wittgenstein uses the word 'grammar' to cover what corresponds in language to the existence of these various 'spaces'. (ibid., 200)

Although he does recognize the importance of possibility in the concepts "space" and "grammar", Russell does not see a connection to logical space in the *Tractatus*. Joachim Schulte recognizes this connection but does not specify it (cf. 2018, 319) and elsewhere highlights the disappearance of "spaces" as a central feature of late Wittgensteinian thinking (cf. 2006, 566). Instead, I would like to specify the continuity from early to middle Wittgenstein and, beyond that, argue that the "spatial method" is still present in the later writings even though the terminology changes. To substantiate this, I shall highlight below some of the many occurrences of "spaces" from the period 1929–1933.

(i) As Russell observes, colors are considered a space of possibilities by Wittgenstein. What "you can say significantly" about colors is a matter of *a priori* exclusion or implication: if "This is blue" is true, this excludes all the other colors; it implies that the propositions "This is not red", "This is not green" etc. are also true and their opposites false. This dependency threatens a crucial claim of the *Tractatus*: the independence of elementary propositions. Wittgenstein had thought that color statements are not elementary and hence could be further analyzed until that kind of logical impossibility becomes explicit which is only implicit in the unanalyzed statement: that "A is red and A is green" would, in its analyzed form, be a contradiction such as p & $\sim p$.[8] However, he came to accept that no such analysis can be provided without assuming further rules for statements about colors. These rules

[8] Frank Ramsey had pointed out this problem to Wittgenstein: e.g. "No one could say that inference from 'This is red' to 'This is not blue' was formally guaranteed like the syllogism" (1990, 48).

would require familiarity with color concepts and hence be "in a certain sense a posteriori" (Wittgenstein 1929, 163). But, as Russell rightly points out, this knowledge is distinct from knowledge *that* a certain patch is green. Since we know color space before we actually look at how things are, it can be said to be *a priori* in a modified Kantian sense.

As seen, Wittgenstein had thought about these *a priori* spaces of possibilities in the *Tractatus*. Implicitly the spaces mentioned there already contain the idea that one coordinate in this space can only be assigned once, but he did not elaborate on this, focusing instead on the common logical form. Ten years later he saw the peculiarities of these sub-spaces, many of which contain statements of degree in which one degree excludes all others, just like a stick that is 2 m long cannot be 3 m long at the same time. In 1930 he writes:

> In my old conception of elementary propositions there was no determination of the value of a coordinate. Although my remark that a colored body is in color space etc. could directly have led me there.
>
> A coordinate of reality can only be assigned *once*. (MS108, 53)

So the space of colors is an *a priori* space of possibilities. Another thing Russell noted correctly is that this account of *a priori* possibilities is associated with "grammar". It excludes certain expressions just like Euclidean geometry excludes that two parallels intersect. Moreover, the grammatical space of color statements is also supposed to account for relations such as "orange is closer to red than to blue" and to exclude combinations such as "reddish green". In order to illustrate this color space as it can be known *a priori*, Wittgenstein suggests a "color octahedron" with black and white and the four primary colors at its corners. Each point on its surface would be a meaningful proposition about color and the octahedron itself would form a "grammar" for those propositions:

> Color space is *roughly* represented by the color octahedron with pure colors at its corners and this representation is a grammatical one, not a psychological one. (TS213, 440f.)

(ii) The notion of *grammar*, which occurs only once in the *Tractatus*, is continuously refined in the course of the years 1929–1933. Since dependencies between basic determinations (formerly elementary propositions) are now acknowledged, grammatical spaces cannot be conceived as contained in an all-encompassing logical space. The peculiarities of each grammar have to be accounted for. A tempting way to put it is that the independence of elementary propositions was for Wittgenstein's logical space what the parallel postulate was for Kant's Euclidean space: the supposed universal space proved to be untenable as an *a priori* certainty and other geometries had to be accepted. The difference is, of course, that logical space stands at the end of a development of increasing formalization which started with Euclidean geometry; its universality was based on the unity of logic and not on the mind's forms of intuition. However, both claims for universality were shattered by the insight that some things, which were supposedly guaranteed by the universal space, cannot be known *a priori*.

Wittgenstein does mention different logics and grammars (e.g. MS114, 15v), but he never considers the question how many grammars there are. To follow the analogy to the history of geometry we can put it in the words of Ernst Cassirer who commented on the new plurality of geometries, not only in mathematics, but also in physics: "Instead of speaking ontologically of the being or indeed the co-existence of diversely constituted 'spaces', which results in a tangible contradiction", one should rather, as relativity theory does, speak "methodically of the possibility or necessity to use different measurements, i.e., different geometrical conceptual languages" (1922, 440).

(iii) Wittgenstein repeatedly used examples and analogies from the domain of measurement in general and of relativity theory in particular in order to illustrate his conception of grammar and grammatical space. The analogy is so attractive because it captures two important aspects of his interest in logic: the measure exists *before* anything is actually measured, in this sense it is *a priori*; but unlike Kantian *a priori* it is not *true*, it rather determines *how* something is measured, not the result which can be true or false. The metric system is not true itself, only a statement about the length of an object measured in meters can be true. Wittgenstein considered this focus on the method of verification, on sense instead of truth, as a characteristic of his new method.[9] In his analogies to relativity theory he refers to Einstein's insight that the method of verification is constitutive of the meaning of the concepts used, in this case the concept of simultaneity:

> The concept does not exist for the physicist until he has the possibility of discovering whether or not it is fulfilled in an actual case. We thus require a definition of simultaneity such that this definition supplies us with the method by means of which, in the present case, he can decide by experiment whether or not both the lightning strokes occurred simultaneously. As long as this requirement is not satisfied, I allow myself to be deceived as a physicist (and of course the same applies if I am not a physicist), when I imagine that I am able to attach a meaning to the statement of simultaneity. (Einstein 1920, 19)

When mentioning Einstein, Wittgenstein refers to the constitutive function of grammar which resembles that of the stipulation of a method for judging agreement in time.[10] He transfers this idea to agreement in other fields, notably to phenomenal experience and its agreement with mental anticipation.

> How can I know that I expected *this* if not through the fact that it satisfies my expectation now, meets my expectation now.
>
> A step is needed which resembles that taken in relativity theory. (MS108, 270)

Just like Einstein undermined the idea of an absolute time relative to which events could be said to be simultaneous, Wittgenstein undermines the assumption of a metaphysical realm in which thought and reality meet. Language is the only measure

[9]"It is the specific criterion of structure to find out which propositions have sense in it—not which ones are true. To search this is the method of philosophy" (MS105, 36); "This method is essentially the transition from the quest for *truth* to the quest for *sense*" (MS105, 45).

[10]See also: "Verification is not *one* indication of truth, but *the* sense of the proposition (Einstein: how a magnitude is measured, is what is measured)" (MS107, 142r).

that can serve as a criterion of agreement. If two phenomena can be described by using the same words there is an internal relation between them irrespective of empirical circumstances. The examples Wittgenstein considers are the relations between a wish and its satisfaction, an order and its execution, an expectation and its fulfillment. What all these have in common is that the two internally related statements refer to the same grammatical space, which functions as the measure to make a statement about agreement in the first place; whether the agreement actually occurs, whether the wish gets satisfied, the expectation fulfilled, does not affect the level of sense which is, again, conceived in spatial imagery.

> The expectation is not a fantasy, for if I expect that a light point will appear, then there must be a place *there* at which the point *really* will appear if it appears.
>
> Anticipatory in the expectation as well as retrospectively in the fulfillment and memory of the expectation, expectation and fulfillment are in the same space. (MS108, 228)

That these are related, that I can recognize an experience I had before, is a background assumption which belongs to this method of verification. As "Einstein has taught the world: our method of time measurement belongs to the grammar of our time expressions" (MS119, 116v). And like Einstein Wittgenstein aims to investigate "the clocks with which we measure the phenomena" (MS164, 82). What makes his philosophy so comparable to measurement is that his grammar resembles the stipulation of a measure which constitutes the possibility of truth; it is concerned with sense, not with truth.

(iv) But not everything grammatical can be compared to measurement. Another function of the space metaphor in the middle period is to distinguish between linguistic systems, between different approaches to given situations that require different grammars. A prominent example is the distinction between various degrees of exactitude that are used in different grammatical spaces (cf. PI §88). The treatment of generality is a related case. A general statement accounts for a number of instances without addressing each individual element. To say that a circle is in a square does not logically translate into an infinite disjunction of potential locations for the circle, but rather to an "in" or "out"; it is, as it were, "not in measuring space" (MS111, 37). Speaking of a line as an interval in a coordinate system does not treat of all individual points of this line, but takes this interval out as a whole *in contrast* to the rest of the coordinate system. The grammatical space of such general statements only allows for binary possibilities and is not concerned with the potentially infinite possibilities that may fall under its concepts.

(v) Visual space, too, is not a "measuring space". It features a similar indeterminacy as the space of generality by prescinding from the attempt to determine something with an absolute exactitude that is not achievable. That the edges of the visual field are blurred is a feature of its grammar and not a deficiency. Here Wittgenstein distinguishes between visual space and physical space in terms of exactitude and measurement. In physical space an ever greater (but not an absolute) exactitude is theoretically possible by conducing ever finer measurements. In visual space this is not possible as it is inherently inexact due to its peculiar grammar. Philosophical problems result from confusion of these spaces and from the attempt to capture the inherent inexactness of visual space with apparent exactitude, for example in Mach's picture "where the vagueness of objects at the edges of his visual field is represented by a vagueness (in a completely different sense) of the drawing" (MS107, 284).

Another example for the contrast of different spaces is the distinction of time concepts in two systems: a primary system of immediate experience and a secondary system of linguistic description. Wittgenstein illustrates this with the analogy of a film projector. On the film strip there are distinct pictures and it has sense to call that which is in front of the lamp "present", others "past" or "future" according to their position on the strip. On the screen, however, there is only the continuum of one ever changing picture and the immediacy of experience slides along this continuum so that one is tempted by a presentist interpretation of time: "only the present is real". Wittgenstein's reply is motivated by his grammatical method: "present as opposed to what?". Since in the grammatical space of immediate experience no other time than the present is even thinkable, there is no antithesis to this statement, and hence it has more or less the status of tautologies in the *Tractatus*: it fills out the whole space of possibilities and therefore is no informative proposition in this space (cf. MS108, 3). It does, however, tell us something about the grammatical space in which we are operating and thereby can be called a geometrical statement.

(vi) In the manuscripts from the middle period, the discourse about systems with diverging grammars for time, exactitude or similar concepts is paralleled by a mathematical discourse where, again, spatial analogies and conceptions abound. The very first remark after Wittgenstein's return to Cambridge in 1929, starts with such a comparison:

> Is a space conceivable which contains only the rational points, but not the irrational ones?
>
> And this only means: are the irrational ones not already preconceived by the rational ones? (MS105, 1)

In the following volumes Wittgenstein reflects on the opposition between determinate and indeterminate objects, finite and infinite spaces which we are already familiar with from the other space metaphors in the domains of phenomenal experience, physical measurement and generality. Here the function of the space metaphor is obviously to stress the differences between different kinds of geometries or grammars for different kinds of numbers. Another function related to mathematics is the development of new geometries which plays little role in Wittgenstein's writings as a whole (after all, one of its central maxims is that philosophy ought to be merely descriptive). But in mathematics this issue is touched upon: innovations and understanding in mathematics are conceived in terms of new spaces, whereas the solving of puzzles within one and the same space is declared uninteresting, "the solving of a school exercise" (MS108, 19). He stresses the creative aspects of mathematics, for instance, when he compares the insight into the periodicity of certain fractions to Kant's view that mathematics is synthetic *a priori* (MS114, 11r). We have seen that Wittgenstein rejected synthetic *a priori* truth in general, but what he means here is that such a discovery is prior to experience and yet is something more than what was already known: it is the development of a new space. Likewise, Sheffer's "discovery" that logical operations can be simplified with just one symbol for joint negation, the

Sheffer stroke, was not the discovery of something within a space which was already given, but rather the invention of a new space (MS108, 106f.). The famous Wittgensteinian slogan that "the mathematician is not a discoverer, but an inventor" is to be taken in the same sense; an expedition of mathematical discoverers would be one which "was not sure of *space* itself!" (MS108, 91).—It is significant that these new spaces often involve the development of a new notation. I shall take up this creative understanding of the space metaphor that glimpses through the usage in Wittgenstein's mathematical considerations later when I discuss possible applications of Wittgensteinian "spatial" thinking to other fields.

(vii) In the later manuscripts of the middle period, Wittgenstein often takes up remarks from the earlier ones, from 1929 and 1930, and re-evaluates them in light of his now more robust understanding of the grammatical method. The resolution of the question for a common space for rational and irrational numbers is an indicator for the later fate of the space metaphor:

> Is a space conceivable which contains only the rational points, but not the irrational ones?
>
> And this only means: are the irrational ones not preconceived by the rational ones?
>
> As little as the game of chess is preconceived in the game of draughts. (MS111, 29)

During the 1930s games gradually take up many functions of grammatical spaces whose frequency decreases accordingly. Reflecting on the various forms his analogies to space can take Wittgenstein comes to recognize that these are not based on a single understanding of the word, but are themselves, like games (cf. PI §§66, 67), something like a family of interrelated concepts:

> Our notation, or mode of expression, shows which similarities – and differences – we want to highlight in particular. So sometimes one calls everything a space which has a similar structure as space, and one always wants to stress this analogy. And then again one only wants to shun this analogy because it leads to confusion and one wants to stress the differences between "spaces" [...] (MS108, 135)

14.5 New Concepts, Similar Functions

The first step away from "space" as the preferred metaphor for frameworks of possibilities is the notion of "calculus". It is generally considered a concept that is more on the *Tractatus* side of the "middle Wittgenstein" as it still conceives of language as a set of fixed and static rules as opposed to the more fluid concepts of the *Investigations*. Recently it has been highlighted that the idea of linguistic calculi is not completely abandoned by the late Wittgenstein, but is still considered, along with games, as an important "object of comparison" for philosophical clarification (cf. Kuusela 2008, 42). To better understand this intermediate concept I suggest to see it in the context of the development of grammatical spaces and language-games. First, one feature which calculi share with grammatical spaces and which separates them from the *Tractatus* is their plurality. Wittgenstein repeatedly stresses the differences from one calculus to another in order to highlight different rules, just like he distinguished between different grammatical spaces.[11] Second, calculi are already less static than spaces insofar as the possibilities they constitute are possible *moves,* not points or regions. This is what they share with games with whom Wittgenstein sees multiple connections, e.g. when he acknowledges formalist approaches to mathematics. However, what is less flexible and too static about calculi is that the possible moves are, in a way, "already made" by fixed rules. This leads to the problem of rule-following.

The problem of rule-following can be framed as a problem about space and time. The logical and grammatical spaces we have encountered are *complete* spaces of possibilities which are known *a priori*. They are hence concerned with meaning or sense, not with experience and truth. However, the later Wittgenstein would increasingly conceive of meaning as influenced by the actual use of language, that is, by experience, by something that unwinds in time. The meaning of a word in the early 30s was determined by "its place in grammatical space" (TS213, 34); in the *Investigations* we read: "But we understand the meaning of a word when we hear or say it; we grasp the meaning at a stroke, and what we grasp in this way is surely something different from the 'use' which is extended in time!" (PI §138). The new problem is to conceive of the individual applications of a rule as distinct and *a priori* knowable steps, while at the same time they are determined by something empirical, a posteriori. A comprehensive interpretation of the rule-following considerations would go beyond the scope of this paper, but conceiving of the problem in this way could perhaps lead to interesting new approaches to it.

A related point is the shifting sense of *completeness*. The idea that all steps are already determined by the rule is the idea that the space of possibilities is complete. However, the completeness, the use of the word "all", is different whether we talk about rules or finite sets; a rule is rather similar to the geometrical concept of a triangle which is not defined by a number of real triangles. The distinction between different grammars, for example between generality and measuring space resonates here: completeness does in one case consist in a disjunction of all distinct elements, but in another case this account leads to nonsense. Completeness of an infinite row rather consists in familiarity with the rule, not in an infinite disjunction of applications; and the completeness of visual space is not affected by the fact that I cannot describe its edges in all exactitude.

Now, games, which are as *language*-games important in all of Wittgenstein's later writings, are characterized by being rule-governed. Their advantage over calculi, which are also rule-governed, is that their notion of rules is much more liberal: moves can be part of a game even though they are not determined by the rules (PI §68); the rules of certain games may even be altered "as we go along" (§83). Games share with spaces the fact that they determine a certain domain of possibilities for which they constitute *sense*. What is not in the game does not have sense, such as

[11] He distinguishes, e.g., between the space of generality from the disjunction of individual elements: "It is a different calculus to which the determination of generality belongs, and another one which yields that disjunction" (MS111, 65).

check-mating in draughts; and this case also shows that there may be a *different* game in which the expression does have sense. The discussion of ostensive definitions in the earlier parts of the *Investigations* can be summed up—a bit simplistically—with the requirement of a grammatical space in which I know the *place* of the defined object. Otherwise it would be unclear whether I am pointing to a piece of paper, or to its shape, or its color. Having carved out the functions of grammatical spaces it is easy to see how knowledge of a "space" can serve as such a requirement which Wittgenstein in the *Investigations* compares to the mastery of a game (cf. PI §§27–33). Moreover, the plurality of conceptual frameworks is inherited from the geometrical pluralism we have encountered in the middle period. Unlike spaces, but similar to calculi, "the word 'language-*game*' is used here to emphasize the fact that the *speaking* of language is part of an activity, or of a form of life" (PI §23), i.e. is not something static in which all possibilities stay at their places. Unlike calculi then, language-games are interwoven with our ordinary human activities and forms of life.

With "form of life" a concept recurs that was very close to space in the *Tractatus*, namely the notion of "form". *Lebensform* is the most prominent of its variants, but Wittgenstein also considers concepts like "form of representation" or "form of expression" which have a similar function. In all these cases "form" is a framework of possibilities which *a priori* determines what counts as a location or a move. It is already quite emancipated from the earlier, rather literal, space imagery, but it serves a very similar purpose. Other concepts that similarly determine the way something is perceived are "world view" and "style".[12] While there may be many differences between these concepts in detail, there is sufficient similarity to see them as related to the early concept of "space". That its functions occur with different names at all stages of his work makes it a central characteristic of Wittgensteinian thinking.

14.6 Applications of the Spatial Method

The preceding sections have been restricted to Wittgenstein's writings and his conception of logic and grammar which he often construes in terms of geometrical or space-like structures. Importantly this is not a theory of meaning, but the suggestion of a method (the similarities justify the use of the singular despite the method's inherent plurality and openness which I have pointed out). This method can be applied to various fields. To conclude this chapter, I want to give two examples where methods are applied that are similar to those associated with the use of "spaces" and related concepts. The first is the field of principles in science which are meant to provide an *a priori* or *constitutive* framework. The second is a contribution to the much-debated problem of Wittgenstein's ethics which struggles with the apparent paradox that he refused to acknowledge any ethical propositions, but sometimes considered

[12]Plaud (2011) has highlighted the role of "style" as a kind of *a priori* framework very similar to space in the middle period.

his own philosophical work as, in a sense, ethical and regarded questions of ethics or aesthetics as entangled with logical or mathematical problems.

The debate about principles in science was initiated by Michael Friedman's work on the "relativized *a priori*", an account of the history of science which spans from the formalization of geometry via relativity theory and logical positivism to Thomas Kuhn's theory of scientific revolutions. Along these stages Friedman follows the fate of the Kantian *a priori* as it was questioned by 19th century mathematicians and then reduced to its constitutive function without the claim for universality and necessity. While the connection to "space" is evident in geometry and the logical positivists' conventionalism, it is less obvious in Friedman's reading of Kuhn: paradigms are seen as successors of Carnap's conceptual frameworks which constitute the range of possible knowledge, but are not themselves candidates for truth in the same way statements within the paradigm are verified. We can compare this development with the story of Wittgenstein's spaces: the assumption of a universal space *a priori* was abandoned and replaced with a plurality of formal systems with their own grammar; the strict formalism of these frameworks gets loosened in his later thinking—this is why the literal mentions of "space" decrease—but the basic idea of *a priori* frameworks remains a central characteristic of his philosophy.

In recent contributions to the debate about constitutive principles in science, the central problem is the status of these principles between *a priori* and empirical propositions.[13] This is precisely the problem that is at the heart of Wittgenstein's later philosophy. As meaning is increasingly conceived of as use, logic seems to lose its rigor, *a fortiori* as there is now a plurality of grammatical systems conceivable: what he wanted to avoid in the *Tractatus*, that there are *kinds* of logic, is now accepted. However, these grammars *locally* have the role of constitutive principles which are not questioned while one is playing the specific language-game. If we take again the game of the metric system, we can say that the specific role the prototype meter takes in the game of measuring is that it does not itself have a length that is known by measurement; in this sense it is "neither one meter long nor not one meter long". Instead it constitutes what a meter is and thereby allows measuring other things (cf. PI §50). The emphasis on the role of certain assumptions is something that is shared with the current debates about the *a priori* in science (cf. Friedman 2001, 46; Stump 2015, 167). Likewise, the problem that many of these constitutive principles seem to be empirical and yet are grammatical is repeatedly discussed in Wittgenstein's writings, especially in the last manuscripts which have been published under the title *On Certainty*. The point of propositions that seem to be empirical truths and yet feature an absolute certainty is that they function as grammatical propositions: they are not candidates for contingent truth, but are part of the geometry of our space, or at the basis of our form of life. Their certainty is not absolute, but even though they are only relative to the given system or space, they can nevertheless have a constitutive function.

With regards to ethics, Wittgenstein gives some hints what he could mean when stressing that the ethical (as well as aesthetics) is ineffable in logical space and,

[13]See Stump 2015 for a good summary of the development and current trends in this debate.

like logic, “transcendental” (TLP 6.421) and, later, that philosophy rather deals with difficulties of the will, not of the intellect (cf. MS112, 112r). It is obviously concerned with changing one’s whole perspective towards the world, that is, with the choice of space itself, not with a particular change of the facts *in* space. And in the later writings, although the word ethics does not figure prominently, it is clear that his grammatical investigations, his struggle to clarify spaces and to invent new ones, have an ethical dimension for him (see, for example, the passionate tone of the notebooks that lead to the *Investigations*, MSS152 and 157).

We have seen that Wittgenstein often stresses the creative aspects of mathematics and that this often comes with the use, or invention, of a specific notation. My claim that his ethics has something to do with the choice of conceptual framework—to which both space and notation amount in this case—is supported by his repeated insistence on the importance of notation for his philosophy.

> R. [Ramsey] does not understand the value I place on a specific notation, nor the value I place on a specific word, because he does not see that a whole way to look at things is expressed in it; the angle from which I am now looking at the matter. The notation is the last expression of a philosophical point of view. (MS105, 10).

In the following remarks this sensitivity for stylistic matters is connected to the spatial notions of “form” and “grammar” and the differences between spaces:

> One must not pass over a difference of forms – like one may well pass over a difference between suits if it is a very small difference.
>
> In a certain sense for us – that is in grammar – there are no “small differences”. Anyway, the word “difference” means something completely different here than when we speak of differences between things.
>
> A philosopher feels changes in the style of a derivation which a contemporary mathematician passes over calmly with a blank face. (TS213, 643r)

This strategy, as has been shown, is not restricted to mathematics. A major aim of Wittgenstein’s philosophy is to influence his readers, not in order to convince them of this or that truth or fact, but to change their perspective. The forms of representation, be they conceived in terms of space, calculus, game or style, are responsible for many of the philosophical problems that he was working on, and on that level they have to be cured. In light of the views on ethics Wittgenstein held already in the *Tractatus*, it is possible to see how this can be seen as an ethical activity. It is often quoted that Wittgenstein wants to change the philosopher’s “one-sided diet” of examples (PI §593). In his space phase he expresses a related idea: “He who teaches philosophy today gives dishes to the other, not because he likes them, but in order to change his taste” (MS112, 223).

References

Carnap, Rudolph. 1963. *The Philosophy of Rudolph Carnap*. Ed. by P. A. Schilpp. La Salle, Illinois: Open Court.

Cassirer, Ernst. 1922. *Einstein's Theory of Relativity. Considered from the Epistemological Standpoint*. In: The Monist. Vol. 32, No. 3 (July, 1922). 412–448.

Caygill, Howard. 1995. *A Kant Dictionary*. Oxford: Blackwell Reference.

Einstein, Albert. 1920. *Relativity. The Special and the General Theory*. Tr. by Robert Lawson, London: Methuen & Co. Ltd.

Friedman, Michael. 1999. *Reconsidering Logical Positivism*. Cambridge University Press.

Friedman, Michael. 2001. *Dynamics of Reason*. Stanford: CSLI Publications.

Helmholtz, Hermann von. 1903. *Über den Ursprung und die Bedeutung der geometrischen Axiome*. In: *Vorträge und Reden*, vol. 2. Braunschweig: Vieweg.

Hertz, Heinrich. 1899. *The Principles of Mechanics*. London: Macmillan.

Hyder, David. 2002. *The Mechanics of Meaning: Propositional Content and the Logical Space of Wittgenstein's Tractatus*. Berlin, New York: Walter de Gruyter.

Kant, Immanuel. 1999. *The Critique of Pure Reason*. Edited and translated by Paul Guyer and A.W. Wood. Cambridge University Press.

Kuusela, Oskari. 2008. *The Struggle against Dogmatism*. Harvard University Press.

Lützen, Jesper. 2005. *Mechanistic Images in Geometric Form. Heinrich Hertz's Principles of Mechanics*. Oxford University Press.

Pilch, Martin. 2017. *The Structure of Wittgenstein's Logical Space*. In: Wittgenstein-Studien, 8 (1). 15–60.

Plaud, Sabine. 2011. *From 'Styles of Thinking' to 'World-Pictures': Wittgenstein and Spengler on Style*. In: *Unsocial Sociabilities. Wittgenstein's Sources*. Ed. by Esther Ramharter. Berlin: Parerga. 75–95.

Preston, J.M. 2017. *Wittgenstein, Hertz and Boltzmann*. In: *A Companion to Wittgenstein*. Ed. H.J.Glock and J. Hyman. Chichester: Wiley Blackwell. 110–124.

Ramsey, F.P. 1990. *Philosophical Papers*. Ed. by Hugh Mellor. Cambridge University Press.

Russell, Bertrand. 1968. *The Autobiography, Vol. 2, 1914–1944*. London: George Allen and Unwin.

Schulte, Joachim. 2006. *Wittgenstein on Time (1929–33)*. In: *Time and History. Zeit und Geschichte*. Ed. by F. Stadler and M. Stöltzner. Frankfurt: Ontos Verlag.

Schulte, Joachim. 2018. *Body and Soul*. In: The Cambridge Companion to Wittgenstein. Ed. by H. Sluga and D. Stern. Cambridge University Press.

Stump, David. 2015. *Conceptual Change and the Philosophy of Science. Alternative Interpretations of the A Priori*. New York: Routledge.

Wittgenstein, Ludwig. 1922. *Tractatus Logico-Philosophicus* [TLP]. Tr. by C.K. Ogden and F.P. Ramsey. London: Routledge and Kegan Paul.

Wittgenstein, Ludwig.1929. *Some Remarks on Logical Form*. In: Proceedings of the Aristotelian Society, Supplementary Volumes, Vol. 9, Knowledge, Experience and Realism. 162–171.

Wittgenstein, Ludwig. 2004. *Briefwechsel. Innsbrucker Elektronische Ausgabe*. Ed. by M. Seekircher, B.F. McGuinness and A. Unterkircher.

Wittgenstein, Ludwig. 2009. *Philosophical Investigations* [PI]. Ed. and tr. by P. M. S. Hacker and J. Schulte, Oxford: Wiley-Blackwell.

Wittgenstein, Ludwig. 2015. *Bergen Nachlass Edition*. Ed. by A. Pichler, Wittgenstein Archives, University of Bergen. In: Wittgenstein Source (2009-) [wittgensteinsource.org]. Bergen: WAB.

Chapter 15
Foundations of Mathematics: From Hilbert and Wittgenstein to the Categorical Unity of Science

Yoshihiro Maruyama

15.1 Hilbert's Transcendental Finitism and Wittgenstein's Perspicuous Finitism

It is a common understanding that mathematical truth is absolute and hold forever whereas empirical truth in natural science is subject to change and revision. Why is it so? It might be because there is a Platonic, ideal universe of mathematical entities and that mathematical reality is invariant and universal. Along this line of thinking, the absolute universality of mathematical reality entails the absolute universality of mathematical truth. This is a realist view of mathematics, which was, for example, advocated by Kurt Gödel, one of the greatest thinkers on foundations of mathematics. In this view, the human being just discovers pre-existing mathematical truth, and mathematical truth per se is independent of any human activity.

Yet at the same time there is a more human view of mathematics, which, if not necessarily, tends to be some sort of antirealism when contrasted with the aforementioned, strongly realist view. In a constructivist view of mathematics, such as Luitzen Egbertus Jan Brouwer's intuitionism, mathematical entities are constructed or created by the human mind, and thus not independent of the human activity of mathematical practice. Put another way, mathematical entities do exist as far as they are constructible, that is to say, non-constructible entities are not justified to exist. To Brouwer, mathematical truth is a construction of the human mind, based on the Kantian intuition of time, especially the primordial intuition of two-oneness in time. Yet there are different versions of constructivism, which endorse different views on what mathematical entities do exist and on where mathematical truth is grounded. Constructivism often does not accept highly infinitary entities, thus leading to revisionism in mathematical practice. In a nutshell, constructivism sacrifices the transfinite realm of classical mathematics in order to secure the certainty of mathematics. In

---

Y. Maruyama (✉)
The Hakubi Centre for Advanced Research, Kyoto University, Kyoto, Japan
e-mail: maruyama@i.h.kyoto-u.ac.jp

S. Wuppuluri and N. da Costa (eds.), *WITTGENSTEINIAN (adj.)*,
The Frontiers Collection, https://doi.org/10.1007/978-3-030-27569-3_15

other words, the security of mathematics is kept at the cost of transfinite mathematics. Constructive mathematics, generally speaking, is more immune to inconsistency than classical mathematics allowing for transfinite entities and principles. Constructivists, such as Per Martin-Löf, were concerned with the potential inconsistency of highly infinitary principles in mathematics, and his intuitionistic type theory (aka. Martin-Löf type theory; Martin-Löf 1984) is strictly more certain in terms of consistency strength than classical mathematics. Even if classical mathematics is inconsistent, his intuitionistic type theory can be consistent; note that intuitionistic ZF set theory, referred to as IZF, only has the same consistency strength as classical mathematics (i.e., classical ZF or ZFC). The early impredicative version of Martin-Löf's theory actually turned out to be inconsistent due to the so-called Girard's paradox, and this motivated him to construct a significantly more secure, predicative version of it. Now his theory lies at the heart of Vladimir Voevodsky's univalent foundations programme (Voevodsky 2013), which is strongly influencing both pure and applied mathematics as well as logic, and also supporting the automation of mathematical verification. Some pure mathematicians, such as Leopold Kronecker, Bartel Leendert van der Werden, and Hermann Weyl, actually had some tendency towards what is now called constructivism in mathematics. Yet most mathematicians are not so happy to reject the existence of some infinitary entities on the ground that they are transfinite and go beyond the constructible realm of mathematics. Why would they like to save the infinitary domain of mathematics?

As David Hilbert, Richard Dedekind, and others have shown, infinitary methods are highly effective even for deriving results concerning the finitary domain of mathematics. Hilbert is known for his formalism and finitism, the so-called Hilbert programme aiming at the finitist justification of infinitary mathematics in terms of conservativity (and thus consistency) proofs. Hilbert's programme, even though it failed in the end, was rich in both philosophy and mathematics, giving the Kantian, transcendental justification of finitism per se as well as the finitist justification of infinitary mathematics. From the point of view of mathematical logic, finitism is a necessary prerequisite to express formal systems in the first place, and there seems to be no foundational justification of finitism. Yet Hilbert's idea is that any science or any (rational) thinking presupposes (something equivalent to) finitism at least, and thus finitism must be accepted as the minimal basis of our intellectual investigation, without which thinking as such is not possible in the first place. In a nutshell, finitism is the condition of possibility of thinking or reasoning, and all our scientific enterprises build upon it, that is, science is not possible without it. This is highly reminiscent of Kant's transcendental argument, and Hilbert's position may be named transcendental finitism (or even Kantian finitism). He had a profound understanding of Kantian transcendental philosophy. It should be emphasized that Hilbert's foundational theory was much richer than superficial formalism in both of its philosophical and mathematical content. Hilbert's transcendental finitism may be seen as an attempt to justify platonism by means of constructivism, and thereby to save the infinitary realm of mathematics from the foundational crisis. His formalism is a method to reduce the infinitary realm of mathematics to the finitary one; put another way, it reduces infinitary reality to finitary language. Hilbert concludes his seminal lecture "On the Infinite" as follows:

> Operating with the infinite can be made certain only by the finitary. (Hilbert 1983)

The same idea is actually shared by his algebraic geometry, which reduces infinitary geometric objects to finitary algebraic formulas. This logic-geometry correspondence can even be given a rigorous formulation: while logical completeness shows the (category-theoretical) duality between theories and models, Hilbert's Nullstellensatz shows the (category-theoretical) duality between algebras and varieties (and the two dualities are indeed equivalent for the two-element field of truth values and also for any Galois field of finitary truth values). Hilbert's foundational philosophy, thus, was intertwined with his work in pure mathematics, especially algebraic geometry, in which he proved two major theorems, i.e., Hilbert's basis theorem and Hilbert's Nullstellensatz. To Hilbert, finitism is prior to anything whatsoever as he says in the aforementioned lecture:

> As a further precondition for using logical deduction and carrying out logical operations, something must be given in conception, viz., certain extralogical concrete objects which are intuited as directly experienced prior to all thinking. For logical deduction to be certain, we must be able to see every aspect of these objects, and their properties, differences, sequences, and contiguities must be given, together with the objects themselves, as something which cannot be reduced to something else and which requires no reduction. This is the basic philosophy which I find necessary, not just for mathematics, but for all scientific thinking, understanding, and communicating. (Hilbert 1983)

This passage also clarifies Hilbert's idea that basic arithmetic (dealing with extra-logical concrete objects) is prior to logic. Put another way, logic presupposes certain arithmetic, without which all scientific thinking, understanding, and communication are impossible. Hilbert's ontology of mathematics does not accept infinites as real objects:

> [T]he infinite is nowhere to be found in reality, no matter what experiences, observations, and knowledge are appealed to. Can thought about things be so much different from things? Can thinking processes be so unlike the actual processes of things? In short, can thought be so far removed from reality? Rather is it not clear that, when we think that we have encountered the infinite in some real sense, we have merely been seduced into thinking so by the fact that we often encounter extremely large and extremely small dimensions in reality? (Hilbert 1983)

Yet infinites or ideal elements are highly useful in mathematical practice, and so Hilbert's finitism aims at an epistemological justification of them. This is why Hilbert's philosophy is occasionally referred to as instrumentalism (about infinitary mathematics).

It is a widely accepted view that Wittgenstein had some tendency towards finitism and constructivism, but there are, as a matter of fact, quite some disagreements about how strong his finitism and constructivism were. Some assert that his position was ultrafinitism, which only accepts the existence of numbers up to a certain finite number, that is, rejects any number greater than it. In ultra-finitism, too large numbers are like infinities, and simply do not exist. It is a strongly revisionist view of mathematics, and related to complexity-theoretical foundations of mathematics, such as mathematics feasible with polynomial time algorithms (cf. Buss' bounded arithmetic; Cook's bounded reverse mathematics). Note that computability-theoretical foundations of mathematics, such as recursive mathematics, allow for any computable algorithms with no condition on computational complexity. Wittgenstein's philosophy of mathematics is also labelled as strict finitism. Infinities in mathematics are problematic to Wittgenstein as well as to Hilbert. Referring to proofs relying upon mathematical induction, for example, Wittgenstein remarks as follows:

> We ought not to confuse the infinite possibility of its application with what is actually proved. The infinite possibility of application is not proved! The most striking thing about a recursive proof is that what it alleges to prove is not what comes out of it. (Wittgenstein 1975, p. 193)

He also argues as follows:

> Proof by induction. The most striking fact here is that the proposition to be proved doesn't itself occur in the proof at all. That is to say, induction isn't a procedure leading to a proposition. Instead, induction shows us an infinite possibility, and the essence of proof by induction consists in this alone. (Wittgenstein 1975, p. 328)

According to Wiitgenstein:

> A proof proves what it proves and no more. (Wittgenstein 1975, p. 203)

There is little doubt that Wittgenstein's philosophy of mathematics is some sort of finitism, formalism, and constructivism. His constructivist tendency is clear in his view of the task of the mathematician:

> The mathematician is an inventor, not a discoverer. (Wittgenstein 1978, p. 6)

His formalist tendency may be observed in his grammatical conception of mathematics and its certainty:

> What is unshakably certain about what is proved? To accept a proposition as unshakably certain–I want to say–means to use it as a grammatical rule: this removes uncertainty from it. (Wittgenstein 1978, p. 170)

Yet at the same time the principal challenge in elucidating Wittgenstein's philosophy of mathematics is to identify what kind of finitism, formalism, and constructivism it precisely is. What is then crucial would be Wittgenstein's notion of perspicuity or surveyability:

> Perspicuity is part of proof. If the process by means of which I get a result were not surveyable, I might indeed make a note that this number is what comes out – but what fact is this supposed to confirm for me? I don't know 'what is supposed to come out' (Wittgenstein 1978, p. 95)

To Wittgenstein, proof is not just about justification, but also about understanding. Wittgenstein does not accept proofs too long to be surveyed (cf. Shanker 1987). It is well known that the four colour theorem was proven by the computer, and the proof was too long for the finitary human mathematician to understand its underlying idea (if there is any such thing in it at all). Lengthy proofs have no problem in justifying statements involved, and yet, if they are too long to be surveyed by the mathematician, they cannot confer any proper understanding of mathematical phenomena involved. Understanding is what demarcates formal proofs in logic from actual proofs in mathematics. Formal proofs are good enough to give a logical justification of mathematical statements while they do not make much sense in understanding mathematical phenomena. The significance of perspicuity in mathematical proof lies in its capacity to give a conceptual understanding of what is going on in mathematical proof. The computer is able to check whether a given list of symbols form a logically correct proof or not, and yet it would not be able to judge whether the list of symbols is perspicuous or give a conceptual understanding of mathematical phenomena concerned. Formal proofs are about the justification of propositions concerned, and does not address the understanding of their mathematical content. By contrast, informal proofs are more about the understanding of propositions involved than about their ultimate justification. Formalizations of proofs do not preserve their understanding even if they do preserve or even sharpen their justificatory nature. To Wittgenstein, proof must confer both of justification and understanding. Wittgenstein's philosophy of mathematics, therefore, may be called perspicuous finitism or surveyable finitism. It could also be named transparent finitism, since the transparency of proof is what lies at the heart of mathematical understanding. Marion (2011) emphasizes Wittgenstein's geometrical understanding of proof in his conception of surveyability. To Hilbert, finitism is a foundational basis for justification; to Wittgenstein, finitism is a fundamental basis for understanding or surveyability.

15.2 Wittgenstein's Natural History View and Logical Foundations of Mathematical Truth

Where does the certainty and objectivity of mathematical truth come from? Since Euclid, proof has been considered a crucial process to secure the certainty and objectivity of mathematics. What exactly does count as a proof, nonetheless, varies from one period of mathematical practice to another and from one discipline to another (and even from one mathematician to another in certain rare cases; cf. the highly controversial case of Shinichi Mochizuki's alleged proof of the ABC conjecture). The historicity of proof is almost a brute fact in the history of mathematics.

The historicity of proof may be observed from different angles. Some constructive algebraists, for example, did not accept transcendental or indirect proofs in the early days of abstract algebra. Proofs in 19th century mathematics were quite loosely formulated, and some of them were not really proofs according to the high standard of clarity required in 20th century mathematics. Disciplinary disagreements about proofs can be observed, for example, in different forms of geometry. Modern algebraic geometry by Oscar Zariski, Andre Weil, and Alexander Grothendieck were concerned with the rigorous, formal reconstruction of intuition-driven proofs in the Italian school of algebraic geometry led by Guido Castelnuovo, Federigo Enriques and Francesco Severi. It is generally true even today that the geometer's proofs are involved in more intuition than the algebraist's proofs. Along this line of thinking, the concept of proof is historical and contextual rather than universal. Daston and Galison (2007) argue in their treatise *Objectivity* that scientific objectivity has a history, while providing a lot of evidence from the history of science. Mathematical objectivity as expressed in the notion of proof has a history too.

All this is about proofs in natural language. They cannot avoid certain ambiguities, and there is no criterion to judge whether a given mathematical argument really counts as an absolutely valid proof or not. Mathematics performed via natural language heavily rests upon human intuition in practice and the cognitive capacity of the human mind; it is even left implicit what sorts of principles and arguments are allowed to use in mathematical proofs. This sometimes leads to controversies on the admissible methods of proofs in mathematics, such as the case of the axiom of choice. Lacking a precise criterion for validity of arguments necessitates a restricted conception of objectivity: the objectivity of proofs in natural language boils down to intersubjectivity, i.e., overall agreement within the mathematical community. If there is an explicit criterion for validity of arguments, a single person can check whether a given series of arguments constitute a proof. Yet if not there is no way to allow such a justification. Where there is no explicit criterion, justification can only be made collectively within the relevant community; it is left to judges of the mathematical court as it were.

Proofs in formal language, on the other hand, are involved in less ambiguities, and it is explicated what exactly counts as a proof, if not completely. A formal proof consists of two kinds of ingredients, namely instantiations of axioms and applications of inferential rules. And it is indeed mechanically computable whether a given series of symbols count as a proof or not (as far as the entire system is recursively axiomatized). Formal proofs apparently involve no historicity, contingency, or ambiguity, and thus could serve as a rigorous foundation for the certainty and objectivity of mathematical truth. Formal systems of mathematics as developed in symbolic logic are fully explicated forms of mathematics, the certainty and objectivity of mathematics being pursued to its extreme. So does symbolic logic secure the absolute certainty and objectivity of mathematics? This is precisely where Wittgenstein's distinctive philosophy of logic and mathematics comes into the play. Wittgenstein asserts:

> We incline to the belief that *logical* proof has a peculiar, absolute cogency, deriving from the unconditional certainty in logic of the fundamental laws and the laws of inference. Whereas propositions proved in this way can after all not be more certain than is the correctness of the way those laws of inference are *applied*. (Wittgenstein 1978, p. 174)

Wittgenstein's following remark is also reminiscent of his notion of surveyability discussed above:

> The logical certainty of proofs [...] does not extend beyond their geometrical certainty. (Wittgenstein 1978, p. 175)

The certainty of propositions with formal proofs is not more than the certainty of applications of inferential rules in them. A possible interpretation of this assertion is scepticism about the application of rules, which risks the certainty of mathematics.

Another is an infinite regress argument about logical truth, which bankrupts the foundationalist enterprise of grounding logical or mathematical truth upon formal rules. And the two interpretations are intertwined with each other. In the following we shall discuss the details of the two interpretations whilst elucidating the relationships between logic and mathematics.

The logicist endorses the thesis that mathematics, ultimately, is reduced to logic. To justify the absolute truth of mathematics, it is necessary to think of the ultimate ground of mathematical truth. Where is the ultimate ground of mathematics? The logicist's tenet is that the absolute certainty of mathematics is rooted in the fundamental principles of logic. Axioms and rules of inference in logical systems are conventions set by humans. The logical positivist, Rudolf Carnap in particular, endorses the thesis that mathematical truth is reduced to logical truth, and logical truth, in turn, is reduced to truth by convention. The relationships between logic and mathematics are intricate, and different conceptions of them may be developed. Both Hilbert and Brouwer argue that mathematics, especially arithmetic, is prior to logic; Wittgenstein would agree with this view as well. Indeed, logic does not exist in vacuum, presupposing many things including certain elementary arithmetic, without which formal systems of logic cannot even be expressed. To define what a formula is, what a proof is, and so fourth, it is necessary to use mathematical induction, and the overall structure thus required is rich enough to accommodate elementary arithmetic (basically, what is called PRA, namely Primitive Recursive Arithmetic). We can only make sense of formal logic on the basis of arithmetic.

Put another way, natural numbers are, both ontologically and epistemologically, irreducible; they cannot be reduced to something more primitive. It is possible to refer to Peano axioms to clarify what natural numbers are, and yet it is not a proper reduction because the formal structure of Peano Arithmetic presupposes arithmetic. It is well known that real numbers can be constructed on the basis of rational numbers, which can in turn be constructed on the basis of integers. It is also possible to construct integers using natural numbers. Natural numbers can be constructed in terms of sets (e.g., as finite von Neumann ordinals); set theory, nevertheless, is stronger than arithmetic. The language of first-order logic per se presupposes elementary arithmetic, and thus no proper reduction of natural numbers to logic is possible. A similar argument may be applied to the laws of logic per se. It is possible to explicate what the laws of logic are by means of formal systems, and yet those formal systems presuppose certain elementary logic. Think, for example, of an introduction rule for conjunction, which says that if you have both A and B you may deduce $A \wedge B$, the conjunction of A and B. Taking a careful look at this rule, one notices that it presupposes several concepts such as "if" and "both", which essentially amount to implication and conjunction. Logic, therefore, cannot exist in vacuum, presupposing itself to some extent. Logical and arithmetical laws are so fundamental that any attempt to define them result in essentially circular arguments. It is thus impossible to ground logical and arithmetical truths upon formal rules.

From a contemporary point of view, fundamental proof theory consists of three components: background-theory, object-theory and meta-theory. Background-theory is what must be presupposed to express a formal system in the first place. Object-theory is then formulated on the basis of the background-theory. And meta-theory is what is required to reason about object-theory. Background-theory includes format-theory, which is a theory of symbolic objects like comma, context, and so fourth, as, for example, in typed sequent calculus. In Hilbert's proof theory, the background-theory and the meta-theory coincide, and both of them are a finitist theory such as PRA (Primitive Recursive Arithmetic). Yet they are different in general; it is possible that the background-theory is finitism and the meta-theory is set theory. In categorical semantics for intuitionistic logic, format-theory and object-theory almost coincide, both being cartesian closed categories. Still they are different in general. As we discussed above, it would be impossible to fully explicate our background-theory in view of the fundamental circularity lurking behind the scene (e.g., to spell out the inferential rules of conjunction and implication, you rely upon the theory of comma and if-then). Hilary Putnam says:

> The 'exciting' thesis that logic is true by convention reduces to the unexciting claim that logic is true by convention plus logic. (Putnam 1983, p. 116)

We can still explicate the underlying format-theory, which is indeed crucial for categorical semantics because categories must model the structure of the format-theory (',', '[...]', type contexts, etc.) as well as the structure of the object-theory. The impossibility of grounding logical truth upon formal convention was clearly shown for the first time in Willard Van Orman Quine's argument in his "Truth by Convention", and the idea goes back to Lewis Caroll's "What the Tortoise Said to Achilles."

Hilbert also gives a clear argument on the role of logic in foundations:

> In summary, let us return to our main theme and draw some conclusions from all our thinking about the infinite. Our principal result is that the infinite is nowhere to be found in reality. It neither exists in nature nor provides a legitimate basis for rational thought — a remarkable harmony between being and thought. In contrast to the earlier efforts of Frege and Dedekind, we are convinced that certain intuitive concepts and insights are necessary conditions of scientific knowledge, and logic alone is not sufficient. (Hilbert 1983)

Hilbert considers his idea to be rooted in Kantian philosophy:

> Does material logical deduction somehow deceive us or leave us in the lurch when we apply it to real things or events? No! Material logical deduction is indispensable. It deceives us only when we form arbitrary abstract definitions, especially those which involve infinitely many objects. In such cases we have illegitimately used material logical deduction; i.e., we have not paid sufficient attention to the preconditions necessary for its valid use. In recognizing that there are such preconditions that must be taken into account, we find ourselves in agreement with the philosophers, notably with Kant. Kant taught — and it is an integral part of his doctrine — that mathematics treats a subject matter which is given independently of logic. Mathematics, therefore, can never be grounded solely on logic. Consequently, Frege's and Dedekind's attempts to so ground it were doomed to failure. (Hilbert 1983)

Actually, both Hilbert and Wittgenstein were influenced by Neo-Kantians in their time, especially Heinrich Hertz, who, along with Hermann von Helmholtz known for his Back-to-Kant movement, was a Neo-Kantian thinker as well as a physicist. There is rich historical evidence for the Neo-Kantian roots of Hilbert's axiomatics and of Wittgenstein's philosophy, especially his early picture theory, even though our primary interest here is not of historical nature (see, e.g., Corry 2004 and Baird et al. 1998, respectively).

To Hilbert (and also to Brouwer), Kantian intuition is the ultimate ground of mathematical truth. What is the ultimate ground of truth to Wittgenstein then? He asserts that it is our "natural history." According to Wittgenstein, this natural history is the history of rules:

> I should like to be able to describe how it comes about that mathematics appears to us now as the natural history of the domain of numbers, now again as a collection of rules. (Wittgenstein 1978, p. 230)

The rules that Wittgenstein is concerned with are not formal rules in mathematical logic, but natural rules in our everyday practice, which would not really be formalizable. Wittgenstein says:

> [T]he natural history of the use of a word can't be any concern of logic. (Wittgenstein 1975, p. 59)

In Wittgenstein's philosophy, mathematical activity is understood as part of our larger natural history:

> I go through the proof and then accept its result.–I mean: this is simply what we do. This is use and custom among us, or a fact of our natural history. (Wittgenstein 1978, p. 61)

Ultimately, the laws of logic have their origin in our natural history:

> [L]ogic belongs to the natural history of man. And that is not combinable with the hardness of the logical "must". But the logical "must" is a component part of the propositions of logic, and these are not propositions of human natural history. If what a proposition of logic said was: Human beings agree with one another in such and such ways (and that would be the form of the natural-historical proposition), then its contradictory would say that there is here a lack of agreement. Not, that there is an agreement of another kind. The agreement of humans that is a presupposition of logic is not an agreement in opinions, much less in opinions on questions of logic. (Wittgenstein 1978, pp. 352–253)

"The agreement of humans that is a presupposition of logic" is an agreement in the natural history of following rule. As the issue of rule-following in Wittgenstein's later philosophy poses the so-called Kripkenstein's paradox, Wittgenstein's "conception of arithmetic as the natural history of numbers" (Wittgenstein 1978, p. 11) is involved in much the same type of paradox as follows:

> Suppose someone were to say: Russellian arithmetic agrees with ordinary arithmetic up to numbers less than 10^{10}; but then it diverges from it. And now he produces a Russellian proof that $10^{10} + 1 = 10^{10}$. Now why should I not trust such a proof? How will anybody convince me that I must have miscalculated in the Russellian proof? (Wittgenstein 1978, p. 152)

This is almost the same as Kripke's "quus" argument to account for the rule-following paradox (Kripke 1982). What determines arithmetic facts is not any formal system, which inevitably faces the rule-following paradox as Kripke (1982) argued, but rather is the natural history of numbers, which amounts to what Kripke calls the sceptical solution to the rule-following paradox. The following is a well-known passage from Wittgenstein's *Philosophical Investigations*:

> [N]o course of action could be determined by a rule, because any course of action can be made out to accord with the rule. (Wittgenstein 1986, p. 81)

Simply saying, any rule is subject to reinterpretation, and any attempt to fix the intended interpretation fails because any such attempt itself can be reinterpreted again and again. It also happens in our everyday life that the meaning of some legal law is debated, and reinterpreted again and again. The same thing happens in any sort of rule, however clear it seems, including the rules of mathematics. As long as we rest upon formal systems to fix the meaning of mathematical rules, it is impossible to single out their intended meaning among a variety of possible interpretations. This is analogous to the fact that it is impossible to characterize the intended model of arithmetic via finitary logic; any theory of arithmetic over finitary logic allows for the existence of (unintended) non-standard models, which encompass strange "infinite" numbers as well as ordinary "finite" numbers. If you use infinitary logic and formulate arithmetic over it, then you can actually obtain the complete system of arithmetic that allows for the intended model only. Yet this is like cheating because in that case you have to presuppose at the meta-level what the totality of natural numbers is. There is no purely logical way out from the rule-following circularity. Wittgenstein thus appeals to our natural history, in which meaning is crystalized historically rather than logically. And this applies to mathematics in particular. To Wittgenstein (in his later philosophy), mathematical truth is grounded upon our natural history rather than formal logic. While Wittgenstein's notion of surveyability is concerned with the understanding of mathematics, his natural history view is concerned with the meaning of mathematics, which does not have an origin in any sort of formal convention, but in our natural history per se.

It should be remarked that the issue of understanding is intertwined with the issue of constructibility:

> But if you say "The point isn't understanding at all. Mathematical propositions are only positions in a game" that too is nonsense! 'Mathematics' is not a sharply delimited concept. Hence the issue whether an existence-proof which is not a construction is a real proof of existence. That is, the question arises: Do I understand the proposition "There is ..." when I have no possibility of finding where it exists? And here there are two points of view: as an English sentence for example I understand it, so far, that is, as I can explain it (and note how far my explanation goes). But what can I do with it? Well, not what I can do with a constructive proof. And in so far as what I can do with the proposition is the criterion of understanding it, thus far it is not clear in advance whether and to what extent I understand it. (Wittgenstein 1978, p.299)

Wittgenstein's constructivism thus stems from his emphasis on understanding in mathematics; the problem of indirect proofs of existence lies in a lack of understanding in it.

15.3 Categorical Foundations of Mathematics and the Categorical Unity of Science

Foundationalism, whether Cartesian or of any other sort, is considered something bad in contemporary philosophy, including the philosophy of logic and mathematics. In foundations of mathematics, some partial realizations of Hilbert's programme have been achieved despite Gödel's incompleteness theorems, for example, in Thierry Coquand's celebrated work applying Andre Joyal's duality theory to the finitist justification of commutative algebra and algebraic geometry (Coquand 2009). Even Grothendieck's concept of schemes in modern algebraic geometry has been given a constructive finitist foundation along the same lines. It is also not impossible to argue that Gödel's theorems do not necessarily undermine Hilbert's programme in its entirety. There are indeed some technical subtleties lurking behind Gödel's theorems: if logic is strong enough then it is possible to axiomatize the complete theory of arithmetic on it (yet Hilbert would not accept such infinitary logic); if the so-called provability predicate (encoding of system's provability within the system; arithmetization of metamathematics) is formulated in a certain manner then the corresponding consistency statement actually gets provable within the system concerned (this is why the first incompleteness is called extensional, and the second incompleteness intensional). Traditional foundations of mathematics, nevertheless, have lost most of its driving force since Gödel's theorems. Different fields in contemporary foundations of mathematics only target at some highly technical problems just as ordinary mathematics does. And anti-foundationalism is prevailing in philosophy in general and in philosophy of logic and mathematics in particular. So is there no proper foundation of mathematics now?

There are several novel developments after Gödel, including category theory and categorical foundations of mathematics, such as Grothendieck-Lawvere's topos theory, which integrates logic and algebraic geometry, and Voevodsky's univalent foundations (aka. homotopy type theory), which integrate type theory and homotopy theory. In both of them, logic concerned is primarily intuitionistic or constructive, even though classical logic obtains as their extenions. Yet at the same time the categorical models they yield allow us to understand intuitionism and constructivism within classical mathematics. An origin of interpreting intuitionism and constructivism within classical mathematics lies in Stephen Kleene's realizability models. In Brouwer's intuitionism, all functions from reals to reals are continuous, and this contradicts the classical existence of discontinuous functions. Yet, for example, any computable function from reals to reals are always continuous, and this sort of computational models of mathematics allow us to understand intuitionism and constructivism within classical mathematics. Topos theory and univalent foundations play the same role of understanding intuitionism and constructivism in terms of classical mathematics such as algebraic geometry and homotopy theory. At the same time, topos theory has been effective in proving (both classical and constructive) independence results, such as the independence of the continuum hypothesis. In both of topos and univalent foundations, there is a close, vital collaboration between foundations and mainstream mathematics; this is something that did not exist before because any of proof theory, set theory, and recursion theory was far apart from mainstream mathematics. In a certain sense they elucidate a special place in which logic and space are one and the same thing.

The meaning of foundations in category theory, nonetheless, is quite different from the meaning of foundations in set theory. Set theory aims at reductive absolute foundations of mathematics, and yet category theory at structural relative foundations of mathematics. While set theory is concerned with global foundations of mathematics, category theory is concerned with local foundations of mathematics and even science in general. There would be yet another sort of foundations of mathematics, that is, conceptual foundations of mathematics. Let us clarify the major differences between them as follows:

- Global Foundations or Reductive Absolute Foundations. They allow us to reduce everything to a single framework, thus giving an absolute, domain-independent environment to work within. In set theory, there is a single universe of sets, into which all mathematical entities are encoded, and there is nothing other than sets. Both set theory (e.g., ZFC) and categorical theories (e.g., Lawvere's ETCS; Lawvere 1964) work for this. Yet category theory may give another sense of foundation for mathematics and science in general. Indeed category theory today is serving as a unifying language for the sciences, especially for physics and informatics as well as logic.
- Local Foundations or Structural Relative Foundations. There is no single universe; there are plural categories, each of which provides a local foundation rather than global one. While set-theoretical foundations are monistic (except for the multiverse view of sets), categorical foundations are inherently pluralistic. They change a framework according to our structural focus (and see what remains invariant, and what does not). Base change is a fundamental idea of category theory (e.g., Grothendieck's relative point of view). They give a relative, domain-specific context to work in, such as ribbon categories for knot theory, model categories for homotopy theory, and dagger-compact categories for quantum mechanics and information, to name but a few. The foundational significance of categories in mathematical practice would be with respect to relative foundations rather than absolute foundations.
- Conceptual Foundations. They involve no (absolute or relative) foundationalism, unlike the two kinds of foundations above, and aim at elucidating the nature of fundamental concepts in mathematics (including space, number, continuum, infinitesimal, computation and so fourth), and thus they are compelling for the working mathematician as well as the foundational mathematician. The meaning of "foundation" in the term "foundations of physics" seems to be close to this sense of foundations. In the next section we shall discuss the nature of space in mathematics while interconnecting Brouwer's and Wittgenstein's conceptions of it.

Mathematicians work on different objects, some of which are algebraic and others of which are geometric. In set theory, all that is translated or encoded into the language of sets because there is nothing else in set theory (which does not allow for ur-elements).

This monistic ontology of set theory provides a common, ontological ground for all mathematical entities, and differences between them are understood as differences in additional structures on them. Everything is a set in the first place, and yet one can introduce a richer structure on it, such as algebraic, topological, and order structures, making structural differences even on the same set. Put another way, sets are prior to structures in set theory. In category theory, nonetheless, structures are prior to sets, and structures may even have no underlying sets. Category theory thus allows us to express pure structures rather than structures on sets.

As topos and univalent foundations show, category theory can play the role of foundations of mathematics. It can even play the role of foundations of science in general. A category is a network structure of objects and morphisms, which can arise from science as well as mathematics. In mathematics, there is a category of algebras and homomorphisms, a category of spaces and continuous maps, and so fourth. There are, moreover, different categories in different sciences:

- Physics: objects are physical systems (e.g., quantum systems as Hilbert spaces), and morphisms are physical processes (quantum processes as bounded operators). The category of finite-dimensional quantum systems and processes forms a compact closed category (which is a specific type of monoidal closed category).
- Logic: objects are propositions; morphisms are proofs or deductions (from one proposition to another). It is well known that the category of propositions and proofs in intuitionistic logic forms what is called a bicartesian closed category. It is also known that the category of propositions and proofs in classical linear logic forms what is called a $*$-autonomous category (which is a specific type of monoidal closed category), sharing much the same structure as the category of quantum systems. This correspondence between substructural logic and quantum physics is called the Abramsky-Coecke correspondence.
- Computer Science: objects are data types; morphisms are (functional) programs. The category of types and programs in the standard typed λ-calculus forms a cartesian closed category, sharing almost the same structure as intuitionistic logic. This correspondence between logic and computation is what is called the Curry-Howard correspondence.
- Linguistics: objects are grammatical types, and morphisms are semantic flows of information. The resulting categorical structure is similar to the categorical structure of quantum physics.

In such a way, category theory allows us to analyze analogies and disanalogies between different sciences in a systematic manner, and thereby enables knowledge transfer between different sciences. A case of knowledge transfer enabled by category theory is that an automated reasoning system for quantum mechanics and quantum computation has been developed by applying an automate reasoning system for substructural logic to quantum mechanics through the categorical correspondence between substructural logic and quantum mechanics. This is a sort of axiomatization of quantum mechanics, thus counting as a solution to the Hilbert's Sixth Problem, which was envisaged and yet could not be realized for a long time.

Category theory as foundations of science arguably embodies the unity of science in the present age of disunity. In the following let us unpack what this means in a broader perspective on the issue of unity versus disunity in contemporary science. Today we are, in a sense, living in the age of disunity (or the lack of meaning, i.e., nihilism), and the search for unity, absolute foundations and the like mostly failed in the 20th century (cf. Richard Rorty): there is no "first philosophy" or "grand narrative", and there is no foundation of mathematics, let alone that of knowledge in general. After all, there is no unity but disunity. Put in a broader context, modernization went hand-in-hand with the death of natural philosophy, the fragmentation of science, and the overall disenchantment of the world leading to disunity. For science, the Stanford School (Patrick Suppes, Peter Galison, Ian Hacking, and others) represents this tendency, claiming a positive value of disunity. We can fine yet another type of disunity between humanistic and natural sciences as in Dilthey's distinction between Geisteswissenschaften and Naturwissenschaften. The philosophy of science, too, is disunified into plural philosophies of particular sciences; this, surely, is not necessarily bad. In mathematics, Bourbaki had a clear conception of the tension between unity and disunity as they said in The Architecture of Mathematics:

> [I]t is legitimate to ask whether this exuberant proliferation makes for the development of a strongly constructed organism, acquiring ever greater cohesion and unity with its new growths [...] In other words, do we have today a mathematic or do we have several mathematics? Although this question is perhaps of greater urgency now than ever before, it is by no means a new one; it has been asked almost from the very beginning of mathematical science. Indeed, quite apart from applied mathematics, there has always existed a dualism between the origins of geometry and of arithmetic [...] (Bourbaki 1950)

The quest for unity, absolute foundations, and the like have mostly failed, and hence anti-foundation tendency is pervasive today. Yet mathematics did achieve some unity due to Bourbaki, who considered mathematics to be a single organism as a coherent whole. In physics, Helmoltz, who was also a Neo-Kantian thinker, was aware of the tendency of disunity at an earlier stage of the fragmentation of science:

> No one could oversee the whole of science and keep the threads in one hand and find orientation. The natural result is that each individual researcher is forced to an ever smaller area as his workplace and can only maintain incomplete knowledge of neighbouring areas. (Weingart 2010)

In the accelerating process of fragmentation, Leibniz came to be conceived of as "one-man-embodiment of the unity of science"; his natural philosophy is the paradigmatic example of unity in the sciences. It is well known that the Vienna Circle (Rudolf Carnap, Otto Neurath, Hans Hahn, and others) envisaged unified science in specific form and continued to publish their *The International Encyclopedia of Unified Science* for a while. Simply saying, their unity of science meant monistic reductionism, or the reductive justification of the sciences, which is comparable to Hilbert's programme aiming at the reductive justification of mathematics (aka. reductive proof theory). Galison in the Stanford School claims disunity with binding. Galison says:

> It is precisely the disunification of science that underpins its strength and stability. (Galison 1999, p. 137)

Yet it is disunity with binding or networking, that is, what Galison calls the "binding culture of science whilst referring to Peirce (1974, p. 157):

> It should not form a chain which is no stronger than its weakest link, but a cable whose fibres may be ever so slender, provided they are sufficiently numerous and intimately connected.

Galison thinks that there is no single golden thread in science, which is a cable rather than a chain:

> With its intertwined strands, the cable gains its strength not by having a single, golden thread that winds its way through the whole. No one strand defines the whole. (Galison 1997, p. 844)

We can historicize unity in such a way. All this represents the strong tendency toward pluralism in the age of disunity. Pluralism is inherent in category theory; there is no single golden category for all sciences but different categories for different sciences. Set theory, on the other hand, endorses global ontology and global epistemology: there is the absolute universe of sets and all mathematical knowledge is about sets. There is no single universal language of categories, which is going to be indefinitely extended to richer and richer (e.g., higher and higher) languages, whereas set theory mostly sticks to the language of ZF set theory.

Now we articulate the meaning of unity in three ways from a theoretical point of view. In general, unity is a source of meaning, in particular scientific meaning. There can be three conceptions of unity. One is unity as a transcendental condition in the Hume-Kant tradition: the uniformity of nature is a condition of possibility for (meaningfulness of) scientific knowledge (both theoretical and empirical). Another is unity as a scientific norm (or devastating longing): as the physicists longs for "one single unifying equation that explains everything in the universe" (Stephen Hawking's words in the 2014 film *The Theory of Everything*), laws are of unifying nature in general, and the unifying nature of laws do matter in prediction as well (if laws are highly context-dependent they could not really be applied in practice). The other is unity as existential absurdity (in the sense of Albert Camus): the absurd lies in the discrepancy between the human being longing for unified meaning and the world rejecting it in its irreducible diversity. Unity lies at the heart of the enterprise of science in the two respects. And in both science and existence, we tend to long for unity (and often fail).

How can we attain the ideal of unity in the age of disunity? Category theory gives us a distinct possibility due to its powerful mechanism of interweaving different sciences as discussed above. The idea of categorical unified science still stands whilst taking the disunity of science at face value: we separate the unity problem within the existing sciences from the possibility of creating unified science as a new kind of special science, claiming the latter rather than the former for the following reason. The disunity of science is almost a brute fact: different scientists rely upon different epistemologies and ontologies, disagreeing on conditions of justification allowed or universes of entities posited. Even in the quite uniform realm of mathematics, which enjoys a good amount of unity thanks to Bourbaki's Architecture of Mathematics, there are nonetheless disagreements on what count as proofs (e.g., between algebra and topology) and on what count as real objects of mathematics (e.g., Hilbert only accepted the finitary as the real objects of mathematics; to him, the infinitary were instruments to know about the finitary). There is arguably no monolithic norm to unify different scientists with different backgrounds. There is no such monolithicity even within mathematics. And thus the only possibility of unified science lies in the pursuit of unity as a special science. There is no unity of science as the unity of the existing sciences, which is misguided in the present age of disunity. Yet unified science can still be a new kind of special science to give a unified, systematic understanding of the sciences. And there is no revisionism involved in this enterprise of unified science. Unity was also an issue of absorbing Western scholarship into Japan. Keiji Nishitani regards the modernist divide of world-views as a form of nihilism, arguing in "Overcoming Modernity" as follows: "We are facing three, separated directions to build a world-view" (i.e., God-based, Nature-based, and Human-based ones); hence "the development of a unified world-view is a compelling, fundamental issue in the contemporary world." In further retrospect Amane Nishi had an idea of (non-revisionist, non-foundationalist, non-reductionist) unified science in his programme of "Interweaving a Hundred Sciences." Unified science may be put in such a Japanese context. The history of Western logic in Japan also dates back to Nishi's treatise on logic as the canon of reason leading to knowledge.

Now let us formulate the three conditions of unified science as paradigmatically exemplified by categorical unified science. The three conditions of contemporary unified science (as a special science) are: epistemological and ontological networking; the methodology of networking; the decentred networking of scientific knowledge, that is, networking of scientific theories, which gives an account of both structural analogies and disanalogies between scientific theories. Unified science must support scientific pluralism in an epistemological sense: there is no privileged Archimedean vantage point upon which everything is grounded, and there is no espistemological centre of the network of sciences. This is realized in categorical unified science; there is no foundationalism or reductionism involved in it. Ontologically, it arguably allows the networking of levels of reality; a stream-lined account of matter, life, and mind, and of language, economy, and society can be given in pluralistic unified science. In other words, categorical unified science reconciles the science of facts and the science of artifacts, two separated levels of reality. Note that the idea of levels of reality goes back to Heisenberg's "levels of Reality, corresponding to different objectivity modes", and even to Nocolai Hartmann, a Neo-Kantian philosopher. Categorical unified science is concerned with higher scientism pursuing higher laws to network the sciences, which unify or interconnect first-order laws in different sciences (cf. second-order laws in physics). And it is crucial that unified science ensures productivity and knowledge-transfer in scientific practice, rather than abstract philosophy, and categorical unified science indeed realizes them as argued above. By means of epistemological and ontological networking via higher science, we lead to the local-to-global correspondence: a global world-view to patch together local world-views existing in particular sciences. Several consequences of the unity conditions are the following: there is no universal ontology or epistemology; there is no universal language; and there is no revisionism or uniformization. Rather, the global interweaving of local epistemologies and ontologies in different sciences is the primary task of contemporary unified science, in particular categorical unified science.

It should be remarked that what is at stake here is unity as a matter of knowledge, and not as a matter of activity. Unity qua science comes down to unified science, whereas unity qua scientific activity boils down to interdisciplinary science. The logical positivist attempted to invent or force "unity from above", and failed. Some still do so. Yet it would naturally show up in scientific practice. Categorical unified science indeed embodies "unity from below" (cf. "history from below"); category theory gives a unified treatment of logic as the category of propositions and proofs, informatics as the category of data types and programs, physics as the category of systems and processes (e.g., time evolution), linguistics as the category of words and meanings, economics as the category of items and preferences (with reasons), sociology as the category of agents and communications, and so fourth. The science of physical, computational, and logical networks is especially well-trodden while the others are less so. Although there are some categorical developments in life sciences as well, it is not yet crystal-clear how categorical unified science can get along with life sciences. Yet life is difficult for anyone; the nature of life and mind is to be explored in science per se, rather than unified science, at least at the moment. Categories are structural networks (or graphs with compositional structures on their edges). And category theory is the higher science of structural networks across the sciences. Categorical unified science is the only development of unified science evidenced by broad scientific practice rather than dogmatic delusions. Its achievements are remarkable in physics and mathematics on one hand, and for a few humanities and social sciences on the other, including linguistics and social choice theory. Categorical unified science is indeed in the process of intensive worldwide implementation by categorical scientists across the world. The problem of unity we face today may be characterized as a lack of the local-to-global correspondence: there is no global world-view to patch together local world-views in different sciences. Unity, along this line, means the global networking (binding) of local epistemologies and ontologies in different sciences. And this is the unity that categorical unified science is embodying at the interface of different sciences.

15.4 Conceptions of Space in Foundations of Mathematics: Wittgenstein, Brouwer, and Beyond

Let us come back to Wittgenstein's philosophy of mathematics. Since ancient Greek philosophy, there have been a vast number of debates on whether or not the concept of a point precedes the concept of the space continuum. On the one hand, one may conceive of points as primary entities, and of the continuum as secondary ones to be understood as the collection of points; on the other, the whole space continuum may come first, and then the concept of a point is derived as a cut of it. This is more or less analogous to the well-known dichotomy between Newton's conception of absolute space and Leibniz's conception of relational space. Wittgenstein gives a fresh look at the issue of the relationships between space and points:

> What makes it apparent that space is not a collection of points, but the realization of a law? (Wittgenstein 1975, p. 216)

Wittgenstein's intensional view on space is a compelling consequence of his persistent disagreement with the set-theoretical extensional view of mathematics:

> Mathematics is ridden through and through with the pernicious idioms of set theory. One example of this is the way people speak of a line as composed of points. A line is a law and isn't composed of anything at all. (Wittgenstein 1974, p. 211)

In the following we attempt to examine and articulate Wittgenstein's conception of space in relation to Brouwer's theory of the continuum and its mathematical descendants in modern geometry. Wittgenstein's intensional conception of space is closely related, in its core idea, with "the modernist transformation of mathematics" (Gray 2008), especially the resulting revolutionary change of the concept of space in geometry: space is not a collection of points any more, but a sort of abstract algebraic structure, such as a C^*-algebra, topos, locale, formal space, or scheme (see, e.g., Cartier 2001 and MacLane-Moerdijk 1992). One of the first steps was taken by Brouwer in his intuitionistic conception of the continuum in terms of his peculiar (but indeed natural) notions of spreads and choice sequences, which, as explained below, may be seen as based upon the concept of a law rather than that of a point. Although the enterprise of intuitionism did not succeed so much in convincing mathematicians of its significance, nevertheless, similar ideas on space have won widespread acceptance with the crucial help of duality theory, which elegantly enables us to derive points from an abstract algebraic structure as mentioned above, thereby establishing a tight (or functorial) link between point-set and point-free concepts of space; we could even say that duality justifies to regard an algebra as space, since it tells us that certain algebras carry the same amount of information as space itself. In the following we shall explicate the philosophical significance of such advances in modern geometry.

It is widely believed that Brouwer's intuitionism influenced Wittgenstein. Although the main point here is not on historical discussion, nevertheless, it could still be said that Wittgenstein's philosophy of space in particular was affected by Brouwer to some degree; his *Philosophical Remarks* and *Philosophical Grammar* contain a number of relevant descriptions, and even refers to Brouwer explicitly at several places. In the following we first elucidate the relationships between Wittgenstein's and Brouwer's concepts of space, and move on to the explication of how modern geometry with duality conceptually connects with Wittgenstein's philosophy of space. Wittgenstein's and Brouwer's point-free conceptions of space are not independent of each other. Brouwer's influence on Wittgenstein is almost unquestionable, as for instance Rodych (2011) says, "There is little doubt that Wittgenstein was invigorated by L.E.J. Brouwer's March 10, 1928 Vienna lecture." At the same time, however, connections between their philosophies of space in particular have remained untouched and to be investigated; we hereby embarks upon this project, taking a first step towards a full-fledged account of the relationships. First of all, what Wittgenstein calls a law should be clarified. Just after the first sentence quoted above, he says, "In order to represent space we need–so it appears to me–something like an expansible sign" (Wittgenstein 1975, p.216); here, the concept of a sign already suggests relevance to algebra. What precisely is a sign, then? As he proceeds in the same page, it is "a sign that makes allowance for an interpolation, similar to the decimal system"; he then adds, "The sign must have the multiplicity and properties of space." To elucidate what he means, the following discussion on a coin-tossing game seems crucial:

> Imagine we are throwing a two-sided die, such as a coin. I now want to determine a point of the interval AB by continually tossing the coin, and always bisecting the side prescribed by the throw: say: heads means I bisect the right-hand interval, tails the left-hand one. (ibid., pp.218–219)

The point here is that a point is being derived from the coin-tossing game, a sort of law, which Wittgenstein thinks realises space. The process of tossing the coin, of course, does not terminate within finite time, so Wittgenstein remarks, "I have an unlimited process, whose results as such don't lead me to the goal, but whose unlimited possibility is itself the goal" (ibid., p.219). To put it differently, such a rule for determining a point only gives us the point in infinite time, but still we may regard a rule itself as a sort of point; this idea of identifying points with rules or functions is now prevailing in mainstream mathematics, such as Algebraic and Non-Commutative Geometry. It should be noted here that a shift of emphasis is lurking behind the scene, from static entities like points to dynamic processes like laws. Those who are familiar with Brouwer's theory of the continuum would already have noticed that there is a close connection between Brouwer and Wittgenstein on the nature of space. The above illustration of tossing a coin almost defines the Cantor space in terms of contemporary mathematics: the Cantor space is the space of infinite sequences consisting of zeros and ones only, which in turn correspond to heads and tails of a coin in Wittgenstein's terms; actually, he himself discusses this correspondence (ibid., p. 220). Now let me quote a passage which, together with the quotations above, strikingly exhibits a remarkable link between Brouwer's and Wittgenstein's ideas of space (Brouwer 1918, p. 1; translation by van Atten 2007):

> A spread is a law on the basis of which, if again and again an arbitrary complex of digits [a natural number] of the sequence ζ [the natural number sequence] is chosen, each of these choices either generates a definite symbol, or nothing, or brings about the inhibition of the process together with the definitive annihilation of its result; for every n, after every uninhibited sequence of $n-1$ choices, at least one complex of digits can be specified that, if chosen as n-th complex of digits, does not bring about the inhibition of the process. Every sequence of symbols generated from the spread in this manner (which therefore is generally not representable in finished form) is called an element of the spread.

For Brouwer, a law is a rule to make a sequence of countably many digits. The difference between Brouwer's and Wittgenstein's laws basically lies in which to use two digits only (the Cantor space in modern terms) or all natural numbers (the so-called Baire space). Although this in fact gives rise to a certain technical difference, however, there is no doubt that the underlying conceptual view of capturing the concept of space in terms of laws is fundamentally the same in their thoughts. It may thus be concluded that Wittgenstein's and Brouwer's conceptions of space build upon the same core idea of regarding space as a law to form infinite digital sequences (important differences in the light of Wittgenstein's distinction between arithmetical and geometrical space shall be addressed later); interestingly, their philosophically motivated idea has now become a standard method, in Computer Science, to implement exact computation over continuous infinitary structures.

The modernist transformation of geometry via duality is striking. Modern mathematics has encountered drastic changes in both conceptual and technical senses. Prominent among them are a shift of emphasis from space itself to the structure of functions on it. In Algebraic Geometry, for example, properties of algebraic varieties are often shown through analysis of their associated function algebras; this was already noticed by Riemann in his study of what we now call Riemann surfaces. Duality expressed in terms of category theory is lurking behind the efficacy of function algebras in mathematics; in the case of Riemann surfaces, there is category-theoretical duality between Riemann surfaces and rational function fields, which tells us that Riemann surfaces can be reconstructed from the purely algebraic information of rational function fields on them, and vice versa. The modernist transformation of geometry has naturally led mathematicians to regard algebras themselves as spaces that do not presuppose the concept of points in the first place; points are entities derived as prime ideals of algebras under suitable conditions. In Non-Commutative Geometry (Connes 1994), for example, spaces are indeed defined as certain algebras, and commutative spaces are equivalent, via Gelfand Duality, to locally compact Hausdorff spaces in a classical sense. The same duality phenomenon happens to exist in mathematical logic, and is pursued under the name of Stone Duality, which is duality between syntax and semantics, thus functioning as a strengthened version of logical completeness theorems. In this case, we can regard logical systems on the side of syntax as spaces of their models on the side of semantics; classical logic amounts to the Cantor space in this sense.

Philosophically phrasing, we have both ontology and epistemology of space. As in many philosophies of space, points could be seen as metaphysical ultimate constituents of space, therefore an ontological account of space may be given in terms of points. At the same time, however, we cannot really see points with no extension, and hence the point-based concept of space is not acceptable from an epistemological point of view. In a sense, we can actually recognise regions with some extensions, or properties of space, so that a region-based or property-based notion of space, which has been implemented in algebraic concepts of space, is suitable for an epistemological account of space. Here note that, in the context of topology, regions (i.e., open sets) bijectively correspond to properties or predicates on space (i.e., Boolean-valued continuous functions). Wittgenstein (1974) says, "A line as a coloured length in visual space can be composed of shorter coloured lengths (but, of course, not of points)" (p.211). This gets quite closer to the idea of Locale Theory and Formal Topology, which replace topological spaces by the structures of (basic) open sets. Formal Topology by Sambin and Martin-Löf is a descendant of Brouwer's intuitionism, contriving a predicative framework for constructive topology. Locale

Theory models intuitionistic propositional logic, and closely connected with Topos Theory by Grothendieck and Lawvere, which gives a category-theoretical concept of point-free spaces serving as models of intuitionistic predicate logic. Deligne's theorem in Topos Theory, which guarantees the existence of sufficiently many points of certain toposes, amounts to Gödel's completeness theorem.

Just as intuitionism is an epistemological enterprise, so Wittgenstein's intensional conception of space is arguably epistemological in its nature. In contrast, Cantor's or Hausdorff's set-theoretical extensional concept of space may be conceived of as ontological in the sense above. Duality theory establishes a categorical-theoretical equivalence (or adjunction) between point-free and point-set spaces, and may thus be interpreted as exposing a remarkable link between ontology and epistemology of space in a mathematically profound and effective fashion. Furthermore, it should be noted that the idea of duality between the ontological and the epistemological makes sense in surprisingly diverse disciplines. In Quantum Physics, there is duality between quantum states and observables, which, in a sense, may be seen as ontological and epistemological respectively. In Computer Science, there is duality, due to Samson Abramsky, between computer systems (or programs) and observable properties of them, which are ontological and epistemological respectively. Duality in logic was already mentioned above. What matters here is that all this can indeed be understood as duality between set-theoretical extensional concepts of space and algebraic intensional concepts of space.

Summing up, Wittgenstein's conception of space can be articulated in connection with Brouwer's and other mathematical concepts of space in modern geometry. And the philosophical significance of duality in mathematics and sciences can be elucidated in terms of duality between ontology and epistemology of space. Wittgenstein's disagreement with the set-theoretical view of mathematics led him to the idea that space is not an extensional collection of points, but the intensional realisation of a law. Brouwer's theory of the continuum is arguably a theory of space based upon the point-free conception of space qua law rather than collection of points; this exhibits yet another case of Brouwer's influence on Wittgenstein's philosophy, in particular that of space. We may then consider the conception of space qua law to represent epistemology of space, and the conception of space qua points to represent ontology of space. In this perspective, modern geometry, such as Topos Theory, Algebraic and Non-Commutative Geometry, and Formal Topology, some of which are conceptual ramifications of Brouwer's intuitionism, yields rich instances of duality between epistemology and ontology of space. In the following we shall address links with Husserl, Whitehead, Cassirer, Granger, Lawvere, and Japanese philosophers.

From a historical perspective, Brentano and Husserl, as well as Wittgenstein, rejected to see space as a collection of points (but in their case for phenomenological reasons); a link between Brouwer and Husserl is pursued in van Atten (2007). Merleau-Ponty argued for the phenomenological priority of anthropological space (i.e., lived actual space) over geometric space (i.e., posited ideal space). Whitehead vigorously developed his process philosophy, putting strong emphasis on dynamic processes (like laws) rather than static entities (like points); in other words, he advocated the shift from Being to Becoming. He considered a point to be a bunch of shrinking regions (which converges to the point); it is exactly a prime filter in mathematical terms. Brentano, Husserl, and Whitehead may be seen as origins of what is now called Mereology. Cassirer weaved his philosophy based upon the dichotomy between substances and functions, which roughly amounts to the dichotomy between objects and operations in the more recent case of Granger, a French philosopher who problematised duality between objects and operations. In category-theoretical logic, Lawvere discussed duality between the conceptual and the formal. Concerning Japanese philosophy, it should be remarked that Hajime Tanabe in the Kyoto School of Philosophy was directly influenced, as discovered through Susumu Hayashi's textual analysis, by Brouwer's theory of the continuum, in contriving what he calls logic of species; he worked in the context of German idealism (the Marburg School of Neo-Kantianism, to be more specific), prominently seeing a parallelism between individuals and points, and between societies and regions (or laws). Kitaro Nishida, the leader of the Kyoto School of Philosophy, asserts "it is not that experience exists because there is an individual, but that an individual exists because there is experience" (Nishida 1990; originally 1921; xv in the Introduction). Just as a point is a bunch of observable regions for Whitehead, so an individual is a bundle of actual experiences for Nishida. Wataru Hiromatsu, a Tokyo philosopher of a later generation than the Kyoto School, similarly propounded the shift from things (or objects) to events, which is analogous to the shift from points to laws in philosophy of space by Wittgenstein and Brouwer, and more conspicuously to the aforementioned shift from substances to processes and functions in Whitehead and Cassirer respectively. These philosophers (and some others) seem to share certain ideas in some way or other; however, a coherent and rigorous perspective on them is yet to be explored.

Several remarks are to be made here, including those on future directions. First and foremost, it should be emphasised that Wittgenstein's and Brouwer's conceptions of space are not claimed to be absolutely the same in any sense. There are debates as in Rodych (2011) regarding discrepancies between Brouwer and Wittgenstein on real numbers (and lawlike or lawless sequences; see also Han 2011). We have not much discussed real numbers so far, not just for this reason but more crucially because Wittgenstein carefully distinguishes between real numbers and points (or positions), and between arithmetical space and geometrical visual space (even though they are often identified in set-theoretical modern mathematics). In the light of this distinction between arithmetical space and geometric space, the claim here is concerning geometrical space, and indeed consistent with the orthodox view of discrepancies in terms of numbers or arithmetical space. And it may be that Wittgenstein's true characteristic as opposed to Brouwer's lies in this conceptual articulation of the notion of space.

15.5 A Bird's-Eye View of Mathematical Philosophy and Wittgenstein's Constructive Structuralism

To wrap up, Wittgenstein's philosophy of mathematics may be elucidated in contrast to other major positions in the following way:

- It is well known that Wittgenstein stopped to work on philosophy after having finished his *Tractatus Logico-Philosophicus*, and that he came back again to it under the influence of Brouwer's philosophy. Yet Wittgenstein's philosophy disagrees with Brouwer's psychologism or mentalism that mathematics is a mental activity and exists within the single human mind; his philosophy also disagrees with Brouwer's revisionism about mathematical practice. Wittgenstein's anti-psychologism is obvious in his denial of so-called private language and many other thoughts. Wittgenstein is also well known for his anti-scientism; yet at the same time he did not intend to revise science by any means. That is not the task of philosophy, and everything must be left as it is, according to Wittgenstein (cf. Dawson 2015). Wittgenstein was not a revisionist about scientific practice, including mathematical practice in particular:

  > Philosophy may in no way interfere with the actual use of language; it can in the end only describe it. For it cannot give it any foundation either. It leaves everything as it is. It also leaves mathematics as it is, and no mathematical discovery can advance it. A "leading problem of mathematical logic" is for us a problem of mathematics like any other. (Wittgenstein 1986, p. 49)

  The above passage tells us that Wittgenstein's philosophy of logic and mathematics was seamlessly interconnected with his philosophy of language in general; this is clear in his following saying as well:

  > Logical inference is part of a language game. (Wittgenstein 1978, p. 24)

- Wittgenstein's philosophy occasionally counts as some sort of formalism. His formalism is very different from Hilbert's formalism, however. Both Hilbert and Wittgenstein do talk about symbols and rules governing the use of them. Yet at the same time, Wittgenstein considers symbols and rules to be part of what Wittgenstein calls "natural history" as discussed above; to Hilbert, they are just formal entities with no such history. Wittgenstein himself contrasts the "natural history" view with the "logical must" view as follows:

  > 'That logic belongs to the natural history of mankind' – is not combinable with the logical 'must'." The agreement of human beings which is a presupposition of the phenomenon of logic, is not an agreement in opinions, much less in opinions about questions of logic. (Wittgenstein 1978, p. 21)

  Wittgenstein is sympathetic to some sort of historicism in mathematics; what shapes mathematics is its history of practice, without which the language game of mathematics gets meaningless because what determines meaning is collective human practice as a concrete historical process rather than any abstract formal machinery. Wittgenstein's philosophy also lacks a transcendental perspective on mathematics as observed in Hilbert's finitism and its Kantian justification, which, from a philosophical point of view, would be the deepest part of Hilbert's thought on foundations of mathematics.

- Wittgenstein started his career as a philosopher within the Frege-Russell's tradition of logical philosophy, and his *Tractatus* inspired the movement of logical positivism. The centrality of logic was clear in his early view of logic as reflecting and determining the structure of the world. Yet his later philosophy diverged from it, and his mature philosophy of mathematics in particular endorsed anti-logicism. He indeed says:

  > 'Mathematical logic' has completely blinded the thinking of mathematicians and philosophers. (Wittgenstein 1978, p. 15)

  Needless to say that mathematical logic was of positive value to most major thinkers in foundations of mathematics, a possible exception being Brouwer, to whom mathematics, primarily, was a purely mental activity, and any formalized mathematics was an incomplete representation of the mathematician's creative mind. Yet Wittgenstein's anti-logicism was even more strong:

  > The curse of the invasion of mathematics by mathematical logic is that now any proposition can be represented in a mathematical symbolism, and this makes us feel obliged to understand it. Although of course this method of writing is nothing but the translation of vague ordinary prose. [...] 'Mathematical logic' has completely deformed the thinking of mathematicians and of philosophers, by setting up a superficial interpretation of the forms of our everyday language as an analysis of the structures of facts. Of course in this it has only continued to build on the Aristotelian logic. (Wittgenstein 1978, pp. 299–300)

  He also says:

  > Logic as foundation of mathematics does not work if only because the cogency of the logical proof stands and falls with its geometrical cogency. In mathematics we can get away from logical proofs. [...] Criticism of the conception of logic as the 'foundation' of mathematics. Mathematics is a motley of calculating techniques. [...] Why mathematics is not logic. (Wittgenstein 1978, p. 10)

To Wittgenstein, mathematics is a special type of language game as grounded upon our forms of life, and its certainty, objectivity, and universality must all stem from the linguistic practice of mathematical activity as grounded upon our natural history in his own terms; this would be the most striking idea in his philosophy of mathematics.

In the following several remarks shall be made on Wittgenstein's philosophy of logic and mathematics in order to complement the views articulated so far and to give some outlooks for further investigation. From a contemporary point of view, Wittgenstein's philosophy may also be seen as some sort of structuralism; it could be called constructive structuralism with a distinctive emphasis on the notion of pattern:

> Is mathematics supposed to bring facts to light? Does it not take mathematics to determine the character of what we call a 'fact'? Doesn't it teach us how to enquire after facts? In mathematics there are no causal connexions, only the connexions of the pattern. (Wittgenstein 1978, p. 23)

Whereas natural science is concerned with causal connexions or mechanisms, mathematics is concerned with the connexions of the pattern, which is structural rather than factual. Mathematics is not about facts, but about patterns and structures, which in Wittgenstein's philosophy of mathematics, must be perspicuous and surveyable as we discussed above. This finitary nature of patterns and structures would be what characterizes Wittgenstein's structuralism. Put another way, patterns and structures are what we construct, rather than discover, in our mathematical practice, and thus Wittgenstein's structuralism is constructive structuralism. Structure is what we epistemically impose on the world, and not what the world naturally has.

Another striking feature of Wittgenstein's philosophy of mathematics is his emphasis on application of mathematics, which is not an extra but necessary element of mathematics. Wittgenstein especially emphasizes the autonomy and necessary applicability of arithmetic:

> One always has an aversion to giving arithmetic a foundation by saying something about its application. It appears firmly enough grounded in itself. And that of course derives from the fact that arithmetic is its own application. Arithmetic doesn't talk about numbers, it works with numbers. The calculus presupposes the calculus. Aren't the numbers a logical peculiarity of space and time? The calculus itself exists only in space and time. Every mathematical calculation is an application of itself and only as such does it have a sense. That is why it isn't necessary to speak about the general form of logical operation when giving a foundation to arithmetic. A cardinal number is applicable to the subject-predicate form, but not to every variety of this form. And the extent to which it is applicable simply characterizes the subject-predicate form. On the one hand it seems to me that you can develop arithmetic completely autonomously and its application takes care of itself. (Wittgenstein 1975, p. 130)

Wittgenstein then says:

> You could say arithmetic is a kind of geometry […] You could say it is a more general kind of geometry. […] The point of the remark that arithmetic is a kind of geometry is simply that arithmetical constructions are autonomous like geometrical ones, and hence so to speak themselves guarantee their applicability. (Wittgenstein 1975, pp. 131–132)

Wittgenstein also argues that applied mathematics does exist even in the absence of pure mathematics:

> People can be imagined to have an applied mathematics without any pure mathematics. […] Thus these people have rules in accordance with which they transform the appropriate signs (in particular, e.g., numerals) with a view to predicting the occurrence of certain events. […] The centre of gravity of their mathematics lies for these people entirely in doing. (Wittgenstein 1978, p. 232)

To Wittgenstein, mathematics is more about the dynamic act of reason than about the static judgment, ceaselessly producing what he calls the natural history of humankind, which alone allows for the determination of meaning in the rule-following problem or Kripkenstein's paradox.

It must be remarked that Wittgenstein is a finitist of some sort, and yet he does not totally reject infinities. Wittgenstein actually gives a Kantian view of infinites:

> We all of course know what it means to say there is an infinite possibility and a finite reality, since we say space and time are infinite but we can always only see or live through finite

> bits of them. But from where, then, do I derive any knowledge of the infinite at all? In some sense or other, I must have two kinds of experience: one which is of the finite, and which cannot transcend the finite (the idea of such a transcendence is nonsense even on its own terms), and one of the infinite. And that's how it is. Experience as experience of the facts gives me the finite; the objects contain the infinite. Of course not as something rivalling finite experience, but in intension. Not as though I could see space as practically empty, with just a very small finite experience in it. But, I can see in space the possibility of any finite experience. That is, no experience could be too large for it or exhaust it: not of course because we are acquainted with the dimensions of every experience and know space to be larger, but because we understand this as belonging to the essence of space.–We recognize this essential infinity of space in its smallest part. Where the nonsense starts is with our habit of thinking of a large number as closer to infinity than a small one. As I've said, the infinite doesn't rival the finite. The infinite is that whose essence is to exclude nothing finite. [...] Space has no extension, only spatial objects are extended, but infinity is a property of space. [...] And the same goes for time. (Wittgenstein 1975, pp. 157–158)

Concerning time, Wittgenstein argues as follows:

> That we don't think of time as an infinite reality, but as infinite in intension, is shown in the fact that on the one hand we can't imagine an infinite time interval, and yet see that no day can be the last, and so that time cannot have an end. We might also say: infinity lies in the nature of time, it isn't the extension it happens to have. We are of course only familiar with time–as it were–from the bit of time before our eyes. It would be extraordinary if we could grasp its infinite extent in this way (in the sense, that is to say, in which we could grasp it if we ourselves were its contemporaries for an infinite time.) We are in fact in the same position with time as with space. The actual time we are acquainted with is limited (finite). Infinity is an internal quality of the form of time. (Wittgenstein 1975, p. 164)

In mathematical logic, infinities are packed into finite language, just as higher infinities in set theory boil down to finitary formulas characterizing them. Does this allow us to eliminate infinites from mathematics? Wittgenstein asks a similar question:

> 'We only know the infinite by description.' Well then, there's just the description and nothing else. Does a notation for the infinite presuppose infinite space or infinite time? [...] Time contains the possibility of all the future now. The space of human movement is infinite in the same way as time. [...] Empty infinite time is only the possibility of facts which alone are the realities.–If there is an infinite reality, then there is also contingency in the infinite. (Wittgenstein 1975, pp. 26–27)

The same question is asked again in the following passage:

> 'We only know the infinite by description.' Well then, there's just the description and nothing else. To what extent does a notation for the infinite presuppose infinite space or infinite time? Of course it doesn't presuppose an infinitely large sheet of paper. But how about the possibility of one? We can surely imagine a notation which extends through time, not space. Such as speech. Here too we clearly find it possible to imagine a representation of infinity, yet in doing so we certainly don't make any hypothesis about time. Time appears to us to be essentially an infinite possibility. Indeed, obviously infinite from what we know of its structure. Surely it's impossible that mathematics should depend on an hypothesis concerning physical space. And surely in this sense visual space isn't infinite. And if it's a matter, not of the reality, but of the possibility of the hypothesis of infinite space, then this possibility must surely be prefigured somewhere. (Wittgenstein 1975, pp. 155)

Wittgenstein's idea that infinity is an internal quality of the form of time or space clearly sounds Kantian. However his entire philosophy of infinity is yet to be explored and articulated further. Wittgenstein's comparison of finitism with behaviorism is interesting as well:

> Finitism and Behaviourism are similar trends. Both deny the existence of something with a view to getting out of a muddle. (Wittgenstein 1978, p. 9)

The relationships between propositions and proofs are intriguing in Wittgenstein's philosophy. He says as follows:

> We might also put it like this: the completely analysed mathematical proposition is its own proof. Or like this: a mathematical proposition is only the immediately visible surface of a whole body of proof and this surface is the boundary facing us. A mathematical proposition–unlike a genuine proposition–is essentially the last link in a demonstration that renders it visibly right or wrong. [...] A mathematical proposition is related to its proof as the outer surface of a body is to the body itself. We might talk of the body of proof belonging to the proposition. Only on the assumption that there's a body behind the surface, has the proposition any significance for us. We also say: a mathematical proposition is the last link in a chain of proof. (Wittgenstein 1975, p. 192)

Wittgenstein's thesis of meaning as use is well known. The following conception of meaning as proof may count as a mathematical instantiation of the idea of meaning as use:

> Its meaning must derive from its proof. What the proof proves-that is the meaning of the proposition (neither more nor less)...(Wittgenstein 1975, p. 144)

The meaning of a proposition is its proof, and this implies that a proof is indispensable to make sense of a proposition. He says:

> [T]he inference isn't expressed by a proposition. (Wittgenstein 1975, p. 151)

And then Wittgenstein proceeds as follows:

> The essential point in all these cases is that what is demonstrated can't be expressed by a proposition. (Wittgenstein 1975, p. 152)

To Wittgenstein, propositions only touch upon the surface of mathematics, and the depth and meaning of it are rooted in proofs, demonstrations, and constructions. The gist of mathematics consists in constructions themselves rather than propositions thus established (and those constructions must be surveyable). Martin-Löf, known for intuitionistic type theory, had a similar idea too. The conception of meaning as proof is further pursued in contemporary developments such as proof-theoretic semantics. Yet what is peculiar about Wittgenstein's conception of proof is the surveyability condition as discussed above. He calls it "the 'geometrical' conception of proof" (Wittgenstein 1978, p. 9). "Proof as a memorable picture" (Wittgenstein 1978, p. 9) expresses a similar idea. The memorable picture is essential in mathematical proof:

> The queer thing is that the picture, not the reality, is supposed to prove a proposition. (Wittgenstein 1978, p. 15)

Surveyability or memorability is crucial for "identity and reproducibility of a proof" according to Wittgenstein. It is not that memorable proofs are better than other proofs, but that surveyability is a necessary condition for something to count as a proof in the first place:

> The proof must be a procedure plain to view. (Wittgenstein 1978, p. 10)

This is why Wittgenstein's philosophy of mathematics may be understood as surveyable finitism. Wittgenstein argues that it is the genuine creation of a new proof to make too long a proof a memorable proof plain to view, and surveyability or memorability is indispensable in mathematical proof. He also thinks of "proof as part of an institution" (Wittgenstein 1978, p. 10). In a nutshell, proofs must exhibit constructions plain to view, and yet all this just lives within our institution or natural history. Wittgenstein says that convictions given by mathematical proofs are rooted in the natural history as well:

> I have read a proof–and now I am convinced.–What if I straightway forgot this conviction? For it is a peculiar procedure: I go through the proof and then accept its result.–I mean: this is simply what we do. This is use and custom among us, or a fact of our natural history. (Wittgenstein 1978, p. 61)

Proofs, under the surveyability condition, are sources of mathematical understanding and conviction:

> The proof conceived as a movement from one concept to another. Understanding a mathematical proposition. The proof introduces a new concept. The proof serves to convince one of something. (Wittgenstein 1978, p. 14)

Wittgenstein also says that a proof is a pattern and a (cinematographic) picture (Wittgenstein 1978, p. 84), which must be surveyable and memorable, thus serving us as the source of mathematical understanding as well as transparent justification. The geometric or pictorial nature of proof is essential in mathematical understanding.

Let us finally have a look at Wittgenstein's philosophy of contradictions, which is relevant to his understanding of Gödel's incompleteness theorems. Wittgenstein says:

> If a contradiction were actually found in arithmetic, this would show that an arithmetic with such a contradiction can serve us very well. (Wittgenstein 1978, p. 25)

He also says as follows:

> By permission and prohibition, I can always only define a game, never the game. What Hilbert is trying to show by his proof is that the axioms of arithmetic have the properties of the game, and that's impossible. It's as if Hilbert would like to prove that a contradiction is inadmissible. (Wittgenstein 1975, p. 321)

According to Wittgenstein, a contradiction is basically no problem, being no obstacle to actual practice at all. As long as practice goes well, contradictions cannot do any harm to it. This view strikingly distinguishes Wittgenstein's philosophy from any other major positions in foundations of mathematics, such as logicism, intuitionism, and formalism. Wittgenstein's understanding of potential contradictions in mathematics is even more clear in the following conversation recorded by Waismann (1979, p. 125):

> [S]uppose among the rules there were two that contradicted each other, but I had such a bad memory that I never noticed this and always forgot one of these two rules or obeyed alternately the one and then the other. Even then I would say, Everything is all right. After all, the rules are instructions for playing the game, and as long as I can play, they must be all right. It is only when I notice that they contradict each other that they cease to be all right, ant that manifests itself only in this: that I cannot apply them any more. For the logical product of the two rules is a contradiction, and a contradiction no longer tells me what to do. Thus the conflict appears only when I notice it. There was no problem as long as I was able to play the game.

This is a strongly pragmatic view of mathematics and its consistency. To Wittgenstein, mathematics is not really a theory; rather it is pure practice. Wittgenstein's pragmatism on mathematical inconsistencies is reminiscent of the physicist's use of mathematics, which occasionally involve mathematical inconsistencies, as for example in quantum field theory, and yet cannot really do any harm to the significance of physics itself. This may even generalized to our everyday life. Although our natural language is full of paradoxes, such as the Liar paradox, we do not really care about them, leading quite consistent lives with our inconsistent languages, which are highly useful despite their logical inconsistencies. Wittgenstein just applies the same idea to mathematical language. To Wittgenstein, the foundation of mathematics is mathematics itself. The certainty, objectivity, universality, and applicability of mathematics all stem from the natural history of down-to-earth mathematical practice, and not from any idealized foundational theory in which mathematics is jailed.

Acknowledgements I am grateful to Samson Abramsky for discussions about the unity of science, to Makoto Kikuchi for discussions about Wittgenstein's philosophy of logic and mathematics, and to Norihiro Yamada for discussions about logic and foundations of mathematics. I hereby acknowledge financial support from JSPS (grant number: 17K14231) and JST (grant number: JPMJPR17G9).

References

Baird, David, et al. (eds.), 1998, *Heinrich Hertz: Classical Physicist, Modern Philosopher*, Netherlands: Springer.

Bourbaki, Nicole, 1950, "The Architecture of Mathematics", *The American Mathematical Monthly*, vol. 4, pp. 221–232.

Brouwer, L.E.J., 1918, "Begründung der Mengenlehre unabhängig vom logischen Satz vom ausgeschlossenen Dritten. Erster Teil, Allgemeine Mengenlehre", *KNAW Verhandelingen* 5, pp.1-43.

Cartier, Pierre, 2001, "A Mad Day's Work: From Grothendieck to Connes and Kontsevich: The Evolution of Concepts of Space and Symmetry", *Bull. Amer. Math. Soc.* 38, pp. 389–408.

Connes, Alain, 1994, *Noncommutative Geometry*, San Diego: Academic Press.

Coquand, Thierry, 2009, Space of valuations, *Annals of Pure and Applied Logic*, vol. 157, pp. 97–109.

Corry, Leo, 2004, *David Hilbert and the Axiomatization of Physics (1898-1918)*, Netherlands: Springer.

Daston, Lorraine, and Peter Galison, 2007, *Objectivity*, NewYork: Zone Books.

Dawson, Ryan, 2015, *Leaving Mathematics As It Is: Wittgenstein's Later Philosophy of Mathematics*, Norwich: University of East Anglia.
Dummett, Michael, 1977, *Elements of Intuitionism*, Oxford: Clarendon Press.
Galison, Peter, 1999, "Trading Zone: Coordinating Action and Belief", in: *Science Studies Reader*, pp. 137–160, NewYork: Routledge.
Galison, Peter, 1997, *Image and Logic: A Material Culture of Microphysics*, Chicago: Univesity of Chicago Press.
Gray, Jeremy, 2008, *Plato's Ghost: The Modernist Transformation of Mathematics*, Princeton: Princeton University Press.
Han, Daesuk, 2011, "Wittgenstein and the Real Numbers", *History and Philosophy of Logic* 31, pp. 219–245.
Hilbert, David, 1983, "On the Infinite", in: *Philosophy of Mathematics: Selected Readings*, Cambridge: Cambridge University Press.
Kripke, Saul, 1982, *Wittgenstein on Rules and Private Language*, Massachusetts: Harvard University Press.
Lawvere, William, 1964, "An elementary theory of the category of sets", *Proceedings of the National Academy of Science of the U.S.A.*, vol. 52, pp. 1506–1511.
MacLane, Saunders, and Ieke Moerdijk, 1992, *Sheaves in Geometry and Logic*, New York: Springer-Verlag.
Marion, Matthieu, 2011, Wittgenstein on surveyability of proofs, in: Marie McGinn and Oskari Kuusela (eds.), *The Oxford Handbook of Wittgenstein*, Oxford: Oxford University Press.
Martin-Löf, Per, 1984, *Intuitionistic type theory*, Napoli: Bibliopolis.
Nishida, Kitaro, 1990 (original publication in 1921), *An Inquiry into the Good*, New Haven: Yale University Press.
Peirce, Charles Sanders, 1974, *Collected Papers of Charles Sanders Peirce*, Massachusetts: Harvard University Press.
Putnam, Hilary, 1983, *Realism and Reason*, Philosophical Papers vol. 3, Cambridge: Cambridge University Press.
Rodych, Victor, 2011, "Wittgenstein's Philosophy of Mathematics", *Stanford Encyclopedia of Philosophy*, Stanford: Stanford University.
Shanker, Stuart, 1987, *Wittgenstein and the Turning Point in the Philosophy of Mathematics*, New York: SUNY Press.
van Atten, Mark, 2007, *Brouwer meets Husserl: on the phenomenology of choice sequences*, Dordrecht: Springer.
Voevodsky, Vladimir, et al., 2013, *Homotopy Type Theory: Univalent Foundations of Mathematics*, Princeton: IAS.
Waismann, Friedrich, 1979, " Ludwig Wittgenstein and the Vienna Circle", Oxford: Basil Blackwell.
Weingart, Peter, 2010, "A short history of knowledge formations", in: *The Oxford Handbook of Interdisciplinarity*, pp. 3–14, Oxford: Oxford University Press.
Wittgenstein, Ludwig, 1974, *Philosophical Grammar*, Oxford: Basil Blackwell.
Wittgenstein, Ludwig, 1975, *Philosophical Remarks*, Oxford: Basil Blackwell.
Wittgenstein, Ludwig, 1978, *Remarks on the Foundations of Mathematics*, Oxford: Basil Blackwell.
Wittgenstein, Ludwig, 1986, *Philosophical Investigations*, Oxford: Basil Blackwell.

Chapter 16
Wittgenstein on Physics

Chon Tejedor

In this paper, I explore Wittgenstein's philosophical approach to physics, an approach that crystallises in the *Tractatus* and is then polished—rather than replaced—in his later writings. The question of Wittgenstein's attitude towards science has been the subject of much scholarly debate. Wittgenstein maintained throughout his life that philosophy, ethics and religion should be kept separate from the natural sciences. In his view, any attempt to apply scientific methodologies to philosophical, ethical and religious discussions is both dangerous and futile. Some interpreters have read this aspect of Wittgenstein's thinking as expressing a hostility to science:[1] Wittgenstein, they suggest, views science in an negative light and, for this reason, wishes to protect other areas from scientific encroachment. Wittgenstein certainly conveys on occasion a deep frustration with what he regards as the increasing dominance of science over other forms of culture. In *Culture and Value*, we find him writing:

> In order to marvel human beings–and perhaps peoples–have to wake up. Science is a way of sending them off to sleep again. (CV: 7)
>
> Our civilization is characterized by the word progress. Progress is its form, it is not one of its properties that it makes progress. Typically it constructs. Its activity is to construct a more and more complicated structure. And even clarity is only a means to this end & not an end in itself.
>
> For me on the contrary clarity, transparency, is an end in itself
>
> I am not interested in erecting a building but in having the foundations of possible buildings transparently before me.
>
> So I am aiming at something different than are the scientists & my thoughts move differently than do theirs. (CV: 9)

[1] See notably Phillips (1993).

C. Tejedor (✉)
Philosophy Faculty, Universitat de València, Valencia, Spain
e-mail: Chon.Tejedor@uv.es

S. Wuppuluri and N. da Costa (eds.), *WITTGENSTEINIAN (adj.)*,
The Frontiers Collection, https://doi.org/10.1007/978-3-030-27569-3_16

Whilst the cultural dominance of scientific mindsets is certainly a matter for concern to Wittgenstein, it is, in my view, mistaken to read this as an expression of hostility to science *as such*. For Wittgenstein's hostility is not so much aimed at science, but at *any* ill-formed effort to bring together methodologies and practices whose very integrity requires separation from each other. In this respect, Wittgenstein is as opposed to scientific thinking encroaching into philosophy, ethics and religion as he is opposed to philosophy (notably; metaphysics), ethics and religion encroaching in scientific thinking. Indeed, both forms of encroachment represent dangerous versions of *scientism* to his mind: scientism poses both a threat from science to philosophical, ethical and religious practice and a threat from metaphysics, ethics and religion to science.[2] Far from being driven by a hostility to science, much of Wittgenstein's philosophical thinking stems from a fascination with science and from the conviction that shedding light on its aims is crucial to safeguarding its integrity.

As we know, Wittgenstein embarked on Engineering studies before turning to philosophy and moving to Cambridge in 1911. His experience as a budding engineer enriched his philosophy in many ways and I think it is fair to say that he never entirely shook the suspicion that, at a dispositional level at least, he remained partly a scientist. Even in the midst of his 1929 "Lecture on Ethics" (LOE), when he famously imagines witnessing a human being miraculously growing a lion's head, he thinks of himself as standing in the shoes of the scientist:

> Now suppose such an event happened. Take the case that one of you suddenly grew a lion's head and began to roar. Certainly that would be as extraordinary a thing as I can imagine. Now whenever we should have recovered from our surprise, what I would suggest would be to fetch a doctor and have the case scientifically investigated and if it were not for hurting him I would have him vivisected. (LOE: 5)

The use of the first person in this passage is striking: "what I would suggest [...is to] have the case scientifically investigated". Perhaps Wittgenstein continued to feel, even by this stage, that his instincts remained those of empirical experimentation, the instincts of the natural scientist.

Be this as it may, it is clear that one key lesson learnt by Wittgenstein during his brief stint as an Engineering student was that he was far less interested in applying scientific methodologies than in exploring them from a philosophical perspective. This philosophical preoccupation with the natural sciences emerges in his very first writings: some of his earliest philosophical remarks focus on science and, significantly, they are the only ones to survive virtually unchanged into his first major philosophical work, the *Tractatus Logico-Philosophicus,* published in 1921.[3] Since so much of Wittgenstein's thinking on science ends up surviving into the post-*Tractatus* period, we would do well to conclude that this aspect of his thinking represents one of the strongest lines of continuity in his entire philosophy. When, in addition, we consider that Wittgenstein's early discussions of scientific systems and scientific laws

---

[2]Tejedor (2017).

[3]The *Tractatus'* remarks on natural science systems (notably, *TLP* 6.341–6.343) first emerge in the *Notebooks* in as early as December 1914, at which point Wittgenstein comments 'This I have felt for a *long* time' (*NB* 6.12.14). I am grateful to John Preston for drawing my attention to this.

are the precursors to his remarks on, respectively, language-games in *Philosophical Investigations* and hinges in *On Certainty*, it seems reasonable to conclude that Wittgenstein's early work on science is a central part of the backbone underpinning his entire philosophy.

In light of all this, I shall begin with a discussion of Wittgenstein's treatment of physics in the *Tractatus* and continue with an examination of how it evolves in his later writings. In the *Tractatus*, Wittgenstein presents physics as a natural science centrally concerned with physical causation and associated to the mechanics.

> 'Law of causality'—that is a general name. And just as in mechanics, for example, there are 'minimum-principles', such as the law of least action, so too in physics there are causal laws, laws of the causal form. (*TLP* 6.321)

The *Tractatus* remarks on causation and mechanics are organised into two separate sections: TLP 5.133–5.1362 and TLP 6.32–6.3611. The first of these sections is mostly negative or critical: in it, Wittgenstein seeks to undermine a traditional metaphysical understanding of physics and of causation which he regards as self-defeating, and, for this reason, as nonsensical. The second of these sections is more positive: in the TLP 6.3ff, Wittgenstein defends an understanding of physics and of causal principles which is, in his view, devoid of distorting metaphysics and far more in line with actual scientific practice.[4] I shall consider each of these two sections from the *Tractatus* in turn, before turning to the evolution of Wittgenstein's thinking on physics in his later writings.

(1) The *Tractatus* on Metaphysic's Self-Defeating Approach to Physics

In TLP 5.133–5.1362, Wittgenstein aims to dissolve a confused understanding of physics and of physical causation, one he views as distinctly philosophical—or, more specifically, as *metaphysical*. The view criticised by Wittgenstein—which I call the Natural Necessity View—suggests that causal relations hold necessarily between causes and effects. Consider the example of a flame causing a piece of wax to melt. According to the Natural Necessity View, when a flame causes a piece of wax to melt, the connection between the flame and the melting of the wax and is not accidental but necessary: given the presence of the flame near the wax, it is impossible (in the existing conditions) for the piece of wax not to melt. That is to say: the presence of the flame renders the melting of the wax necessary, in those circumstances. In this view, causal relations are necessary relations that hold in the physical world by virtue of certain laws of nature or laws of physics being in place: that the wax necessarily melts in the presence of the flame is determined by these laws. These laws could conceivably have been different; they are themselves, in this respect, akin to *falsifiable generalisations*. However, given that they do obtain, it is impossible for the wax not to melt in the circumstances described. And so it goes for all instances of causation. Consider the case of a magnet *causing* an iron pin to move. According to the Natural Necessity View, the magnet causes the iron pin to move *in that* the presence of the magnet *necessarily* results in the movement of the iron pin,

[4] Both of these are discussed in further detail in my: Tejedor (2015: Chaps. 4 and 5).

by virtue of the fact that certain laws of physics happen to obtain. In this view, the laws in question justify and ground the necessary causal connection between cause and effect.

For Wittgenstein, this view of physics and of causation—the Natural Necessity View—is problematic: far from capturing genuine scientific practice in physics, it is a metaphysical pseudo-theory about physics generated by philosophers, one which defeats itself and which is, for this reason, nonsensical. Let us consider his reasons for suggesting this in TLP 5.133–5.1362. Wittgenstein begins by noting that the Natural Necessity View presents us with a particular understanding of entailment. In this view, the presence of the flame *necessarily entails* the melting of the wax; similarly, the presence of the magnet *necessarily entails* the movement of the iron pin. This form of entailment is supposed to be necessary for the reasons discussed above: since causal connections are not accidental, advocates of the Natural Necessity View conclude that they must be necessary—necessary by virtue of the fact that certain laws of nature obtain (hence the expression "Natural Necessity"). This understanding of entailment contrasts with our entailment practices in ordinary thinking and language. Ordinarily, when I am presented with a thought or with proposition in language, there are quite a few things I can tell about it simply by attending to it. For instance, when someone says to me "It is raining", I can tell immediately, simply by attending to the proposition, that it sets out to *inform* me about the state of reality (namely: that it is raining out there) and that it may be either true or false. In other words, I can tell that the proposition purports to be *informative* and that it is *contingent* merely by attending to it. This may not seem like much. After all, in order to establish whether the proposition is true, I need to go beyond simply attending to the proposition: I need to empirically check outdoors, with my senses, whether it is indeed raining. Still, there are key aspects of the proposition that I can be confident about simply by attending to it, notably, that it purports to be informative and that it is contingent. Similarly, I can tell that the proposition "Either it is raining or it is not raining" can never be false (i.e. that it is necessarily true) simply by attending to it: no matter what that is happening outdoors, "*Either* it is raining *or* it is not raining" is true—it is true necessarily. "Either it is raining or it is not raining" is necessarily true by virtue of its internal characteristics, not by virtue of the facts that happen to obtain outdoors, in reality. For this reason, I can establish that it is necessary, rather than contingent, simply to attending to its internal characteristics.

In ordinary language and thought, I can tell whether a proposition or a thought is contingent (and thus in principle verifiable) or necessary, simply by attending to its internal characteristics, without having to check any facts external to it. This, for Wittgenstein, is a crucial aspect of ordinary linguistic communication: if someone tells me "it is raining", I need the proposition to signal clearly to me that something here stands in need of verification, so that I can decide whether or not to take an umbrella. In ordinary language and thinking, this is a central aspect of propositions and thoughts of all types, including those featuring relations of entailment. Compare for instance the following two propositions: "If Alex is a bachelor, Alex is a man" and "If it is raining, there is a traffic jam". Both of these propositions feature an entailment relation, in so far as they are of the form "if… then". The first of these propositions is necessarily true. I can tell this simply by attending to its internal characteristics: the proposition is true under all circumstances, irrespective of the facts, simply by virtue of the internal or analytic relation between the concept of bachelor and that of man, since a bachelor just *is* an unmarried man. Similarly, I can tell that the proposition "If it is raining, there is a traffic jam" is contingent rather than necessary simply by attending to its internal characteristics, without having to check any facts external to it. For instance, I may reflect on the proposition and note that, since rain may or may not be accompanied by traffic jams (i.e. since there is no internal or analytic connection between the concept of rain and that of traffic jam), "If it is raining, there is a traffic jam" is a contingent proposition, rather than a necessarily true one. When it rains and there is a traffic jam "If it is raining, there is a traffic jam" is true; when it rains and the roads are clear "If it is raining, there is a traffic jam" is false.

The understanding of entailment that emerges from the Natural Necessity View is subtly but importantly different from this. Consider the proposition "If the magnet is present, the iron pin moves". Ordinary linguistic and thinking practice suggests that we view this as contingent (that is, as amenable to verification) simply on the grounds of its internal characteristics: since there are no internal or analytic connections between the concept of magnet and that of iron pin, ordinarily we would confidently conclude that the proposition is contingent. If the Natural Necessity View were correct, however, this confidence would not be warranted. For, according to this view, there will be occasions—namely when the relevant laws of nature and associated circumstances obtain—when the proposition will be necessarily true: necessarily true, not by virtue of characteristics internal to the proposition (such as analytic connections between its concepts), but by virtue of the obtaining, as a matter of *fact*, of particular physical laws of nature. One of the upshots of this is that we can no longer treat any of the propositions or thoughts subject to such laws of nature (e.g. propositions such as "If the magnet is present, the iron pin moves" or "the magnet causes the iron pin to move") as we do ordinarily: we can no longer treat them as if we can confidently establish their logical status (i.e. whether they were contingent or necessary) simply by attending to them. For, in order to ascertain their status, we need, in this view, to know certain facts of the matter *external* to the proposition, namely, whether or not a certain law of nature happens to obtain.

This may not strike us as an unsurmountable problem. We may note that whereas language and thought operate in the ordinary manner in the majority of cases, in a small number of cases (i.e. those governed by obtaining laws of nature) they do not. And we may come to conclude that, so long as the exceptional cases are indeed rare enough, accommodating them need not disrupt our prevailing ordinary linguistic and thinking practices. This is indeed a conclusion that the advocates of the Natural Necessity View would do well to angle for. For it is crucial to the credibility and internal coherence of their view that it should not present itself as *undermining* our entire ordinary linguistic and thinking practices, but only as *restricting* these practices in a few, carefully circumscribed areas. In other words, the credibility and coherence of the Natural Necessity View depends on its leaving *most* of our ordinary thinking and linguistic practices intact. Wittgenstein suggests however that it does not. For, in effect, accepting the Natural Necessity View means *accepting that we can no longer be confident as to the status of a proposition merely by attending to its internal characteristics*. Imagine that I am presented with a proposition I have never encountered before. Say for instance that a Parisian friend tells me "Whenever you are in London, it rains in Paris". Ordinarily, I would take this to be a joke. "When you are in London, it rains in Paris" is a contingent proposition, since there is no analytic connection between rain in Paris and my presence in London; so "Whenever you are in London, it rains in Paris" is an apt joke. If the Natural Necessity View were correct, however, I would not be entitled to this confidence. For my confidence as to the logical status of "Whenever you are in London, it rains in Paris" derives from my having attended, in the ordinary manner, to the internal characteristics of the proposition. One of the implications of the Natural Necessity View, however, is that *there might just be some law of nature at work here*, perhaps one not yet discovered even by science, that might refute that confidence. For laws of nature, in this view, are akin to facts: they are contingent and are empirically discoverable by science. This means, however, that it is conceivable for a law of nature to obtain without it having been (as yet) discovered. A similar scenario emerges with more basic propositions and thoughts, such as the proposition "It is raining in Paris". Normally, proper attention to this proposition would entitle me confidently to conclude that it is contingent: "It is raining in Paris" is either true or false and in principle verifiable. If the Natural Necessity View were correct, however, I would no longer be entitled to this confidence in my conclusion. For the Natural Necessity View implies that there could be some as yet undiscovered law of nature to the effect that it was *always* in fact raining in Paris, regardless of whether people noticed this or not.

The moment we allow for the logical status of *some* propositions to become dependent on discoverable facts of the matter (here: on scientific laws of nature), we are surrendering the *normative authority* we ordinarily possess, by virtue of our mastery of language and thought, to determine the logical status of propositions and thoughts. In particular, we can no longer confidently appeal to our having paid proper attention to the propositions and thoughts in question in order to determine whether or not they are contingent. The Natural Necessity View thus inadvertently undermines the entire ordinary linguistic and thinking practice it relies upon in order to get off the ground. This, however, destabilises the view: for if we can no longer confidently proceed in language and thought as we ordinarily do, how can anyone be in a position to defend, understand or justify the Natural Necessity View? By undermining ordinary language and thinking, the Natural Necessity View ends up subverting itself. The view is self-defeating and, for this reason, nonsensical, according to Wittgenstein.

For Wittgenstein, the Natural Necessity View presents an understanding of causation and of the laws of physics that ultimately reveals itself to be a piece of *philosophical* nonsense. The Natural Necessity View emerges as part of a philosophical tradition immersed in metaphysics which is the source of a number of apparently intractable philosophical problems. Consider, for instance, the sceptical problem expounded by Hume in the following passage from the *Enquiry Concerning Human Understanding*:

> There are no ideas, which occur in metaphysics, more obscure and uncertain, than those of power, force, energy or necessary connexion [...].
>
> To be fully acquainted, therefore, with the idea of power or necessary connexion, let us examine its impression; and in order to find the impression with greater certainty, let us search for it in all the sources, from which it may possibly be derived.
>
> When we look about us towards external objects, and consider the operation of causes, we are never able, in a single instance, to discover any power or necessary connexion; any quality, which binds the effect to the cause, and renders the one an infallible consequence of the other. We only find, that the one does actually, in fact, follow the other. The impulse of one billiard-ball is attended with motion in the second. This is the whole that appears to the outward senses. The mind feels no sentiment or inward impression from this succession of objects: Consequently, there is not, in any single, particular instance of cause and effect, anything which can suggest the idea of power or necessary connexion. (EHU VII)

For Wittgenstein, apparently intractable philosophical problems such as that described by Hume result from adopting confused starting positions. These problems are not in fact intractable at all, but *non-existent*: they don't require *resolution*, but *dissolution*. In this particular case, dissolving Hume's problem involves exposing the understanding of causation on which it rests—i.e. the Natural Necessity View—as a self-defeating piece of nonsense.

For Wittgenstein, the Natural Necessity View therefore presents us with a nonsensical treatment of the role played by causation and laws of nature in physics. Nonsense, in the *Tractatus*, is associated with using sentences in ways that defeat themselves and thereby fail to serve any communicative purpose whatsoever. Nonsense is not always negative or to be avoided, however. Indeed, Wittgenstein suggests that generating nonsense can be entirely unobjectionable when, for instance, it is done in full awareness, as in certain forms of humour. This is exemplified by Wittgenstein's "nonsense" postcards to Gilbert Pattison: Wittgenstein would send his friend postcards (for which he used the term "nonsense") bearing photos of austere-looking patriarchs which he claimed—obviously falsely—to be photos of himself in various situations.[5] Wittgenstein's nonsense postcards are humorous partly because they pretend to have a communicative purpose which they evidently fail to fulfil: the postcards purport to inform Pattisson about Wittgenstein's appearance and character, whilst clearly failing to have any such aim. Their self-defeating aspect—i.e. their *nonsensically*—is very much part of what makes them humorous. For Wittgenstein, nonsense of this type is not in any way problematic. Nonsense only becomes problematic—indeed, pernicious—when it is not presented or recognised clearly for what it is, that is, when it arises from and promotes confusion.[6] The problematic type of nonsense emerges when we are under the illusion of being engaged in a purposeful activity, whereas in fact we are not. For Wittgenstein, this particular type of nonsense is central to a number of traditional philosophical practices—notably in metaphysics—and needs to be overhauled. Metaphysics poses a problem in that it attempts to bring together methodologies from philosophy and the natural sciences in

[5]Monk (1991: 265, Fig. 42).

[6]This is further supported by McGuinness (2002: 359) and by Monk (1991: 282).

ways that are self-defeating. For Wittgenstein, approaches such as this pose a threat to philosophy (properly understood) but also, significantly, a threat to science itself.

The question remains at this stage: why did the notion of a necessary causal relation seem appealing in the first place? For there is an undeniable surface attractiveness to the Natural Necessity View. After all, causation *is* meant to differ importantly from accidental correlation or mere coincidence; and positing causal relations as necessary does seem to account for this difference. If causal relations are not necessary, how do we account for our sense that they are not merely accidental? And how are we to understand the role played by laws of nature in physical causation, if not in the way proposed by the Natural Necessity View? As we will now see, these are precisely the questions that Wittgenstein seeks to address in the second, more positive section of the *Tractatus* dedicated to the natural sciences, causation and physics—namely, TLP 6.32–6.3611.

(2) A Positive Conception of Physics: Causal Laws as Instructions in Optional Systems

Interpreters of the *Tractatus* (*cf.* McGuinness 2002, 116–123) have for the most part held that, for the early Wittgenstein, all scientific propositions are truth-assessable and characterised by at least one of the following:

1. They represent possibilities, possible states of reality (e.g. the possibility that it might be raining).
2. They purport to inform us of those possibilities that actually obtain in reality—i.e. of the facts.
3. They give empirical descriptions of facts.
4. They give causal explanations or make statements about the causal relations—or causal mechanisms—that hold between possible states and/or facts.

For the sake of concision, I shall refer to truth-assessable propositions characterised by at least one of 1–4 as "truth-assessable propositions".

Contrary to what is commonly claimed, however, Wittgenstein does not regard the natural sciences as restricted to truth-assessable propositions in the *Tractatus*. For, already in this early text, Wittgenstein's interest in science is far broader than this: it is both an interest in the *subject matter* of science—(captured by truth-assessable propositions) and also, to a far greater extent, an interest in the scientific laws or *principles* constitutive of specific scientific approaches to this subject matter. This becomes evident as soon as one examines TLP 6.32–6.3611 at close range and considers Wittgenstein's positive remarks on physics. In *Tractatus* 6.321, Wittgenstein writes:

> [...] in physics, there are causal principles, principles of the causal form (*TLP* 6.321 – my translation).

Although this remark may at first sight seem cryptic, the term "form" is a prominent and familiar one in the *Tractatus*. It aims partly to capture the notion of a unified set of instructions (or "principles")—the instructions that enable us to produce *properly formed* truth-assessable linguistic propositions and thoughts within particular systems. For instance, English-language grammatical principles on the conjugation of verbs (i.e. the grammatical conventions relating to verbs) are *instructions* that enable us to say certain things in the English-language system, such as, for instance, "It is raining".

> The tacit conventions on which the understanding of everyday language depends are enormously complicated (TLP 4.002)

Wittgenstein suggests that, just as there are different linguistic systems stemming from different natural languages (English, French, etc.), there are different scientific systems, constituted by different sets of principles or forms. In connection to mechanics, for instance, he writes:

> Mechanics determine a form of description by saying: All propositions in the description of the world must be obtained in a given way from a number of given propositions—the mechanical axioms. It thus provides the bricks for building the edifice of science, and says: 'Any building that you want to erect, whatever it may be, must somehow be constructed with these bricks, and with these alone' (TLP 6.341)

The *Tractatus* suggests that causal laws should be understood as causal principles in this way, that is as instructions that circumscribe what can be said causally within particular systems of physics. Causal laws—the causal principles associated with physics—are instructions that demarcate what kind of proposition counts as *causal* within a given system of physics, just as the English-language conventions determine what counts as the appropriate use of a verb in this language. For Wittgenstein, it is part of the remit of physics to come up with such instructions or causal principles and—crucially—these unified sets of principles (these systems) can change over time. Consider for instance the shift in physics from the view that causation involves physical contact to the view that two objects at a distance from each other can still causally interact with each other: the latter allows for causation in the absence of contact, an idea sometimes captured by the phrase "action at a distance"; the former does not. Let us call the former system the Causal Contact system and the latter the Action at a Distance system. In the Causal Contact system of physics, "The magnet causes the iron pin to move" would not be a properly formed causal proposition, since this system does not include an instruction (a principle, a law) to the effect that causation can occur in the absence of physical contact. In contrast, "The flame touching the iron pin causes the iron pin to get hot" would be a properly formed causal proposition, in this system. The move from the first system to the second is in effect a move from one optional set of instructions to another.

The laws of physics, like those of other natural sciences, are optional, in so far as the systems to which they belong can be replaced by others. And they are at the same time a priori in that, being instructions for the proper use of signs, they are constitutive of the truth-assessable propositions and thoughts (including the beliefs and experiences) that we can construct with those signs. Hence:

> And just as in mechanics, for example, there are 'minimum-principles', such as the law of least action, so too in physics there are causal laws, laws of the causal form. (TLP 6.321)

> All such propositions, including the principle of sufficient reason, the laws of continuity in nature and of least effort in nature, etc. etc.—all these are a priori insights about the forms in which the propositions of science can be cast. (TLP 6.34)
>
> The form is optional. (TLP 6.341)

What, we may wish to ask, leads physicists to move from one causal system to another? Although Wittgenstein does not explicitly discuss this question in the *Tractatus*, he hints at a possible answer in his "Lecture on Ethics", a text written in 1929 but still in many respects close to the Tractarian position. In 'Lecture on Ethics', Wittgenstein repeatedly aligns the scientific approach with a particular form of relative evaluative judgement. The 'scientific book' (LOE: 2), he writes, would be such as to[7]:

> contain all relative judgments of value and all true scientific propositions and in fact all true propositions that can be made. (LOE: 2)

'Relative judgements of value', in this context, are judgement as to what is valuable in instrumental (means-ends) or pragmatic terms. He writes:

> If for instance I say that this is a good chair this means that the chair serves a certain predetermined purpose and the word good here has only meaning so far as this purpose has been previously fixed upon. In fact the word good in the relative sense simply means coming up to a certain predetermined standard. Thus when we say that this man is a good pianist we mean that he can play pieces of a certain degree of difficulty with a certain degree of dexterity. And similarly if I say that it is important for me not to catch cold I mean that catching a cold produces certain describable disturbances in my life and if I say that this is the right road I mean that it's the right road relative to a certain goal. (LOE: 1)

It is likely that Wittgenstein regarded it as part of the remit of physics to come up with those instructions or causal laws that best serve the discipline's instrumental purposes *qua* natural science and, more generally, the instrumental purposes of human beings.

(3) Physics and Causation in the post-*Tractatus* works.

Although Wittgenstein's philosophical thinking undergoes substantial and well documented changes in the post-*Tractatus* period, the most striking aspect of his treatment of the natural sciences—and of physics in particular—is in fact its continuity. Part of the reason for this is that some of the key notions that emerge in the *Tractatus*' discussion of science are retained and further refined—rather than replaced—in his later writings. The Tractarian notion of a linguistic system, which emerges both in the context of Wittgenstein's early discussion of natural languages (under the guise of "representational form"—cf. *TLP* 2.173 and 4.002) and in the context of his positive discussion of scientific systems (TLP 6.32–6.3611) is a clear forerunner of the concept of "language-game" which takes full shape in the *Philosophical Investigations*. In this respect, the move from Wittgenstein's early approach to scientific systems, in particular to causal systems in physics, to his *Philosophical Investigations* position

[7]This notion of the 'scientific book' is a clear echo of the *Tractatus*' notion of a book entitled *The World as I Found It* (*TLP* 5.631).

does not mark so much a conceptual change as a change in the level of detail with which Wittgenstein is now looking at the differences between (rather than the aspects in common to) such systems.

> But how many kinds of sentence are there? Say assertion, question, and command?—There are *countless* kinds: countless different kinds of use of what we call "symbols", "words", "sentences". And this multiplicity is not something fixed, given once for all; but new types of language, new language-games, as we may say, come into existence, and others become obsolete and get forgotten. (We can get a *rough picture* of this from the changes in mathematics.)
>
> Here the term "*language-game*" is meant to bring into prominence the fact that the *speaking* of language is part of an activity, or of a form of life.
>
> Review the multiplicity of language-games in the following examples, and in others:
>
> Giving orders, and obeying them—
>
> Describing the appearance of an object, or giving its measurements –
>
> Constructing an object from a description (a drawing)—
>
> Reporting an event—
>
> Speculating about an event—
>
> Forming and testing a hypothesis—
>
> Presenting the results of an experiment in tables and diagrams—
>
> Making up a story; and reading it—
>
> Play-acting—
>
> Singing catches—
>
> Guessing riddles—
>
> Making a joke; telling it—
>
> Solving a problem in practical arithmetic—
>
> Translating from one language into another—
>
> Asking, thanking, cursing, greeting, praying. —
>
> It is interesting to compare the multiplicity of the tools in language and of the ways they are used, the multiplicity of kinds of word and sentence, with what logicians have said about the structure of language. (Including the author of the *Tractatus Logico-Philosophicus*.) (PI 23)
>
> Misleading parallel: psychology treats of processes in the psychical sphere, as does physics in the physical.
>
> Seeing, hearing, thinking, feeling, willing, are not the subject of psychology *in the same sense* as that in which the movements of bodies, the phenomena of electricity etc., are the subject of physics. You can see this from the fact that the physicist sees, hears, thinks about, and informs us of these phenomena, and the psychologist observes the *external reactions* (the behaviour) of the subject. (PI 571)

Like in the *Tractatus*, in the *Philosophical Investigations* we find Wittgenstein arguing that no metaphysical moves, facts of the matter or empirical experiences ground or justify the rules (in the *Tractatus*: the instructions, the principles) that demarcate possible moves within scientific and other linguistic systems. For language is simply not founded upon metaphysical or empirical *justification*. The rules of grammar can only be expounded, if at all, pragmatically (PI §492). This idea emerges with force in Wittgenstein's discussion of the role of reasons as forms of justification:

> Why is it that if I burn my hand when I put my hand in a flame I am reluctant to do so again? Is it because I reason to myself 'it burned me last time so it *must*, or is at any rate very *likely*, to burn me again'? [...] Whether the earlier experience is the cause of the certainty depends on the system of hypotheses, of natural laws, in terms of which we are considering the phenomenon of certainty.
>
> Is such confidence justified? – What people accept as justification shows how they think and live. (PI §325; see also PI §§466, 472-4)

In *Philosophical Investigations*, the rules (the instructions, the principles) that constitute our scientific practices are arbitrary to the extent that it would be wrong to seek a justification for their existence outside those broadly arising from instrumental or pragmatic evaluation (cf. PI §466; see also PI §§467-85; cf. LOE).

Whereas, in *Philosophical Investigations*, the rules constitutive of language-games are often imagined as explicitly expressed in language, from 1946 onwards Wittgenstein rekindles an interest in what he calls the "tacit presuppositions" implicit in our language-games (PPF §§31-3)—an idea strongly reminiscent of the *Tractactus* notion of a "tacit convention" mentioned above (cf. TLP 4.002). This culminates towards the end of *On Certainty* in the notion of a *hinge certainty*—a notion similar to that of a heuristic maxim for interpreting phenomena.[8] By this stage in his philosophical life, the level of attention paid by Wittgenstein to the differences (rather than commonalities) between language-games has rendered his view maximally pluralistic: hinge certainties may by now take the form of action (OC §204), dispositions (OC §337), expectations, laws, rules, propositions (OC §608) and even of belief (in the sense of "faith", see OC §602). These hinge certainties are not inferred from experience and do not provide justifications or grounds, but rather function as the constitutive "*scaffolding*" (OC §211) of our language-games – another *Tractarian* metaphor that we see returning in *On Certainty*.[9] In connection to physics in particular, Wittgenstein writes:

> Is it wrong for me to be guided in my actions by the propositions of physics? Am I to say that I have no good ground for doing so? Isn't precisely this what we call a 'good ground'? (OC §608).
>
> Supposing we met people who did not regard that as a telling reason. Now, how do we imagine this? Instead of the physicist, they consult an oracle. (And for that we consider them primitive). Is it wrong for them to consult an oracle and be guided by it? – If we call this "wrong" aren't we using our language-game as a base from which to *combat* theirs? (OC §609).
>
> I said I would 'combat' the other man, – but wouldn't I give him *reasons*? Certainly; but how far do they go? At the end of reason comes *persuasion*. (Think of what happens when missionaries convert natives). (OC §612)

[8] On this, see notably Moyal-Sharrock (2005).

[9] Moyal-Sharrock (2005: 80–2).

Pragmatic considerations and instrumental evaluations may of course guide us towards attempting such persuasions. Nevertheless and crucially, our hinge certainties are emphatically not akin to true beliefs or to veridical perceptions. On the contrary: they consists in our *acting without further justification*.

> Giving grounds, however, justifying the evidence, comes to an end; - but the end is not certain propositions striking us immediately as true, i.e. it is not a kind of seeing on our part; it is our acting, which lies at the bottom of the language-game. If the true is what is grounded, then the ground is not true, not yet false. (OC §§204-5)

In *Philosophical Investigations*, we find Wittgenstein repeatedly insisting that knowledge presupposes the possibility of doubt. Without abandoning this view, in *On Certainty,* Wittgenstein comes to highlight the multifarious ways in which new doubts and certainties lead to revisions of our language-games themselves, rather than just to revisions of the claims we make within such games (OC §617). By this stage, Wittgenstein has fully surrendered not only to the conclusion that there are a multiplicity of scientific systems, including a multiplicity of causal systems of physics, but also to the idea that shifts from one such system to another themselves follow a multiplicity of patterns and cannot be subsumed under one single account.

References

McGuinness, B. *Approaches to Wittgenstein: Collected Papers*, 2002.
Monk, R. *Wittgenstein: The Duty of Genius* (Penguin) 1991.
Moyal-Sharrock, D. 2005, *Understanding Wittgenstein's "On Certainty"*. Palgrave Macmillan.
Phillips, D. Z. 1993, *Wittgenstein and Religion* (Palgrave Macmillan).
Tejedor, C. 2017, "Scientism as a Threat to Science: Wittgenstein on Self-Subverting Methodologies." J. Beale and I. J. Kidd (eds.), *Wittgenstein on Scientism* (Routledge): 14–33.
Tejedor, C. *The Early Wittgenstein on Natural Science, Metaphysics, Language and Value* (Routledge) 2015.

Chapter 17
The Physics of Miniature Worlds

Susan G. Sterrett

After his father died, in early 1913, Ludwig Wittgenstein spent some time with various friends, family, and acquaintances. In late 1913, he began making plans to withdraw to Norway, away from Cambridge [UK],[1] where he had been working very closely with Bertrand Russell, and away from people, to work on solving the problems of logic. There was no looking back at aeronautics as an alternative career after that, it seems. The age-old problem of controlled, heavier-than-air flight on a scale that permitted humans to fly had [just recently] been solved, albeit by others (Fig. 17.1). There was still exciting and important work to do in aeronautics, but he had by then made the agonizing decision to become a philosopher, and, in working with Russell, he had found the age-old problem that he felt he was meant to solve instead: finding a correct theory of symbolism [105].[2]

Wittgenstein's own investigations into logic were bringing him around to notions of mirroring and corresponding. In notes expressing his views as of April 1914, he concludes "thus a language which *can* express everything *mirrors* certain properties of the world by these properties which it must have" (Wittgenstein 1979b, p. 107). And he struggles to accommodate his observation of the problematic fact that "in the case of different propositions, the way in which they correspond to the facts to which they correspond is quite different" (Wittgenstein, 1979b, p. 113). A year earlier, he had said there was no such thing as the form of a proposition ["the form

[1]Text in brackets indicates addition or change from original as it appeared in the First Edition of Sterrett (2005/2006).

[2]Numerals in brackets from 105 through 153 refer to the pagination in the original book, *Wittgenstein Flies A Kite* (Sterrett 2005/2006).

Adapted from Chapter 6 of *Wittgenstein Flies A Kite: A Story of Models of Wings and Models of the World* by Sterrett (2005/2006, pp. 105–153).

S. G. Sterrett (✉)
Department of Philosophy, Wichita State University, Wichita, KS, USA
e-mail: susangsterrett@gmail.com

S. Wuppuluri and N. da Costa (eds.), *WITTGENSTEINIAN (adj.)*,
The Frontiers Collection, https://doi.org/10.1007/978-3-030-27569-3_17

THE TIMES, MONDA

AERIAL NAVIGATION.

SUCCESS OF THE WRIGHT AEROPLANE.

(FROM OUR CORRESPONDENT.)

PARIS, Aug. 9.

The Special Correspondent of *The Times* at Le Mans telegraphed late last night :—

" Mr. Wilbur Wright has made a remarkable flight this evening, lasting 1 minute 45 seconds, over a course of about 2,500 feet. He will resume his experiments on Monday. The average height maintained during to-day's flight was 30ft."

The news of this remarkable achievement, which took place in the presence of some of the leading members of the Aéro Club, well-known aviators like M. Blériot, and aeronauts like M. Archdeacon, MM. Paul and Edmond Zens, and M. Peyrey, has been received with enthusiasm in the French Press. Such secrecy had been maintained with regard to the Wright aeroplane that a large number of Frenchmen were sceptical even as to Mr. Wright's seriousness. All accounts, however, published in this morning's papers from the correspondents on the spot attest the complete triumph of the American inventor. All present affirm that, after yesterday's experiments, there can be no doubt that the Wrights possess a machine capable of remaining an hour in the air and almost as manageable as if it were a small toy held in the hand.

It was at half-past six

Fig. 17.1 Several months after Ludwig Wittgenstein moved to England to pursue graduate study in aeronautical research, it was revealed in a public exhibition in Paris, France, that the problem of controlled, sustained, heavier-than air flight had been solved. From *The London Times*, August 9, 1908

of a proposition is not a thing" (Wittgenstein, 1979b, p. 105)]; his resolution of the issue of how propositions correspond to facts now is in terms of the general form of a proposition—something, he has decided, that all propositions do have in common. Then he says, "In giving the general form of a proposition, you are explaining what kind of ways of putting together the symbols of things and relations will correspond to (be analogous to) the things having those relations in reality" (Wittgenstein, 1979b, p. 113).

These exploratory thoughts about the notion of correspondence were the beginning steps toward an answer to one of the puzzles raised much earlier by musical scores and the gramophone records that had been such a striking arrival on the scene the year Wittgenstein was born: "What is the relationship between the symbols in the score and the patterns of grooves in the gramophone record?" That there was a mechanical process that could be used to make a gramophone record and one that could be used to play sound from it was well known. What about the process of creating a musical score, and the process by which a symphony could be imagined or [106] produced by a musician reading the score? Were these just as straightforward? Wittgenstein had already steered clear of simplistic accounts of a symbol as "sign of thing signified" a year earlier, at least in the case of words, in deciding that "Man possesses an innate capacity for constructing symbols with which some sense can be expressed, without having the slightest idea what each word signifies" (Wittgenstein 1979c, p. 100). But how did mirroring work, if not by a straightforward correspondence?

Wittgenstein was not alone in pondering how items of language could mirror a situation and how propositions could correspond to the world. The specific suggestion that equations function like pictures or models was made by Boltzmann in his *Lectures on the Principles of Mechanics*, a work in which he strove for an accurate exposition of mechanics that would be accessible to members of the general public. In explaining the role of pictures in physical theories, Boltzmann had there explained that even those who thought their approach had dispensed with pictures had not really done so: "[Partial differential equations] too are nothing more than rules for constructing alien mental pictures, namely of series of numbers. Partial differential equations require the construction of collections of numbers representing a manifold of dimensions" (Boltzmann, 1974, p. 226). Thus, he said, at the bottom Maxwell's equations "like all partial differential equations of mathematical physics...are likewise only inexact schematic pictures for definite areas of fact" (Boltzmann, 1974, p. 226). Boltzmann's suggestion, however, went just as far as claiming that symbolic equations could function like scientific models or pictures—it did not purport to explain exactly how either worked. It does seem, though, that picturing involved some imagined entities that may, but need not, correspond to something in reality. Boltzmann speaks of pictures almost interchangeably with mental pictures (Fig. 17.2).

Boltzmann became an extremely popular lecturer in Vienna around 1903, when, as mentioned earlier, Wittgenstein would have been about fourteen years old and would have known of Boltzmann's lectures. These lectures were so popular that the lecture hall in Vienna could not accommodate the audience, and Boltzmann was invited to give them at the palace instead. Boltzmann was present in Wittgen-

Fig. 17.2 Ludwig Boltzmann (1844–1906), with whom Wittgenstein had hoped to study. *Photo credit* University of Vienna, 1898. Public Domain

stein's youth through his prolific writing as well as through these lectures [which he] delivered a stone's throw from [Wittgenstein's family] home. The second volume of Boltzmann's *Lectures on the* [107] *Principles of Mechanics* was published in 1904, and a collection of his writings was published as *Popular Writings* in 1905, when Wittgenstein was sixteen. As we have seen, at that crucial time in his life, he was so interested in Boltzmann that, at least as he later recounted things to his friend and colleague von Wright, he had originally planned to study physics with him (McGuinness 1988, p. 54).

Boltzmann's *Popular Writings* anthology (Fig. 17.3) included an essay republished from a physics journal, "On the Indispensability of Atomism in Natural Sci-

Inhalt.

Fig. 17.3 The Table of Contents to Boltzmann's Popular Writings. (1905) Essay 6 is "On Aeronautics" and Essay 10 is "On the Indispensibility of Atomism in Natural Science"

ence," in which he emphasizes the remarks about equations quoted earlier: "The differential equations of mathematico-physical phenomenology are evidently nothing but rules for forming and combining numbers and geometrical concepts, and these in turn are nothing but mental pictures from which appearances can be predicted". Here he refers the reader to Ernst Mach's *Principles of the Theory of Heat,* remarking that, as far as the use of models goes, there is no essential difference in the approach he takes and approaches such as energetics, in which equations rather than models of material points, are central: "Exactly the same holds for the conceptions of atomism, so that in this respect I cannot discern the least difference. In any case it seems to me that of a comprehensive area of fact we can never have a direct description but always only a mental picture". He had a rather precise criticism specific to differential equations and the corresponding assumption of a continuum, in that differential equations relied on the notion of a limit, and that observationally there was no distinguishing between systems of large numbers of finite particles and actual continuums. Thus, he said "those who imagine they have got rid of atomism by means of differential equations fail to see the wood for the trees" (Boltzmann, 1974, p. 45).[3]

Elsewhere, in his encyclopedia article on "Model," which was reprinted in the same anthology, Boltzmann again described the method he referred to as the theory of "mechanical analogies," remarking that, unlike in earlier days, "nowadays Philosophers postulate no more than a partial resemblance between the phenomena visible

[3] A number of commentators have pointed out the relevance of these and similar passages to Wittgenstein's early thoughts: McGuinness (1988), Janik and Toulmin (1973) and Barker (1980). "Hertz and Wittgenstein".

in such mechanisms and those which appear in nature" (Boltzmann 1974, p. 214). Looking closely at his remarks, though, it is clear he had run into a brick wall with this approach. He had to except from the models to which his remarks applied the kind of model that was used in [108] experimental engineering scale models. On the approach in which physical models constructed with our own hands are actually a continuation and integration of our process of thought, he says, "physical theory is merely a mental construction of mechanical models, the working of which we make plain to ourselves by the analogy of mechanisms we hold in our hands". In contrast, in his discussion of mental models, Boltzmann had explicitly described experimental models as of a different sort than the kind with which he was comparing mental models. Boltzmann even explained why they must be distinguished:

> A distinction must be observed between the models which have been described and those experimental models which present on a small scale a machine that is subsequently to be completed on a larger, so as to afford a trial of its capabilities. Here it must be noted that a mere alteration in dimensions is often sufficient to cause a material alteration in the action, since the various capabilities depend in various ways on the linear dimensions. Thus the weight varies as the cube of the linear dimensions, the surface of any single part and the phenomena that depend on such surfaces are proportionate to the square, while other effects — such as friction, expansion, and condition of heat, etc., vary according to other laws. Hence a flying-machine, which when made on a small scale is able to support its own weight, loses its power when its dimensions are increased. The theory, initiated by Sir Isaac Newton, of the dependence of various effects on the linear dimensions, is treated in the article UNITS, DIMENSIONS OF. (Boltzmann 1974, p. 220)

Thus, the experimental models represent a challenge: for experimental models, the relationship between model and what is modeled is in some ways unlike the relationship between a mental model and what is modeled by it.

Boltzmann committed suicide in the year 1906, the year after the anthology of his popular writings appeared, and Wittgenstein never did get to study with him. This remark of Boltzmann's might well have resonated with Wittgenstein's personal experience, even though he did not get to do experimental work under Boltzmann, for we know that Wittgenstein had built and played with a toy airplane (Spelt and McGuinness 2000), and these toys were quite serious affairs technically.[4] Boltzmann's remarks about experimental models, and his specific mention of a model [109] of a flying machine as a model that does not behave like the full-size machine it models, could scarcely fail to command Wittgenstein's attention. If an airplane design would only work the same way when enlarged, the problem of sustained, controlled heavier-than-air flight would have pretty much already been in the hands of countless children in Europe and America. As we saw earlier (Sterrett 2005/2006, pp. 11–15), Penaud had developed a rubber band-powered model airplane that was capable of sustained, stable flight (Fig. 17.4), and some of Penaud's designs were available as toys even before Wittgenstein was born. Boltzmann was especially aware of the fact that, in England, Maxim had shown that it was possible to design a full-

[4]Chanute's compendium (Chanute 1894) surveys an astounding number and variety of different proposed aeroplane designs, and identifies which of the designs were implemented as model aeroplanes that had undergone experimental trials.

In 1872 *Pénaud*, who had already succeeded (1870 and 1871) in compassing flight with the superposed screws and

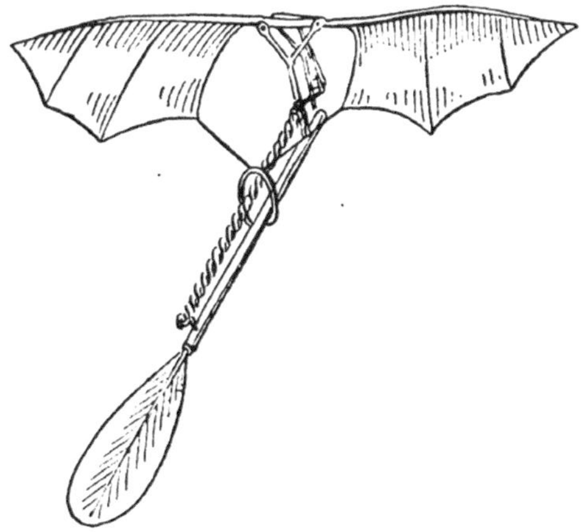

FIG. 17.—PÉNAUD—1872.

Fig. 17.4 Penaud had success with a rubber-band powered model airplane. Though some of his designs were intended as toys, larger versions were meant for researching specific full-size designs. The photograph of one of Penaud's model planophores being held by a man viewable online at the Science and Society Picture Library, Image No. 10438807 provides a sense of scale of these models

size steam-powered airplane capable of getting off the ground. In doing so, Maxim showed that an airfoil or kite could be powered—that is, that the power an engine produced could be large enough in proportion to its weight to get an airplane off the ground, which is all he was trying to establish at that point.

The unsurmounted obstacle was to get the sustained flight that Penaud had already achieved in a small-scale model, in a full-size airplane. Penaud's work had been the most promising, but as we saw earlier, he too had committed suicide (Fig. 17.5). In fact, he had done so upon receiving news that construction of the full-size model he had designed would not receive the funding he had been expecting. In 1905, the air was full of the promise of controlled heavier-than-air-flight, and there were some who believed the stories that two Americans had achieved it. Thus, Boltzmann's remarks describing significant and essential differences between mental models and models of flying machines would have had the effect of diminishing interest in mental models, because they made mental models seem like less robust representations of the world. And they may well have piqued interest in understanding whether and how models of flying machines could represent larger ones.

The puzzlement about the effect of size on the ability of machines to fly was common to just about anyone who played with toy flying machines. We saw earlier that the Wright Brothers recalled very clearly their puzzlement as children that the larger-sized models they built of exactly the same design didn't perform like the wonderful toy did. Even had Wittgenstein at age sixteen not recalled similar experiences with his childhood toys when reading this passage in Boltzmann's 1905 anthology,

Fig. 17.5 Alphonse Penaud (1850–1880) committed suicide in despair when funding for building a full-size version of his successful model was not granted. *Photo* Public Domain

Boltzmann's point about the effect of size on the strength and performance [110] of machines would almost certainly be remembered, given that Wittgenstein soon found himself enrolled in an engineering certificate program, and especially interested in aeronautics.

Certainly by the time he left his position as a research student in aeronautical engineering at Manchester in the fall of 1911 to show up at Cambridge asking to study logic with Bertrand Russell, Wittgenstein would have been familiar with the use of experimental scale models for specific types of engineering problems. Two wind tunnels were already in use for aeronautical research in England by that time. There was the tunnel that Wenham had convinced the Aeronautical Society of Great Britain to fund before anyone fully understood in general how to use the results on small-size models to predict the behavior of a full-size model (Baals and Corliss 1981, p. 3). Then there was the privately-funded wind tunnel that the inventor Maxim had built, which was constructed after Reynolds had shown how to use experiments to predict behavior in similar flow situations—at least for fluid flow in pipes. The significant thing about Reynolds' work was that it provided a way to determine similarity of flow regime for different-sized pipes, as well as for flow at different velocities and viscosities. Reynolds' work used liquids to investigate fluid flow, but air is a fluid, too, so Reynolds' work also bore on the questions of how wind tunnel results on models of aerodynamic surfaces could be used to predict the behavior of larger versions of the surfaces tested, and how that behavior would vary with different air velocities.

This is not to say that Wittgenstein would have then known exactly how to pick up where Boltzmann left off and fill in the story for experimental models, for it is not clear that there was an account of a general methodology of experimental models at that time. The way things stood with the practice of using engineering scale models might well have evoked almost exactly the puzzlement Wittgenstein had about propositions: "in the case of different propositions, the way in which they correspond to the facts to which they correspond is quite different". Given the approach then to experimental scale models, where the rules about how to scale from results on a model to results on a full-size object depended on whether you were talking about experiments in a towing canal or fluid flow in a pipe, one could just as well say that, in the case of different models, "the way in which they correspond to the things they model is quite different" [111].

What would the analogous point he had made for propositions be for experimental models? Wittgenstein's view in early 1914 about propositions was that "In giving the general form of a proposition you are explaining what kind of ways of putting together the symbols of things and relations will correspond to (be analogous to) the things having those relations in reality". What could it mean to give the general form of a model? Or, on Boltzmann's view that equations are really models of a sort, what does it mean to give the most general form of an equation?

In early 1914, Wittgenstein was asking these questions for propositions. Curiously, as we shall see, by the end of 1914, there would be a paper in the field of physics addressing analogous questions about empirical equations. The investigation in that physics paper involved finding the general form of an empirical equation, and it ended up addressing the question of what a universe built on a smaller scale would be like. There was more to be said in answering the question about the relationship between empirical equations and models than Wittgenstein was able to say about propositions in early 1914, the extra twist having to do with the fact that empirical equations involve measurement. The answer given for such equations would appear in late 1914 in a paper that also presented a formal basis for the methodology of experimental models. Though its author, Edgar Buckingham, was American, he had studied in Leipzig with Wilhelm Ostwald for his doctorate and had written a book on the foundations of thermodynamics. Thus, Buckingham's discussion was informed by the debates between Boltzmann and Ostwald about energetics, the kinetic theory of gases, and statistical thermodynamics (Figs. 17.6, 17.7 and 17.8).

Boltzmann had tried to tone down the strident claims of supporters of energetics such as Ostwald, who was antagonistic to the use of models. Ostwald's view, at least as Boltzmann understood Ostwald's emphasis, was not only that the use of models in thermodynamics and the kinetic theory of gases were so much extraneous and distracting baggage, but also that the use of models at all was suspect. In defending the use of models against such strident claims, we saw, Boltzmann pointed out that even proponents of energetics used models of a sort, inasmuch as they used equations as a sort of model—a model made of symbols. Boltzmann's suggestion that equations function like models may well have prodded Wittgenstein to think of a proposition as a model, and it may have even been implicit in some of Wittgenstein's statements in the manuscripts on logic he was [112] working on in 1913 and 1914. At any rate,

Fig. 17.6 The physicist Edgar Buckingham (1867–1940) as an undergraduate at Harvard. He later studied with W. Ostwald at Leipzig, earning a Ph.D. in 1893. *Photo* Harvard University Library

the notion of a proposition being like a model in some way was not explicit then. Wittgenstein just did not talk about propositions being models or pictures during his stay in Norway—that would come only after the crucial insight in late 1914 (Wittgenstein 1979a).

However, in early 1914, Wittgenstein *was* talking about propositions in terms of the facts to which they correspond, as was Russell. In the first manuscript on logic he produced in 1914, he writes "Proposition [which are symbols having reference to facts] are themselves facts: that this inkpot is on this table may express that I sit in this chair" (Wittgenstein 1979c, p. 97). Wittgenstein's move here about propositions and facts is at least vaguely reflective of Boltzmann's move in saying that manipulating symbols in an equation is using the equation like a model. Likewise, as indicated in the preceding quote from his manuscript, Wittgenstein had already, during his time

AN
OUTLINE OF THE THEORY
OF
THERMODYNAMICS

BY
EDGAR BUCKINGHAM, PH.D. (LEIPZIG)

New York
THE MACMILLAN COMPANY
LONDON: MACMILLAN AND CO., LIMITED
1900

Fig. 17.7 Title page of Buckingham's *An Outline of the Theory of Thermodynamics* (1900). *Image credit* Google Books

in Norway, made the move that propositions not only are symbols that correspond to facts, but are themselves facts—that is, they are the same sort of thing that they correspond to. Thinking of propositions as facts was not new. Frege had spoken of the marks on paper associated with a written sentence, and Russell [had spoken] of the varieties of fact that correspond to a proposition, including the example of the acoustic fact associated with a spoken sentence (Russell and Whitehead 1997, p. 402). Frege and Russell, though, tended to de-emphasize this kind of fact, to mention it

Fig. 17.8 W. Ostwald (1853–1937) Ostwald was Buckingham's Ph.D. advisor in Leipzig, Germany. He received the Nobel Prize in 1909. In 1919, Ostwald accepted Wittgenstein's manuscript for publication in a journal he edited, *Annalen der Naturphilosophie*. *Photo* Public Domain

only by way of contrasting the kind of fact that a proposition is with the kind of fact to which a proposition *corresponded*. But, in all fairness, even if Wittgenstein was tentatively exploring the possibilities of the observation that a proposition is a fact before the war (rather than merely mentioning it as a contrast to the kind of fact to which a proposition does correspond), he was not, pre-war, exploring the idea of a proposition being a fact *in terms of picturing or modeling* then. The key notion regarding propositions that shows up in the work Wittgenstein did during his stay in Norway just before the outbreak of World War I is the notion of correspondence rather than the notion of picture or model.

It was almost as though something in the atmosphere was stimulating people's appetites for a satisfying understanding of correspondence, similarity, and form. For, while Wittgenstein was living in Norway pondering problematic issues in logic such as the fact that "it seems as if, in the case of different propositions, the way in which they correspond to the facts to which they correspond is quite different," interest in similarity, [113] correspondence, and similarity transformations was appearing in a wide variety of contexts, especially in Britain.

Someone familiar with Boltzmann's *Popular Writings* might well find that these discussions of different kinds of correspondence, and, especially, the accompanying explorations of the consequences of similarity, brought to mind Boltzmann's remark about the methods of theoretical physics: "The new approach compensates the abandonment of complete congruence with nature by the correspondingly more striking appearance of the points of similarity. No doubt the future belongs to this new method" (Boltzmann 1974, p. 11). Boltzmann had written that in 1892, in the wake of Hertz's spectacular successes in electrodynamics, which in turn (according to Boltzmann) owed much to Maxwell's ingenious mechanical analogies for his equations describing electromagnetic phenomena.

Boltzmann's point here is that, although the analogies Maxwell came up with were crucial to Hertz's advances, Maxwell did not intend the analogies to be taken literally. Maxwell did not mean them to be regarded as hypotheses; Boltzmann felt that, just as with Maxwell's attitude toward his own equations, so things had become in general with the equations of theoretical physics: as he put it there, science speaks "merely in similes". Two decades after Boltzmann penned these remarks about the new methods in physics, we find that in the contexts of discussing thermodynamics, hydrodynamics, and biology (morphology), many other thinkers in many other fields were seeking definitive statements about similarity, too. Not all the ideas that sprang up were unprecedented, but ideas about the use of similarity, whether old or new, were now being explicitly reflected upon, talked about, and written about. In the years 1913–1914 in particular, simultaneous activity of this sort occurred in a number of very different disciplines. Looking back, the activity in the few years just prior to 1914 portends a convergence of ideas about similarity and correspondence.

Certainly some notion of correspondence was already familiar, from pure mathematics as well as from theoretical mechanics. Bertrand Russell's *Principia Mathematica*, published in England in three volumes from 1910 to 1913, reflected the approaches of pure mathematicians in Germany, such as Richard Dedekind and Georg Cantor. Dedekind and Cantor, friends and colleagues, became friends and published very different definitions of the real numbers in 1872, but both had employed the notions of [114] correspondence and similarity in formalizing and defining numbers and other mathematical concepts. Dedekind, for instance, had set up an analogy between the set of rational numbers and a straight line, which he then used to explicitly define a correspondence between rational numbers and points on the line (O'Connor and Robertson 1998). Rational numbers are numbers that can be expressed as the ratio of two whole numbers; irrational numbers cannot. Constructing a line segment corresponding to a rational number on a line uses only the most basic methods of geometrical construction, so Dedekind considered the rational numbers a logical starting point from which to define the rest of the real numbers. This correspondence was not merely illustrative, but was put to good use, for the fact that there were some points on the straight line to which no rational numbers corresponded motivated a definition of irrational numbers. Numbers, as Dedekind defined them, corresponded to "cuts" of the rational numbers that were analogous to "cuts" of the straight line. So analogy and correspondence were not of merely heuristic value in discovering the definition of number, but vestiges of them actually appear in the definition. Dedekind also showed how to use the definition of real numbers to construct a definition of complex numbers (numbers of the form a + bi, where a and b are real numbers and i is the square root of −1).

Cantor took a totally different approach. He did not presume the rational numbers, or any kind of numbers, as already familiar and known, and he aimed for a notion of number even more general than real numbers and complex numbers. Cantor's approach was to start with what he thought one of the most basic of mental activities—abstracting from individual properties of objects in a collection. This resulted in a definition of what he called "transfinite" numbers, which contrast with "finite" numbers. He gave a definition of infinite that distinguished infinite sets and finite sets, and in doing so discovered a whole world of infinite numbers. His definitions are based on two main notions: the notion of one-to-one correspondence and the notion of an order-preserving mapping. Roughly put, the notion of one-to-one correspondence captures the idea of how many, whereas the notion of an order-preserving mapping captures the notion of similarity of order structure. He found that these notions led to a way to delineate infinite sets. His definition of infinite says that infinite sets are sets that can be put in one-to-one correspondence with a proper [115] subset of themselves, something that is impossible for finite sets. On this definition of infinite, the set of whole numbers is easily seen to be infinite because it can be put into one-to-one correspondence with the even numbers (just map each number n to the number 2n), and the even numbers are contained within the set of whole numbers. He defined cardinal type, which expresses the informal idea of how many things are in a collection, and ordinal type, which expresses the informal idea of how things are ordered in a collection as well as how many things are in it. As he defined cardinal number and ordinal number, the two concepts coincide for finite numbers but are forced apart for transfinite (infinite) numbers. Russell's third volume of *Principia Mathematica*, published in 1913, dealt with transfinite cardinal and ordinal numbers and incorporated many of Cantor's ideas. Thus, these ideas became more widely known in England in 1913, especially among philosophers and people interested in philosophy and logic.

The ideas were the horizon-expanding kind, provoking the kind of exhilaration in thought that ballooning had for the senses. They made people feel that, using only simple, familiar ideas, they were transported to a realm where they were suddenly freed of things that had bound the before. Similarity was shown to be a very powerful, if unassuming, idea. Cantor argued that not only had he extended the notion of number, but that the notion of a transfinite ordinal number reflected the most general notion of number possible. The notion of "similar" aggregates (he used the term "aggregates" to refer to collections formed by the mind by abstracting from individual characteristics of the things in the collection) turned out to be exceedingly fruitful. It required a notion of something more structured conceptually than a mere aggregate. Thus Cantor was led to define ordered aggregates (where different notions of "less than" induce different orders) and then an even more important concept: a well-ordered aggregate, which has a "least element"—a starting point or end point. We can think of one aggregate being "transformed" into an aggregate to which it is similar by a similarity transformation, just as we can think of an aggregate being transformed into one to which it is equipollent (meaning that its elements and the transformed one can be put into a one-to-one correspondence) by a simple replacement of each element of the aggregate, one by one. The mathematical notion of a transformation can be used to discuss a [116] mapping. Instead of describing a mapping, which involves specifying a rule by which each element in one aggregate or object is paired with an element in the aggregate or object to which it is similar, we can talk about an aggregate or object being transformed into one to which it is similar.

Similarity is not just a mathematical notion, however. The notion of similarity is entwined in the thought and practice of just about any discipline you can think of, although it is not always talked about per se. Then, as now, notions of similarity were essential to much of scientific reasoning and engineering practice, even though discussion of the topic itself appears infrequently in scientific papers. However, in the years just prior to and including 1914, there was a cultural precipitation of papers that did explicitly reflect on the use of similarity.

In 1912, when James Thomson's *Collected Papers in Science and Engineering* appeared (twenty years after his death), it contained a paper about similar structures (Thomson and James 1875, 1912). These are structures in the most literal sense—structures such as bridges and columns. In that paper, "Comparison of similar structures as to elasticity, strength, and stability" he distinguishes two kinds of similarity between structures: similarity with respect to elasticity and bending, and similarity with respect to stability. The paper was written in 1875, just after Dedekind and Cantor published their accounts of number. James Thomson's style of reasoning illustrates that, in practical engineering, the method of similarity, though based on reasoning from principles of natural science, was still conceived of in terms of *specific kinds* of similarity, specific kinds of loads (wind on a surface versus attached weight), and specific disciplines (hydrodynamics versus mechanics of materials). The kinds of things that were then called "similarity principles" were statements covering a certain class of cases. The point of the "principle" was usually to state how one variable—the weight, size, elasticity—was to be varied as the linear dimension was varied—that is, as an object was increased or decreased in size but kept the same shape. James Thomson's examples are often about how to vary some quantity such that two structures of different sizes are similar in one of these respects. Here is one example of what is meant by a "similarity principle" taken from that work: "Similar structures, if strained similarly within limits of elasticity from their forms when free from applied forces, must have their systems of applied forces, similar in arrangement and of amounts, at homologous [117] places, proportional to the squares of their homologous linear dimensions" (Thomson 1875, p. 362) (Figs. 17.9 and 17.10).

Sometimes the reasoning is based on equations, but often it is not. Rather, some arguments from physical intuition are used: that weight increases as the cube of the linear dimension, and a cross-sectional area of a rope increases as the square of the linear dimension. They are very much like the statements, cited earlier, that Boltzmann used in describing the kind of model he called "those experimental models which present on a small scale a machine that is subsequently to be completed on a larger, so as to afford a trial of its capabilities".

At about the same time, the polymath D'Arcy Wentworth Thompson, then a professor of biology at the University of St. Andrews, was working out ideas that would soon appear in a compilation titled *On Growth and Form*. It is most well-known for its illustrations showing sketches of animals and animal parts that are "morphed" into others. Each sketch is overlaid with a grid; in the transformed sketch, the grid is stretched or slanted in some way so that the form of one species of animal looks as though it is obtained from another via a transformation mapping the points on the lines of one sketch to another by a mathematical function. Some commentators today regard this work as putting forth an alternative to Darwin's theory of natural selection, but this obscures the nature of D'Arcy Thompson's masterwork. Certainly it is true that D'Arcy Thompson wanted to put the brakes on the tendency of his contemporaries to use Darwin's theory of natural selection to explain everything, to the exclusion of other kinds of explanations. But careful readers of Darwin know that scientific explanations of animal forms according to Darwin's theory did not exclude the role of physics. Like Darwin, D'Arcy Thompson was a wonderful naturalist; unlike Darwin, he was a mathematician. D'Arcy Thompson wanted especially to ensure that the role of physics was not overlooked in explaining biological form. Likewise, the mathematical aspects of his work, which have to do with similarity and transformation, do not of themselves conflict with Darwin's theory of natural selection, either. D'Arcy Thompson's deeper mission was the mathematization of biology. The spirit of mathematics of the day was similarity, and it was reflected in his work on what might be called mathematical biology.

In a lecture he gave in 1911 to the British Association for the Advancement of Science titled "Magnalia Naturae: of The Greater [118] Problems of Biology," D'Arcy Thompson spoke of a tendency in recent biological work: "the desire to bring to bear upon our science, in grater measure than before, the methods and results of the other sciences, both those that in the hierarchy of knowledge are set above and below, and those that rank alongside our own" (Thompson 1911, p. 419).

Fig. 17.9 James Thomson (1822–1892), Professor of Engineering, was the older brother of Lord Kelvin, and they collaborated on thermodynamics. He also wrote on different kinds of similarity in mechanics and on dimensional equations. *Image* Google Books (Thomson 1912)

He spoke of the unifying influence of physiology, with its focus on the living rather than the dead organism, and its amenability to being treated by the methods of the physical sciences, remarking "Even mathematics has been pressed into the service of the biologist, and the calculus of probabilities is not the only branch of mathematics to which he may usefully appeal". He spoke of the personal appeal that problems about morphology that were related to "mechanical considerations, to mathematical laws, or to physical and chemical processes" held for him. he also laid out reasons

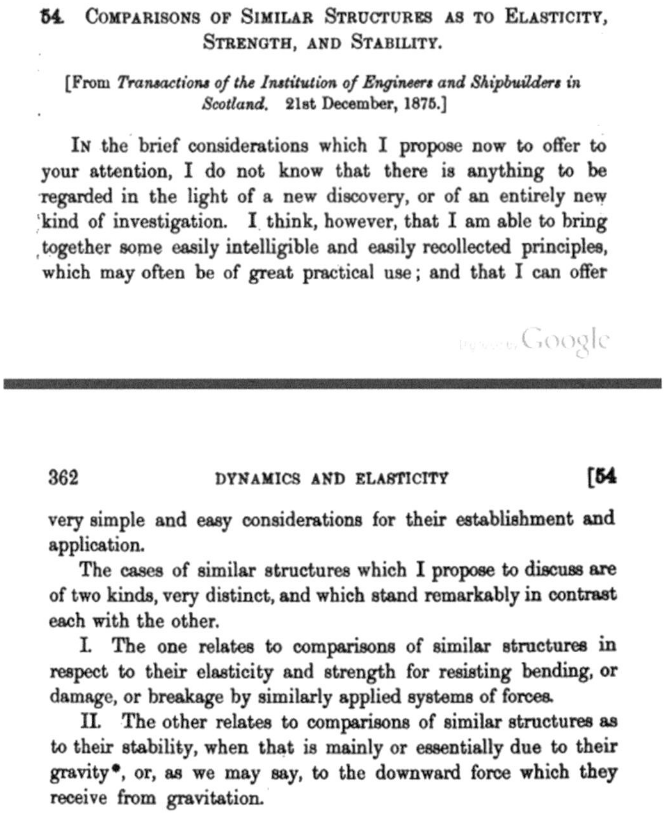
54. COMPARISONS OF SIMILAR STRUCTURES AS TO ELASTICITY, STRENGTH, AND STABILITY.

[From *Transactions of the Institution of Engineers and Shipbuilders in Scotland.* 21st December, 1875.]

IN the brief considerations which I propose now to offer to your attention, I do not know that there is anything to be regarded in the light of a new discovery, or of an entirely new kind of investigation. I think, however, that I am able to bring together some easily intelligible and easily recollected principles, which may often be of great practical use; and that I can offer

362 DYNAMICS AND ELASTICITY [54

very simple and easy considerations for their establishment and application.

The cases of similar structures which I propose to discuss are of two kinds, very distinct, and which stand remarkably in contrast each with the other.

I. The one relates to comparisons of similar structures in respect to their elasticity and strength for resisting bending, or damage, or breakage by similarly applied systems of forces.

II. The other relates to comparisons of similar structures as to their stability, when that is mainly or essentially due to their gravity*, or, as we may say, to the downward force which they receive from gravitation.

Fig. 17.10 From James Thomson's 1875 "Comparison of Similar Structures as to Elasticity, Strength, and Stability" which is included in his *Collected Papers in Physics and Engineering* published in 1912. *Image* Google Books

supporting the possibility of "so far supporting the observed facts of organic form on mathematical principles, as to bring morphology within or very near to Kant's demand that a true natural science should be justified by its relation to mathematics" (Thompson 1911, p. 426).

On the first page of the compilation of his ideas into the large compendium *On Growth and Form*, which was published in 1917 and is even now regarded as a masterpiece, he opens the work with a reference to Kant's declaration, as he put it, "that the criterion of true science lay in its relation to mathematics" (Thompson 1992, p. 1). He goes on to say that:

> As soon as we adventure on the paths of the physicist, we learn to weigh and to measure, to deal with time and space and mass and their related concepts, and to find more and more our knowledge expressed and our needs satisfied through the concept of number, as in the dreams and visions of Plato and Pythagoras. (Thompson 1992, p. 2)

D'Arcy Thompson recognized Newton, too, of course, as a prime example of someone whose work had shown the tremendous fruitfulness of mathematizing a class of phenomena. The continuity D'Arcy Thompson saw between his own project of mathematizing biology and Newton, though, was in the use of similarity. "Newton did not shew the cause of the apple falling, but he shewed a similitude ('the more to increase our wonder, with an apple') between the apple and the stars" (Thompson 1992, p. 9). As in physics, so in the life sciences: "The search for differences or fundamental contrasts between the phenomena of organic and inorganic, of animate [119] and inanimate things, has occupied many men's minds, while the search for the community of principles or essential similitudes has been pursued by few" (Thompson 1992, p. 9). He compared the "slow, reluctant extension of physical laws to Tycho Brahe, Copernicus, Galileo and Newton (all in opposition to the Aristotelian cosmology), that the heavens are formed of like substance with the earth, and that the movements of both are subject to the selfsame laws" (Thompson 1992, p. 11). D'Arcy Thompson did not go so far as to claim [that] physics and mathematics were comprehensive, nor even to know how much they could explain. He recognized there were limits: "... nor do I ask of physics how goodness shines in one man's face, and evil betrays itself in another" (Thompson 1992, p. 13).

He did look to physics, though, for an explanation of lots of different kinds of behavior in creatures both living and nonliving—including an explanation of heavier-than-air flight. He reiterated the point Boltzmann made "that various capabilities depend in various ways on the linear dimensions," citing Helmholtz's 1873 lecture on similar motions and dirigibles for the reasoning supporting the by-then familiar conclusion that "the work which *can be done* varies with the available weight of muscle, that is to say, with the weight of the bird; but the work which *has to be done* varies with mass and distance; so the larger the bird grows, the greater the disadvantage under which all its work is done" (Thompson 1992, p. 42). But D'Arcy Thompson goes further and points out that, while this is true for a specific machine or animal form, it is not the whole story. Not all flight is powered by the sources assumed in these analyses. There is also, he says, "gliding flight, in which ... neither muscular power nor engine power are employed; and we see that the larger birds, vulture, albatross or solan-goose, depend on gliding more and more" (Thompson 1992, p. 42). This is just one illustration of the fact that many factors other than size are involved in comparing flight capabilities of various birds. These other factors, he says, "vary so much in the complicated action of flight that it is hard indeed to compare one bird with another". In living things, we find that "Nature exhibits so many refinements and 'improvements' in the mechanism required, that a comparison based upon size alone becomes imaginary, and is little worth the making" (Thompson 1992, p. 44).

What can be said is that, in both how the fish swims and how the bird flies, streamlining is important. In properly streamlined wings, "a partial vacuum is formed above the wing and follows it wherever it goes, so long [120] as the stream-lining of the wing and its angle of incidence are suitable, and so long as the bird travels fast enough through the air" (Thompson 1992, p. 43). Here the kind of reasoning based on the observation Boltzmann had made in his "Models" essay ("that various capabilities depend in various ways on the linear dimensions") is informative; it tells us how the speed required to stay aloft increases with size. D'Arcy Thompson refers to this as a "principle of *necessary speed*," which he describes as "the inevitable relation between the dimensions of a flying object and the minimum velocity at which its flight is stable" (Thompson 1992, p. 45). That is, "in flight there is a certain necessary speed—a speed (relative to the air) which the bird *must attain* in order to maintain itself aloft, and which must increase as its size increases" (Thompson 1992, p. 41). This principle explains the qualitative differences between large and small birds. Large birds "must fly quickly, or not at all," whereas insects and very small birds such as hummingbirds, are capable of what appears to be "stationary flight," since, for them, "a very slight and scarcely perceptible velocity relatively to the air [is] sufficient for their support and stability" (Thompson 1992, p. 46).

Thus, a proper understanding of the significance of Boltzmann's observation about capabilities varying with dimension, in conjunction with finer distinctions about what is involved with flight, does permit conclusions based on size: "The ostrich has apparently reached a magnitude, and the moa certainly did so, at which flight by muscular action, according to the normal anatomy of the bird, becomes physiologically impossible. The same reasoning applies to the case of man" (Thompson 1992, p. 48). However, this is not to say that flight is impossible above a certain size—rather, that "gliding and soaring, by which energy is captured from the wind, are modes of flight little needed by the small birds, but more and more essential to the large". So the proper lesson to be drawn from considerations of the dependence of capabilities on size is not the one that had been drawn during the eighteenth century—that humans should not try to fly at all—but rather that humans should learn to *glide*. Thus, he observes, "It was in trying to glide that the pioneers of aviation, Cayley, Wenham and Mouillard, Langley, Lilienthal and the Wrights—all careful students of birds—renewed the attempt; and only after the Wrights had learned to glide did they seek to add power to their glider" (Thompson 1992, pp. 48–49) (Fig. 17.11).

What D'Arcy Thompson stressed in all this variety of phenomena was a principle that explained the variety: *the principle of similarity*. Recall that [121] he had mentioned that Newton's insight had involved discerning the similarity underlying the two very different cases of the apple's fall to the earth and the moon's hanging in the sky. In the case of flight, the hummingbird's hanging in the air and the difficulty large birds have in becoming airborne is explained by a principle as well: the principle of similarity. Galileo, too, had introduced the principle of similarity by remarking on the difference in performance of large and small creatures and machines Thompson draws out Galileo's point as it pertains to differences in animal form and to engineering design:

> But it was Galileo who, wellnigh three hundred years ago, had first laid down this general principle of similitude; and he did so with the utmost possible clearness, and with a great wealth of illustration drawn from structures living and dead. [citing 1914 translation, p. 130] He said that if we tried building ships, palaces or temples of enormous size, yards, beams and bolts would cease to hold together; nor can Nature grow a tree nor construct an animal beyond a certain size, while retaining the proportions and employing the materials which suffice in the case of a smaller structure. The thing will fall to pieces of its own weight unless

Fig. 17.11 Otto Lilienthal in a gliding experiment. Boltzmann commended the approach, and corresponded with him about the experiments. D'Arcy Wentworth Thompson notes the significance of the approach in his *On Growth and Form*. *Photo* Public Domain

> we either change its relative proportions, … or else we must find new material, harder and stronger than was used before. Both processes are familiar to us in Nature and in art, and practical applications, undreamed of by Galileo, meet us at every turn in this modern age of cement and steel". (Thompson 1992, p. 24)

To "change its relative proportions" is to change an animal's form. Thus, the form of animals are dependent on, or conditioned on, not only the material properties of the stuff of which they are made, but also on the force of gravity. Form is an effect of scale, but, in turn, "The effect of scale depends not on a thing in itself, but in relation to its whole environment or milieu" (Thompson 1992, p. 24). If this is so, the form of a land-based animal reflects the strength of the gravitational force. D'Arcy Thompson illustrates the point by asking what things would be like were the gravitational force different:

> Were the force of gravity to be doubled our bipedal form would be a failure, and the majority of terrestial animals would resemble short-legged saurians, or else serpents. Birds and insects would suffer likewise, though with some compensation in the increased density of the [122] air. On the other hand, if gravity were halved, we should get a lighter, slenderer, more active type, needing less energy, less heat, less heart, less lungs, less blood. Gravity not only controls the actions but also influences the forms of all save the least of organisms". (Thompson 1992, p. 51)

For very tiny organisms, the same general principles—that the effect of scale on form is a matter not only of the features of the organism itself, but of its whole environment – applies. However, for motions of such tiny animals, it is not gravity, but surface tension, that tends to be the dominant feature of the environment. "The small insects skating on a pool have their movements controlled and their freedom limited by the surface tension between water and air, and the measure of that tension determines the magnitude which they may attain". There are other constraints on their size due to their form, too. In the respiratory system of insects, "blood does not carry oxygen to the tissues, but innumerable fine tubules or tracheae lead air into the interstices of the body". There are natural limitations on the size of such a system; if they grew too much larger, "a vast complication of tracheal tubules would be necessary, within which friction would increase and fusion be retarded, and which would soon be an inefficient and inappropriate mechanism" (Thompson 1992, pp. 51–52).

Besides the limitations on size for insect forms, we can, conversely, see the insect's form as constrained by its size: "we find that the form of all very small organisms is independent of gravity, and largely if not mainly due to the force of surface tension" (Thompson 1992, p. 57). One of D'Arcy Thompson's well-known phrases comes from his point that the form of an object is a "diagram of forces"; the immediate context in which that phrase occurs is as follows:

> The form, then, of any portion of matter, whether it be living or dead, and the changes of form which are apparent in its movements an din its growth, may in all cases alike be described as due to the action of force. In short, the form of an object is a 'diagram of forces,' in this sense, at least, that from it we can judge of or deduce the forces that are acting or have acted upon it. (Thompson 1992, p. 16)

The point here is the effect on form of the environment, not the importance of forces. D'Arcy Thompson was clear that he was using forces only as a sort of shortcut expression: "... force, unlike matter, has no [123] independent objective existence. It is energy in its various forms, known or unknown, that acts upon matter" (Thompson 1992, p. 15). Here, we recognize his awareness of the view of energeticists (such as Ostwald and Hertz), of the problematic status of forces, and their tendency to replace explanations made in terms of force with explanations in terms of mass and energy. Throughout *On Growth and Form*, we find many explanations of animal behavior and form given in terms of energy available and expended. The reason D'Arcy Thompson used the notion of force in describing form was because form is abstract, rather than material, and he justifies his use of the term "force" as appropriate here *without* reifying force:

> But when we abstract our thoughts from the material to its form, or from the thing moved to its motions, when we deal with the subjective conceptions of form, or movement, or the movements that change of form implies, then Force is the appropriate term for our conception of the causes by which these forms and changes of form are brought about. When we use the term force, we use it, as the physicist always does, for the sake of brevity, using a symbol for the magnitude and direction of an action in reference to the symbol or diagram of a material thing. It is a term as subjective and symbolic as form itself, and so is used appropriately in connection therewith. (Thompson 1992, p. 16)

He elaborates on the interrelations of magnitude, ratio, and picture: "When we deal with magnitude in relation to the dimensions of space, our diagram plots magnitude in one direction against magnitude in another—length against height, for instance, or against breadth" (Thompson 1992, p. 78). What we get, he says there, is "what we call a picture or outline, or (more correctly) a 'plane projection' of the object". His emphasis on ratio is striking, and in fact he sums up the whole idea of form in terms of it: "what we call Form is a ratio of magnitudes referred to direction in space". This particular ratio is dimensionless, since it is a ratio of like magnitudes. A length, a height, a breadth, are all measured in dimensions of linear length, whatever units are used. Hence, whatever units are used to measure these magnitudes – inches, feet, millimeters, or centimeters—so long as the same units are used for both the magnitudes in the ration, the units cancel, and the resulting ratio has no units at all. So any ratio of like magnitudes is dimensionless [124].

However, Thompson does not restrict the ratios of interest to such ratios. When considering the variation of a length over time, as in studying growth, he points out that the ratio involved there has the dimensions of velocity: "We see that the phenomenon we are studying is a velocity (whose 'dimensions' are space/time, or L/T) and this phenomenon we shall speak of, simply, as *rate of growth*" (Thompson 1992, p. 78). The symbols L and T denote that the dimensions of the magnitudes being measured are length and time, respectively. They do not specify units of measurement, just what kind of measurement is being taken. Constructing a ratio of length to time gives rise to another kind of quantity, and thus we say that the quantity velocity has dimensions of L/T. So ratios of unlike magnitudes give rise to additional kinds of quantities, or kinds of magnitudes. D'Arcy Thompson's graphical representations also used contour-lines, or "isopleths," to represent a third dimension or magnitude on a two-dimensional surface. The contour-lines can show depth, or the third dimension, of a three-dimensional form.

So far, time is represented only insofar as each of these representations represents a form or configuration at a particular time. Then, Thompson explains, the outlines of an organism as it changes over time can be set out side by side (or, alternatively, overlaid on each other), and this series represents the organism's gradual change over time. Such a representation—somewhat like a series of comic strip frames—exhibits both the form of the organism and the growth of the organism. In addition, it shows how an organism's growth and form are interrelated: "it is obvious that the form of an organism is determined by its rate of growth in various directions" (Thompson 1992, p. 79).

As mentioned earlier, D'Arcy Thompson's goal was to mathematize biology—to treat it the way a physicist treats his subject. He had earlier remarked that "physics is passing through an empirical phase into a phase of pure mathematical reasoning," (Thompson 1992, p. 17n) and certainly the energeticists' emphasis on equations and energy balances was an example of the newer style of mathematical reasoning. Thompson had, however, identified the "old-fashioned empirical physics" as the one "which we endeavour, and are alone able, to apply [when we use physics to interpret and elucidate our biology]" (Thompson 1992, p. 15). That remark was made in the context of explaining the sense in which it was still appropriate to speak of forces as determining biological form, in spite of his recognition that forces do not exist [125].

The approaches he mentions in which the variables tracked are velocities seems to reflect at least some features of the Lagrangian approach. Lagrangian mechanics is a reformulation of Newton's formulation of mechanics. In Lagrangian mechanics, energy conservation principles, rather than force balances, are used to solve equations of motion. By that time, Lagrangian mechanics was generalized so that the quantities did not even need to be velocities and spatial coordinates; they were instead called "generalized velocities" and "generalized coordinates". It then became possible to express the equations of motion of a system using only variables for generalized velocities and time. The generalized velocities bore the same kind of relationship to generalized coordinates as velocities do to coordinates in classical formulations: they expressed the change in coordinates with respect to time. What was important was that the kinds of quantities that are arguments, or inputs, into the functions that express a system's equations of motion were independent of each other and together characterized the system.

The methods proposed for biology in On Growth and Form are easily generalized, so that biology appears as a case of more general principles that apply in physics as well. For, he said, many other things in the world can be seen as cases of the phenomenon of growth, if growth is seen as change in magnitude over time: "since the movement of matter must always involve an element of time, … in all cases the *rate of growth* is a phenomenon to be considered" (Thompson 1992, p. 81). If, as he also said, rate of growth is velocity, what he is saying here is akin to approaches in Lagrangian mechanics, where generalized velocities, rather than forces, are the variables considered important in addition to coordinates of position. We shall see these notions appear elsewhere, such as the use of side-by-side depictions of forms changing over time representing the dynamics of a situation, and the notion of form as consisting of ratios of magnitudes.

Thompson emphasizes the effects of magnitude or size: because different forces are predominant at different scales (gravity at one scale, and surface tension at another), animals on different-sized scales [have] very different kinds of forms. The mechanical principles that describe these forms and how the growth of these forms is constrained are different at different scales; the different mechanical principles are responsible for the [126] difference in form we observe in animals of very different sizes. In overview, he remarks:

> We found, to begin with, that "scale" had a marked effect on physical phenomena, and that increase or diminution of magnitude might mean a complete change of statical or dynamical equilibrium. In the end we begin to see that there are discontinuities in the scale, defining phases in which different forces predominate and different conditions prevail … [the range of magnitude of life] is wide enough to include three such discrepant conditions as those in which a man, an insect and a bacillus have their being and play their several roles. (Thompson 1992, p. 77)

He describes what life is like on three different scales, each such "world" smaller than the other:

> Man is ruled by gravitation, and rests on mother earth. A water-beetle finds the surface of a pool a matter of life and death, a perilous entanglement or an indispensable support. In a

> third world, where the bacillus lives, gravitation is forgotten, and the viscosity of the liquid, the resistance defined by Stokes's law, the molecular shocks of the Brownian movement, doubtless also the electric charges of the ionised medium, make up the physical environment and have their potent and immediate influence on the organism. The predominant factors are no longer those of our scale; we have come to the edge of a world of which we have no experience, and where all our preconceptions must be recast. (Thompson 1992, p. 77)

However, when Thompson looked for an underlying common principles of which these differing phenomena are illustrative, he found it in the principle of similitude, the same one he credited to Galileo in his work on mechanics of materials, and which he says is also recognizable in Newton's explanation of his discovery of the theory of gravitation.

The compilation of D'Arcy Thompson's works into a massive masterwork unified around the theme of growth and form was not published until 1917, but it is based on lots of scientific and engineering work done prior to 1914. He seems to have been especially interested in artificial and natural flight, even citing technical works from the years following the Wright Brothers' 1908 demonstrations in Europe, such as G. H. Bryan's 1911 *Stability in Aviation*, F.W. Lanchester's 1909 *Aerodynamics*, and [127] George Greenhill's 1912 *The Dynamics of Mechanical Flight*. He cited works more directly relevant to the biological emphasis of the work, too, such as E. H. Hankin's 1913 *Animal Flight*, and many, many scientific papers about insects and other animals.

In *On Growth and Form*, he cited papers from those who had been thinking about aviation well before 1900, too, including Helmholtz's 1873 paper on similar motions and dirigibles, mentioned earlier. Recall that, in that paper, Helmholtz had shown how, even when the differential equations governing the motions of dirigibles could not be solved, one could rewrite the equations in a form such that the coefficients were all dimensionless parameters. Then he showed that any two situations in which these dimensionless parameters were the same would have the same solutions. This gives a mathematically sound basis on which one can infer the motions of dirigibles (which were extremely large and unwieldy) from a model or from other observed cases (Fig. 17.12).

However, in citing this paper, D'Arcy Thompson does not draw from it anything more general than the kind of reasoning used for a specific case; recall that he was interested in showing that Helmholtz's conclusion held only assuming that the energy keeping a bird aloft came from muscular energy. He did not take issue with Helmholtz's method, only with his assumptions.

Still, it is telling that D'Arcy Thompson does not seem to be interested in the general theory of dimensions and similarity, or even in hydrodynamical similarity, which is rather presciently laid out in the remarkable paper by Helmholtz to which he refers, in which Helmholtz's methodology outpaces both his assumptions and conclusions. It was a paper ahead of its time. Helmholtz had directed his analysis of similar motions to the practical problem of steering air balloons, not gliders or airplanes, though it applied equally to both problems. When research into heavier-than-air flight was pursued in earnest, Helmholtz's paper was resurrected and recognized as containing the basis for all the important dimensionless numbers in hydrodynamics.

The attitude we see in Thompson's treatment of dimensional analysis in *On Growth and Form*—being interested in, even inspired by, the principle of similitude found in Galileo and Newton, (Thompson 1992, p. 79) yet being content to be led by that interest only so far as necessary to solve a problem at hand—seems to be representative of scientists of that era. Reasoning based on dimensional analysis was used to reach conclusions, and these conclusions were [128] considered basic principles of a general class of problems. Hence, we have various "laws" for specific kinds of situations, such as liquid flow in pipes, boats being towed in canals, streams flowing into lakes, and so on, with a corresponding rule about how a measurement taken on one scale has to be transformed to yield the corresponding value in the situation on another scale. So, for instance, in discussing the speed of aquatic animals, wherein the resistance is provided not by gravitational forces, but by "skin-friction," he reasons:

> Now we have seen that the dimensions of W are l^3 and of R are l^2; and by elementary mechanics $W \alpha RV^2$, or $V^2 \alpha W/R$. Therefore $V^2 \alpha l^3/l^2$ and V α [square root of l]. This is what is known as Froude's Law, of the correspondence of speeds – a simple and most elegant instance of 'dimensional theory'. (Thompson 1992, p. 31)

He goes on to say that sometimes such questions about the effect of scale are "too complicated to answer in a word".

He points out that, depending on an engine's design, the design work can instead depend on the square, rather than the cube, of linear dimensions, and he mentions a different law in such a case: Froude's law of steamship comparison. In a footnote, he cites with approval Lanchester's wry remark that "the great designer was not hampered by a knowledge of the theory of dimensions," which reflects a respect for practical knowledge above this kind of theoretical principle (Thompson 1992, p. 31n) (Fig. 17.13).

Thompson goes on to show that there are subtleties involved that complicate such simplified generalizations. They usually have to do with details of the mechanisms by which different functions are accomplished, so they actually tend to be criticisms of the assumptions used rather than of the methodology of dimensional analysis. One

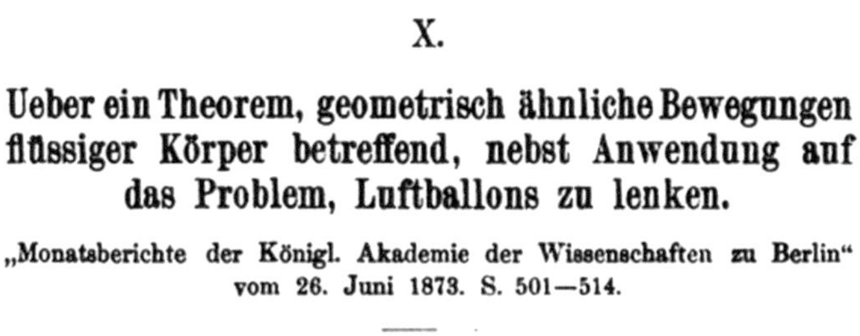

X.

Ueber ein Theorem, geometrisch ähnliche Bewegungen flüssiger Körper betreffend, nebst Anwendung auf das Problem, Luftballons zu lenken.

„Monatsberichte der Königl. Akademie der Wissenschaften zu Berlin" vom 26. Juni 1873. S. 501—514.

Fig. 17.12 Helmholtz's remarkable 1873 paper on similar motions and the steering of air balloons, cited by D'Arcy Wentworth Thompson's *On Growth and Form* for its method. The paper is now recognized for having identified the dimensionless parameters important in meteorological research, and hence to the work Wittgenstein and Eccles were doing at the Kite Flying Station in Glossop in 1908

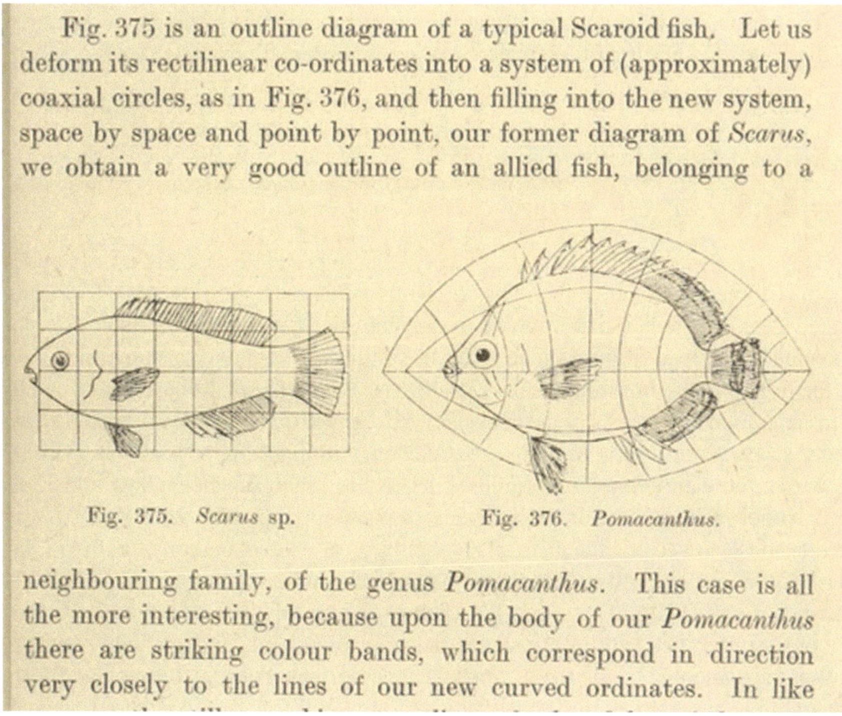

Fig. 375 is an outline diagram of a typical Scaroid fish. Let us deform its rectilinear co-ordinates into a system of (approximately) coaxial circles, as in Fig. 376, and then filling into the new system, space by space and point by point, our former diagram of *Scarus*, we obtain a very good outline of an allied fish, belonging to a

Fig. 375. *Scarus* sp. Fig. 376. *Pomacanthus.*

neighbouring family, of the genus *Pomacanthus*. This case is all the more interesting, because upon the body of our *Pomacanthus* there are striking colour bands, which correspond in direction very closely to the lines of our new curved ordinates. In like

Fig. 17.13 An example of an illustration of the transformation of one form into another from D'Arcy Wentworth Thompson's *On Growth and Form*. Published in 1917, it reflects on and discusses many works on similarity and flight research around the same time Wittgenstein was writing the *Tractatus*

example of such a criticism is the point he had made about analyses based on dimensional reasoning that neglected the importance of gliding flight, which completely reversed the previous conclusion about the possibility of humans achieving heavier-than-air flight. The subtleties, anecdotes, and considerations Thompson brings up are meant to temper looking to any specific derivation using the principle of similitude as the arbiter of effects of scale—that is, to warn against regarding such laws of correspondence as themselves principles of nature. The validity of the particular laws of nature that are [129] consequences of the principle of similitude are very dependent on correct insight into the functions and forces relevant to the behavior of the machine or organism. The principle of similitude, though, is simply a principle about the behavior of forces and functions.

So despite the reservations expressed, the principle of similitude is called out as the underlying principle of D'Arcy Thompson's book, and with appropriate justification:

> In short, it often happens that of the forces in action in a system some vary as one power and some as another, of the masses, distances or other magnitudes involved; the "dimensions" remain the same in our equations of equilibrium, but the relative values alter with the scale. This is known as the "Principle of Similitude," or of dynamical similarity, and it and its consequences are of great importance. (Thompson 1992, p. 25)

Thompson's conclusion here about the significance of the principle of similarity, or dynamical similarity, follow his more specific observations that:

> A common effect of scale is due to the fact that, of the physical forces, some act either directly at the surface of the body, or otherwise in proportion to its surface or area; while others, and above all gravity, act on all particles, internal and external alike, and exert a force which is proportional to the mass, and so usually to the volume of the body. (Thompson 1992, p. 25)

Thus, the principle of similitude is a very general principle of science—of any science using measurement, in fact. The criticisms and qualifications D'Arcy Thompson raised were directed at inappropriate uses of the specific "laws" and "laws of correspondence" derived from considerations of similarity and theory of dimensions, not at the principle of similitude itself. Usually, the principle of similarity states conditions of similarity in terms of the constancy of a certain dimensionless parameter, and hence states when similar motions are expected, within a certain "scale" or "world". However, there are often discontinuities in behavior, such as when a flow transitions from smooth to turbulent, or as a substance transitions from a liquid to a gas. D'Arcy Thompson seems to cite the principle of similarity as an explanation of both the similarities that can be drawn within a particular scale and the discontinuities that exist between scales [130]. In fact, the way Thompson has stated the principle, it explains not only similarities and differences in behavior of animals and artificial machines within the same one of the three "worlds" he describes, but it also accounts for the discontinuities and very different kinds of forms and forces encountered between those three "worlds".

Physical similarity, dynamical similarity, and principles of similarity or similitude became more prominent topics in 1913 and early 1914. For one thing, an English translation of Galileo's *Dialogue Concerning Two New Sciences* was published in February 1914, after being practically unobtainable (Galileo 1914) (Fig. 17.14).

It was bound to be of great interest to anyone interested in the history or philosophy of science. According to the Translator's Preface by Henry Crew and Alfonso De Salvio, copies of the previous translation of Two New Sciences into English, done in 1730 by Thomas Weston, had become "scarce and expensive" by 1914 (Galileo 1914, p. vi). An even earlier English translation had "issued from the English press in 1665". But fate intervened and, they said, "It is supposed that most of the copies were destroyed in the great London fire which occurred in the year following, ... even [the copy] belonging to the British Museum is an imperfect one" (Galileo 1914, pp. v–vi). The drama of the long struggle for this work to become available in English to twentieth-century readers made it a publishing event. At least, that is how the translators saw it:

> For more than a century English speaking students have been placed in the anomalous position of hearing Galileo constantly referred to as the founder of modern physical science, without having any chance to read, in their own language, what Galileo himself has to say. Archimedes has been made available by Heath; Huygens' Light has been turned into English by Thompson, while Motte has put the Principia of Newton back into the language in which it was conceived. To render the Physics of Galileo also accessible to English and American students is the purpose of the following translation. (Galileo 1914, p. v)

Fig. 17.14 The 1914 publication of a new English translation of Galileo's *Dialogues Concerning Two New Sciences* by the physicist Henry Crew was a major publishing event. English translations of the work had been "practically unobtainable"; however, Wittgenstein owned a rare and expensive 1730 edition translated by Thomas Weston. It is known for discussing the problem of scaling in mechanics and in biology

Crew lectured on the topic of the things to be learned from Galileo in 1913, and he mentioned the "theory of dimensions" in passing, even before the English translation of *Two New Sciences* was published (Crew 1913).

According to the recent study of Wittgenstein's books left with Russell [mentioned in a previous chapter of *Wittgenstein Flies A Kite*], which were books Wittgenstein owned during the period from 1905 to 1913, Wittgenstein owned one of the [...] [131] editions of Galileo's *Discourses Concerning Two New Sciences* published in London in 1730. [Taking into account what we have just seen, i.e., that the edition would have been "scarce and expensive," this entry on the list takes on a special significance. A perusal of the list shows that] it is the *only* book by Galileo in the collection of Wittgenstein's books surveyed. Oystein Hide, the author of the study, remarks that "Wittgenstein later drew comparisons between his philosophical activity and the work of, for example, Galileo within his scientific field" (Hide 2004, p. 74).

The collection of books owned by Wittgenstein that Hide describes does not contain many other books of this sort, with two striking exceptions: a six-volume set of facsimilies of notebooks of "Leonardo De Vinci" published in Paris in 1891, and a copy of *Principia Philosophia* by "Isaco Newtono" published in 1728. Galileo and Newton stand out as the early scientists who wrote on similarity and similitude. We saw that D'Arcy Thompson cited Newton's use of similitude in discovering the law of gravitation, and that Boltzmann cited Newton in his encyclopedia article on models just after mentioning models of flying machines, in mentioning "The theory, initiated by Sir Isaac Newton, of the dependence of various effects on the linear dimensions". And we shall see that the physicist Heike Kamerlingh Onnes mentions Newton's use of mechanical similitude in his Nobel lecture in 1913. Leonardo da Vinci [was, and] is known not only for his work on flying machines, including parachutes and helicopters, but for his work in hydrodynamics and his explicit discussions and illustrations of the use of proportion in art as well as in science. His notebooks used both his artistic capabilities and his mechanical genius. His sketch of the human body with geometric proportions overlaid on it is well-known, but he investigated many topics, and his work in hydrodynamics was striking. His sketches of fluid flows, especially turbulent flows, show a remarkable talent for observation of the phenomena that were being investigated with such intensity in the years prior to 1914. He is also known for his use of proportion in art as well as in science. It's possible that the theft of the Mona Lisa from the Louvre in 1911 and its recovery in 1913 put Leonardo's works in the public eye during this time. At any rate, the fact that among the books Wittgenstein owned prior to 1914, we find expensive copies of works by Newton and Galileo in which they discuss similitude may reflect interests stimulated by the resurgence of interest in their works on similitude in England (and perhaps elsewhere) in the years preceding 1914. The interest in Leonardo's notebooks can be accounted for by the aeronautical work in [132] his notebooks alone, especially his invention of a helicopter, but it also fits with an interest in ratio and proportion.

The publishing event of a new English translation of Galileo's *Two New Sciences* in 1914, which was being written about and discussed in 1913 prior to its publication, is significant to our story because of the prominence of the discussion of similarity it contains and the accessible and memorable way the principle is explained and its consequences depicted. One of the translators, Henry Crew, was a physicist, a professor of physics at Northwestern University, and had studied with Hermann von Helmholtz in Berlin in 1883–1884; (Cahan 2004) this would have been after Helmholtz had already developed the criteria for similar motions, and perhaps it accounts for Crew's interest in translating this particular work of Galileo's. In a nutshell, the dialogue begins with a conversant using a limited notion of similarity (geometric similarity), puzzling over the invalidity of the consequences one can draw from it, and then proceeds to a discussion in which the same conversant comes to use a more general notion of similarity, from which the conclusions and experimental predictions drawn are in fact valid. There are not, as in D'Arcy Thompson's treatment of the principle, examples about flight, but there is more emphasis on correct, rather than incorrect, uses of the principle.

Galileo's dialogue begins with Salviati, usually taken to be the voice of Galileo, recounting numerous examples of a large structure that has the same proportions and ratios as a smaller structure but that is not proportionately strong. In these opening pages of the dialogue, the wise and seasoned Salviati explains to the earnest but puzzled Sagredo that "if a piece of scantling [corrente] will carry the weight of ten similar to itself, a beam [trave] having the same proportions will not be able to support ten similar beams". The phenomenon of the effect of size on function of machines of similar design holds among natural as well as artificial forms, Salviati explains: "just as smaller animals are proportionately stronger and more robust than the larger, so also smaller plants are able to stand up better than larger". Perhaps the most well-known of Salviati's illustrations is about giants:

> ... an oak two hundred cubits high would not be able to sustain its own branches if they were distributed as in a tree of ordinary size; [...] and nature cannot produce a horse as large as twenty ordinary horses [133] or a giant ten times taller than an ordinary man unless by miracle or by greatly altering the proportions of his limbs and especially his bones, which would have to be considerably enlarged over the ordinary. Likewise the current belief that, in the case of artificial machines the very large and the small are equally feasible and lasting is a manifest error. (Galileo 1914, pp. 52–53)

It is this point about scale we saw reflected throughout D'Arcy Thompson's work, applied especially to the case of flight. Yet it was well known in 1914 that engineers used scale models. The question was whether a certain use was valid, and the problem was that determining this often seemed a matter of engineering knowledge and skill, not a matter of pure science Recall that there was some ambivalence in D'Arcy Thompson's discussion; he criticized a number of analyses based on dimensional considerations, although he is clear that his reservations were not about the validity of the principle of similitude itself. He credited Galileo as the first to articulate the principle of similitude, or dynamical similarity.

Although Galileo's work opens with the wise participant in the dialogue reminding the others of the reasons for the lack of giant versions of naturally occurring life-forms, it soon proceeds to a case of a valid use of a small (artificial) machine to infer the behavior of a large (artificial) machine. As in Helmholtz's reasoning in his paper on similar motions and dirigibles, the basis for similarity is found not in mere geometric similarity, but, more deeply, in dimensional considerations drawn from an equation of motion. At a later point in the dialogue, Sagredo makes use of Salviati's statement that the "times of vibration" (period of oscillation) of bodies suspended by threads of different lengths "bear to each other the same proportion as the square roots of the lengths of the thread; or one might say the lengths are to each other as the squares of the times" (Galileo 1914, p. 139). From this, Sagredo uses one physical pendulum to infer the length of another physical pendulum:

> Then if I understand you correctly, I can easily measure the length of a string whose upper end is attached at any height whatever even if this end were invisible and I could see only the lower extremity. For if I attach to the lower end of this string a rather heavy weight and give [134] it a to-and-fro motion, and if I ask a friend to count a number of its vibrations, while I, during the same time-interval, count the number of vibrations of a pendulum which is exactly one cubit in length, then knowing the number of vibrations which each pendulum makes in the given interval of time one can determine the length of the string. Suppose, for example, that my friend counts 20 vibrations of the long cord during the same time in which I count 240 of my string which is one cubit in length, taking the squares of the two numbers, 20 and 240, namely 400 and 57600, then, I say, the long string contains 57600 units of such length that my pendulum will contain 400 of them; and since the length of my string is one cubit, I shall divide 57600 by 400 and thus obtain 144. Accordingly I shall call the length of the string 144 cubits. (Galileo 1914, p. 140)

The basis on which Sagredo infers the length of the larger from the smaller is a fundamental relationship—the relationship between length and period that describes the behavior of any pendulum, which can be expressed in terms of the constancy of the value of a certain ratio containing them. What he derives from it is a *law of correspondence* telling him how to find the corresponding length in the large pendulum from the length of the small. Salviati (the voice of Galileo) responds approvingly to his claim that this method will yield the length of the string: "Nor will you miss it by as much as a hand's breadth, especially if you observe a large number of vibrations" (Galileo 1914, p. 140). Henry Crew, the physicist who co-translated the work, thought the emphasis on experiment in *Two New Sciences* was one of the important contributions of Galileo's work.

This work of Galileo's, which is generally credited with giving not only the first, but probably the best ever, exposition of physical similarity, appeared just as scientists from Britain's National Physical Laboratory (NPL) presented a compendium of their own about similarity (Stanton and Pannell 1914). The paper, "Similarity of Motion in Relation to the Surface Friction of Fluids," by T. E. Stanton and J. R. Pannell, was submitted to the Royal Society of London in December 1913 and was read to the Society in January of 1914. Stanton was superintendent of NPL's Engineering Department and was interested in the possibilities of using small-scale models in wind tunnels for engineering research. Except for the work of Osborne Reynolds, experimental study of similar motions in fluids was, according to the authors, only done since 1909: [135]

> Apart from the researches on similarity of motion of fluids, which have been in progress in the Aeronautical Department of the National Physical Laboratory during the last four years, the only previous experimental investigation on the subject, as far as the authors are aware, has been that of Osborne Reynolds .... (Stanton and Pannell 1914, p. 200)

Osborne Reynolds was the celebrated professor of engineering at Manchester who had retired a few years prior to Wittgenstein's arrival and enrollment there as an engineering student. What Stanton and Pannel cite as Reynolds' major discoveries were that there was a critical point at which fluid flow suddenly changed from "lamellar motion" to "eddying motion" (today we would say "from laminar flow to turbulent flow"), that the critical velocity was directly proportional to the kinematical viscosity of the water and inversely proportional to the diameter of the tube, and that for geometrically similar tubes, the dimensionless product:

$$\text{(critial velocity)} \times \text{(diameter)}/\text{(kinematic viscosity of water)}$$

was constant (Stanton and Pannell 1914, p. 200). They also noted that, no matter what the conditions of flow, whether above or below the critical velocity, whenever the values of the dimensionless product:

$$\text{(velocity)} \times \text{(diameter)}/\text{(kinematic viscosity of water)}, \text{ or } vd/\nu$$

were the same [in two different setups], so were the corresponding values of another dimensionless parameter:

$$(\text{density}) \times (\text{diameter})^3/(\text{coefficient of viscosity of water})^2 \\ \times (\text{rate of fall of pressure along the length of the pipe})$$

As is often the case, there was a complication: experiment showed that the surface roughness of the pipe wall (or of whatever surface forms the flow boundary) needed to be taken into account as well. This is a matter of geometry on a much smaller scale making a difference. However, the overall approach of the use of dimensionless parameters to establish similar situations was still seen to be valid, as their experiments illustrated:

> From the foregoing it appears that similarity of motion in fluids at constant values of the variable vd/ν will exist, provided the surfaces relative to which the fluids move are geometrically similar, which similarity, as Lord RAYLEIGH has pointed out, must extend to those [136] irregularities in the surfaces which constitute roughness. In view of the practical value of the ability to apply this principle to the prediction of the resistance of aircraft from experiments on models, experimental investigation of the conditions under which similar motions can be produced under practical conditions becomes of considerable importance … By the use of colouring matter to reveal the eddy systems at the back of similar inclined plates in streams of air and water, photographs of the systems existing in the two fluids when the value of vd/ν was the same for each, have been obtained, and their comparison has revealed a remarkable similarity in the motions. [ref: Report of the Advisory Committee for Aeronautics, 1911–1912, p. 97] (Stanton and Pannell 1914, p. 201)

The authors here refer to the dimensionless parameter "vd/ν" as a variable. This variable is a product of several measurable quantities and is a dimensionless parameter. That is, the units all cancel out, and its value is independent of the choice of units of measurement. To see this, we can talk about the dimensions of each contributor to the dimensionless parameter, also known as a dimensionless product: the dimensions of v are length divided by time; the dimensions of d are length, and the dimensions of ν are (length time length) divided by time. What Stanton and Pannell meant in referring to it as a variable was that their equation for the resistance R includes a function of this dimensionless parameter:

$$\text{resistance R} = (\text{density}) \times (\text{velocity})^2 \times (\text{some function of vd}/\nu)$$

Or, as they put it, $R = \rho v^2 F(vd/\nu$, where "F (vd/ν" indicates some unspecified function of vd/ν Hence, vd/ν is a variable in the sense that the relation for resistance includes an unspecified function of vd/ν It is also a variable in the more practical sense: it can be physically manipulated.

Stanton and Pannell present this relation as a consequence of the Principle of Dynamical Similarity, in conjunction with assumptions about what "the resistance of bodies immersed in fluids moving relatively to them" depends on (Stanton and Pannell 1914, p. 201). Evidently, it was Rayleigh who suggested the generalization; they cite Rayleigh's contribution [on p. 38] of the *Report of the Advisory Committee for Aeronautics, 1909–1910*. Rayleigh had there spoken of the possibility of taking a more general approach than current researchers were taking in applying "the principal of dynamical similarity". He explained his 'more general' approach as follows [137].

> … We will commence by supposing the plane of the plate perpendicular to the stream and inquire as to the dependence of the forces upon the linear dimension (l) of the plate and upon the density (ρ), velocity (v) and kinematic viscosity (ν) of the fluid. Geometrical similarity is presupposed, and until the necessity is disproved it must be assumed to extend to the thickness of the plate as well as to the irregularities of surface which constitute roughness.
>
> If the above-mentioned quantities suffice to determine the effects, the expression for the mean force per unit area normal to the plate (P), analogous to a pressure, is
>
> $$P = rv^2 \times f(n/vl) \tag{A}$$
>
> where f is an arbitrary function of the one variable n/vl.
>
> It is for experiment to determine the form of this function, or in the alternative to show that the facts cannot be represented at all by an equation of form (A). (Rayleigh 1910, pp. 532–533)

Rayleigh does not here say exactly how the relation (A) is obtained from the assumptions he lists, which was not uncommon at that time when invoking the principle of similarity. Referring to his other papers, and papers by others who invoked it, such as Reynolds and Helmholtz, it is fairly clear that invoking the principle of similarity meant using dimensional considerations, sometimes using the specific fact that a scientific equation required consistent units.

Stanton and Pannell present the results obtained at the National Physical Laboratory in the paper. It is interesting to note that the results are presented in graphs where one of the variables plotted is the term $R/\rho v^2$. [Since, as explained above, "$R = \rho v^2\ F(vd/\nu)$, where "$F(vd/\nu)$" indicates some unspecified function of vd/ν", the term $R/\rho v^2$ is equal to $F(vd/\nu)$] and so it just another expression for the unspecified function, and is dimensionless. What this implies is that the laboratory experiments are not conceived of in terms of the values of individual measurable quantities such as velocity but are classified in terms of the value of a dimensionless parameter.

Rayleigh, too, published a kind of survey paper in early 1914 (Rayleigh 1914). Here he actively campaigned for wider appreciation and use of the principle, which he credited Stokes with having "laid down in all its completeness".

Rayleigh wrote: "There is a general law, called the law of dynamical similarity, which is often of great service. In the past this law has been unaccountably [138] neglected, and not only in the present field. It allows us to infer what will happen upon one scale of operations from what has been observed at another" (Rayleigh 1914). He then discussed the example of a sphere moving uniformly through air, remarking that, if the kinematic viscosity can be assumed to be the same across the cases considered:

> When a solid sphere moves uniformly through air, the character of the motion of the fluid round it may depend upon the size of the sphere and upon the velocity with which it travels. But we may infer that the motions remain similar, if only the product of diameter and velocity be given. Thus if we know the motion for a particular diameter and velocity of the sphere, we can infer what it will be when the velocity is halved and the diameter doubled. The fluid velocities also will everywhere be halved at the *corresponding* places. (Rayleigh 1914)

Fig. 17.15 Britain's National Physical Laboratory in 1914, as featured in an article in *Nature* magazine. In 1908, Thomas Stanton, Superintendent of NPL's Engineering Department since 1901, was directed to apply techniques of wind research to the study of flight, especially the efficiency and safety of the aeroplane. Stanton had gone to study with Osborne Reynolds at Owens College in Manchester in 1888, the same institution where Wittgenstein would later enroll as a student in aeronautical research in 1908 (though it later became Victoria University, then the University of Manchester). In 1914, Stanton and Pannell published the results of a major study in hydrodynamics, "Similarity of Motion in Relation to the Surface Friction of Fluids"

So we see one use of the principle is to be able to use one observation or experiment as representative of a whole class of actual cases: all the other cases to which it is similar, even though the cases may have very different values of measurable quantities such as velocity. The important fact of the situation is the dimensionless parameter just mentioned: "It appears that similar motions may take place provided a certain condition be satisfied, viz., that the product of the linear dimension and the velocity, divided by the kinematic viscosity of the fluid, remain unchanged" (Figs. 17.15 and 17.16).

The important feature of the situation is the value of this dimensionless parameter; that would mean that, even in cases of a *different fluid*, so long as this dimensionless product is the same, the motions will be similar! Can that be right? Yes, and Rayleigh points out that it is a particularly useful application of the principle: "If we know what happens on a certain scale and at a certain velocity in *water*, we can infer what will happen in *air* on any other scale, provided the velocity is chosen suitably".

There is a qualification he adds here, one that the reader familiar with the qualifications needed since the advent of special relativity (that relativistic phenomena do not appear only so long as the velocities involved are much smaller than the velocity of light) might see as an analogue for sound: "It is assumed here that the compressibility of the air does not come into account, an assumption which is admissible so long as the velocities are small in comparison with that of sound" (Rayleigh 1914, p. 246). Neglecting the compressibility of air in certain velocity ranges and not others is another illustration of [139] the discontinuity of scale D'Arcy Thompson emphasized. It does not mean that the method of similarity requires neglecting the compressibility of air, just that in the low-velocity range, establishing similarity of motions between two different situations does not require accounting for it.

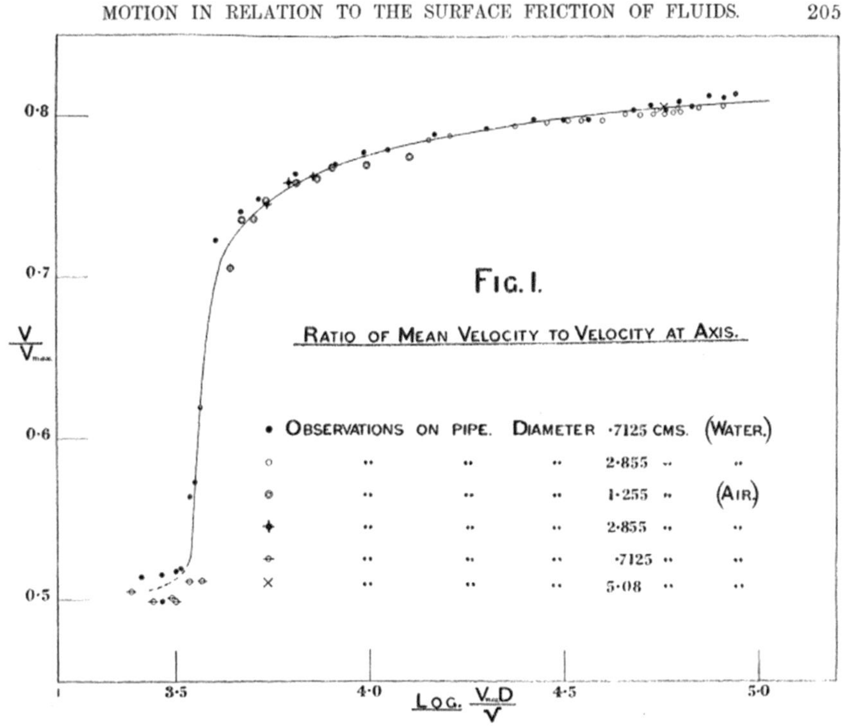

Fig. 17.16 Figure I from Stanton and Pannell's 1914 "Similarity of Motion…" paper

However, Rayleigh also pointed out that, in contrast to the compressibility of air, it appeared that viscosity *was* important in many cases where it was so small that it seemed improbable that it should matter (Rayleigh 1914, p. 237). When viscosities were low, as in water, one would not expect that the actual value of viscosity would be a significant factor in water's qualitative behavior Osborne Reynolds' results on fluid flow in pipes had shown that it is; Reynolds began to suspect that viscosity was important even in water when he observed unexpected changes in fluid flow as the temperature was varied. Since viscosity varies with temperature, he investigated the effect of viscosity and found that it was indeed important for fluid flow through pipes, even for nonviscous fluids such as water. He also investigated cases where viscosity was the "leading consideration," as Rayleigh put it, in remarking that "It appears that in the extreme cases, when viscosity can be neglected and again when it is paramount, we are able to give a pretty good account of what passes. It is in the intermediate region, where both inertia and viscosity are of influence, that the difficulty is greatest (Rayleigh 1914, pp. 245–246).

These unexpected experimental results showing that viscosity needed to be taken into account in these intermediate regions were as unwelcome as they were unexpected, since there was already on hand a well-developed theory of hydrodynamics that neglected viscosity—on that lent itself to closed-form mathematical solutions of the equations. This discipline, hydrodynamics, had produced many beautiful mathematical theorems and solutions. If fluid friction were included, though, the mathematical equations became intractable, as Helmholtz had remarked in his papers on hydrodynamics. Helmholtz had lain out and responded to this problem in his 1873 paper on similar motions of air balloons; his response had been to show how to use the intractable equations to find similarity conditions – the relevant dimensionless parameters for dynamical similarity—to permit inferring the behavior of a large balloon from experiments on smaller ones. This was an experimental alternative to the impossible task of finding a solution of the intractable hydrodynamical equations that accounted for fluid friction, or viscosity. As the authors of an historical survey put it: "Helmholtz, in fact, presented to the Berlin Academy of Sciences in 1873 (Fig. 17.12) a [140] dimensional analysis of the equations of fluid motion that already encompassed what we know today s the Froude, Reynolds, and Mach criteria for model-prototype similarity" (Rouse and Ince 1957, p. 200). [The guidance about dynamical similarity in Stokes' (Fig. 17.17) 1850 paper and Helmholtz' 1873 paper] may be what Rayleigh was referring to when he added that, even in the intermediate region, where the difficulty is the greatest, "we are not wholly without guidance," lamenting the neglect of "the law of dynamical similarity" (Rayleigh 1914, p. 246).

One might think that, by 1914, when the use of wind tunnels had become recognized as essential to practical aeronautical research, this principle would have become accepted and would no longer be in question, at least among aeronautical researchers. But if Rayleigh's estimation of the state of the profession is correct, even as late as 1914 this wasn't so! He writes that "although the principle of similarity is well established on the theoretical side and has met with some confirmation in experiment, there has been much hesitation in applying it, … (Rayleigh 1914, p. 246). He especially mentions problems in its acceptance in aeronautics due to skepticism that viscosity, which is extremely small in air, should be considered an important parameter: "In order to remove these doubts it is very desirable to experiment with different viscosities, but this is not easy to do on a moderately large scale, as in the wind channels used for aeronautical purposes".

Rayleigh tries to persuade the reader of the significance of the effects of viscosity on the velocity of fluid flow by relating some experiments he performed with a cleverly designed apparatus in his laboratory. The apparatus consisted of two bottles containing fluid at different heights, connected by a tube with a constriction, through which fluid flowed due to the difference in "head," or height of fluid, in the two bottles. The tube with the constriction contained fittings that allow measurement of pressure head at the constriction, and on either side of it. To investigate the effects of viscosity, Rayleigh varied the temperature of the fluid, which changes the fluid viscosity, and he observed how the velocity of the fluid flowing between the two bottles was affected. The kind of relationship he establishes and uses is of the form Galileo employed in reasoning from one pendulum to another. In other words, he worked in terms of ratios (ratios of velocities, ratios of viscosities, ratios of heads), and he employed the fact that some ratios are the square root of others. He took the

Fig. 17.17 Sir George Gabriel Stokes (1819–1903) Rayleigh credited Stokes for the Principle of Dynamical Similarity. In 1850 Stokes had read a paper "On the Effect of the Internal Friction of Fluids on the Motion of Pendulums" to the Cambridge Philosophical Society, which identified the dimensionless parameters important for hydrodynamical similarity. *Image* Google Books

experimental results he reported in this 1914 paper to conclusively settle the question of the relevance of viscosity to fluid motions [141].

Thus, in early 1914, there was first a report form Britain's National Physical Laboratory on the experimental work performed there recently on similarity of fluid motions, presented in January; then, in March, Rayleigh's paper on Fluid motions appeared. Rayleigh's paper explained the need fulfilled by such experimental work, which employed the principle of dynamical similarity and reported his own experiments meant to eradicate any remaining skepticism about the principle. In these works, there is a move toward generalization of the specific relationships then in use for specific applied problems in hydraulics. The work at the National Physical

Laboratory was aimed at carrying out a “systematic series of experiments” for “the purpose of establishing a general relations which would be applicable to all fluids and conditions of flow”. The dual purpose of both establishing the relation and exploring its limits is seen in how Stanton and Pannell stated their purpose in the paper submitted in December 1913:

> The object of the present paper is to furnish evidence confirming the existence, under certain conditions, of the similarity in motions of fluids of widely differing viscosities and densities which has been predicted, and further by extending the observations through a range in velocity of flow which has not hitherto been attempted to investigate the limits of accuracy of the generally accepted formulae used in calculations of surface friction. (Stanton and Pannell 1914, pp. 199–200)

The immediate motivation for performing experiments to furnish this evidence was practical, though there was no hope of producing an equation from which predictions could be directly generated. Recall that the general relation the paper proposed contained an undetermined function. So the goal was to find out how to find situations similar to the one you needed to know about that could be carried out in the laboratory and to use the correspondence between them to determine what you wanted to know. It was the question of similar motions that was of special interest in the investigation: “experimental investigation of the conditions under which similar motions can be produced under practical conditions becomes of considerable importance” because of “the practical value of the ability to apply this principle to the prediction of the resistance of aircraft from experiments on models” (Stanton and Pannell 1914, p. 201). To be brief, what they were really [142] interested in was providing a methodology for experimental engineering scale models in hydrodynamics and aerodynamics. Somehow the key lay not in a fully detailed equation, but in an equation that enabled them to determine when certain situations were similar and told them how to determine corresponding motions in one situation from observations made on a similar situation.

Adding to this swirl of ideas gathering in late 1913 and early 1914 was the attention being paid to a wonderful experimental result based on similarity and correspondence in the area of thermodynamics: the successful liquefication of helium. The same month that Stanton and Pannell submitted their paper from the National Physical Laboratory, December 1913, Heike Kamerlingh Onnes delivered his Nobel lecture. Again, the ideas were not brand-new in 1913, but the worldwide public attention being paid to them was. Onnes had published a paper in 1881 called “General Theory of Liquids,” in which he argued that van der Waals’ “Law of Corresponding States,” which had just been published the prior year, could be derived from scaling arguments, in conjunction with assumptions about how molecules behaved (Onnes 1881). Van der Waals was impressed with the paper, and a long friendship between the two ensued. Van der Waals won the Nobel Prize in 1910 for “The equation of state for gases and liquids,” and he mentioned in his lecture that the “law of corresponding states” had become “universally known,” though he said nothing more about it there (van der Waals 1910). The Nobel Prize awarded to Onnes in 1913 brought wider attention to Onnes’s account of the foundations of the law of corresponding states—though you might not guess it from the title of Onnes’s award, given for

"Investigations into the properties of substances at low temperatures, which have led, amongst other things, to the preparation of liquid helium".

Onnes's lecture highlighted the connection between his investigations into properties of substances at low temperatures and similarity principles:

> [F]rom the very beginning … I allowed myself to be led by Van der Waals' theories, particularly by the law of corresponding states which at that time had just been deduced by Van der Waals.
>
> This law had a particular attraction for me because I thought to find the basis for it in the stationary mechanical similarity of substances and from this point of view the study of deviations in substances of simple chemical structure with low critical temperatures seemed particularly important". [143] (Onnes 1913, p. 306)

What's special about low temperatures, of course, is that, according to the kinetic theory of gases on which van der Waals' equation of state was based, there would be much less molecular motion. Onnes's approach in looking for the foundation of the law of corresponding states has a slightly different emphasis than the kinetic theory of gases. Boyle's Law (often called the ideal gas law) and van der Waals' equation were based on investigating the relationship between the microscale (the molecular level) and the macroscale (the properties of the substance, such as temperature and density.) but Onnes was instead looking at the foundation for the similarity of states. Like van der Waals, he looked to mechanics and physics for governing principles, but Onnes pointed out that it was also useful to look at principles of similarity. At low enough temperatures, where motion of the molecules was not the predominant factor, the relevant principles of similarity would be principles of static mechanical similarity, as opposed to dynamical similarity.

As had happened in Osborne Reynolds' work in hydrodynamics, a criterion for similarity had arisen out of investigations into the transition from one regime to another. For Reynolds, it as the critical point at which fluid flow underwent a transition from laminar to turbulent flow (or, in his terminology, from "lamellar" to "eddying" flow) that led to the identification of the dimensionless parameter that later became known as Reynolds Number. The Reynolds Number is in a way a criterion of similarity: fluid systems with the same Reynolds Number will be in the same flow regime, regardless of the fluid. So it was with thermodynamics: the critical point at which a substance undergoes a transition from the gaseous to liquid state led to the identification of a criterion of similarity of states that held for all substances.

Van der Waals was very interested in the continuity of states, particularly the continuity between the gas and liquid states. Although his prize was awarded for his work on "The equation of state for gases and liquids," the doctoral dissertation in which he presented the equation had been titled "On the continuity of the gas and liquid state," and he emphasized the important role that the idea of continuity had played in developing the equation:

> … I conceived the idea that there is no essential difference between the gaseous and the liquid state of matter — that the factors which, apart from the motion of the molecules, act to determine the [144] pressure must be regarded as quantitatively different when the density changes and perhaps also when the temperature changes, but that they must be the very

> factors which exercise their influence throughout. And so the idea of continuity occurred to me. (Onnes 1913, p. 255)

Van der Waals thought that the force of attraction between molecules would be a big part of the story about the continuity of the gas and liquid states, so he wanted to revise the equation to account for it. He added two more variables to Boyle's Law, which he called "a" and "b" and which are characteristic of a particular substance. Instead of Boyle's ideal gas law—(pressure) × (volume) = (gas constant) × (temperature)—van der Waals' equation can, as Johann Levelt Sengers points out, be written as a cubic equation—that is, as a third-degree polynomial in volume, with coefficients depending on temperature and pressure. There are many variations on the equation he proposed, but the specific form of the equation and the details about the various ways in which it was revised are not important to us here. What is significant about it for us here is that van der Waals could use the equation to explicitly solve for the values of volume, temperature, and pressure at the critical point (Levelt-Sengers 2002). The solution of the equation at the critical point is thus for the trio and is in terms of the parameters a and b introduced by van der Waals:

$$P_c = a/\left(27b^2\right)$$

$$V_c = 3b$$

$$RT_c = 8a/(27b)$$

The specific expressions [above] are not important to our story here; the important thing to notice about them is that the critical values of pressure, volume, and temperature [indicated by the subscript 'c' in the expressions above] can be expressed in terms of the parameters a and b that van der Waals introduced, and that [the specific values of] a and b are characteristic of a [particular] substance. Once these are in hand, the next conceptual step is to use these expressions to eliminate the constants a, b, and R from his equation of state he does this by using the critical values of these measurable quantities as units of measurement. Or, as we might say today, he normalizes pressure, volume, and temperature using their critical values, just as Mach number is a ratio of the velocity of something (such as a bullet) in a medium (such as air) to the celerity [(velocity of sound in)] in the medium. It is only because their critical values are given as functions of the constants a [145] and b that his reduction is possible. Thus, reduced pressure, denoted by P*, is the pressure of the gas or liquid divided by P_c, and V* and T* are similarly defined. This yields an equation of state in which neither a, b, nor R appears. Levelt Sengers writes, "This is a truly remarkable result". The equation:

> … is universal: all characteristics of individual fluids have disappeared from it or, rather, have been hidden in the reduction factors. The reduced pressures of two fluids are the same if the fluids are in corresponding states, that is, at the same reduced volume and pressure.

Fig. 17.18 Paul Ehrenfest, Hendrik Lorentz, Neils Bohr, and Kamerlingh Onnes. Onnes' "Principle of Corresponding States" was a remarkable result about fluids, allowing scientists to deduce behavior of one liquid from experiments on another. Onnes was inspired by Newton's use of mechanical similarity in developing it. *Photo credit* nobelprize.org

In his presentations to the Academy, van der Waals deduces straightforwardly that in reduced coordinates, the vapor pressure curve and the coexistence curve must be the same ("fall on top of each other") for all fluids (Levelt-Sengers 2002, p. 25) (Fig. 17.18).

The principle of corresponding states allowed scientists to produce curves representative of all substances from experiments on a particular substance:

> The principle of corresponding states … frees the scientist from the particular constraints of the van der Waals equation. The properties of a fluid can now be predicted if only its critical parameters are known, simply from correspondence with the properties of a well characterized reference fluid. Alternatively, unknown critical properties of a fluid can be predicted if its properties are known in a region not necessarily close to criticality, based on the behavior of the reference fluid. (Levelt-Sengers 2002, p. 26)

If we reflect on how this method of prediction works, we see that the same could be said of Reynolds' discoveries about the critical point of transition from laminar to turbulent flow. Although people today tend to think in terms of a critical Reynolds Number—that is, the value the dimensionless parameter has at the point of transition—in their 1914 paper, Stanton and Pannell put Reynolds' discovery in terms of a critical velocity. By using v_c as an abbreviation for critical velocity, this can be conveniently abbreviated by saying, as he put it, that "Reynold's discovery was that for geometrically similar tubes $v_c d/\nu$ was constant," where d denotes a diameter or other chosen distance in the situation, and ν denotes viscosity. This is stating Reynold's discovery as the kind of constraint [146] provided in thermodynamics by the statement that the dimensionless parameter $(P_c \times V_c)/(R \times T_c)$ is constant. Don't be misled by the fact that these look like the kind of equations you are familiar with using to plug in values and predict the [value of the] one chosen as the unknown. As in thermodynamics, so in hydrodynamics; their value lies not in using the equation directly, but in telling how experimental curves for one substance can be used to predict the behavior of another. The curves for fluid flow were meant to apply to any fluid—hence Rayleigh's comments that experiments on oil as a reference fluid could be used to predict the critical velocity of another fluid from its properties in a noncritical region—just as the law of corresponding states allows one to make predictions about the critical points of other substances from experiments performed on a reference substance.

Onnes used this insight about corresponding states to set up an experimental apparatus to liquefy helium, which has an extremely low critical temperature. What is so exciting about his story is that he had to rely on the law of corresponding states to estimate the critical temperature so that he would know where to look—that is, so that he would know what condition to create in order for the helium to liquefy. What is especially relevant to our story is that he did more than just use van der Waals' law of corresponding states. He also gave a foundation for it that was independent of the exact form of van der Waals' equation and did not depend on results in statistical mechanics. Instead, he used *mechanical* similarity:

> Kamerlingh Onnes's (1881) purpose is to demonstrate that the principle of corresponding states can be derived on the basis of what he calls the principle of similarity of motion, which he ascribes to Newton. he assumes, with Van der Waals, that the molecules are elastic bodies of constant size, which are subjected to attractive forces only when in the boundary layer near a wall, since the attractive forces in the interior of the volume are assumed to balance each other … He realizes this can be valid only if there is a large number of molecules within the range of attraction … [Onnes] considered a state in which N molecules occupy a volume v, and al have the same speed u (no Maxwellian distribution). The problem is to express the external pressure p required to keep the system of moving particles in balance, as a function of the five parameters. He solves this problem by deriving a set of scaling relations for M, A, v, u and p, which pertain if the units of length, mass, and time are changed". [147] (Levelt-Sengers 2002, p. 30)

The 'scaling relations' Onnes developed are another way of bringing in dimensional considerations, or the "theory of dimensions" that we saw earlier as key to D'Arcy Wentworth Thompson's work in biology on the importance of size, to Galileo's arguments about similarity in mechanics of materials, to Helmholtz's paper on similar motions and dirigibles, to the work by Reynolds, Stanton, and Pannell on similar motions in fluids, and to Rayleigh's crusade for a proper appreciation and more widespread use of the principle of dynamical similarity. As Sengers notes, scaling relations are supposed to hold no matter what units are used for the measureable quantities involved. Onnes provides a criterion for corresponding states based on these scaling relations, along with assumptions abut what the molecular-sized objects are like. Sengers remarks:

> Two fluids are in corresponding states if, by proper scaling of length, time and mass for each fluid, they can be brought into the same "state of motion". It is not clearly stated what he means by this, but he must have had in mind an exact mapping of the molecular motion in one system into that of another system if the systems are in corresponding states.

Then she gives her own suggestion of what it means to be in the same "state of motion":

> In modern terms: suppose a movie is made of the molecular motions in one fluid Then, after setting the initial positions and speed of the molecules, choosing the temperature and volume of a second fluid appropriately, and adjusting the film speed, a movie of the molecular motion in a second fluid can be made to be an exact replica of that in the first fluid. (Levelt-Sengers 2002, p. 30)

Shortly after Onnes's Nobel lecture, Richard Chase Tolman, a physicist at the California Institute of Technology, published a paper titled "The Principle of Similitude" in *Physical Review*, a major physics journal in the U.S. What it suggested sounded a lot like the idea of being able to make movies of one situation that were replicas of movies of other situations except for film speed. Tolman's paper proposed the following:

> The fundamental entities out of which the physical universe is constructed are of such a nature that from them a miniature universe could be constructed exactly similar in every respect to the present universe. [148] (Tolman 1914, p. 244)

Tolman then proceeded to show that he could derive a variety of laws, including the ideal gas law, from the principle of similitude he proposed. He proceeded in much the same way as Onnes had done in showing that the principle of corresponding states was a consequence of mechanical similarity. Tolman first developed scaling laws, laying out a transformation rule for how each quantity from a short list he had constructed—length, time, velocity, acceleration, electrical charge, and mass—should be scaled from the present universe to the miniature universe. He laid out and justified a transformation rule for each quantity individually, but some quantities were dependent on others (the transformation equation for velocity is a consequence of the equations for length and time). His justifications seem to be based on the criterion that the two universes would be observationally equivalent from the standpoint of an observer located in one of them:

> ... let us consider two observers, O and O', provided with instruments for making physical measurements. O is provided with ordinary meter sticks, clocks and other measuring apparatus of the kind and size which we now possess, and makes measurements in our present physical universe. O', however, is provided with a shorter meter stick, and correspondingly altered clocks and other apparatus so that he could make measurements in the miniature universe of which we have spoken, and in accordance with our postulate obtain exactly the same numerical results in all his experiments as does O in the analogous measurements made in the real universe. (Tolman 1914, p. 244)

Examples are that if O measures a length to be I, O' measures it to be xl, where O' has a meter stick that is shorter than O's and x is a number less than 1. In obtaining the transformation equation for time, however, Tolman appeals to the physical fact that "the velocity of light in free space must measure the same for O and O'," (Tolman 1914, p. 245) and he concludes that if O measures t, O' will measure xt. So the equations for length and time are $l' = xl$ and $t' = xt$. In obtaining the transformation equation for mass, he appeals to Coulomb's law as setting a constraint that must be satisfied, in conjunction with the requirement that O and O' measure the same charge, to obtain $m' = m/x$. Once the equations for the "fundamental magnitudes, length, time and mass" have been obtained, he says, we "can hence obtain a whole series of further equations for force, temperature, etc., by merely [149] considering the dimensions of the quantity in question" (Tolman 1914, p. 247). He then tries to show how various physical relations, such as the ideal gas law, can be deduced from simple physical assumptions and his proposed principle of similitude. for relations about gravitation, however, a contradiction arises, which he embraces and uses to propose new criteria for an acceptable theory of gravitation. He ends feeling triumphant about his proposed principle. It's a new relativity principle, he concludes: "the principle of the relativity of size"!

> … in the transformation equations which we have developed we have shown just what changes have to be made in lengths, masses, time intervals, energy quantities, etc., in order to construct such a miniature world. If, now, throughout the universe a simultaneous change in all physical magnitudes of just the nature required by these transformation equations should suddenly occur, it is evident that to any observer the universe would appear entirely unchanged The length of any physical object would still appear to him as before, since his meter sticks would all be changed in the same ratio as the dimensions of the object, and similar considerations would apply to intervals of time, etc. From this point of view we see that it is meaningless to speak of the absolute length of an object, all we can talk about are the relative lengths of objects, the relative duration of intervals of time, etc., etc. The principle of similitude is thus identical with the principle of the relativity of size. (Tolman 1914, p. 255)

Einstein had shown that the principle of relativity of uniform motion led to the conclusion that it was meaningless to speak of two events occurring simultaneously—that one could talk only about relative simultaneity, never absolute simultaneity. Tolman structures his claim along the same lines, in saying that the principle of relativity of size leads to the conclusion that it is meaningless to speak of absolute length (Fig. 17.19).

[Tolman's claim that it is meaningless to speak of absolute length] seems at least on the face of it at odds with D'Arcy Thompson's view about the importance of size—that different laws govern at different size scales We shall see that Tolman did not have the last word on the issue of the principle of similarity he proposed. [A later chapter of *Wittgenstein Flies a Kite*, "Models of Wings and Models of the World", relates how, later that year, Buckingham published a paper in *The Physical Review* in which he develops the background needed for a comprehensive response to the problem of observationally indistinguishable universes that Tolman raised. He shows where Tolman's reasoning errs, and presents a correct statement of the relevant similarity criterion.] Still, we may ask why [Tolman] addressed the topic of the significance of size in physics at all. It is clear why a physicist would be interested in principles of relativity of motion, for relative velocities [150] were essential to analyzing any dynamic situation in physics. But why an interest in size and scale?

Fig. 17.19 Richard C. Tolman (1881–1948) and Albert Einstein (1879–1955) at the California Institute of Technology, where Tolman was Professor of Physical Chemistry and Mathematical Physics from 1922 to 1948. *Photo credit* Los Angeles Times Photographic Archive, UCLA Library

It does seem that the equation of the effect of size in scientific works was on people's minds even earlier, and when answered in one context, seems to rise in another. The question of size had been raised by Newton, who endorsed the notion of replica miniature worlds, at least in terms of their dynamic similarity, and by Galileo, who pointed out that, although there are certainly rules that inform us how to build replicas of different sizes, sometimes there are so many things you have to take into account in making a replica of a different size that you might not be able to address all the considerations: the rules are more complicated than just keeping relative magnitudes of the same kind the same. This is the point D'Arcy Thompson picks up on, and so there is at least an apparent tension between, on the one hand, Thompson's insight about how different life is for organisms that live in worlds at different size scales, and on the other hand, his conviction that physical laws are universal, and that universality has to extend across worlds of different size scales.

Actually, in Osborne Reynolds' most famous work, the 1883 "An Experimental Investigation of the Circumstances Which Determine Whether the Motion of Water Shall be Direct or Sinuous, and of the Law of Resistance in Parallel Channels," he expresses his conviction that physical laws are universal, when faced with an apparent challenge to it. In a subsection of the paper he called "Space and Velocity," he expresses this conviction and hints at an unexpected resolution to the seeming paradox:

> As there is no such thing as absolute space or absolute time recognised in mechanical philosophy, to suppose that the character of motion of fluids in any way depended on absolute size or absolute velocity, would be to suppose such motion without the pale of the laws of motion. If then fluids in their motions are subject to these laws, what appears to be the dependance of the character of the motion on the absolute size of the tube and on the absolute velocity of the immersed body, must in reality be a dependance on the size of the tube *as compared with* the size of some other object, and on the velocity of the body *as compared with* some other velocity. What is [151] the standard object and what the standard velocity which come into comparison with the size of the tube and the velocity of an immersed body, are questions to which the answers were not obvious. (emphasis added) (Reynolds 1883, p. 937)

Reynolds considered this the philosophical, as opposed to the practical, aspect of his "experimental investigation" and the primary result of it (Reynolds 1883, p. 935). An idea about how the apparent paradox might be resolved had come to him only after he had figured out that the law of transpiration depends on an unusual relation of the magnitude: "the size of the channel [the opening the gas has to flow through] and the *mean range* of the gaseous molecules". This discovery arose when he realized that a change in temperature affected the rate of transpiration, or gas flow, through the pores. Then he looked at Stokes' equation for further clues, upon which he realized that the form of Stokes' equation did contain the information that there was a relation of a sort that had been hitherto overlooked: a relation between what he called "dimensional properties"— properties that did depend on size (velocity, size of the tube) and "the external circumstances of motion". Deriving and then equating different expressions for fluid acceleration yielded what later became known as the Reynolds Number. He reported his result at the outset of the paper as follows:

> In their philosophical aspect these results related to the fundamental principles of fluid motion; inasmuch as they afford for the case of pipes a definite verification of two principles, which are – that the general character of the motion of fluids in contact with solid surfaces depends on the relation between a physical constant of the fluid and the product of the linear dimensions of the space occupied by the fluid and the velocity. (Reynolds 1883, p. 935)

This seems to be saying that the way to resolve the paradox between the conviction that size can only be relative, and our experience with phenomena that do seem dependent on size, is to expand the notion of "relative size" to include relations of linear magnitudes to other magnitudes. The crucial ratio is still dimensionless, as a ratio of two linear magnitudes would be, but it involves relations [to quantities] such as the viscosity of a fluid, not just the geometry of the situation. This still doesn't answer the question of [152] whether there can be a miniature universe or not, for the question of what is meant by a miniature universe becomes more complex. Preserving the [magnitudes of] things that a length is relative to is no longer a matter of size, no longer [merely] a matter of geometry. Thus it is natural that the question of whether a miniature universe can be a replica universe was bound to be asked again—as Tolman did, in 1914.

There may be some historical reasons for the intensified interest in scale, similarity, and transformation equations (correspondence rules) among physicists in the years leading up to 1914. Certainly there are different scales of magnitudes described by the kinetic theory of gases. Onnes's 1913 Nobel lecture in which he credits use of the law of corresponding states for his success in liquefying helium may have brought attention to the power of using scaling laws and principles of mechanical similarity, since they were involved in Onnes's derivation of that law. That alone would explain Tolman's interest: although he worked in a number of areas, he had a special interest in foundations of statistical mechanics, and he later authored [textbooks] on it (Tolman 1927, 1938).

An impetus for the interest in scale and similarity may have come from other areas of physics as well. For instance, did the discovery in 1911 that there was something much, much smaller than an atom—its nucleus—raise questions about what things were like inside the nucleus, or about what sorts of laws governed at such tiny scales? That is, did it raise the sort of image for physics that D'Arcy Thompson had evoked for biology: the image of how different things are in the world within a world within "our" world experienced by a bacillus? Or, alternatively, might the discovery of atomic nuclei have spurred reflection on, and helped renew commitment to, the universality of physical laws at all scales, as in Tolman's paper? Einstein's special theory of relativity emphasized the difference between cases in which velocities were very small in comparison to the velocity of light and those that were not. It was interesting that Einstein presented that theory as continuous with, rather than an overthrowing of, classical mechanics, and, emphatically, as a consequence of universal laws. The time around 1914 was also a time when there were new techniques allowing measurement of distances between molecules, meaning that assumptions that previously could only be debated as theoretical or inferred could be more directly determined experimentally (AIP 1990). [153]

Whatever the historical reasons for the interest, principles of similarity were in fact relevant to discussions and active research programs in fields as diverse as zoology and atomic physics, and to endeavors as practical as applied aerodynamics. Another historical fact in 1914 was that the U. S. lagged behind almost every European country in aerodynamic research. Some in the U. S. were trying to change that. The ensuing bureaucratic struggle set the stage for the writing of Edgar Buckingham's remarkable paper on physically similar systems that had something to say in response to all these claims about similarity that were swirling around in 1914 (Fig. 17.20).[5]

[5]Later chapters of the book from which this paper is excerpted delineate the structure of the explanation given in Buckingham's 1914 paper describing how a physical system can be constructed so as to model another physical system, which follows in part from examining the logical consequences of the "most general form" of a "physical equation". I then show how it can be seen as analogous to the structure of the explanation presented in the *Tractatus*, on which a proposition can be regarded as "a model of reality as we imagine it".

ON PHYSICALLY SIMILAR SYSTEMS; ILLUSTRATIONS OF THE USE OF DIMENSIONAL EQUATIONS.

By E. Buckingham.

1. *The Most General Form of Physical Equations.*—Let it be required to describe by an equation, a relation which subsists among a number of physical quantities of n different kinds. If several quantities of any one kind are involved in the relation, let them be specified by the value of any one and the ratios of the others to this one. The equation will then contain n symbols $Q_1 \cdots Q_n$, one for each kind of quantity, and also, in general, a number of ratios r', r'', etc., so that it may be written

$$f(Q_1, Q_2, \cdots Q_n, r', r'', \cdots) = 0. \qquad (1)$$

Let us suppose, for the present only, that the ratios r do not vary during the phenomenon described by the equation: for example, if the equation describes a property of a material system and involves lengths, the system shall remain geometrically similar to itself during any changes of size which may occur. Under this condition equation (1) reduces to

$$F(Q_1, Q_2, \cdots Q_n) = 0. \qquad (2)$$

If none of the quantities involved in the relation has been overlooked, the equation will give a complete description of the relation subsisting among the quantities represented in it, and will be a complete equation. The coefficients of a complete equation are dimensionless numbers, *i. e.*, if the quantities Q are measured by an absolute system of units, the coefficients of the equation do not depend on the sizes of the fundamental units but only on the fixed interrelations of the units which characterize the system and differentiate it from any other absolute system.

Fig. 17.20 First page of Edgar Buckingham's 1914 *Physical Review* paper "On Physically Similar Systems: Illustrations of the Use of Dimensional Equations". A brief summary of the work appeared in May 1914 in the Journal of the Washington Academy of Sciences, as "Physically Similar Systems" (pp. 347–353)

References

AIP 1990. American Institute of Physics. Interview with Linus Pauling, Ph.D. November 11, 1990. Bigsur, California. http://www.achievement.org/achiever/linus-pauling/#interview

Baals and Corliss (1981), Wind Tunnels of NASA (NASA SP-440). Washington, D.C.: Scientific and Technical Information Office.

Barker, Peter (1980). "Hertz and Wittgenstein," *Studies in the History and Philosophy of Science*, 11: 243–256.

Boltzmann, L. (1974). *Theoretical Physics and Philosophical Problems: Selected Writings* (*Vienna Circle Collection*). (B. McGuinness, Ed.) Boston, MA: D. Reidel.
Cahan, David (2004) "Helmholtz and the shaping of the American physics elite in the Gilded Age." Hist Stud Phys Biol Sci, Vol. 35 No. 1, September 2004; (pp. 1–34).
Chanute, Octave (1894) *Progress in Flying Machines*. New York: The American Engineer and Railroad Journal.
Crew, Henry. (1913) "Galileo, the Physicist." *Science*, New Series, Vol. 37, No 952 (March 28, 1913), 463–470.
Galileo G. (1914) *The Two New Sciences by Galileo Galilei*. (Tr. by Henry Crew and Alfonso de Salvio.) Evanston IL: Northwestern University Press.
Hide, Oystein (2004) "Wittgenstein's Books at the Bertrand Russell Archives and the Influence of Scientific Literature on Wittgenstein's Early Philosophy" *Philosophical Investigations* Vol 27 No. 1. (January 2004), pp 68–91.
Janik, Allan and Stephen Toulmin (1973) *Wittgenstein's Vienna*. New York: Simon & Schuster.
Levelt-Sengers, Johanna (2002) *How Fluids Unmix: Discoveries By the School of Van der Waals and Kamerlingh Onnes*. Amsterdam: Edita, KNAW.
McGuinness, Brian F. (1988). *Wittgenstein: A Life: Young Ludwig 1889–1921*. Berkeley: University of California Press.
O'Connor, J. J. and E. F. Robertson (1998). "Georg Ferdinand Ludwig Philipp Cantor". http://www-history.mcs.st-andrews.ac.uk/Biographies/Cantor.html.
Onnes, H. Kamerlingh (1881) *Algemeene theorie der vloeistoffen* (General theory of liquids) Verhandelingen der Koninklijke Akademie van Wetenschappen: Afdeeling Natuurkunde.
Onnes, H. Kamerlingh (1913) "Investigations into the properties of substances at low temperatures, which have led, amongst other things, to the preparation of liquid helium." Nobel Lecture, December 11, 1913.
Rayleigh (1910) "Note as to the Application of the Principle of Dynamical Similarity." in Rayleigh, Sir John William Strutt, *Scientific Papers*, No. 340, pp. 532–533. Cambridge, England: Cambridge University Press (Six volumes).
Rayleigh, Sir John William Strutt (1914) "Fluid Motions" *Nature*, Vol. XCIII, p. 364 (Also appeared in *Proceedings of the Royal Institute*, March 1914 and is in *Scientific Papers*, No. 384, Volume VI; page numbers cited are from this work).
Reynolds, Osborne (1883) "An Experimental Investigation of the Circumstances Which Determine Whether the Motion of Water Shall be Direct or Sinuous, and of the Law of Resistance in Parallel Channels".
Rouse, Hunter and Simon Ince (1957) *History of Hydraulics*, 1st ed. Iowa City: Institute of Hydraulic Research.
Russell, Bertrand and Alfred North Whitehead (1997). *Principia Mathematica*, 2nd ed. Cambridge, England: Cambridge University Press.
Spelt, P. D. M. and B. F. McGuinness (2000) "Marginalia in Wittgenstein's copy of Lamb's Hydrodynamics." *Wittgenstein Studies* 2, 131–148.
Stanton, T. E. and J. R. Pannell (1914) "Similarity of Motion in Relation to the Surface Friction of Fluids." *Proceedings of the Royal Society of London, Series A, Containing Papers of a Mathematical and Physical Character*. Vol. 90, No. 619 (July 1, 1914).
Sterrett, Susan G. (2005/2006) *Wittgenstein Flies A Kite: A Story of Models of Wings and Models of the World*. New York: Pi Press (Pengin).
Thompson, D'Arcy Wentworth (1911) "Magnalia Naturae: of the Greater Problems in Biology." In Science, October 6, 1911. New Series, Vol. 34, No 875, p. 417–428.
Thompson, D'Arcy Wentworth (1992) *On Growth and Form*, complete revised edition. New York: Dover Publications.
Thompson, James (1875, 1912) "Comparison of similar structures as to elasticity, strength, and stability." In Thompson, James (1912) *Collected papers in physics and engineering*, 1st ed. 1912. 361–372.
Tolman, Richard Chace (1914) "The Principle of Similitude." *Physical Review* 3.

Tolman, Richard C. (1927) *Statistical Mechanics with applications to physics and chemistry*. New York: The Chemical Catalog Company.
Tolman, Richard C. (1938) *The Principles of Statistical Mechanics*. Oxford: Claredon Press.
van der Waals (1910) "The equation of state for gases and liquids." Nobel Lecture, December 12, 1910.
Wittgenstein, L. (1979a). *Wittgenstein, Ludwig. Notebooks 1914–1916.* (G. H. Anscombe, Ed., & G. Anscombe, Trans.) Chicago: University of Chicago Press.
Wittgenstein, L. (1979b). *"Notes Dictated to G. E. Moore April 1914." Appendix II to Wittgenstein, Ludwig. Notebooks 1914–1916.* (G. v. Anscombe, Ed., & G. Anscombe, Trans.) Chicago: University of Chicago Press.
Wittgenstein, L. (1979c). *"Notes on Logic." Appendix I to Wittgenstein, Ludwig. Notebooks 1914–1916.* (G. H. Anscombe, Ed., & G. Anscombe, Trans.) Chicago: University of Chicago Press.

Chapter 18
Wittgenstein's Thought Experiments and Relativity Theory

Carlo Penco

BT	*The Big Typescript*, TS 213. Edited and translated by G. G. Luckhard and M. A. E. Aue, Oxford: Blackwell 2005
LFM	*Lectures on the Foundation of Mathematics*, Cambridge, 1939. Edited by Cora Diamond, Ithaca: Cornell U.P. 1976
LW	*Last Writings on the Philosophy of Psychology*, vol. 1, Blackwell, Oxford, 1982
OC	*On Certainty/Über Gewissheit*. Oxford: Blackwell, 1974
PR	*Philosophical Remarks/Philosophische Bemerkungen*. Oxford: Blackwell, 1964. Engl. transl. Oxford: Blackwell, 1975
PI	*Philosophical Investigation/Philosophische Untersuchungen*. Edited and translated by G.E.M. Anscombe. Oxford: Blackwell 1968
RFM	*Remarks on the Foundations of Mathematics*. Oxford: Blackwell, 1978 (Third edition)
RPP	*Remarks on the Philosophy of Psychology*, 2 voll., Oxford: Blackwell, 1980
VB	*Vermischte Bemerkungen*. Edited by Von Wright, Frankfurt aM, 1977
Z	*Zettel*. Oxford: Blackwell, 1967

Different drafts of this paper have been presented—in 2012 at the EPILOG Seminar in Genoa and at the "In Wittgenstein's footsteps" Conference at the University of Iceland in Reykjavìk; in 2014 at the University of East Anglia and at a conference on "Wittgenstein and Physics" at St. Cross College, Oxford. I thank all participants for their useful comments and criticism, especially Joanna Maria Antonia Ashbourn, Mikael Karlsson, Oskari Kuusela, Ray Monk, Valeria Ottonelli, Silvia Panizza, Rupert Read, Mauricio Suarez. A special thanks to Diego Marconi, Brian McGuinness and Nino Zanghì for their suggestions and encouragement.

C. Penco (✉)
Philosophy of Language, Department of Philosophy, University of Genova, Genoa, Italy
e-mail: penco@unige.it

S. Wuppuluri and N. da Costa (eds.), *WITTGENSTEINIAN (adj.)*,
The Frontiers Collection, https://doi.org/10.1007/978-3-030-27569-3_18

18.1 Kinds of Mental Experiments in Wittgenstein

In his famous paper "On Thought Experiment"[1] Mach distinguished between those thought experiments that require confirmation and those that are not to be conceived as anticipation of real experiments but can work only in imagination. Probably thinking of Mach's second kind of thought experiment Wittgenstein comments: "What Mach calls a thought experiment is of course not an experiment at all. At bottom it is a grammatical investigation" (PR: 1). The reference to Mach follows a discussion on how an octahedron with pure colours is a "grammatical representation" of colour-space. This example suggests that we can give analogous "grammatical representations" not only of colours but also of other semantic fields: we need a logic that goes beyond the classical truth-conditional logic, a logic or grammar that is given by describing different *Bedeutungskörper*. Although critical of Mach, Wittgenstein probably found in that paper two suggestions very close to his own sensibility: (a) the idea that thought experiments are kinds of "variations" on a situation, and (b) the idea that "paradoxical" thought experiments "help thoughts not to come to rest". In what follows I propose to consider all or most of Wittgenstein's language games as thought experiments that typically take the form of counterfactual reasoning involving a concrete scenario, although a clear conclusion does not always follow in Wittgenstein's text.[2]

Language games may be grossly[3] classified as follows: (i) "constructive": aimed at arguing central claims in Wittgenstein's view. These are typically presented as a

[1] Mach's paper "Über Gedankenexperimente" was first published in 1897 in a little known journal of chemistry and later included in his book *Erkenntnis und Irrtum* (1905; 2nd ed. 1906) where Wittgenstein might have read it. It is probable that Wittgenstein was implicitly referring to Einstein's thought experiments, given the time of the entry in the late twenties, when Wittgenstein used to discuss with Waismann and Schlick, one of the main interpreters of Einstein's work (we find an echo of the discussions on Einstein's simultaneity problem in the first chapter of Waismann 1965)

[2] I will speak interchangebaly of "thought experiment" and "mental experiment", keeping in mind that they are not intended as analysis of our psychological and biological activities but as the analysis of concepts. Wittgenstein was not studying the biological development of our capacity of categoriziation, but had an idea of concept more similar to the Fregean one. See Marconi (2015) for the contraposition between cognitive and metaphysical views of concepts. Roux 2011 defines the characteristic features of mental experiments as (i) counterfactual reasoning (ii) referring to concrete scenarios (ii) linked to a specific cognitive intention. Wittgenstein's thought experiments can be framed as counterfactual reasoning, with a concrete scenario (e.g. what if thing were different from what they normally are, for instance if a rigid rod shrinks? RFM I: 119). This does not imply a metaphysical viewpoint on possible worlds (as in Williamson 2007; see Marconi 2017 for a criticism) but a style of reasoning.

[3] From Horowitz-Massey (1991) to Ierodiakonou-Roux (2011) the literature on mental experiments has incredibly increased. Here I will simply use the basic classification given by Brown (1986) between constructive and destructive mental experiments, which I think helps understanding different roles of mental experiments in Wittgenstein's work. Wittgenstein's mental experiments are liable of being represented as arguments, although not always easy, given that we often need to reconstruct his arguments from scattered remarks. However, reconstructing arguments seems to be the best way to avoid "bad intuition pumps" (cfr. Brendel 2004).

chain of variations[4] that purports to give, in the end, not a theory but a perspicuous representation of the working of language. Variations on a situation are especially developed in *the Blue and Brown Books* and in *Philosophical Investigations*: for instance, the initial situation of builders with the progressive insertion of different kinds of words for colours, numbers and so on (see PI: 2, 8, 21, 48, 64, 86, but also OC: 566) or the situation of the pupil learning to follow a rule (see PI: 143, 151, 179, 185, 201); (ii) "destructive": almost exclusively designed to challenge erroneous "theories" on language, such as the case of human beings as reading machines on §157, or the diary of private sensations on §258 or the beetle in the box on §293; their main origin is probably to be found in the strategy prominent in the first part of the *Big Typescript*, where a series of theories is presented to be challenged and dismantled; (iii) "paradoxical": they present bizarre cases whose point is not easily understood, and are sometimes called "alien language games" because of the awkward situations they depict. Paradoxical mental experiments are of two kinds:

(iiia) mental experiments in which general psychological or physical facts are different from the usual ones, such as the community of deaf people or people without dreams (Z: 371, 530), or situations in which objects suddenly disappear (RFM I: 37,137). Wittgensgein gives a clear description and purpose of (iiia) at the end of *Philosophical Investigations* II, xxxi:

> if anyone believes that certain concepts are *absolutely* the correct ones, and that having different ones would mean not realizing something that we realize—then let him imagine certain very general facts of nature to be different from what we are used to, and the formation of concepts different from the usual ones will become intelligible to him. (PI II, xii)

(iiib) mental experiments in which the general facts of nature are unchanged, but concepts entertained are so different from ours that it is difficult to understand their proper concern. Examples: wood sellers who exchange money with goods independently of the amount of money or value of goods (RFM I: 149–153), people who use elastic rulers (RFM I: 5, 19, 140), people that follow oracles instead of "our" knowledge of physics (OC: 609–12), people who have nothing like our concept of "knowing" (OC: 562), people who see the pattern "I I I I I" as "I I II I I" (RFM I: 169),—or people who use the notion of "pain" to laugh when pain is expressed (Z: 380). Wittgenstein seems to implicitly refer to (iiib) of mental experiments when he says:

> Suppose you came as an explorer into an unknown country with a language quite strange to you. In what circumstances would you say that the people there gave orders, understood them, obeyed them, rebelled against them, and so on? Shared human behaviour is the system of reference by means of which we interpret an unknown language. (PI 206)

---

[4]Picardi (2009) sees primitive language games as having the function "to enable us to arrange phenomena along a scale", using different extensions or expansions of it, as happens with the language game of the builders. One of the main claims of this remarkable paper is that "the conceptual tools required in certain extensions of the language depend on the availability of conceptual resources acquired at a lower level. Thus, in Dummett's terminology, there is a relation of relative priority among language-games; there is (partial) order in the motley".

In this paper I will deal mainly with (iiib): here the focus is not on different facts of nature but on behaviours and forms of life that appear different from our own (where here for "our own" we may mean the form of life of western culture).

Some "alien" mental experiments occur in *Remarks on the Foundation of Mathematics*, as if they were an answer to Frege's challenge presented in the introduction of *Grundgesetze*. Here Frege claims that tribes who held different logical laws contradicting ours would represent "a hitherto unknown kind of insanity". Wittgenstein's comment was: "… but he never said what this 'insanity' would really be like" (RFM I: 152). Some examples of alien language games may be seen as an attempt to answer Frege, as if Wittgenstein wanted to describe this kind of insanity; however we are left with the question as to what the point of these examples is. Wittgenstein's answer to Frege's challenge therefore demands an explanation that goes beyond his reaction to Frege's thought (see later at §18.3).

18.2 A Missing Background: The Analogy Between Wittgenstein and Relativity Theory

Discussion of alien mental experiments is typically framed in the contrast between relativistic interpretations (e.g. Winch 2014; Rorty 1979; Marconi 1987; Hacker 1996; Grayling 2001; Boghossian 2006; Glock 1996, 2007; Kusch 2013) and anti-relativistic ones (Stroud 1965; Williams 1974; Broyles 1974; Anscombe 1976; Lear 1986; Conant 1991; Blackburn 2004; Williams 2007; Coliva 2010; Tripodi 2013; Kuusela 2017; Pritchard 2018).[5]

Relativists usually disregard the more constructive examples of language games that rely on similarities with our normal concepts, while anti-relativists[6] have a more

[5]The distinction is not so sharp, and many interpretations are more nuanced than it may appear. I make just one classical example: Winch (2014) attempts to do with Edward Evans Pritchard and his judgment on Zande culture what Wittgenstein did with James Frazer. Winch criticizes Evans Pritchard for thinking that Western scientific explanations are "in accord with the objective facts", while we cannot define a reality which is independent of social practices and institutions (e.g. Winch 1976: 12). He seems therefore to be a strong cultural relativist, and to support the idea that what is real depends on human mind (see also Burley 2012). Against the standard interpretation of Winch as a relativist, Hutchinson et al. (2008) claim that a correct rendering of Winch ideas is not as a relativist, but as a proponent of a more subtle and complex stance that would be not so different from what I propose in this paper.

[6]It is worth remarking that the anti-relativistic interpretation has been supported by Anscombe's translation of the passage of *Philosophical Investigation* 206 quoted above, where the phrase "*die gemeinsame menschliche Handlungsweise*" is translated as "the common behaviour of mankind". Accepting Anscombe's translation implies that there is a clear reference to something like universal features of human behaviour. Savigny (1991) was the first to point out the problem of an ambiguity in the German text. In a later edition (2009) the English translation has been changed into "shared human behaviour", permitting also a relativistic interpretation of the passage, where we *might* think of specific features of behaviour and reasoning shared in a local community. However, the German text permits both interpretations and is of no use in deciding which of the two is the right one. For a discussion of the development of the different editions of Wittgenstein's texts see Erbacher (2014).

complex strategy: according to them the first kind of language games (constructive ones) helps us in exploring our conceptual structure; the second kind (bizarre ones) is a way to show that we cannot follow deviant concepts (we cannot conceive the inconceivable).

In what follows I suggest that a better consideration of Wittgenstein's discussion of relativity theory could help clarify the strategy behind his use of "alien" language games. The relevance of Relativity theory for Wittgenstein has been discussed by many authors, among whom (Hintikka and Hintikka 1986; Wright 2002; Brebber 2003; Penco 2010; Kusch 2011, 2015; Mion 2019). It is reasonable to think that Wittgenstein was strongly influenced by the debate on Relativity Theory after his return to Austria in 1919, the year in which even newspapers, especially German and Austrian, were discussing the confirmation of Relativity Theory, and Einstein himself wrote in December 1919 in the *Berliner TageBlatt*[7]:

> But the truth of a theory can never be proven. For one never knows if future experience will contradict its conclusion; and furthermore there are always other conceptual systems imaginable which might coordinate the very same facts.

Such a view, a forerunner of the Duhem-Quine thesis,[8] may have influenced Wittgenstein, together with other aspects of Einstein's revolution that seem to have had an impact on Wittgenstein's thought since the early Twenties. Relativity theory seems to constitute an red thread, linking Wittgenstein's early remarks on colour to his latest remarks on certainty. First, Wittgenstein's idea of a special "logic" of colour words, after his criticism of *Tractatus* in "Some remarks on logical form", may be prompted by a comparison with the idea of a system of coordinates as a necessary framework for the grammar of colour words:

> The written sign without the coordinate system is senseless. Mustn't we then use something similar for representing colours? (PR: 46).

[7]Quoted from Einstein (1919), also reported in Einstein (2002): Volume n. 7, ch. 28. It is reasonable to imagine that Wittgenstein, coming back to Wien at the end of the war in 1919, could not avoid having a look at the many remarks on the topic in newspapers available in most Viennese cafes (and at home). On the presence in German newspapers of news about the confirmation of relativity theory see Elton (1986). Unfortunately there is no research on the presence of articles on relativity theory in the Austrian newspaper *Neue Freie Presse*, that Brian F. McGuinness suggested me to consult. As far as I have verified, Einstein himself was requested to publish a paper by the *Neue Freie Presse*, but he had to answer (December 6, 1919) that he could not give them the content of a paper already published in *The Times* on November 28 1919; however he would later publish an article on the Theory of Relativity for the *Neue Freie Presse* in 1920 (July 24: "Die hauptsächlichen Gedanken der Relativitätstheorie") and in 1921 (January14). The *Berliner Tageblatt* and the *Vossische Zeitung* had already published articles on the confirmation of Einstein's Theory by the solar eclipse (October 6, 1919 in *Berliner Tageblatt*: "The sun brings it to light"; October 15, 1919 in *Vossische Zeitung*: "solar Eclipse and Relativity Theory", where specifications of the bending of the light beam were given in details); the 14th of December 1919 the *Berliner Illustrierte Zeitung* published the famous picture of Einstein on its front page; The *Neue Freie Presse* continued to publish articles on Relativity theory (e.g. by Oskar Kraus on September, 11, 1920 and on October 27, 1923). German newspapers can be consulted at http://zefys.staatsbibliothek-berlin.de; Austrian newspapers at: http://anno.onb.ac.at.

[8]Views of the kind were not absent from the Vienna Circle, where Duhem's work was well known and discussed, given the translation into German, with a preface by Mach, in 1908.

The topic of colours is taken up more than once in this regard until the very effective discussion on rule-following in RFM VI 28:

> The certainty with which I call the colour "red" is the rigidity of my measuring-rod, it is the rigidity from which I start. (…) Following according to the rule is FUNDAMENTAL to our language game.

We cannot enter here the debate on rule-following considerations.[9] I gave just two of many quotations[10] to suggest the reasonable assumption that we may find implicit links with Relativity theory in many of Wittgenstein's examples, especially when discussing the idea of coordinate system or frame of reference in a very general way, which is sometimes to be understood as "conceptual system".[11] Since the Twenties Wittgenstein often uses the terms "coordinate system" and "system of reference" ("*Bezugssystem*", literally "reference system", is a term typically used in physics also by Einstein to refer to a coordinate system). The previous quotation on the idea that "the written sign without the coordinate system is senseless" (PR: 46, see also Z: 432) seems an echo of Einstein's idea that a statement of simultaneity of events without a specified coordinate system is meaningless: "unless we are told the reference-body to which the statement of time refers, there is no meaning in a statement of the time of an event" (Einstein 1916, p. 31).

This is a reminder of the verificationist account, which is explicitly connected, both by the Vienna Circle and Wittgenstein, to Einstein's view: "The verification is not one token of the truth, it is the sense of the proposition. (Einstein: How a magnitude is measured is what it is)." (PR: 166; see also BT I: 60; PG II: 39). The discussion of Wittgenstein's use of the metaphor of measurement has been much developed in two papers by Kusch (2011, 2015), according to which metrology has a central role in Wittgenstein's critique of the *Tractatus* and beyond. Kusch relies on the comparison between the rigidity of rods and clocks in Einstein theory and the rigidity of "grammatical" rules in Wittgenstein – where for "grammatical" we mean also our mathematics: "mathematics as such is always measure, not thing measured" (RFM III: 75). Kusch (2011) claims that "just as Einstein brings the rulers

[9]The debate on rule-following is almost endless. There are at least two dilemmas: Platonist versus communitarian justifications of rule-following (truth conditions vs. assertibility conditions) and individualistic versus communitarian mastering of rule-following (Kusch's communitarian view is dealing with the latter). This duplicity makes it difficult to give a proper assessment of the contrasts among interpretations. For a discussion see also, Baker-Hacker (2009), §IV and Horwich (2010: 133). In RFM VI 49 Wittgenstein suggests the following anti-consensualist view: "If what a proposition of logic said was: Human beings agree with one another in such and such ways (…) then its contradictory would say that there is here a lack of agreement. Not that there is an agreement of other kind. The agreement of humans that is a presupposition of logic is not an agreement *in opinions*, much less in opinions on questions of logic". An anti-consensualist view is developed by Crispin Wright who claims that Wittgenstein's analogy between his philosophy and Relativity theory suggests a more *constructive* kind of response to the rule-following considerations, that partly undermines the interpretation of Wittgenstein's quietism (see also Footnote 15).

[10]For a more complete list of quotations we can find in different phases of Wittgenstein's philosophy see Penco (2010).

[11]While Wittgenstein often uses the term "coordinate system" (since PR 46, 83, 206) or "system of reference" he also uses, although rarely, the term "system of concepts" (e.g. RPP I 47).

and clocks into the empirical world, so the later Wittgenstein rejects the transcendental status that the *Tractatus* accorded to rules, and situates them firmly inside our real and empirical language-games." Somebody might object that Wittgenstein also insists on non-empirical language games (for Wittgenstein "the limit of empirical is concept formation" RFM IV: 29). However there is something in this interpretation of Wittgenstein, which is also connected with the discussion on scepticism in *On Certainty*, where we find Wittgenstein's last explicit remark on Relativity Theory:

> Here once more there is needed a step like the one taken in relativity theory. (OC: 305)

The passage should be read in the context of the problem of scepical doubt[12] discussed in *On Certainy*, which has a starting point on mathematical certainty: "The mathematical proposition has, as it were officially, been given the stamp of incontestability. I.e.: Dispute about other things; *this* is immovable—it is a hinge on which your dispute can turn." (OC: 655) How do these remarks help in facing sceptical doubts? Kusch would say that mathematics is measure and we agree in putting it "in the archive". But, concerning scepticism in particular, Mion (2019) gives a further suggestion, grounded on the basic tenet by Wittgenstein according to whom.

> Every empirical proposition may serve as a rule if it is fixed, like a machine part, made immovable, so that now the whole representation turns around it and it becomes part of the coordinate system, independent of facts. (RFM VII: 74)[13]

The suggestion is that empirical propositions like the ones presented by Moore against the sceptic are not actual empirical propositions, but belong to the background, to the coordinate system with which we describe reality. As Mion (2019: 6) says: "both Einstein and Wittgenstein seek a solution to their problems by rising mere empirical data to the role of apriori postulates. So, for Einstein, the fixity of the speed of light becomes an assumption of his theory of relativity; while, for Wittgenstein, mere empirical propositions such as 'there are physical objects' become assumptions that are generally presupposed in all empirical inquiries".

While this point may help to understand Wittgenstein's analysis of the problem of scepticism in *On Certainty* we still need an answer about the role Wittgenstein gives to "our" frame of reference (although subject to changes). We have to be cautious in the use of the term "frame of reference", and not to assume that Wittgenstein were always concerned with "our" frame of reference as the only one we have (OC 83). It is true that the scientific and technological view of the world seems to unify the world as a whole, as Wittgenstein claims in one of his most pessimistic and apocalyptic passages: "science and industry with infinite misery will unite the world" (VB: 14.7.1947). However, in Wittgenstein's view, a frame of reference may

---

[12]The following analysis is coherent with the general idea of "hinge epistemology" (Coliva 2016). My suggestion is that we may view the idea of "hinge" as a metaphor of a coordinate system. What is a coordinate system if not a theoretical hinge on which we rely to organize our descriptions?

[13]See also OC 321: "any empirical proposition can be transformed into a postulate—and then becomes a norm of description". Mion (2019) remarks that this entry has been written immediately after the claim about Relativity theory in OC 305. See also Mounce 2012.

also be intended as one of different possible choices, as discussed in a late remark of 1947:

> A religious belief could only be (something like) passionately committing oneself to a system of coordinates. Hence although it's belief, it is really a way of living, or a way of judging life. Passionately taking up *this* interpretation. And so instructing in a religious belief would have to be portraying, describing that system of reference and at the same time appealing to the conscience. And these together would have to result finally in the one under instruction himself, of his own accord, passionately taking up that system of reference. (VB: 21.12.1947)

In this last case, it seems apparent that speaking of "system of coordinates" for religious beliefs at least implies that there may be different systems of coordinates or systems of reference. Our frame of reference may be one among many.

18.3 Alien Mental Experiments and Relativity Theory

Crispin Wright suggests considering Wittgenstein's remarks on Einstein as as something central to Wittgenstein's project of philosophy and as the seeds of a constructive perspective, which gives philosophy a theoretical aspect beyond mere quietism.[14] In what follows I try to develop the suggested similarity between Wittgenstein and Relativity, partly expanding what was very shortly hinted at in Penco (2010), and partly going beyound a mere textual interpretation. I attempt to make my earlier remarks more explicit, trying to answer the following more specific questions:

(1) How do time dilation and light contraction fit into Wittgenstein's discussion on alien language games?

[14]Wright (2002: 32–35) presents the comparison with Relativity theory as follows: "Just as, from the relativistic standpoint, we abandon the idea that our measurement practices answer to absolute and independently constituted determinations of spatiotemporal values, instead regarding those very practices as grounding of our concepts of spatio-temporal parameters and of the content of statements concerning them (…) so we should abandon the idea that in basic rule-following, our moves and judgments answer to independently constituted determinations of correctness, instead regarding our propensities to convergence as grounding our concepts of what it is to follow such rules correctly and the content of the judgments on which we converge." At first sight it might appear that these remarks could be interpreted as a move from an absolute Platonist account towards a communitarian or consensualist account of rule-following (as claimed, following Kripke, by Kusch 2013). However, as Wright insists, rule-following still cannot be identified, as a matter of conceptual necessity, with what most human beings converge on. Although *arising from* our primitive dispositions to converge, we cannot identify convergence or consensus as *what constitutes or guides* our *basic* rule-following, e.g. judgments about colours (see RMF VI, 28 where Wittgenstein connects his analogy to Relativity theory to rule-following considerations). In these basic cases we are not really "guided" by anything. It is correct to say that "basic rule-following, like all rule-following, is rational in the sense that it involves intentionality and a willingness to accept correction in the light of error." (Wright 2009: 497–8). But this does not mean that it requires a justification on some further level of analysis, e.g. that it depends on consensus. We rely on the rigidity of our basic judgments on colours just as we rely on the rigidity of clocks and rods. This rigidity permits us understanding what is error and what is correct and incorrect application.

(2) Where can we look in Wittgenstein for invariants and transformations that permit the comparison among different coordinate systems – or conceptual systems?
(3) How should we interpret the similarity between describing measuring tools (clocks and rods) and describing concepts or judgments?

I will try to answer these three questions with the help of one of the most debated Wittgensteinian thought experiments: the wood sellers.[15] The case of the wood sellers at first appears as an answer to Frege's worry as to what kind of insanity it would be not to follow our logical rules.[16] Wood sellers, at first sight, *seem* not to be following basic rationality: they sell wood depending on the extension of the wood on the ground, independently of the height of a pile of wood, so that a smaller quantity of wood would be sold for a higher price than a larger quantity if it were laid out with a larger base. Their system appears, therefore, illogical or incoherent. Wittgenstein suggests that we might try to help them and show that a high pile of wood with a smaller base could be transformed into a larger base and then be shown to be more wood than the other pile. This presentation would help the wood sellers to realize the real quantity of wood of different piles. This solution might work, but Wittgenstein imagines also a possible reaction: "perhaps they would say: 'Yes, now it's a *lot* of wood and costs more'—and that would be the end of the matter." (RFM I: 150) What is the point of this example? Is Wittgenstein trying to show that there are systems of beliefs incommensurable with ours (relativist answer)? Or that our inability to describe the incoherent system of wood sellers will help us to clarify, by contrast, "our" concepts (anti-relativist answer)? Probably neither answer is right, and the analogy with Relativity Theory may provide some new suggestions.

18.3.1 Time Dilation and Concept Deformation

Einstein considered the variations of measuring systems according to different frames of reference (coordinate systems); his mental experiments concerning clocks and rods are first developed in his treatment of Special Relativity, and it is therefore in connection with Special Relativity that we may first look for the putative analogy

---

[15]Kusch (2015) abundantly uses the mental experiment of wood sellers to verify the compatibility of Wittgenstein with a "standard model" of relativism defined according to a model of "metrological relativism". However I will avoid this kind of comparison, in order to find a fresh start on Wittgenstein's similarity with Relativity theory independent of the debate on relativism.

[16]Speaking of logical madness Frege (1893) appears to hint at some *way of thinking* in which we cannot participate. However, as Cerbone (2000) has shown, it is not as simple as that. There is in Frege both a *normative* strand (logical laws are a norm of correct thinking; there may be people that do not follow them, and therefore are wrong in their judgments—and we might not be able to think the way they think) and a *constitutive* strand (logical laws do not prescribe how we *should* think but form the background that constitutes thinking; in principle there cannot be people who *think* and at the same time disagree with logical laws). According to Cerbone, Wittgenstein would follow the second strand, concluding that if we imagine beings who infer and argue, then we cannot imagine them as if they were outside the realm of logic. However which logic are we speaking about when speaking of language games is not clearly defined.

between Wittgenstein and Einstein. In Special Relativity if two Coordinate Systems (CS) approximate the speed of light, from the point of view of one CS, in the other CS clocks slow down and rods shrink.[17] Following the analogy in the conceptual realm, we might ask what happens to our concepts when they are placed in a different conceptual framework. We should expect that in an alien conceptual system our concepts would shrink or be somehow deformed. For instance, our concepts of "selling" and "money" should not be applied as such to the wood seller's conceptual system. Anti-relativistic interpretations claim that the "distance" from our normal concepts is given just to make us understand that our concepts are the only ones we have. But we might also say that the concepts we normally use are analogous to the tools provided by Newtonian principles in dealing with problems in classical mechanics. As Newton's principles don't work properly in a coordinate system that nears the speed of light, our usual concepts don't work properly in an "alien" conceptual system.

We would then suggest that our basic concepts are rigid, like Einstein's measuring rods, but are deformed or bent in alien conceptual systems, as rods shrink relative to a different coordinate system. A limitation of the anti-relativistic interpretation is that we cannot apply our concepts as if they were "absolute": as Einstein showed that time is not an "absolute" measure, but depends on systems of coordinates, Wittgenstein criticises the belief "that certain concepts are absolutely the correct ones, and having different ones would mean not realizing something that we realize" (PI II: 12). Following this quotation, we are led to think that our concepts of "quantity" depending on our paradigm of economic exchange are not the absolutely correct ones, and therefore we could *not* claim that, if the wood sellers did not accept them, they simply did not realize something that we realize. Perhaps we could get them to understand our viewpoint, but they might be not interested.

What is wrong then in applying our concepts (of selling, quantity, money…) in the context of the wood sellers? Barry Stroud and others anti-relativists claim that trying to understand a Wittgensteinian alien society is almost impossible, but this attitude—as Marconi (1987) suggests—may simply reflect a lack of imagination. Monk (2005: 79) claims that "what is required to free us from the picture that holds us captive is an enriched imagination…" I like the idea and, against Monk's own interpretation of wood sellers, I assume a properly "enriched" imagination would imply to imagine something new, which is not our traditional commercial exchange. We have no need to go very far, and we may be content to think of phenomena like the institution of *Potlach*, a kind of economic system where people give away goods for nothing in exchange.[18] Anticipating Quine's criticism of the idea of Lévi-Bruhl's

[17]It is difficult to understand how much Wittgenstein knew about Relativity theory. Probably his remarks fundamentally concern Special Relativity and Wittgenstein maybe influenced by the idea of preferred global space-time coordinates to which Einstein gave a direct physical and intuitive reading. On this limitation of Special Relativity see Dieks (2018).

[18]This is obviously a very simplified vision of the institution of *potlach*, in which natives of the northwestern coast of America gave away things of great value as gifts. The institution of potlach was prohibited because it was considered irrational. However it was much studied after the work by Durkheim's nephew, Marcel Mauss in his *Essay sur le done* (1925), until it became apparent that the

"primitive logic"—partly depending on an erroneous translation of the speech of the inhabitants of the Trobriand Islands—Wittgenstein remarks that we might have *mistaken* the translation for "selling". If the reactions of the natives seem absolutely contradictory and irrational to us, maybe we did not get what was going on. In his '39 *Lectures on the foundation of Mathematics* Wittgenstein suggests translating "selling" as "giving away", and justifies the wood sellers behaviour by saying that, although "we might call this a kind of logical madness", in fact, "there is nothing wrong with giving wood away" (LFM: 202): we may see no point in it, but we do a great many things without a clear point. Therefore, attribution of insanity is not well placed, and other reasonable explanations of an awkward behaviour relative to a specific community are always possible with enriched imagination: perhaps the wood sellers don't care how much money they possess, but rather about the quality of the interaction between them. And this way of exchanging wood for money may be seen as a good way of not accumulating too much money (something perhaps considered unfair and inelegant). Or the time used to distribute pieces of wood in large areas may be considered a kind of artistic endeavour that is very much appreciated. And so on.

Following the analogy with relativity theory, the application of our concepts to an alien society will produce a kind of regular deformation that has to be taken into account. What is wrong in the application of our concept of "selling" is the inability to detect the deformation that this concept undergoes when applied in a different frame of reference. In order to use our concept of "selling" we have to take into consideration their frame of reference, the surrounding circumstances, institutions, habits, and principles of their form of life. Only by this careful anthropological analysis, we may find a different interpretation of what we call "selling", "value", "more", using our concepts in a new way and at the same time giving a more correct translation of the native lexicon that will make the natives more coherent than it appeared. Reasoning on the influence of Relativity Theory, therefore, converges towards the same conclusion reached by Cerbone: "the initial appearance of error, when pressed, reveals instead a lack of understanding on our part. The problem we confront in thinking about this community is not a problem in their practices, but rather a problem for us in interpreting just what it is they are doing".[19]

Therefore, contrary to what anti-relativists think, alien mental experiments are not necessarily devoted only to making our basic concepts more salient, but also to showing that our basic concepts cannot be used *as such* in distant conceptual

---

economy of *potlach* is of a completely different kind from ours, and its prohibition from Western colonialism was based on misunderstanding (see for instance Bracken 1997). Another example of correcting a bad interpretation derived by not realizing that our concept cannot be applied as such in alien systems comes from Malinowsky's critique of Frazer and Durkheim. They claimed that Australian aborigines just "lacked" an economy, as if they were a primitive stage of human society; contrary to this idea Malinowsky shows that totemism amounts to a "magical economy", very different from ours (e.g. Thornton et al. 2006, p. 45). Wittgenstein's critique of Frazer is probably an application to anthropology of his idea that our normal concept cannot be applied as such to an alien system.

[19]Cerbone (2000).

systems, but require some adjustment[20]: we need to make the right transformations in order to adapt our concepts to the conceptual frameworks of people with different habits. Most basic concepts are family resemlance concepts; we need to make the right transformation to use our concepts in a different frame of reference. What seem to be totally different concepts then will be revealed to be a possible non standard application of the concepts we use in our typical environment. Our concepts of "value" or of "selling" may be reconstructed, with some imagination, in order to be adapted to the particular setting in which the alien system is organised (and the fact that we may describe the economic system of *Potlach* using our concepts is an evidence of this possibility). Here we are not speaking of radically different concepts, as it may happen when facts of nature are different, but we are studying different systems where our concepts can be adapted if we recognize the difference of frameworks. Or maybe Wittgenstein wants to say that what we see as radically different concepts are not so different after all. Maybe alien wood sellers have just "a quite different system of payment from us"(PI: 150).[21]

We could summarize the analogy with time dilation and light contraction in the following claim:

> CONCEPT DEFORMATION – Our concepts are like our clocks and rods. When placed in a reference system of an alien society, *our* concepts – to be properly applied – will be deformed in their system, like clocks and rods in a coordinate system slow down and shrink in relation to another coordinate system. We should therefore understand the new applications of our concepts to compensate the difference uses in these systems of reference, in analogy with the transformations needed to compensate the deformation of clocks and rods in coordinate systems that approximate light speed.

While Einstein worked on measuring tools and Coordinate Systems to compare events and verify when two events can be said to be simultaneous, Wittgenstein worked on "grammatical" judgments and conceptual frameworks, to verify whether two concepts belonging to two different conceptual frameworks can be said to be the same concept. The discussion of the wood sellers seems to be an application of this attitude, which amounts to saying that our concepts of "quantity", "selling" or "value" cannot be applied as they are applied in our frame of reference, but need to be modified up to the point in which they may match the alien frame of reference, permitting a right translation and arriving at understanding a sameness of meaning between our concepts and theirs. Analogously, when the two events in the two different coordinate systems are not simultaneous from the perspective of one of the two coordinate systems, we need the right transformations to check what we may define as simultaneous. Simultaneity is sameness of time like synonymy is sameness

---

[20]We may interpret in this way what Tripodi (2013) claims regarding the point of these experiments: to see *how far our concepts can be stretched*. This is not done without a purpose; we try to stretch them in order to find out how we can understand others.

[21]Marconi (2017: 115) shows that the difference with wood sellers and us may be interpreted both as a semantic difference (we have different concepts) and as difference of facts (we have different habits). The thought experiment may be intended to show something about difference concepts of selling and buying, measuring, "more of" and "less of", etc., but also that for them "sales are not necessarily what we had assumed them to be".

of meaning. To reach sameness of meaning we need to adequate our lexicon (our use of words) to the new frame of reference.

The problem is that, in Relativity theory, transformations are definitely stated by a precise calculus, while it seems impossible to give fixed rules of transformation in Wittgenstein's "deep grammar". Our concepts may be deformed in extremely different reference frameworks like the one of the wood sellers, but it seems we have no precise way to do so, no invariant that helps us in doing the right transformation. Where to find in his view the supposed invariants that contribute to transformation rules allowing to pass from one system to another, and to show the different behaviour of concepts in different coordinate systems to make them comparable?

18.3.2 An Invariant Across Different Conceptual Systems

Abandoning the idea of absolute space and absolute time, Relativity Theory is a search for invariants in our description of physical reality, a unifying way of describing physical events in our space-time continuum through systems of transformations. The problem then becomes, for Wittgenstein, to find invariants that permit the right transformations for using our basic conceptual tools in other ("alien") conceptual systems. But our common judgments ("they sell goods for value") cannot be invariants, given that they are "deformed" in their use in alien systems. And invariants like physical laws or the speed of light would be of no use in the conceptual realm.

Here we should put the quotation on shared human behaviour from PI 206 in focus. Wittgenstein, after having concluded his master argument on rule-following considerations (PI: 201–2), seems to try an application of this argument in treating the problem of understanding an alien society. He suggests looking for the surrounding environment and habits—that is the "shared human behaviour" considered as "*the* system of reference (*Bezugssystem*) by means of which we interpret an unknown language." (PI: 206) Given that this passage follows the conclusion of the discussion on rule-following, and begins with the remark "Following a rule is analogous to obeying an order", we are led to think that rule-following behaviour is a central feature of shared human behaviour. It differentiates humans from other animals, and—besides basic features like walking, chasing, eating, and other usual behaviours—the central feature of shared human behaviour is rule following, which is connected with speech acts. We may thereby understand why Wittgenstein speaks of shared human behaviour, after the argument of rule following, as an answer to the question: "In what circumstances would you say that the people there gave orders, understood them, obeyed them, rebelled against them, and so on?" (PI: 106; cfr. RFM VI: 42) The answer is that the right circumstances are those in which the behaviour shared by the alien community shows coordination of sounds and intentional action, which amounts to basic rule following behaviour.

The thought experiment that follows the remark on shared human behaviour may provide a further suggestion in this direction: what if we met people behaving in an apparently rational way, but with sounds that are totally disconnected from their behaviour? "Are we to say that these people have a language: order, reports and the rest? There is not enough regularity for us to call it 'language'" (PI: 207). The conclusion of this thought experiment seems to claim that we need to recognize some kind of *regularity* in the relations between actions and sounds in order to properly speak of language and language games. Only with the recognition of such regularity, we may say that we have a proper language (and therefore the possibility of translating a conceptual system). The point of the thought experiment of PI 207 also seems a reminder of the difference between *rule-following* and *natural regularity* or *expression of a uniformity*, more widely discussed in RFM (VI: 26, 41–49), as well as a distinction between a process *involving* a rule and a process being *in accordance* with a rule, discussed in BB (p. 10). We may describe natural regularities (e.g. rotation of planets or behaviour of animals, or our neurophysiological structure for preceiving regular patterns), but natural regularity is not rule-following. Rule-following *presupposes* the recognition of uniformity, but we have to distinguish rule-following from the natural expression of uniformity. Rule-following characterizes linguistic behaviour, whose mark is the distinction between correct and incorrect (following and disobeying a rule), which grounds the possibility of making a mistake (PI: 54).

Therefore, even rejecting the idea of a kind of behaviour common to all mankind, and considering the "human common behaviour" of PI 206 as behaviour shared in an alien society, we should always look for something invariant, those aspects of behaviour that permits to distinguish what is considered correct or incorrect. Even relativists, therefore, should accept that there is a universal invariant according to which we have to interpret alien societies: rule-following behaviour, that is the central core of human behaviour that permits the organisation of socially shared activities through the use of language. It is therefore this invariant aspect of human behaviour that permits us to compare different conceptual systems. This does not mean that there are *specific* rules common to all societies, but that the analysis of rule-following behaviour is a universal lead to interpret an alien society and an alien language: comparing alien criteria of correctness with ours, we may begin to understand how our concepts may be bended to be used in different conceptual systems. Understanding rule-following behaviour helps us to adapt *our* concepts to an alien frame of reference, as in Relativity theory we use transformations to adapt our measuring tools to different frames of references. We may summarize these remarks in the following claim:

> CONCEPT ADAPTATION – If we meet people that behave in a way that does not match our normal use of concepts, in order to understand their frame of reference we need to adapt our concepts according to their criteria of correctness, according to the way they follow rules; rule-following behaviour is the invariant feature that permits us to make the right transformations from our system of reference to theirs.

This claim does not imply that we have to *accept* any alien frame of reference: we may understand it as we may understand Aristotle's physics or the Ptolemaic system, without adhering to it. But we also have to accept, conversely, that they may understand us and will not accept our frame of reference. Here it is reasonable to refer to an idea of Wittgenstein, for which, when the chain of reasons comes to an end, only persuasion remains. The famous passage in OC: 612 ("at the end of reasons comes persuasion") continues as: "Think what happens when missionaries convert natives". This comment hints at the idea of a commitment to a coordinate system by a religious believer presented above. The missionary is a religious believer with a strong system of coordinates and the only way to convert an alien native cannot be logical reasoning, but the presentation of his own system of coordinates, through examples, methods, habits, and rules. And it may happen that the missionary changes his mind and accept the native viewpoint.

The problem with the analogy between Wittgenstein's thought and Relativity theory is that in a scientific theory invariants are fixed and clear (physical laws and light speed permit the right transformations). On the contrary, in Wittgenstein's view of philosophy we cannot have any *fixed* rules that apply universally to all different conceptual frameworks. We are left with few hints or examples of variations among different forms of life that all are characterized by rule-following. But rule-following is not a fixed set of rules, but a general ability that may be expressed in many different ways. The degree of indefiniteness and flexibility of rule-following behaviour make it impossible to give a formal method for devising how our concepts have to be modulated to adapt to other conceptual systems. It is therefore not very clear *how* we can use rule-following considerations in specific cases of comparison with an alien society, and we are left with the indication of an attitude.

There is a similarity with Quine's radical translation: a kind of charity principle emerges from Wittgenstein's discussion of the wood sellers. If we cannot give a rational explanation of their rule-following, maybe we made a wrong translation: internal logic or coherence will emerge from the careful description of alien language games. But there are differences: according to Wittgenstein, translation is not just grounded on assent and dissent in the face of empirical observational sentences, but on a wider variety of observations on shared behaviours in the community, on what counts as obeying and disobeying orders, of what counts as accepting or rejecting patterns of behaviour. Besides, far from Quine's attitude of continuity, Wittgenstein poses a strong hiatus between philosophy and science: philosophy should not aim—as science does—at explaining what is hidden, but should restrict itself to describing what is in plain view. In order to better understand the analogy with Relativity Theory we should, therefore, take into account the specific character of Wittgenstein's philosophy, whose central core is not explanation, but description (PI: 109, 140).

18.3.3 Describing Descriptions

A suggestion connecting the concept of description and relativity theory is expressed in a remark written in the latest years (1944–45) that already prompted the attention of many scholars. The passage is worth citing:

> Someone asks me: What is the colour of this flower? I answer: "red".—Are you absolutely sure? Yes, absolutely sure! (…) The certainty with which I call the colour "red" is the rigidity of my measuring-rod, it is the rigidity from which I start. When I give descriptions, *that* is not to be brought into doubt. This simply characterizes what we call describing. (…) Following according to the rule is FUNDAMENTAL to our language game. It characterizes what we call description. This is the similarity of my treatment with relativity theory, that it is so to speak a consideration about the clocks with which we compare events. RFM VI: 28

On first reading, it seems that rule-following characterizes what we call "description" because, in order to describe, we need to follow rules, to have standards, to distinguish measurement devices and the measured objects, and this sounds coherent with the recognition of rule-following as an invariant. The analogy Wittgenstein proposes between his work in philosophy and Einstein's work in physics seems therefore to confirm a standard reading: as Einstein analyses the tools with which we measure events in coordinate systems, and the description of the working of these tools is part of the overall theoretical system, philosophy analyses the tools we use in describing facts, tools that belong to the grammar of description, or "logic" in a most general view. This reading is confirmed by a famous passage of *On Certainty*:

> Everything descriptive *of* a language game is part of logic (OC: 56)[22]

On the other hand, Wittgenstein makes a specification that requires clarification:

> Philosophy is not a description of language use (LW I: 121).

The claim in the second quotation seems in sharp contrast with the first. It might be made coherent with the first, saying that philosophy is never a description of the use of words, but always a description of systems, of language games, in which words get their sense.[23] But this answer is not yet clear enough.

Wittgenstein says that there are many *kinds* of descriptions[24] (PI: 24, 291); in fact, saying that rule-following characterizes what we call "description" we have to distinguish at least two kinds of descriptions, concerning different objects of description:

(1) descriptions of language games

---

[22]And later: "what counts as an adequate test of a statement belongs to logic. It belongs to the description *of* the language game" (OC 82).

[23]In this concern the latest Wittgenstein is no more certain of his grammatical systems of the intermediate period: "What am I after? The fact that the description of the use of a word is the description of a system, or of systems.—But I don't have a definition for what a system is" (RPP I: 294).

[24]In Wittgenstein's work there are many examples of different kinds of descriptions (internal or external, verbal descriptions or conventional descriptions,…) but the main difference seems to rely on the kind of objects being described: geometrical constructions (PR 131, 135), logical forms (PG 40), visual space (PR 87), states of mind (LW 1,20–51), colour patches (PR 213), language uses (LW 121, 294…969), feelings (LW I, 400–1), perceptions (LW I, 540–553, PI II; xi), moods (LW I.614), etc. But what is to be taken as a description depends on the language game in which words are used; e.g.: "are the words 'I am afraid' a description of a state of mind? It depends on the game they are in" (PI II, ix, pp. 187f).

(2) descriptions of alternative descriptions of language games.

In the first, we describe how the same expression may have different roles in different language games.[25] In the second we describe which descriptions are used in describing language games. The first is concerned with the question: "How does a language game work? Which is the logic—the grammar—of a language game?" The second is concerned with the question: "How do we compare different descriptions of a language game?" This last question, for instance, requires to describe the wood sellers' descriptions of their actions, and to compare them to our initial description, checking which kind of presuppositions they reveal.

According to (1), in describing alien language games, we may try to extract their background logic or their reference system, and thereby we should be ready to accept the possibility of alternative logics or alternative worldviews. In this case what Wittgenstein calls "logic" belongs to the language games being described. This sounds not so distant from the perspective held by Mikel Burley. Burley (2012: 8), speaking of "the language game of describing", suggests that a description of an event as a manifestation of Azande witchcraft or in terms of classical mechanics reveals important conceptual differences: "they are all playing the language game of describing or reporting an event, yet they are doing so in a way that evinces important conceptual differences." This would amount to saying that we have different kinds of descriptions of an event, equally relevant and possibly incompatible. This apparently seemingly relativistic trend is compatible with the idea that there *might be* different theoretical systems with logics that would be different from ours,[26] putting some doubts on interpreting Wittgenstein as stating the fact that we need to rely on *our* logic as it has been developed—contingently—as the only logic we have—and we cannot conceive of another one.[27]

However, if we follow the distinction between "describing a language game" and "describing a description of a language game", we may gain some further insight. I was brought to underline this difference by a paper by Heather J. Gert.[28] Think of

---

[25]The debate on the boundary between semantics and pragmatics (from Travis 2000; Recanati 2011; Carston 2013; Unnsteinsson 2015 and others) seems an attempt to define these kinds of systems. Donnellan's distinction between attributive and referential uses may be a paradigmatic application of the kind of work Wittgenstein was aiming to (Penco 2017).

[26]This attitude seems more supported from the kinds of alien mental experiments where Wittgenstein imagines different physical laws. If things disappeared irregularly should we have a different mathematics? For instance, a probabilistic mathematics? Something like Quantum Logic? But also, if things disappeared irregularly, our classical mathematics might help to detect how many things disappeared while calculating.

[27]This is basically the typical anti-relativistic conception from Stroud (1965) to Coliva (2010).

[28]Gert (1997) gives a clear analysis of this attitude: "…in the quotation from *Philosophical Investigations* §24 he [Wittgenstein] does not ask us to describe facial expressions, sensations, or moods, or the use of terms for facial expressions, sensations, or moods. He asks us to compare descriptions of facial expressions, sensations, and moods. And in *Zettel* §204 he does not ask for a description of an attitude, nor does he ask how attitude is used. He asks what a description of an attitude is like. In all of these passages the topics are the descriptions themselves." Besides a criticism of Kripke's view on rule-following, Gert seems to derive conclusions not dissimilar to ours, although given in a different context.

the way Wittgenstein describes the descriptions of a state of mind that characterizes a certain game (LW I: 20–51), or, better, the way in which he compares different descriptions of a perceptual report (LW I: 180). What is Wittgenstein doing in these cases? He is not just describing a phenomenon (a language game). He is describing descriptions of phenomena, descriptions of a state of mind or descriptions of perceptual reports. Philosophy is a description of descriptions. Even the paradigmatic mental experiment of the builders (PI: 2) may be interpreted in this way: the remarks in the *Philosophical Investigations* are not important because they describe what two builders do when passing a brick or a slab. The point of Wittgenstein is not just describing the builders' use of words, but describing different descriptions of the builders' interaction: at the beginning Wittgensein presents the builders' as "a language for which the *description* given by Augustine is right" (PI: 2), but soon he criticises the standard description of the meaning of words as evoking images, as it is supposed to be in the Augustinian picture: "in the language of §3 it is *not* the purpose of the words to evoke images" (PI: 6). He then gives a different description of the linguistic interaction of the builders, as "one of those games by which children learn their native language" (PI: 7), and eventually he put doubts on the possibility that "on can reduce the *description* of the use of the word 'slab' to the statement that this word signifies this object" (PI: 10). This seems to me evidence to claim that, at the beginning of the *Philosophical Investigations*, Wittgenstein is describing and comparing different descriptions of the uses of words.

Therefore, according to (2), we might say that rule-following characterizes the activity of describing descriptions of language games. Wittgenstein often describes everyday descriptions[29] (describing a picture, an impression, a sensation, an economic exchange), but often these descriptions are descriptions of language games, especially when presenting alien thought experiments. We might still claim that, trying to describe alternative descriptions of alien language games, we refer to *our* standards of description: our frame of reference supporting the description is presented as something like a universal logical or grammatical structure, the only one we have. However, it is not always mandatory to interpret Wittgenstein's use of "we", or "us", as referring to Western culture or to logic as it has been developed in Western society. Although sometimes he seems to uses "we" referring to Western culture, it is reasonable to understand some central uses of "we" in his remarks on rule-following as referring to us as humans and not as Western people.[30] Therefore the reference to "shared human behaviour" seems to open the way to considering the activity of describing descriptions as something that can possibly be shared among human beings, although very distant in their customs, like native Australians, wood

[29]Cf. Gert (1997: 226).

[30]This remark does not oblige the reader to accept the interpretation of the Wittgensteinian use of "we" as a *transcendental* "we" intended to supersede the transcendental ego of the *Tractatus* (as in Williams 1974 or Lear 1982). Actually we (scholars) would need some more evidence about Wittgenstein's *different* uses of the term "we" in different contexts, and a search in the electronic data may offer space for this work. I didn't have time to make an extensive search, and I rely on my limited understanding in reading the texts, with the help of Andronico (1998, 2013) who discusses different trends in the Wittgensteinian uses of "we".

sellers and experts in quantum mechanics (although apparently there would be a lot of work on both sides to make the right translation, at least a master in physics and a course in witchcraft). Is there a unique logic behind describing descriptions? Something like a chomskyan innate module that we share as human beings? *Philosophical Investigations* were published a few years before Chomsky's *Syntactic Structures*, and we have no idea how Wittgenstein would have reacted. But it seems apparent from our suggestions that the logic behind describing descriptions is derived by our being compelled to make continuous adjustments of our descriptions, confronting different descriptions of language games. Therefore, we may try to find a logic, or a grammar, which we might share or build together with other human beings (or, in the future, with non-human beings who follow rules).

Which of the two perspectives should we choose? Probably there is no clear answer, as often with Wittgenstein. Concerning the two kinds of descriptions (1) and (2), the point of the former is to recognize that every society (or components of a society) has different language games, and our descriptive work of alien systems consists in finding which deep grammar is lying in their conceptual systems, which concepts perform the role of norms of grammar. The point of the latter is to see what happens when we try to describe not simply the language games of an alien society, but what happens when we try to describe their descriptions of their language games and compare theirs to ours. This project seems to point to a more general universalistic perspective, where we humans share a common ability that guides our translations

In this connection the comparison with Relativity theory makes the work of philosophy analogous to the work of theoretical physics: while theoretical physics describes the laws (the systems of measurement) with which we compare events, applied physics deals with using such laws to devise new applications in other areas of science and technology. Analogously, philosophy describes different possible descriptions of language games, leaving to applied philosophy of language (and linguistics) the work of choosing the right description (the right theory) of how words have different meanings in different contexts.

18.4 Conclusions

This paper puts forth two main claims: first, that we have to distinguish among different kinds of thought experiments in Wittgenstein and, second, that the particular thought experiments concerning people who behave in a different way from us can be better understood on the background of the similarity with Relativity theory that Wittgenstein often advocates. After some reminders on how this similarity has been discussed in the critical literature, I tried to define three possible directions in which this discussion may help to illuminate Wittgenstein's attitude towards "alien" people, who apparently think differently from "us".

Wittgenstein was a philosopher, and the similarity between his philosophy and Relativity theory cannot be strict: when Einstein speak of transformations guiding us in comparing events in different reference frameworks he can guide us with specific mathematical formula; on the contrary, Wittgenstein's idea of comparing concepts in different conceptual frameworks cannot afford precise instructions, but only a very general method: describing descriptions, leaving to applied philosophy to define more specific strategies to describe the working of language in different contexts, be they the most usual ones, or very far from the usual surroundings.

References

Andronico, M. 1998. *Antropologia e metodo morfologico.* Napoli: Istituto Italiano per gli studi filosofici.

Andronico, M. 2013. "Ethics and Aesthetics are one: How to escape the Myth of the Ordinary." In *The darkness of this time*, edited by L. Perissinotto, 23–37. Milano: Mimesis.

Anscombe, G. E. M. 1976. "The Question of Linguistic Idealism." *Acta Philosophica Fennica*, 188–215.

Baker, G., and P. M. S. Hacker. 2009. *Wittgenstein - Rules, Grammar and Necessity. Philosophical Books*. Volume 2 of an *Analytical Commentary on the Philosophical Investigations, Essays and Exegesis* 185–242. New York: John Wiley & Sons.

Blackburn, P. 2004. "Relativism and the Abolition of the Other." *International Journal of Philosophical Studies,* no. 12: 245–58.

Boghossian, P. 2006. *Fear of Knowledge. Against Relativism and Constructivism.* Oxford: Oxford University Press.

Bracken, C. 1997. *The Potlatch Papers: A Colonial Case History*. Chicago: University of Chicago Press.

Brendel, E. 2004. "Intuition Pumps and the Proper Use of Thought Experiments." *Dialectica* 58 (1): 89–108.

Brenner, W.H. 2003. "Dialogues on Causality and the Limits of Empiricism Philosophical Investigations." *Philosophical Investigations* 26: 1.23.

Brown, J. R. 1986. "Thought Experiments since the Scientific Revolution." *International Studies in the Philosophy of Science* 1: 1–15.

Broyles, J.E. 1974. "An Observation on Wittgenstein's Use of Fantasy." *Metaphilosophy* 5: 291–97.

Burley, M. 2012. "Mounce and Winch on Understanding (or Not Understanding) an Indigenous Society." *Philosophical Investigations* 35 (3–4): 350–72. https://doi.org/10.1111/j.1467-9205.2012.01473.x.

Carston, R. 2013. "Word Meaning, What is Said and Explicature." In F. Domaneschi, C. Penco, What is Said and What is not. Stanford: CSLI.

Cerbone, D.R. 2000. "How To Do Things With Wood Wittgenstein, Frege and the Problem of Illogical Thought." In *The New Wittgenstein*, edited by Alice Crary and Rupert J. Read, 293–314. London: Routledge.

Coliva, A. 2010. "Was Wittgenstein an Epistemic Relativist?" *Philosophical Investigations*. https://doi.org/10.1111/j.1467-9205.2009.01394.x.

Coliva, A. 2016. "Which Hinge Epistemology?" *International Journal for the study of scepticism* 6 (2–3): 69-76.

Conant, J. 1991. "In Search for a Logically Alien Thought: Descartes, Kant, Frege, and the Tractatus." *Philosophical Topic* 20: 115–80.

Dieks, D. 2018. "Time Coordinates and Clocks: Einstein's Struggle".

Einstein, A. 1916. *Uber Die Spezielle Und Allgemeine Relativitätstheorie, Engl. Transl. Relativity; the Special and the General Theory*. New York: Random House 1961.

Einstein, A. 1919, "Induction and Deduction in Physics." *Berliner TagenBlatt,* December, 25, 1919.

Einstein, A. 2002. *The Collected Papers of Albert Einstein*. Princeton: Princeton University Press.

Elton, L. 1986. "Einstein, General Relativity, and the German Press: 1919-1920." *Isis* 77 (1): 95–103.
Erbacher, C. 2014. "Editorial Approaches to Wittgenstein's Nachlass: Towards a Historical Appreciation." *Philosophical Investigations*. https://doi.org/10.1111/phin.12064.
Frege, G. 1893. Grundgestze of Arithmetik. Vol.1. Jena: Pohle . Engl. Transl. G., Ebert, P. A., & Cook, R. T. (2013). Gottlob Frege: Basic laws of arithmetic (Vol. 1). Oxford University Press (UK).
Gert, H. J. 1997. "Wittgenstein on Description." *Philosophical Studies*. https://doi.org/10.1023/A:1004276819300.
Mion, G. 2019. On Certainty: Wittgenstein and Einstein. Philosophical Investigations 42 (2):163–170.
Glock, H.J. 1996. *A Wittgenstein Dictionary*. Oxford: Blackwell.
Glock, H. J. 2007. "Relativism, Commensurability and Translatability." *Ratio* 20: 377–402.
Grayling, A.C. 2001. "Wittgenstein on Scepticism and Certainty." In *Wittgenstein, A Critical Reader*, edited by H.J. Glock. Oxford: Blackwell.
Hacke, r P.M.S. 1996. "On Davidson's Idea of a Conceptual Scheme." *Philosophical Quarterly* 46: 289–307.
Hintikka, J., and M. B. Hintikka. 1986. *Investigating Wittgenstein*. Oxford: Blakwell.
Horowitz, T., and Massey, G. 1991. Thought experiments in science and philosophy. Philosophy and Phenomenological Research 54 (2):479–482.
Horwich, P. 2010. "Regularities, Rules, Meanings, Truth Conditions and Epistemic Norms." In *Wittgenstein: Mind, Meaning and Metaphilosophy*, edited by P. Frascolla, D. Marconi, and A. Voltolini, 113–39. London: Palgrave-McMillan.
Hutchinson, P., Rupert R., and W.Sharrock. 2008. *There Is No Such Thing as a Social Science*. Aldershot: Ashgate.
Ierodiakonou, K., and S. Roux, eds. 2011. *Thought Experiments in Methodological and Historical ContextsIerodiakonou K., Roux S.* Leiden: Brill.
Kusch, M. 2011. "Wittgenstein and Einstein's Clocks." In *Unsocial Sociabilities: Wittgenstein's Sources*, edited by E. Ramharter, 203–18. Berlin: Parerga.
Kusch, M. 2013. "Annalisa Coliva on Wittgenstein and Epistemic Relativism." *Philosophia,* 41: 37–49.
Kusch, M. 2015. "A Branch of Human Natural History: Wittgenstein's Reflection on Metrology." In O. Schlaudt, and L. Huber (eds.) *Standardization in Measurement: Philosophical, Historical and Sociological Issues*, 11–24. London: Routledge.
Kuusela, O. 2017. "Wittgenstein, Ethics and Philosophical Clarification." In *Wittgenstein's Moral Though*, edited by R Agam-Segal and E. Dain, 37–65. London: Routledge.
Lear, J. 1982. Leaving the world alone. The Journal of Philosophy, 79(7), 382–403.
Lear, J. 1986. "Transcendental Anthropology." In *Subject, Thought and Context*, edited by Philip Pettit and John McDowell, 267–98.
Mach E. 1905. "Über Gedankenexperimente." In *Erkenntnis Und Irrtum,* 181–97. Leipzig: Verlag von Johann Ambrosius Barth. Translated by J. McCormack, in *Knowledge and Error*, 134-147. Dordrecht: Reidel.
Marconi, D. 1987. *L'eredità di Wittgenstein*. Rome: Laterza.
Marconi, D. 2015. "Concepts: too heavy a burden." In *Wittgenstein: Mind, Meaning and Metaphilosophy*, edited by P. Frascolla, D. Marconi & A. Voltolini, 140–166. London: Palgrave Macmillan.
Marconi, D. 2017. "Philosophical Thought Experiments: The Case for Engel." *Philosophia Scientiae*, 21 (3): 111–124.
Monk, R. 2005. *How to Read Wittgenstein*. London: Granta.
Mounce, H. 2012. "Response to Mikel Burley." *Philosophical Investigations* 35 (3/4): 373–76. https://doi.org/10.1111/j.1467-9205.2012.01476.x.
Penco, C. 2010. "The Influence of Einstein on Wittgenstein's Philosophy." *Philosophical Investigations* 33 (4): 360–79.

Penco, C. 2017. "Donnellan's Misdescriptions and Loose Talk." In *Reference and Representation in Language and Thought*, edited by Kepa Korta and Maria De Ponte, 104–25. Oxford: Oxford University Press.

Phillips, D. n.d. *Wittgenstein and Scientific Knowledge*. London: Macmillan.

Picardi, E. 2009. "Concepts and Primitive Language Games." In *Wittgenstein's Enduring Argument*, edited by E. Zamuner and D. Levy, 109–34. London: Routledge.

Pritchard, D. 2018. "Wittgensteinian Hinge Epistemology and Deep Disagreement." *Topoi*, 1–9.

Putnam, H. n.d. "Rethinking Mathematical Necessity." In *H. Putnam, Words and Life*, edited by James Conant, 245–63. Harvard University Press.

Recanati, F. 2011. *Truth Conditional Pragmatics*. Oxford: Oxford University Press.

Rorty, R. 1979. *Philosophy and the Mirror of Nature*. Princeton: Princeton University Press.

Savigny, E. von. 1991. "Common Behaviour of Mankind: Philosophical Investigations Section 206." In *Wittgenstein's Philosophical Investigations: Text and Context,* edited by R.L. Arrington and H. J. Glock, 105–19. London: Routledge.

Stroud, B. 1965. "Wittgenstein and Logical Necessity." *Philosophical Review* 68: 504–18.

Thornton, R.J., P. Skalnik, and L. Kryzanowski. 2006. *The Early Writings of Bronislaw Malinowsk*. Cambridge: Cambridge University Press.

Travis, C. 2000. *Unshadowed Thought*. Cambridge (Mass): Harvard University Press.

Tripodi, P. 2013. "Necessity, Imagination and Meta-Philosophy." *Philosophical Inquiries* 1: 55–78.

Unnsteinsson, E. G. 2015. "Wittgenstein as a Gricean Intentionalist." *British Journal for the History of Philosophy* 24 (1): 155–72. https://doi.org/10.1080/09608788.2015.1047735.

Waismann, F. 1965. The Principles of Linguistic Philosophy. London: Macmillan.

Williams, B. 1974. "Wittgenstein and Idealism." In *Understanding Wittgenstein*, edited by G. Vesey, xxii, 285. London: Macmillan.

Williams, M. 2007. "Why (Wittgensteinian) Contextualism Is Not Relativism." *Episteme* 4 (01): 93–114. https://doi.org/10.3366/epi.2007.4.1.93.

Williamson, T. 2007. *The Philosophy of Philosophy*. Malden: Blackwell.

Winch, P. 1976. "Language, Belief and Relativism." In *Contemporary British Philosophy*, Edited by H.D. Lewis, 322–37. London: Allen & Unwin.

Winch, P. 2014. "Understanding a Primitive Society." *American Philosophical Quarterly* 1 (4): 307–24.

Wright, C. 2002. "What Is Wittgenstein's Point in the Rule-Following Discussion?" Manuscript for Boghossian/Horwich Language and Mind Seminar. New York.

Wright, C. 2009. "Rule-Following without Reasons: Wittgenstein's Quietism and the Constitutive Question." In *Wittgenstein and Reason*, 20:123–44. John Wiley & Sons, Ltd (10.1111). https://doi.org/10.1002/9781444307092.ch7.

Chapter 19
Wittgenstein's Solution to Einstein's Problem: Calibration Across Systems

Susan Edwards-Mckie

19.1 Introduction

Firstly, I will share with you significant *Nachlass* discoveries that I have made over the past few years which reconstruct an early version of the *Philosophical Investigations* (*Philosophische Untersuchungen*). It is the version for which the Wittgenstein scholarly community has been searching as the missing text promised to the Cambridge University Press in 1938 with the projected title *Philosophische Bemerkungen*, and for which the translation that Rhees and Wittgenstein worked on together was made. It is relevant as it is crucially important for us as we consider Wittgenstein as a philosopher of mathematics and physics because this early version spans the well-known TS 220 which constitutes the first part of the published *Philosophical Investigations*, Remarks 1–189 *and* the mathematical section of TS 221, which was later published (Wittgenstein 1956) as the first section of *Remarks on the Foundations of Mathematics*.

In this version we have a seamless numbering across these texts, 1–192 in the first TS and 193-316 in the latter. I shall call this the Hidden Revision as it was literally partially erased and hidden within the later TSS 239, 238 and 237. In addition, cross-referencing of numbers which Wittgenstein promised in the *Vorwor*t to his projected 1938 book, I have been able to reconstruct from the renumberings of the Hidden Revision. These focus on central contemporaneous issues in the philosophy of physics.

Four forms of triangulation became significant. Firstly, the Hidden Revision provided the projected title *Philosophische Bemerkungen* rather than *Philosophische Untersuchungen*, which became fully visible through colour filters and enhance-

I have written elsewhere about similarities between aspects of Wittgenstein and Einstein. In this paper I will explore some of their differences.

S. Edwards-Mckie (✉)
Darwin College, Cambridge, UK
e-mail: susan.edwards-mckie@cantab.net

S. Wuppuluri and N. da Costa (eds.), *WITTGENSTEINIAN (adj.)*,
The Frontiers Collection, https://doi.org/10.1007/978-3-030-27569-3_19

ment. Secondly, when reconstructed it is shown to correlate faithfully with the Rhees translation and numbering. Thirdly, as the consecutive numbering has been established across all of TS 220 before the cuts for TS 239 occur and are visible in internal numberings in both TS 220 and the Rhees translation, the Hidden Revision and the TS 239 project become indisputably distinct with the latter dated 1944, for which the Smythies translation provides corroboration. Fourthly, the Hidden Revision's cross-referencing of remarks provides a meaningful cluster of ideas which are of philosophical significance for understanding Wittgensteinian mathematics and cosmology. Also, Fragment 178e, which I have elsewhere called the Cosmic Fragment, after Heraclitus, is used to explore these ideas.

Particularly significant key concepts fall along lines of a developed action-at-a-distance which operates in proof, the internal following of a rule, infinite divisibility and the implications which I will suggest could provide insights into the working at quantum and nano levels.

19.2 Placing of the Hidden Revision Mathematically and Cosmologically

We have a strong nexus of intra-related Wittgenstein texts during Turing's publications on computable numbers and Goedel's Incompleteness proofs. Wittgenstein's unrelenting criticism of Cantor, Euler and Goedel falls within a larger strategy to disarm a philosophy of mathematics which relies on completed infinite sets. Because transfinite numbers are seen to resolve the Zeno paradoxes of motion, Wittgenstein's philosophy of mathematics began to be seen as backward looking, particularly in the period of Turing's work (Turing 1936) on the *Entscheidungsproblem*. However, Wittgenstein offered consistent criticism and alternative approaches to paradoxes of the infinitely large and small through a consistent systems approach to space, time and generality.

The logical and mathematical material which forms the later published *Philosophische Bemerkungen* (*Philosophical Remarks* 1975) and *Philosophische Grammatik* (Philosophical Grammar 1974) had been written by the 1936/1937 period. The Big TS, 213, with its correlative projects TSS 208–212, had been completed by 1933, including the substance of the mathematically important short essay "Unendliche Möglichkeit" (TS 215). The C-series notebooks (MSS 145–152) are composed 1933–36, with topics *of* space, infinite division, intervals, mathematical aspect perception in constructive proofs and constructions of polygons addressed. The first half of 157a is composed in 1934, with the remainder of 157a and 157b written in Skjolden in the autumn and winter of 1937. This small pocket notebook 157a has passages about action-at-a-distance, and Wittgenstein stated that it is wrong to conceptualise how things work as "push and pull".

During the 1937–38 period and further into the lecture series, Wittgenstein continued to think about infinity, space and time, as evidenced in the catalogued entries of his Whewell's Court short lectures: "Achilles and the Tortoise" and "Absolutely Determinate". MS 121, begun 26.4.1938 is a critique of the diagonal method as a platonising of Cantor's transfinite numbers, with Wittgenstein at 41verso, 12.7.1938 voicing the absurdity that one infinity could be greater than another.

Furthermore, the great distinction between experiment and proof, which he so thoroughly explored in the Vienna Circle years, is firmly present in these Whewell's Court lectures, which began in Lent term 1938 and continued while he held the chair of Philosophy at Cambridge until 1941. At one level, Einstein's move from his earlier empirical position to that of a realism in which, as Einstein writes to Ehrenfest, 5 January 1916, space and time "thereby lose the last vestige of physical reality" (Einstein et al. 1998) has a related taxonomic differentiation to Wittgenstein's differentiation between experiment and proof. However, Wittgenstein's systemic position, while not ruling out realism, does not rely on it. In addition, Wittgensteinian systemic time, with an internalisation of the conception of time, much like an internal rule distinguished his position from what remained for Einstein, even in general relativity, an external point of view.

The Second Conference of the Epistemology of the Exact Sciences, jointly sponsored by the Vienna Circle and Berlin Circle, held at Koenigsberg in 1930 witnessed the important papers of Goedel, Von Neumann on Hilbert formalism (Von Neumann 1930), Heyting on intuitionism, Carnap on logicism, and Waismann's presentation "The Nature of Mathematics: Wittgenstein's Standpoint", the latter an account which allows Wittgenstein a developing mathematics which distinguishes between set theoretic *totalities* and Wittgensteinian *systems*. See Waismann (1979).

A similar congress in Copenhagen in 1936 focussed on causality and quantum physics. Wittgenstein's thinking and writing of the early 1936–38 version of the *Philosophical Investigations*, MS 142 and the transition to TS 221, was influenced by concerns to counter Turing's set-theoretic arguments. Indeed, the references in MS 117, which von Wright initially interpreted to be to TS 221, actually looked back to the earlier Big TS, with extensive hand-written remarks in the margins and on the verso pages, asking what it is to think and if there could be an artificial thinking, Can a machine think?

These questions are intimately tied to conceptions of set theory and the actual infinite as opposed to a type of constructivist mathematics and the potential infinite, which Wittgenstein supported, and for which I have argued elsewhere (Edwards-McKie 2012, 2014a, b, 2015a, b, 2017, 2018). With the Nachlass discovery of the Hidden Revision we find another important theme which parallels the philosophy of physics, this time using his systemic argument to counter Einstein causality through Wittgensteinian systems.

Nedo (1993) suggests a Machian connection in Wittgenstein's early work of the Vienna Circle years, thus we perceive a pattern of holism, a feature which Einstein embraced from Mach's system and which Wittgenstein very much endorsed. There is both no preferential frame of reference in general relativity and all frames of reference are, by definition, referencing all others. As Schlick points out: "…if we wish to maintain the postulate of general relativity, we must refrain from describing metrical and positional relation of bodies by Euclidean methods. This does not mean that in place of Euclidean geometry we are now to use some other definite geometry, such as that of Lobatschewsky or Riemann, for the whole of space, but that all types of measure-determination are to be used: in general, a different sort at every place" (Schlick 1920: 48).

However, during the 1936–38 period the way in which holism *works*, whether it is envisaged as causal, as Einstein, Podolsky and Rosen held in the EPR Paradox of Einstein et al. (1935) or systemically as Wittgenstein held, was the focus of the research community and remains an entrenched research issue in contemporary physics and cosmology. In the Whewell's Court lectures, Wittgenstein addressed the issue of causality (Munz 2010), criticising Russell's views, and when talking about the two seeds in his lecture of 8.5.1938, as recorded by Rhees, Smythies, Taylor and other students of these lectures, Wittgenstein brings in the concepts of action-at-a-distance, indeterminacy and indeterminism as that which shocked and *revolutionised* science (Klagge and Nordmann 1993: 411; Munz and Ritter 2017: 18; Taylor 1938: 24–25).

19.3 The Opportunity of Unending Division and the Idea of Infinity as a Property of Space

Wittgenstein was always fascinated with aspects of divisibility. Before looking closely at the Cosmic Fragment, I want to look at the earlier essay "*Unendliche Moeglichkeit*". This essay successfully draws together several mathematical issues. Firstly, space is characterised not as itself extended, but that "*Der Raum gibt der Wirklichkeit eine unendliche Gelegenheit der/ Teilung*" (TS 215: 19). Secondly, everything must already be contained in the sign/1, x, x + 1/. The unending possibility is not introduced by a mythical element in logic or grammar. "*Es muss also in dem Zeichen/1, x, x + 1/"—dem Ausdruck der Bildungsregel-/schon alles enthalten sein. Ich darf mit der unendlichen Möglichkeit nicht/ wieder ein mythisches Element in die Logik//Grammatik//einführen*" (TS 215: 20).

The idea of infinity as a property of space is rehearsed in one of the C Series notebooks, MS 149, written in 1935–36, and acts as one of the precursor drawings to our 1937 Fragment drawings of the constructions of series (Fig. 19.1):

Given Wittgenstein's conception of space, this conception places infinite divisibility as an *internal* property of space, aligned to an internal rule—not an external rule which produces predetermined outcomes in a mechanical or causally determined way. Within the 1937 drawings of the Cosmic Fragment, the machine-as-symbol-of-its-own functioning has developed as a system which can be compared with the construction of an infinite series because the *logischen Zwang* is now emphasised to be essentially connected with unending possibility. The inevitability of "*eine zwangslaeufige Fuehrung*" (MS 157a: 39) is clearly seen as a hardness which is to be confined to a poorly conceived mysterious mental mechanism or a rigid machine part in time, unlike proof which is not time bound. And relatedly, in the first section of

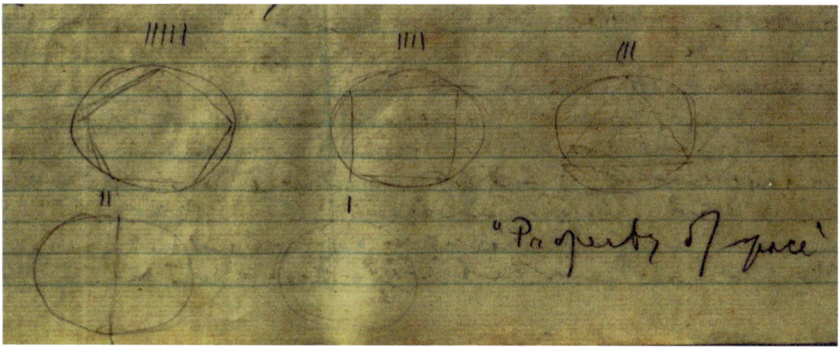

Fig. 19.1 MS 149: 47 verso: property of space

MS 117 which forms part of the Hidden Revision the internal properties of a series, alternative divisions of groups, mathematical aspect perception and the Einsteinian remark "How we measure is what we measure" are present.

19.4 Cosmic Fragment: Placing of the Fragment Physically and Topologically

This particular fragment of notes is unnumbered, written quickly in a particularly loose hand, the content of which sprawls across four pages of a large English diary volume, printed with the dates August 1937, 5–18. The notes and several drawings cover the pages as if the dates are not there. The entries, both at beginning and end, show that there was more text either side of the Fragment, thus is has been torn away and saved.

It is accepted in the literature that MS 142 is crucial as source in the genesis of *PI* 1-188 and 189a. Relatedly, these *PI* remarks are the locus of the widely conjectured connection of TSS 220 and 221 (Baker and Hacker 2005; McGuinness 2001/2002; Pichler 2004; Schulte 2001). Therefore, with its text spanning this juncture, this Fragment is positioned pivotally.

The Fragment's key bridging concepts of *Härte des logischen Zwangs*, *unendliche Reihe*, *Maschine als Symbol ihrer Funktion* and his question "*Aber sind denn die Übergänge also durch die Algebraische Formel bestimmt*?" function pivotally to allow a consistent and developing philosophy of mathematics for Wittgenstein. Significantly, the importance of these concepts within Wittgenstein's thought, paves the way for the radical reordering of material in the *Zwischenfassung*, the Intermediate version of *Philosophical Investigations*. A deep level consideration which I hope will filter through in this paper is that the *Philosophical Investigations*, particularly when viewed through the Hidden Revision, has far more to say about mathematics and physics than hitherto thought.

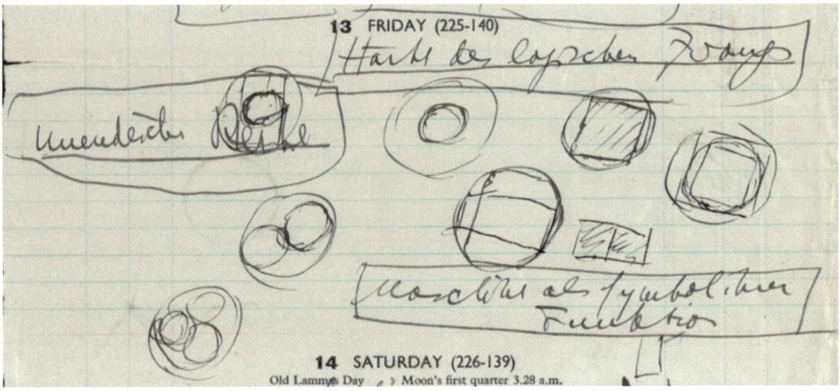

Fig. 19.2 Fragment 178e, page 3: cosmic counting

And cosmologically, the diary entry of 24.2.37 resonates with the mathematical conception of the construction of an infinite series: "It is strange that one says God created the world and not: God is creating, continually, the world" (Fig. 19.2).

The *Maschine als Symbol ihrer Funktion* is conceptually important, because once the symbol can in some sense offer infinite possibility, the idea of infinite-movement-as-symbolised also becomes infinite constructive possibility. Infinite construction symbolised as the machine-as-symbol-of-its-own-function is very like the construction of the infinite series, which are two bridge concepts of this important Fragment.

19.5 Eine Allgemeine Geometrik: The Internal Rule in Iteration and Calibration

Relatedly, we should consider the geometric construction of the triangle which is not the construction of merely a particular triangle. This is why it does not matter to *Verstehen* if the triangle is drawn imperfectly; indeed it is unclear what a perfectly drawn triangle would be or what its use might be. This is important and relevant today because within the machine learning research there is a recalcitrant problem with the gap between the general and the particular. Within Wittgensteinian mathematics the construction is a general proof of triangles. As early as MS 108 Wittgenstein had conceptualised "*eine allgemeine Geometrik*" (MS 108: 117), with space as the possibility of movement (PR: 140). And to bring this idea back to divisibility, we can view constructive divisibility as: The possibility of application projecting/ shining through space: "*Die Anwendungsmöglichkeit strahlt durch den Raum...*" (MS 108: 63).

With the pivotal strength of the combined concepts of the Fragment, by MS 126 which is a highly mathematical notebook, construction becomes proof of the infinite in virtue of its form. This resonates with the early work begun in the 1929 article,

"Some Remarks on Logical Form", in which two values of a continuous spectra being in the same logical space is precluded. The idea that two people cannot occupy the same chair is the homespun example Wittgenstein provides to the theoretical physics conception that time, space, temperature, colour, tone all admit of *unendliche* divisibility while retaining a sense of integrity of value within interval considerations. It is not the clear metric of point coincidence, but these are precursor ideas to entanglement as an indeterminate *system*. This does not mean that is merely a random walk.

In MS 117: 97–98, which is a critique of Cantor, no "*allgemeinen Begriff*" of an unending quantity/set, *Menge*, is appealed to. In other words, there is no need for an infinitely large (or small) *number* to exist for a perfectly good conception of the infinite to be possible. These considerations show that for Wittgenstein mathematics and use as unending possibility of application are conceptually seamless, and in the cross-referencing of the Hidden Revision action-at-a-distance is linked and explained through application of a rule. MS 121 focuses on infinity through exploration of the concept of *Entwicklung* of a series as an internal rule-bound technique, with this technique able to demarcate various systems of *Entwicklungen*.

19.6 Wittgenstein's Fascination with the Logical Properties of Measurement: The Interval Versus the Point; The Heap Versus the Set

As Turing's position moved towards defining a thinking machine as a learning machine (Shanker 1987), I suggest we are able to understand the shift to rule-following and games as a *mathematical response*, in its elucidation of the internal rule. These considerations, of the internal rule, are very important in considerations of iteration and calibration.

In the early notebooks after Wittgenstein's so-called 'return' to philosophy, there is a pronounced interest in intervals, in many types of division, in the characterisation of pi and other irrationals as in different systems in that they are not merely extensions of the rationals. Yet it is clear that the various number systems are tangentially related, in this motley of mathematics, as Wittgenstein called it. The Cosmic Fragment (Edwards-McKie 2015a, b) is the pictorial version of this important idea. This picture is a stage further than entanglement simpliciter because there is calibration across systems rather than just particles.

Within the clustered drawing there is a series of the natural numbers represented by the smaller circles within a larger circle 1, 2, 3, presumably as an infinite series, or as a finite part of an infinite series which is *seen as* part of a continuing series, taking into consideration the prose signs which inform this drawing. The next level of reading this diagram is more interesting. The circle with one smaller circle only which represents 1 in the first series acts pivotally to unlock other possible series.

We could construct 2 in a variety of ways in other series, moving from 1 to the left we have a small circle inside a rectangle inside the larger circle. This is a perfectly good number 2. It could also be the start of an infinite series itself with the rectangle acting as 1, the small circle as 2, ad infinitum, through infinite division, while the first series begun at our first 1 could continue on its path. This would represent an embedding of one series in another, according to which aspectival construction we use.

Another group within this drawing are the circles with squares/rectangles (above "*Maschine als Symbol ihrer Funktion*"). Here we have number 1 represented by 1 large square in a circle at the apex of a triangular positioning, with number 2 represented by 2 rectangles in a circle at the left base position and then 3 represented by a square with a circle inside it with a square inside it, all within a large circle at the right base position. This last more intricate representation could also be an infinite series of its own, which is embedded in the original series. There are several other paths we could construct in this analysis, or we could use modular arithmetics to increase the diversity and flexibility of these systems of numbers. As Wittgenstein states: "Systems are certainly not all in *one* space…" (Wittgenstein 1975, PR: 162).

19.7 Conclusion

There is a growing tendency, at least amongst a few people, to conceptualise the infinite as that with no defined least upper limit and no defined lower bound. I have been pleased to be informed of this fact by the cosmologist Arthur Gibson, as this is how I have conceptualised the infinite and it is certainly in line with Wittgenstein's conceptualisation. So there are a few more people, both living and dead, who join this small group. I was loath to begin my paper with such a conception, lest it be truly unruly, but I encountered exactly the same idea at a conference in Krakow at which I was invited to speak. The context then was Wittgenstein's early, unpublished first version of his symbol for the general proposition. This was changed by the time of publication of the *Tractatus* of 1921, but it is a clear record of a deep problem and a deep thought, during the years of Einstein's move from special to general relativity.

The problem with such a picture is that, initially, we appear to have just one big universe, with no window into its functioning or parts, a heap from which no single atom can be selected. However, we know that Wittgenstein was comfortable with the heap. In a similar vein, he was comfortable with the path or "*eine Weg*", placing emphasis on generality as a direction, on the infinite as a direction, in such a way that the refined conception of generality is entwined with the potential infinite and the operation rather than the function: "Generality in mathematics is a direction, an arrow pointing along a series generated by an operation…It is possible to speak of things which lie in the direction of the arrow but nonsense to speak of all possible positions for things lying in the direction of the arrow as an equivalent for this direction itself" (PR: 142). This moves generality firmly away from the disjunctive set interpretation which Wittgenstein had held earlier.

For our purposes today, the differences between classical and quantum computing is relevant, with the classical computational programme unable to create the type of memory which would allow the conception of a direction in quite the way Wittgenstein envisaged. It is interesting that Wittgenstein spoke on the topic of infinity at the Meeting of the Aristotelian Society in which he had planned to use "Some Remarks on Logical Form", as this article sets the stage for the later entanglement ideas of the Hidden Revision, the Whewell's Court lectures, and many of the passages within the *Nachlass* that we are now reading with greater understanding.

19.8 An Overlay of Intervals, Each with the Unending Possibility of Division

We could almost say that, counterintuitively, we cannot understand directionality without indeterminacy and indeterminism. What this opens up for us is that calibration across systems rather than a geometry of point-coincidences is a real possibility in thinking about how the universe works. Wittgenstein's system is never point based. Rather it is interval based, with each interval both a system unto itself and calibrated across other systems. But since number cannot be thought of as equinumerity, according to Wittgenstein, calibration does not work Russellian style with number/classes or act in any sort of bijective correspondence. Rather, the systems 'hook' onto one another through some form of similarity, family resemblance. This really is a picture worth exploring in our understanding of construction at the nano level.

That the machine-as-symbol remains within Wittgenstein's way of thinking is evidenced by remarks in MS 130 May-August 1946 in which he talks of language as a machine, and the proposition as the machine-part. The proposition is at the level of constructed symbolic meaning and value. And I suppose that if we thought of language as the universe, bounded but unbounded, iterative but creative, then a universal language machine would be just the oscillating-universe-as-information that was proposed in the late 20th century and developed in 21st century theoretical physics and cosmology.

Acknowledgements I wish to thank the Master and Fellows of Trinity College, Cambridge for their kind permission to reproduce images from the Wittgenstein collection. This paper was written as a companion paper to three others: "Wittgenstein's Wager: mathematics, culture and human action", lecture and publication on decision models in our construction of cultures, with considerations on justice and public policy, Krakow, Poland, 2014, 2017; "Wittgenstein's Wager: Quantum Decision Theory Revisited", invited contribution on decision modelling for high risk systems, based on Wittgensteinian philosophy of science and mathematics, Austrian Defence Academy, Vienna, 2015; "On Infinity: in-potentia", invited paper International Wittgenstein Symposium "Philosophy of Logic and Mathematics", Kirchberg, Austria, 2018.

References

Baker, G.P & Hacker, P.M.S. 2005: Wittgenstein. *Understanding and Meaning: Volume 1 of an Analytical Commentary on the Philosophical Investigations*, Part II, 2nd Edition, Oxford: Blackwell.

Edwards-McKie, S. 2012: Some remarks arising from Hacking's 'Leibniz and Descartes: Proof and Eternal Truths', *Contributions to the 34th Wittgenstein Symposium*, Kirchberg, Austria.

Edwards-McKie, S. 2014a: Looking through a microscope. *Philosophische Bemerkungen* 1938: the hidden revision which fits the 1938 Vorwort, the Rhees translation and what this tells us about Wittgenstein's philosophy of mathematics and cosmology", *Contributions to the 36th Wittgenstein Symposium*, Kircherg, Austria.

Edwards-McKie, S. 2014b Wittgenstein's Solution to Einstein's Problem: Calibration across systems, an inaugural lecture HAPP, University of Oxford. Available St Cross College, Oxford website and WAB, University of Bergen website.

Edwards-McKie, S. 2015a: Quantum Decision Theory Revisited; paper presented Austrian National Defence Academy, Vienna; forthcoming 2019.

Edwards-McKie, S. 2015b: The Cosmic Fragment: *Härte des Logischen Zwangs* and *Unendliche Möglichkeit. Nachlass* discoveries and Wittgenstein's conception of generality and the infinite, *Wittgenstein Studien*, Band 6, 51–81.

Edwards-McKie, S. 2017: Wittgenstein's Wager: Mathematics, Culture and Human Action, in: *Wittgenstein, Philosopher of Cultures*, ed. Carl Humphries & Walter Schweidler. Sank Augustin, Germany: Academia Verlag; paper presented symposium *Wittgenstein Philosopher of Cultures*, Krakow, Poland.

Edwards-McKie, S. 2018: On Infinity: in-potentia, in *Philosophy of Logic and Mathematics*, eds. Gabriele M. Mras, Paul Weingartner & Bernhard Ritter, Berlin: Walter de Gruyter (forthcoming).

Einstein A, Podolsky B, Rosen N., 1935: Can Quantum-Mechanical Description of Physical Reality Be Considered Complete?, *Physics. Review* 47(10): 777–780.

Einstein A, Podolsky B, Rosen N., 1998: Collected Papers of Albert Einstein, eds. Robert Schulmann, A.J. Kox, Michael Janssen and Jozsef Illy, Princeton, New Jersey: Princeton University Press, Vol. 8.

McGuinness, B., 2002: Manuscripts and works of the 1930s, in: *Approaches to Wittgenstein*, London, New York: Routledge, 270–286; first published 2001, in: *Proceedings of the Logopesole Workshop October* 2000, ed. Pasquale Frascolla, Annali della Facolta di Lettere e Filosofia dell Universita della Basilicata, Potenza.

Munz, Volker A., 2010 : The Whewell Court Lectures: A Sketch of a Project, in: *Wittgenstein after his Nachlass*, ed. Nuno Venturinha, Basingstoke: Palgrave Macmillan.

Nedo, Michael, 1993: *Wiener Ausgabe Einführung* , Wien, New York: Springer-Verlag.

Pichler, Alois, 2004: Wittgensteins Philosophische Untersuchungen: Vom Buch Zum Albumn, Amsterdam and New York: Brill/Editions Rodopi.

Schlick, Moritz, 1920: *Space and Time in Contemporary Physics: An introduction to The Theory of Relativity and Gravitation*, Oxford: Oxford University Press.

Shanker, Stuart, 1987: Wittgenstein versus Turing on the Nature of Church's Thesis, *Notre Dame Journal of Formal Logic*, 28(4), 615–419.

Turing, Alan, 1936: On Computable Numbers, with an Application to the Entscheidungsproblem, *Proceedings of the London Mathematical Society* (Series 2), 42, 230–265.

Von Neumann, J., 1930: The Formalist Foundations of Mathematics, in: *Philosophy of Mathematics. Selected Readings*, eds. R. Benacerraf & H. Putman 1964, Englewood Cliffs, NJ: Prentice-Hall, 50–54.

Waismann, Friedrich, 1979: Totality and System, in: *Ludwig Wittgenstein and the Vienna Circle*, ed. Brian McGuinness, Oxford: Basil Blackwell.

Wittgenstein, Ludwig, 1993: Cause and Effect: Intuitive Awareness, in: *Philosophical Occasions: 1912–1951*, eds. J. Klagge and A. Nordmann, Indianapolis: Hackett, 368-426. See also: Taylor's Notes: Lectures on Knowledge, Lecture 5, 24-25, cited in Klagge & Nordmann.

Wittgenstein, Ludwig, *The Wittgenstein Papers*, Trinity College, Cambridge, Great Britain.
Wittgenstein, Ludwig, 1997: *Ludwig Wittgenstein. Denkbewegungen. Tagebuecher 1930–1932/1936-1937*, ed. Ilse Sommavilla, Innsbruck: Hayman.
Wittgenstein, Ludwig, 1974: *Philosophical Grammar*, ed. Rush Rhees, Oxford: Basil Blackwell.
Wittgenstein, Ludwig, 1975: *Philosophical Remarks,* ed. Rush Rhees, Oxford: Basil Blackwell.
Wittgenstein, Ludwig, 2001: *Philosophishe Untersuchungen. Kritisch-genetische Edition*, ed. Joachim Schulte, Frankfurt: Suhrkamp.
Wittgenstein, Ludwig, 1956: *Remarks on the Foundation of Mathematics*, eds. G.H. von Wright, Rush Rhees & Elizabeth Anscombe, Oxford: Basil Blackwell.
Wittgenstein, Ludwig, 1986: The Nature of Mathematics. Wittgenstein's Standpoint, in: *Ludwig Wittgenstein: Critical Assessments*, Volume Three, ed. Stuart Shanker, London, Sydney and Dover, NH: Croom Helm.
Wittgenstein, Ludwig, 2017: *Wittgenstein's Whewell's Court Lectures: Cambridge, 1938–1941,* eds. Volker A. Munz & Bernhard Ritter, Oxford: Wiley Blackwell; from the notes taken by Yorick Smythies.

Chapter 20
Quantum Physics and Cognitive Science from a Wittgensteinian Perspective: Bohr's Classicism, Chomsky's Universalism, and Bell's Contextualism

Yoshihiro Maruyama

20.1 The Wittgensteinian Turn and Philosophy Beyond the Analytic-Continental Divide

Wittgenstein's philosophy is trans-disciplinary, and has influenced diverse fields of science. Yet its relationships with ideas in physics and cognitive science (including artificial intelligence) are not so much elucidated or recognized. In the following we shall shed light on how Wittgenstein's philosophy may be placed in such a broad spectrum of contemporary science. Before that, let us revisit his philosophy itself and explicate what was truly innovative in his thought in order to articulate its broad significance in contemporary science. What is particularly intriguing about Wittgenstein's philosophy is its striking transition from the earlier philosophy in *Tractatus Logico-Philosophicus* to the later philosophy in *Philosophical Investigations*. There is some continuity between them, for example, in terms of the nature of logical and mathematical truths. Yet at the same time there is a clear contrast between his earlier and later conceptions of meaning in language. In the following we shall use the term "theory of meaning"; it should be remarked that Wittgenstein himself was clearly against the very idea of theory or theorizing. The same remark applies to the term "science"; Wittgenstein, in his later philosophy, was inclined towards anti-scientism. Having said these, it still holds that Wittgenstein's philosophy is highly relevant to what is called the theory of meaning in contemporary philosophy and also to developments of contemporary science as we shall see in the following.

Wittgenstein's earlier theory of meaning in *Tractatus* is based upon what is called the correspondence theory: linguistic expressions are given their meaning with reference to outer reality, that is, to give meaning in language is to give the correspondence between language and reality. By contrast, Wittgenstein's later theory of meaning has no outward reference to reality, and propounds the autonomy of language, that

Y. Maruyama (✉)
The Hakubi Centre for Advanced Research, Kyoto University, Kyoto, Japan
e-mail: maruyama@i.h.kyoto-u.ac.jp

S. Wuppuluri and N. da Costa (eds.), *WITTGENSTEINIAN (adj.)*,
The Frontiers Collection, https://doi.org/10.1007/978-3-030-27569-3_20

is, meaning exists within the internal structure of language rather than via the correspondence with reality. Wittgenstein's famous thesis of meaning as use tells that the meaning of a word is completely determined by its use in language without referring to any entity outside language. In Wittgenstein's own words:

> The words are not a translation of something else that was there before they were. (Wittgenstein 1970, 33e)

We are naturally inclined to think that there is something in the first place, and then a word comes into existence to refer to it; put another way, language works via its correspondence with reality. According to Wittgenstein's later philosophy, this is a misconceived idea; language is autonomous on its own, being independent of reality or the world outside language. What determines meaning is the internal structure of language per se. This is actually comparable to Jacques Derrida's idea that "there is no outside-text." The autonomy of language and the priority of language over reality as the outside world are the most striking feature of philosophy in the mid-twentieth century that transgresses the boundary between the analytic tradition and the continental tradition. The centrality of language in philosophy of both traditions was a driving force for what is called the Linguistic Turn.

The idea of the Linguistic Turn in philosophy was popularized by Rorty (1967); Richard Rorty started his career in analytic philosophy, and then moved to continental philosophy or post-analytic philosophy. He says that philosophy turned from the philosophy of things (e.g., Aristotle) to the philosophy of ideas (e.g., Kant), and then from the philosophy of ideas to the philosophy of words (e.g., Wittgenstein):

> [T]he dialectical progression which in the course of three centuries has led philosophers away from things to ideas, and away from ideas to words. (Rorty 1962)

The first turn is known as the Copernican Turn; simply saying, ideas are prior to things. The second turn is the Linguistic Turn; words are prior to things and ideas. In Wittgenstein's terms, words are not the translation of any pre-existing ideas or things. Why did these fundamental shifts happen in philosophy? Rorty's account is as follows:

> Given the initial Cartesian sundering of cognition from the object of cognition, it was only to be expected that epistemology should claim priority over metaphysics. Once this happened, it was only a matter of time before metaphysics shrivelled away and died. In due course, epistemology, finding itself unable to function as an autonomous discipline when deprived of metaphysical support, was pushed aside by its ungrateful child, linguistic analysis. (Rorty 1962)

The shifts from things to ideas, and from ideas to language may be enriched with another shift from language to information; the informational view of the world, such as Chalmers's double aspect theory of information (Chalmers 1996) is getting more and more supported by developments of science per se, such as information physics, artificial intelligence, and computational biology. The shifts from things to ideas and from ideas to language have the following mathematical counterpart: mathematics moved away from things to infinities, and from infinities to symbols.

Infinitary mathematics is the mathematics of ideas, and structuralist mathematics is the mathematics of symbolic forms. The same tendency towards symbolic formalism may be observed in quantum physics, which makes the naive, thing-based view of the world untenable, as we shall discuss in more detail later.

Rorty is known for his rejection of the "mirror of nature" view in his influential *Philosophy and the Mirror of Nature*; as a matter of fact he had essentially the same idea already in 1962:

> One can identify the original sin as the representative theory of perception, as the confusion of noetic being with being as such (Rorty 1962)

The abandonment of reality as the mirror of nature is actually the defining characteristic of modernist thought. Jeremy Gray says in his *Plato's Ghost* as follows:

> [M]odernism is defined as an autonomous body of ideas, having little or no outward reference, placing considerable emphasis on formal aspects of the work and maintaining a complicated – indeed, anxious – rather than a naive relationship with the day-to-day world. (Gray 2008)

Axiomatic foundations of mathematics also make the departure from reality as the mirror of nature, blurring the naive correspondence between mathematical structure and natural structure. The same idea applies to a broad variety of cultural developments. In modern art, for example, Paul Klee says, "Art does not reproduce what we see; rather it makes us see." Modern art departed from the mirror of nature or the naive conception of reality as well as modern mathematics. Ernst Cassirer's philosophy of substance and function presents the view that modernization is functionalization as the departure from substantivalism. Cassirer's genetic view replaces the copy theory with a functional theory (Friedman 2000):

> Myth and art, language and science, are in this sense shapings toward being; they are not simple copies of an already present reality, but they rather present the great lines of direction of spiritual development, of the ideal process, in which reality constitutes itself for us as one and many – as a manifold of forms, which are nonetheless finally held together by a unity of meaning. (Cassirer 1955, p. 107; translation by Friedman 2000)

In Cassirer's philosophy, reality or truth is a "never-completed X" as just reached in the ideal limit. Cassirer's shift from the copy theory to the functional theory is in conceptual parallel with Wittgenstein's shift from the picture theory in *Tractatus* to the use theory in *Investigations*. The theory of modernization as functionalization applies to a broad variety of sciences. Developments of structuralist mathematics (Bourbaki, Grothendieck, etc.) clearly exhibit the modernist tendency towards functionalization. The concept of space has been functionalized in mathematics, shifting from the point-based view (e.g., topological space) to the relational view (e.g., locales, schemes, and toposes), or from the Newtonian conception to the Leibnizian conception of space. Quantum physics may be conceived of as the functionalization of reality. Artificial intelligence counts as the functionalization of the mind. Cassirer actually had an explicit idea to understand the modernization of science as functionalization, and yet the scope of functionalization theory is even going beyond his expectation in contemporary developments of science. And it can actually encompass aspects of modern culture in general. Cassirer now counts as a precursor of structuralism in both analytic and continental traditions, including structural realism as we shall discuss below. Indeed, Charles Parsons in the analytic tradition says as follows: "Cassirer takes a position that could be called structuralist" (Parsons 2014, p. 14). Steve Lofts in the continental tradition says more eloquently as follows:

> [T]he philosophy of symbolic forms is presented as a *type* of "structuralism" *avant la lettre* resulting from a fusion and critical transformation of Kant's transcendental critical philosophy, on the one hand, and Hegel's phenomenology of spirit, on the other. (Lofts 2000, p. 21)

Cassirer's structuralism stretches over different disciplines, including mathematics and physics on one hand, and ontology and epistemology on the other. Cassirer had a clear idea of mathematical structural realism in the first place, that is, the view that all there is is structure in mathematics:

> The relational structure as such, not the absolute property of the elements, constitutes the real object of mathematical investigation (Cassirer 1923, p. 93)

This idea is close to the fundamental philosophy of category theory in mathematics, the most sophisticated form of structuralist mathematics. This mathematical structuralism concerning the ontology of mathematics extends to ontological structuralism in general form:

> [W]e can only reach the category of thing through the category of relation (Cassirer 1923, p. 306)

Cassirer also argues that relative properties are "the first and positive ground of the concept of reality" (Cassirer 1923, p. 306). Cassirer's ontological structuralism is, in a sense, a consequence of his epistemological structuralism, the view that all we can know is about structure:

> "Whatever we may know of matter," here, too, we can cite the Critique of Pure Reason, "is nothing but relations, some of which are independent and permanent and by which a certain object is given us" (Cassirer 1923, p. 386)

Cassirer's early work was concerned with Leibniz's relational philosophy, and he combined it with Kantian philosophy, having eventually produced structuralist epistemology and ontology, which allow us to account for various aspects of modern science, including in particular mathematical logic and foundations of mathematics on one had, and relativity theory and quantum physics on the other. Another origin of Cassirer's structuralism is Göthe's morphology. In his time the term *structuralism* was not very popular; yet in his later work he published an article *Structuralism in Modern Linguistics* (1945), in which he argues: "language is neither a mechanism nor organism"; "we must say that language is 'organic', but that it is not an organism"; "the program of structuralism developed by Brondal is, indeed, very near to Humboldt's ideas."

Gödel gives a conceptual perspective on modernist thought in his (never delivered) lecture notes entitled *The modern development of the foundations of mathematics in the light of philosophy*:

> [T]he development of philosophy since the Renaissance has by and large gone from right to left [...] Particularly in physics, this development has reached a peak in our own time, in that, to a large extent, the possibility of knowledge of the objectivisable states of affairs is denied, and it is asserted that we must be content to predict results of observations. This is really the end of all theoretical science in the usual sense. (Gödel 1961)

Some equate Gödel's position with naïve realism, but he is not a simple realist as the lecture notes show. Indeed, he attempts to reconcile the "right" thought (metaphysics, idealism, theology, etc.) and the "left" thought (positivism, materialism, scepticisim, etc.) via the transcendental philosophy of Kant and Husserl in the lecture notes. He illustrates the dynamic history of human thought through the conceptual shift of worldview leading from the metaphysics of substance to the antimetaphysics of different sorts. Gödel's "right" and "left" are a matter of degree; for example, scepticism is an extreme left thought, compared with positivism, which is less lefty than scepticism. He also says that "Schopenhauer's pessimism is a mixed form, namely a pessimistic idealism." Hilbert's transcendental finitism is a left position in mathematics, and yet at the same time, it is a foundation for the right, that is, the infinitary metaphysics of mathematics. The right wing mathematics suffered from Russell's and other paradoxes, and then the left wing foundation of mathematics arose to save the right wing mathematics from the paradoxes. In a nutshell, Hilbert's finitism is about the justification of the right via the left.

Philosophy within the intellectual history of humanity may be characterized by four conceptual turns:

- The Copernican Turn from the philosophy of things to the philosophy of ideas (e.g., Kant);
- The Linguistic Turn from the philosophy of ideas to the philosophy of words (e.g., Wittgenstein);
- The Informational Turn from the philosophy of words to the philosophy of information (e.g., Chalmers and Floridi);
- The Realist Turn (aka. Speculative Turn), a reactionary return to the philosophy of things or realism (e.g., Meillassoux and Ladyman).

The traditional understanding of modernization is given by the Max Weber's concept of disenchantment, which is driven by the modernist propensity for the left in terms of Gödel. The nature of our after-modern society may be characterized via the concept of re-enchantment, a case of which is the Realist Turn as exemplified by developments of speculative realism and materialism in continental philosophy and of structural realism and scientific metaphysics in analytic philosophy. Wittgenstein's use theory of meaning may be applied at every stage of these turns: the meaning of a thing is its use; the meaning of an idea is its use; and the meaning of information is its use. Meaning does not stem from any depth as theory presupposes, bur rather from the very surface of our everyday practice. This idea lies at the heart of Wittgenstein's later philosophy.

Meillassoux's speculative materialism follows the logic of justifying the right (realist metaphysics) via the left (materialism). Structural realism and scientific metaphysics also follow the same logic, justifying the right (realist metaphysics) via the left (structure and science, respectively). Hilbert's transcendental finitism was an earlier case of this logic. Why had the right to be justified via the left in such a cross-disciplinary manner? Realist metaphysics has suffered from the sceptic atmosphere of modernity since the Renaissance, and the naive conception of reality is not tenable any more, both philosophically and scientifically, as Gödel's quote above illustrates the antirealist tendency of modernist thought in both science and philosophy. The indirect justification of the right via the left is the only way to save the right when the right cannot justify itself on its own. When realist metaphysics loses its direct foundation in face of scepticism, it is the only way to save reality to justify it upon the ground even the sceptic has to accept or presuppose.

Modernist (and post-modernist) disenchantment has eventually led us to the nihilistic end of various things: the end of grand narrative (Lyotard); the end of classical realism (Bell); the end of the first philosophy (Quine); the end of meaning (Derrida); the end of history (Fukuyama); and the end of God (Nietzsche). The end of classical realism is almost a fact in quantum physics, mathematically shown by Bell-type theorems as we shall discuss later on. The end of humanity or post-humanity is relevant to the rise of artificial intelligence. Yet the idea itself has an older origin as in Nietzsche's notion of the last man and in Foucault's saying "Man is an invention of recent date." The present concept of human or individual is a social construct of the modern civilization; the meaning of being human in the pre-modern society was different from that in the modernized society. The modernist conception of the human being begins and ends at some points of history. Arguably, the problem of modernity is the problem of nihilism, which can be of scientific nature as well as of human nature, the fragmentation of science being a distinctive case of scientific nihilism. Overcoming modernity is overcoming nihilism. Along the same lines, Keiji Nishitani, a philosopher of the Kyoto School, propounds the overcoming of modernity via modernity, in particular the overcoming of nihilism via nihilism.

Morris Berman illustrates the modernist process of disenchantment as follows:

> The view of nature which predominated in the West down to the eve of the Scientific Revolution was that of an enchanted world. A member of this cosmos was not an alienated observer of it but a direct participant in its drama. His personal destiny was bound up with its destiny, and this relationship gave meaning to his life. This type of consciousness – what I shall refer to in this book as "participating consciousness." (Berman 1981)

The loss of meaning or nihilism is a compelling consequence of the Scientific Revolution that made the mechanistic view of nature prevailing and dominating:

> And historically, our loss of meaning in an ultimate philosophical or religious sense – the split between fact and value which characterizes the modern age – is rooted in the Scientific Revolution of the sixteenth and seventeenth centuries. (Berman 1981)

And finally Berman propounds the re-enchantment of the world:

> [B]ecause disenchantment is intrinsic to the scientific world view, the modern epoch contained [...] an inherent instability that severely limited its ability to sustain itself for more than a few centuries. For more than 99The complete reversal of this perception in a mere four hundred years or so has destroyed the continuity of the human experience and the integrity of the human psyche. It has very nearly wrecked the planet as well. The only hope, or so it seems to me, lies in a reenchantment of the world. (Berman 1981)

Disenchantment and re-enchantment, according to Berman, are tied with the issue of meaning:

> This emphasis stems from my conviction that the fundamental issues confronted by any civilization in its history, or by any person in his or her life, are issues of meaning. (Berman 1981)

Both disenchantment and re-enchantment may be observed in diverse fields of human culture. Analytic philosophy is the disenchantment of traditional philosophy. Axiomatic mathematics is the disenchantment of traditional mathematics. Modern art is the disenchantment of traditional art. How about re-enchantment? The Realist Turn in contemporary philosophy as discussed above is the re-enchantment of the sceptic, antirealist modernism. In mathematics, the twentieth century, by and large, saw the rise of structuralist mathematics, and yet the twenty-first century mathematics is returning to the mathematics of the concrete rather than abstract structure. This return to the concrete may be comparable to the movement of *musique concrète* in contemporary art. Contemporary art puts an emphasis on the contingencies of concrete reality, as seen in John Cage's (contingent) chance music (as opposed to Pierre Boulez's controlled chance music), which is conceptually comparable to the contemporary science of chance such as data science and machine learning. The contingency of Being has been a vital theme in the tradition of contemporary philosophy that leads from Heidegger and Wittgenstein to Derrida and Rorty. Process philosophy such as Whitehead's is also relevant to contemporary developments of re-enchantment (Griffin et al. 1992; Griffin 2001); Cassirer's philosophy may count as process philosophy as well. In Asia, the Kyoto School of Philosophy, while facing the modernist crisis of the fragmentation of worldview, propounds the ideal of overcoming modernity by virtue of constructing a unified view of the world; their project would be an early case of re-enchantment. They argue that we are facing the fragmentation of worldview in three different directions (that is, Human-, Nature-, and God-centered views), and we have to restore the lost unity of worldview again in our time.

From another angle, we can elucidate developments of philosophy in light of the ontic-epistemic duality. Broadly saying, developments of philosophy have centred around a tension between realism and antirealism (the right and the left in Gödel's terms) which may be illustrated by asking the nature of a variety of fundamental concepts. What is Meaning? The realist asserts it consists in the correspondence of language to reality (the early Wittgenstein), whilst the antirealist contends it lies in the autonomous system of language or linguistic practice (the later Wittgenstein). What is Truth? To the realist, it is the correspondence of assertions to facts (Aristotle); to the antirealist, it has no outward reference, constituted by the internal coherency of assertions (Francis Herbert Bradley). What is Being? To the realist, it is persistent substance (Aristotle); to the antirealist, it emerges within an evolving process, cognition, network, environment, or context (Cassirer, Heidegger, Whitehead). What is Intelligence? To the realist, it is more than behavioural simulation, characterised by the intentionality of mind; to the antirealist (John Searle), it is fully conferred by copycatting (Alan Turing). What is Space? To the realist, it is a collection of points with no extension (Newton); to the antirealist, it is a structure of regions, properties, or information (Leibniz).

Everything, from Truth and Meaning to Being and Mind, has dual facets. Duality in turn unites them together, thereby telling us they are the two sides of the same coin. Duality thus conceived is a constructive canon to deconstruct dualism; this has been a blueprint idea underpinning what is called duality theory as a whole. Realism and antirealism are grounded upon the ontic and the epistemic, respectively. The ultimate form of duality would then be duality between ontology and epistemology. An easier ontology of something often makes its epistemology more difficult, and vice versa; this negative correlation is an indication of duality. For instance, a platonic realist ontology of mathematics makes it difficult to account for the possibility of epistemic access to mathematical objects; this is the so-called Benaceraff's dilemma. Conversely, a constructive antirealist epistemology yields ontological difficulties in warranting the existence and objectivity of transfinite entities; this may be called the converse Benaceraff's dilemma. In general, realism gets troubled by our accessibility to entities unable to exist in our world; antirealism faces a challenge of securing their existence and objectivity. To conclude, the hard problem of philosophy is to give a stream-lined account of how the epistemic and the ontic can interact with each other, which is arguably the very task of duality.

20.2 Truth and Meaning in Symbolic AI and Statistical AI: The Earlier versus Later Wittgenstein

The Wittgenstein's earlier conception of meaning as correspondence with reality is now standard in logical semantics. The Wittgenstein's later conception of meaning as use in linguistic context is actually standard in statistical semantics. The relevance of Wittgenstein's later philosophy in Natural Language Processing (as a subfield of Artificial Intelligence) is much less known than the relevance of his earlier philosophy in Natural Language Semantics (as a subfield of Applied Logic and Type Theory). What is called Symbolic AI (aka. GOFAI, i.e., Good Old-Fashioned AI) is based upon the methods of mathematical logic. What is called Statistical AI is based upon the methods of statistical machine learning. There is an interesting debate between Noam Chomsky, the father of modern linguistics, and Peter Norvig, Google's research director, concerning the nature of language. It is a battle between Symbolic and Statistical AI, but not just that. It is concerned with the deeper understanding of science in general, especially the ultimate purpose of the endeavor of science. From a conceptual point of view, it is actually a battle between Wittgenstein's earlier and later conceptions of language and meaning, or between the realist and antirealist conceptions. In the following we shall first review some history of Artificial Intelligence, and then discuss the debate between Chomsky and Norvig, or between the earlier and later Wittgenstein.

The history of Artificial Intelligence may be summarized as follows:

- The Birth of AI at Dartmouth College (1956). Even older origins in Turing, Gödel, Russell, ..., Leibniz. Mathematicians, logicians, and philosophers created AI.
- The Golden Age or Good Old-Fashioned AI (1956–1974). Symbolic AI: Automated Reasoning, Theorem Prover, etc. Novel programming paradigms such as Lisp and Prolog.
- The First AI Winter (1974–1980). Inferior computing power; unable to run some algorithms. Commonsense Knowledge Problem, Frame Problem, etc.
- The First AI Bubble (1980–1987). Expert Systems (with fixed rules on small domains). Japan's Fifth Generation Computer Project (and its failure).
- The Second AI Winter (1987–1993). Funding had stopped. Towards the age of Information Superhighway.
- The Victory of the Neats (1993–2011). Deep Blue defeated a Chess champion. Neat mathematical techniques had been discovered.
- The Big Data Bubble (2011–20??). IBM Watson defeated Quiz Champions. Deep Mind (AlphaGo) defeated a Go champion. Statisitcal AI based upon Machine Learning. Learning rather than Reasoning.
- The Technological Singularity (2045?). AI outperforms human beings in every respect. AI civilizations might replace human civilizations. Humans might become slaves of clever AI. Post-human civilization.

The battle between Chomsky and Norvig is basically the battle between Symbolic AI and Statistical AI. Wittgenstein's early philosophy laid a foundation for symbolic logic, which gave rise to Symbolic AI. Wittgenstein's later philosophy laid a foundation for what is now called distributional semantics in the Statistical AI paradigm of Natural Language Processing. Stephen Clark remarks in his *Vector Space Models of Lexical Meaning* as follows:

> Finally, one classic piece of philosophical work that is often mentioned in the context of distributional semantics is Wittgenstein (1953). The link here is somewhat tenuous, since Wittgenstein was not concerned with the contexts of words in corpora, but rather the conventional, social nature of a whole language. However, Wittgenstein's slogan that "meaning is use" is certainly applicable, under some interpretation, to distributional semantics.

The gist of distributional semantics is as follows: words ($w_1, w_2, \ldots$) are associated with vectors ($v_1, v_2, \ldots$), which are called meaning vectors. There is usually a fixed basis of space, i.e., a set of basic meaning vectors, weighted sums of which give arbitrary meaning vectors. Weights are given by the distribution of words in different contexts. Similarity between words is given by $\langle v_1|v_2\rangle/||v_1||||v_2||$ where $\langle v_1|v_2\rangle$ denotes inner product. Put simply, the relative angle θ between v_1 and v_2 determines their similarity value as $\cos\theta$. If meaning vectors are parallel, for example, the similarity value is 1, which means they have the same meaning. Distributional semantics thus gives the geometry of meaning. The guiding principle lurking behind distributional semantics is the Distributional Hypothesis:

- The Distributional Hypothesis: words that occur in similar contexts have similar meanings. The Distributional Hypothesis may be articulated as different ways to compute aforementioned weights. The Distributional Hypothesis is often said to go back to Wittgenstein's later philosophy, especially his thesis of meaning as use: the meaning of an expression is internally given by the use of it in linguistic practice, with no outward reference to reality (to Wittgenstein, language constitutes the world, and not the other way around).

The more contexts, the less possibilities of meaning. It is duality between meaning and context. It should be remarked that Wittgenstein's philosophy of language does not really admit any formal account; to be even worse, he rejected scientism per se. Yet there is still a conceptual similarity between the Distributional Hypothesis and Wittgenstein's philosophy of language. VSM (Vector Space Model) based on the Distributional Hypothesis has made a great success in Natural Language Processing. For example, Turney and Pantel (2010) say:

> The success of the VSM for information retrieval has inspired researchers to extend the VSM to other semantic tasks in natural language processing, with impressive results. For instance, Rapp (2003) used a vector-based representation of word meaning to achieve a score of 92.5% on multiple-choice synonym questions from the Test of English as a Foreign Language (TOEFL), whereas the average human score was 64.5%.

Let us review the distinction between sense and reference. Reference is about what words denote. Denotational semantics in symbolic logic is concerned with reference. It is compositional, accounting for meaning while respecting grammatical structure. Sense is about the mode of presentation, i.e., how they are presented in expressions concerned. Words can have sense without reference. Distributional semantics is concerned with sense, accounting for meaning via statistical distribution while ignoring the underlying, generative structure. "The Evening Star" and "The Morning Star" have the same reference and yet their senses are different. Two cornerstones of linguistic analysis are: the compositionality principle and the context principle. The principle of compositionality is as follows: "the meaning of a complex expression is a function of the meaning of its parts and of their syntactic mode of combination" (Kamp and Partee 1995; originally due to Frege). "Syntactic mode" means "grammatical mode" of combination, like the mode of the combination between subjects and verbs. The context principle is as follows: "never to ask for the meaning of a word in isolation, but only in the context" (Frege 1884). This is comparable to the distribution hypothesis: words appearing in similar contexts tend to have similar meanings; co-occurrence matrices with each column representing a word and each row a context. Frege says in his *Grundlagen der Arithmetik*:

> In the enquiry that follows, I have kept to three fundamental principles: always to separate sharply the psychological from the logical, the subjective from the objective; never to ask for the meaning of a word in isolation, but only in the context of a proposition; never to lose sight of the distinction between concept and object.

Wittgenstein says in his *Tractatus Logico-Philosophicus* as follows:

> Only the proposition has sense; only in the context of a proposition has a name meaning.

Denotational semantics supports compositionalism: the meaning of the whole is the sum of the meanings of the parts. Distributional semantics supports holism:

the meaning of the parts is determined with reference to the whole (context). Put another way, it supports contextualism: the meaning of the parts is determined with reference to different contexts surrounding them. Chomsky (the early Wittgenstein) is on the side of compositionalism, and Norvig (the later Wittgenstein) on the side of holism or contextualism. There is a tension between the two principles of meaning (or semantics). If meaning is compositional, the meaning of a whole must be determined with reference to the meaning of its parts only, i.e., without reference to anything larger, such as contexts. Put another way, the context principle is a holistic principle. Holism asserts (properties of) a whole cannot be reduced to the mere combination of (properties of) its parts, whereas the central tenet of compositionalism is that this is indeed possible, i.e., the meaning of a whole is composed of that of its parts. It is not clear how we can reconcile the two principles or whether that is possible at all.

Yarden Katz interviews Chomsky at MIT (Katz 2012):

> I want to start with a very basic question. At the beginning of AI, people were extremely optimistic about the field's progress, but it hasn't turned out that way. Why has it been so difficult?

Chomsky answers as follows:

> There's something to that. If you take a look at the progress of science, the sciences are kind of a continuum, but they're broken up into fields. The greatest progress is in the sciences that study the simplest systems. So take, say physics – greatest progress there. But one of the reasons is that the physicists have an advantage that no other branch of sciences has. If something gets too complicated, they hand it to someone else.

And Chomsky continues as follows:

> If a molecule is too big, you give it to the chemists. The chemists, for them, if the molecule is too big or the system gets too big, you give it to the biologists. And if it gets too big for them, they give it to the psychologists, and finally it ends up in the hands of the literary critic, and so on.

Chomsky makes the following remark, clearly against Statistical AI, during the same interview:

> [I]f you get more and more data, and better and better statistics, you can get a better and better approximation to some immense corpus of text […] but you learn nothing about the language.

This is crucial for Chomsky's view of language and science. What Chomsky expects to linguistics or science in general is the systematic account of mechanisms and inner workings that gives us a fundamental understanding of phenomena concerned. A statistical approximation to data, by itself, does not give us any understanding. It may be an excellent simulation, but cannot be a scientific explanation, according to Chomsky. Chomsky says in the same interview:

> [I]t's very different from what's done in the sciences. So for example, take an extreme case, suppose that somebody says he wants to eliminate the physics department and do it the right way. The "right" way is to take endless numbers of videotapes of what's happening outside the video, and feed them into the biggest and fastest computer, gigabytes of data, and do

> complex statistical analysis [...] and you'll get some kind of prediction about what's gonna happen outside the window next. In fact, you get a much better prediction than the physics department will ever give.

Chomsky even asserts that statistical analysis would be superior to physics in terms of future prediction. Still, what he prefers is physics, which is concerned with understanding and explanation, the central features of the enterprise of science in Chomsky's view. Chomsky also says as follows in the same interview:

> These are just two different concepts of science. The second one is what science has been since Galileo, that's modern science. The approximating unanalyzed data kind is sort of a new approach, not totally, there's things like it in the past. It's basically a new approach that has been accelerated by the existence of massive memories, very rapid processing, which enables you to do things like this that you couldn't have done by hand. But I think, myself, that it is leading subjects like computational cognitive science into a direction of maybe some practical applicability.

Gold (2011) summarizes the Norvig vs. Chomsky debate as follows:

> Recently, Peter Norvig, Google's Director of Research and co-author of the most popular artificial intelligence textbook in the world, wrote a webpage extensively criticizing Noam Chomsky, arguably the most influential linguist in the world. Their disagreement points to a revolution in artificial intelligence that, like many revolutions, threatens to destroy as much as it improves. Chomsky, one of the old guard, wishes for an elegant theory of intelligence and language that looks past human fallibility to try to see simple structure underneath. Norvig, meanwhile, represents the new philosophy: truth by statistics, and simplicity be damned.

To Norvig, theory does not mean much, and what is more significant is the proper simulation of actual practice. The emphasis on actual linguistic practice is reminiscent of Wittgenstein's later philosophy, which was strongly against the theorizing tendency, and espoused a nuanced analysis of down-to-earth practice. Gold (2011) seems to be more sympathetic with the Norvig side than with the Chomsky side:

> Norvig is now arguing for an extreme pendulum swing in the other direction, one which is in some ways simpler, and in others, ridiculously more complex. Current speech recognition, machine translation, and other modern AI technologies typically use a model of language that would make Chomskyan linguists cry: for any sequence of words, there is some probability that it will occur in the English language, which we can measure by counting how often its parts appear on the internet. Forget nouns and verbs, rules of conjugation, and so on: deep parsing and logic are the failed techs of yesteryear.

Statistical models do not much take into account the structure of language such as the grammar of language; nonetheless, a purely statistical analysis of co-occurrence of words yields a surprisingly effective model of natural language. Gold (2011) further explains the idea of Statistical AI as follows:

> In their place is the assumption that, with enough data from the internet, you can reason statistically about what the next word in a sentence will be, right down to its conjugation, without necessarily knowing any grammatical rules or word meanings at all. The limited understanding employed in this approach is why machine translation occasionally delivers amusingly bad results. But the Google approach to this problem is not to develop a more sophisticated understanding of language; it is to try to get more data, and build bigger lookup tables.

This is essentially about distributional semantics, which is heavily used by Google as well as many others, and making a great success, especially in information retrieval and language processing. In distributional semantics, words are represented by vectors, and similarities in meaning are represented by angles between them (i.e., what is called cosine measure). Co-occurrence of words determines the meaning (vector) of them: values in vector representations are calculated from co-occurrence frequencies on the ground of the Distributional Hypothesis, which tells that words co-occuring in similar contexts have similar meanings. Cass (2011) summarizes Chomsky's criticism as follows:

> Chomsky derided researchers in machine learning who use purely statistical methods to produce behavior that mimics something in the world, but who don't try to understand the meaning of that behavior. Chomsky compared such researchers to scientists who might study the dance made by a bee returning to the hive, and who could produce a statistically based simulation of such a dance without attempting to understand why the bee behaved that way. "That's a notion of [scientific] success that's very novel. I don't know of anything like it in the history of science," said Chomsky.

Chomsky (2011) himself says at the MIT "Brains, Minds, and Machines" symposium as follows:

> There's a lot work which tries to do sophisticated statistical analysis, you know bayesian and so on and so forth, without any concern for the actual structure of language, as far as I'm aware that only achieves success in a very odd sense of success. There is a notion of success which has developed in computational cognitive science in recent years which I think is novel in the history of science. It interprets success as approximating unanalyzed data.

Chomsky goes on saying:

> If you simply did extensive videotaping of bees swarming, OK, and you did you know a lot of statistical analysis of it, you would get a pretty good prediction for what bees are likely to do next time they swarm, actually you'd get a better prediction than bee scientists do, and they wouldn't care because they're not trying to do that.

Chomsky admits that statistical models are superior, in terms of predictive power, to other models including symbolic ones. Chomsky further proceeds:

> You would get some kind of prediction of what's likely to happen next, certainly way better than anybody in the physics department could do. Well that's a notion of success which is I think novel, I don't know of anything like it in the history of science. In those terms you get some kind of successes, and if you look at the literature in the field, a lot of these papers are listed as successes. And when you look at them carefully, they're successes in this particular sense, and not the sense that science has ever been interested in. But it does give you ways of approximating unanalyzed data, you know analysis of a corpus and so on and so forth.

To Chomsky, data science, perhaps, is a new kind of post-truth science.

There is a well-known argument by Chomsky to show that statistical models cannot capture grammaticalness, which he thinks is of purely symbolic nature. Chomsky (1957) thinks of the following two sentences:

(1) Colorless green ideas sleep furiously.
(2) Furiously sleep ideas green colorless.

Chomsky (1957) then argues as follows:

> It is fair to assume that neither sentence (1) nor (2) (nor indeed any part of these sentences) has ever occurred in an English discourse. Hence, in any statistical model for grammaticalness, these sentences will be ruled out on identical grounds as equally "remote" from English. Yet (1), though nonsensical, is grammatical, while (2) is not grammatical.

Mark (2017) explains Chomsky's idea as follows:

> Chomsky argues that the experiences available to language learners are far too sparse to account for their knowledge of their language.

Chosmky thinks that, without the innate capacity of reason to generate language, we could not even learn language in the first place; this may be called the innate knowledge thesis. Machine learning could be an alleged counterexample to the thesis, but it is not so decisive at the moment. More than 50 years later, Norvig (2011) counters Chomsky (1957) as follows:

> I'm not sure what he meant by "any of their parts," but certainly every two-word part had occurred, for example: [...] Pereira (2001) showed that such a model, augmented with word categories and trained by expectation maximization on newspaper text, computes that (a) is 200,000 times more probable than (b). I repeated the experiment, using a much cruder model [...] and found that (a) is about 10,000 times more probable.

Note that (a) and (b) correspond to (1) and (2) respectively. Concerning this particular point, Norvig can probably win against Chomsky. Yet at the same time, certain deficiencies of present statistical models of language are going to be remedied and overcome with the help of symbolic methods. So the complete theory of language, if any, may require some symbolic methods as well. There is also a huge cost to pay for statistical models, that is, the interpretability of models. Norvig (2011) says:

> I agree that it can be difficult to make sense of a model containing billions of parameters. Certainly a human can't understand such a model by inspecting the values of each parameter individually.

Kuhn-Johnson (2013) assert:

> Unfortunately, the predictive models that are most powerful are usually the least interpretable.

The situation may be comparable to that in quantum mechanics, which comes with strong predictive power and miserably poor interpretation as Feynman-Mermin say "Shut up and calculate!". Kuhn-Johnson (2013) also argue as follows:

> If a medical diagnostic is used for such important determinations, patients desire the most accurate prediction possible. As long as complex models are properly validated, it may be improper to use a model that is built for interpretation rather than predictive performance.

There is a trade-off between predictability and interpretability. A more practical aspect of the cost that complex models have to pay is as follows; Christophe Bourguignat (2014) says:

> In real organizations, people need dead simple story-telling — Which features are you using? How your algorithms work? What is your strategy? etc. ...If your models are not parsimonious enough, you risk to the audience confidence. Convincing stakeholders is a key driver for success, and people trust what they understand. What's more, at the end of the day, the ultimate goal of the data science work is to put a model in production. If your model is too complicated, this will turn out to be impossible or, at least, very difficult.

It should be remarked here that Chomsky simplifies data science to some extent. He contrasts the empirical data-driven approach with the theoretical knowledge-driven approach. Yet data science is not entirely about approximating empirical data. It, for example, attempts to prevent overfitting by introducing domain knowledge or regularization terms or specific priors. It does care about the interpretability of models, and hence the trend of Interpretable Machine Learning. The methodology of data science is not purely empiricist but partly rationalist. By contrast, the resulting knowledge is purely empiricist with no explication of underlying mechanisms.

Norvig (2011) summarizes the points of Chomsky's critique of statistical AI as follows:

- "Statistical language models have had engineering success, but that is irrelevant to science."
- "Accurately modeling linguistic facts is just butterfly collecting; what matters in science (and specifically linguistics) is the underlying principles."
- "Statistical models are incomprehensible; they provide no insight."

The metaphor of butterfly-collecting as non-science is Chomsky's favorite, and has an old origin, as a matter of fact, as Chomsky (1979) says:

> You can also collect butterflies and make many observations. If you like butterflies, that's fine; but such work must not be confounded with research, which is concerned to discover explanatory principles of some depth and fails if it does not do so.

Chomsky's point may be clarified with reference to the shift from Ptolemy's predictive model of the solar system to Copernicus's. In terms of predictive power and approximation of data, Ptolemy's model actually outperformed Copernicus's, and Copernicus nonetheless won the race of science. Machine learning papers often assert like "our method outperforms the state-of-the-art method" in the spirit of what Chomsky calls a new notion of success in science. If the aim of science is at approximating empirical data, the shift was unreasonable in light of the very aim of science. Yet the aim of science in reality is at a rational understanding of Nature as well as approximating or predicting statistical data.

Norvig (2011) characterizes Chomsky's viewpoint "as a Platonist and a Rationalist"; this is, broadly, right to the point. Norvig also says Chomsky is "a bit of a Mystic". Norvig would, then, be an Aristotelian and an Empiricist. Chomsky says in reply to a question in Katz (2012): "It's worth remembering that with regard to cognitive science, we're kind of pre-Galilean, just beginning to open up the subject." So it is a revival of the classic debate between rationalism and empiricism. Indeed, Chomsky (1975) asserts that he has "a rationalist conception of the nature of language." Chomsky (1975) even argues as follows:

> Furthermore, I employed it [...] to support what might fairly be called a rationalist conception of acquisition of knowledge [...] In sharp contrast to the rationalist view, we have the classical empiricist assumption that what is innate is (1) certain elementary mechanisms of peripheral processing (a receptor system), and (2) certain analytical machanisms or inductive principles or mechanisms of association.

As the above quote shows, he is actually aware that what is at stake here is the rationalist versus empiricist conception of language and its aquisition. In the history of philosophy, Kantian epistemology aimed at reconciling the two camps. Kantian AI, rather than Heideggerian AI, may be a solution to the discrepancy between the two camps. There are indeed recent developments to integrate Symbolic AI and Statistical AI via the methods of category theory; the categorical unification of Symbolic AI and Statistical AI may be able to embody Kantian AI, thus making predictive power compatible with interpretability.

If we argue for Norvig, suppose that a war happens between the Chomsky nation and the Norvig nation. Which would win the war at the end of the day? Better predictive power would yield better weapons of mass destruction. The profound understanding Chomsky aims at may be useless to defend his nation. Yet Chomsky could still argue that a deeper understanding of Nature results in revolutionary weapons, just as fundamental physics yielded atomic bombs and Turing broke the German code in the WWII.

Norvig (2011) wraps up the discussion with the following closing remark:

> [L]anguages are complex, random, contingent biological processes that are subject to the whims of evolution and cultural change. What constitutes a language is not an eternal ideal form, represented by the settings of a small number of parameters, but rather is the contingent outcome of complex processes. Since they are contingent, it seems they can only be analyzed with probabilistic models.

In a nutshell, language is contingent, and thus statistical models are necessary for the analysis of language, according to Norvig's idea. By contrast, Chomsky wants to explicate the "eternal ideal form" of language, which is universal rather than contingent. Is the ultimate nature of language contingent or universal? The truth would be that the surface structure of language is contingent, and yet the core structure of it is universal; this would be one way to reconcile the two, opposing views.

Wrapping up, if the primary concern of science lies in empirical prediction, interpretability ought to be compromised in favor of predictive power. If Science is about a rational understanding of Nature, on the other hand, interpretability must be maintained at the cost of predictive power. What is at stake here is the very conception of science and its primary aim as an intellectual endeavor of humanity. The categorically integrated Kantian AI would make empirical performance (Norvig) compatible with systematic understanding (Chomsky). Language is contingent in its surface, and yet universal in its core. The former is the conception of language in Wittgenstein's later philosophy, which sees meaning as rooted in the natural history of linguistic practice (and thus dynamic and contingent), and the latter is the conception of language in Wittgenstein's early philosophy, which sees meaning as arising from the picture-theoretical exact correspondence between language and reality (and thus static and universal).

As a historical remark, James Clerk Maxwell had an interesting idea of the relationships between empiricism and rationalism. On one hand, he says in his 1850 letter to Lewis Campbell as follows:

> [T]he true Logic for this world is the Calculus of Probabilities. (Maxwell 1990)

Maxwell thinks, probably on the basis of the British tradition of empiricism, that human knowledge as a whole rests upon the faculty of sensibility as well as the faculties of reason and understanding, and so it is probabilistic in its nature:

> [A]s human knowledge comes by the senses in such a way that the existence of things external is only inferred from the harmonious (not similar) testimony of the different senses, understanding, acting by the laws of right reason, will assign to different truths (or facts, or testimonies, or what shall I call them) different degrees of probability. (*ibid.*)

Yet at the same time, he admits certainty, immutability, and universality, not in empirical experiments, but in things themselves:

> [O]ur experiments can never give us anything more than statistical information [...] But when we pass from the contemplation of our experiments to that of the molecules themselves, we leave a world of chance and change, and enter a region where everything is certain and immutable. (*ibid.*)

Yet another problem we did not address here is the so-called hard problem of consciousness (Chalmers 1955), which actually goes back to Leibniz. The Leibniz' Mill Argument goes as follows:

> Moreover, we must confess that perception, and what depends on it, is inexplicable in terms of mechanical reasons, that is, through shapes and motions. If we imagine that there is a machine whose structure makes it think, sense, and have perceptions, we could conceive it enlarged, keeping the same proportions, so that we could enter into it, as one enters into a mill. Assuming that, when inspecting its interior, we will only find parts that push one another, and we will never find anything to explain a perception. And so, we should seek perception in the simple substance and not in the composite or in the machine. (Section 17, *Monadology*)

Consciousness is still a big mystery in science, and no one can give a complete answer to the hard problem of consciousness; actually there is no complete answer yet even to the easy problem of consciousness such as the neuroscientific workings of the brain. A fundamental question is: can objective science account for subjective experience, qualia, or the inner self as a whole? We shall touch upon the issue of consciousness later in the cognitive science section.

20.3 The Wittgensteinian Turn in the Foundation of Physics: A Parallelism between Bohr and Hilbert

The essential ingredient of the Linguistic Turn is the departure from reality or the mirror of Nature in Rorty's terms. The departure from reality or the mirror of Nature is the striking characteristic of modernization per se that may be observed in almost all areas of human civilization. It is what Max Weber called "disenchantment." Let us recall Gödel's interesting remarks on developments of contemporary physics:

> "[T]he development of philosophy since the Renaissance has by and large gone from right to left [...] Particularly in physics, this development has reached a peak in our own time, in that, to a large extent, the possibility of knowledge of the objectivisable states of affairs is denied, and it is asserted that we must be content to predict results of observations. This is really the end of all theoretical science in the usual sense".

This is obviously about quantum physics. Quantum physics does not allow for any simple description of reality or what Gödel calls objectivizable states of affairs, as different fundamental theorems in quantum physics indeed show, including Bell's theorem, Kochen-Specker's theorem, and Heisenberg's uncertainty relation.

Bell's theorem is generally understood as showing the end of classical realism in physics. Interestingly, Niels Bohr, one of the founders of quantum physics, nevertheless put a special emphasis on classical concepts, which Bohr even thinks are indispensable in quantum physics. Let us call it the indispensability argument for classical concepts. In the following we shall see a parallelism between Bohr's classicism and Hilbert's finitism. Bohr took strange features of quantum theory at face value, and yet at the same time advocated the necessity of classical concepts in physics. His philosophy may be called the classical-quantum dualism; to him, classical concepts are indispensable in physics. Bohr's indispensability argument goes as follows:

> [O]nly with the help of classical ideas is it possible to ascribe an unambiguous meaning to the results of observation. (Bohr 1934, p. 17)

And he proceeds as follows:

> No more is it likely that the fundamental concepts of the classical theories will ever become superfluous for the description of physical experience. [...] It continues to be the application of these concepts alone that makes it possible to relate the symbolism of the quantum theory to the data of experience". (*ibid.*)

Bohr's philosophy was actually influenced by his Neo-Kantian teacher Harald Hoffding (Faye 1991; Bella 2001). Hoffding, in turn, built his philosophy upon Heinrich Hertz's, an influential Neo-Kantian at that time, as Christiansen (2006) shows. Hertz and Helmholtz were well-known advocates of the so-called Back-to-Kant movement (see, e.g., Baird et al. 1998). Even the early Wittgenstein's picture theory was influenced by Hertz's picture theory (Kjaergaard 2002). Hertz might be the greatest influencer in modern science. Hilbert was another thinker under the Neo-Kantian influence. Hilbert says:

> The axioms are, as Hertz would say, pictures or symbols in our mind. (Hilbert's 1893–1894 lecture notes; see Corry 2004)

Leo Corry also confirms:

> Hilbert became acquainted with Hertz's book on the foundations of mechanics [...] This book seems to have provided a final, significant catalyst for the wholehearted adoption of the axiomatic perspective. (Corry 2004)

We thus have the Bohr-Hoffding-Hertz-Hilbert link. And there is arguably a Neo-Kantian parallelism between Bohr's classicism and Hilbert's finitism. Bohr (1949) says:

> [H]owever far the [quantum] phenomena transcend the scope of classical physical explanation, the account of all evidence must be expressed in classical terms.

Bohr thus aimed at the justification of the quantum by the classical. Hilbert (1983) essentially had an isomorphic idea:

- However far the [infinitary] phonomena transcend the scope of finitistic explanation, the account of all justifications must be expressed in finitistic terms.

Hilbert aimed at the justification of the infinitary (ideal) by the finitary (real). This is the gist of his programme; the conservativity of ideal mathematics (e.g., ZFC) over real mathematics (e.g., PRA) is equivalent to the consistency of the former over the latter. There are even more commonalities between Bohr's classicisim and Hilbert's finitism:

- Hilbert: the source of meaning in mathematics is finitistic concepts (only finitistic statements have meaning on their own; Zach 2003).
- Bohr: the source of meaning in physics is classical concepts; "only with the help of classical ideas is it possible to ascribe an unambiguous meaning to the results of observation."

Note that Bohr is an instrumentalist about quantum theory Faye (2002); Hilbert an instrumentalist about infinitary mathematics (Zach 2007). Philosophically, Bohr's Neo-Kantian thought goes as follows:

- Classical concepts give the condition of possibility of experimental verification, and so of empirical knowledge. The classical is the ultimate (least possible) ground of physical knowledge, without which physics is impossible because any observations and measurements rely upon classical concepts.

On the other hand, Hilbert's Neo-Kantian philosophy goes as follows:

- Finitism gives the condition of possibility of mathematical verification (esp. consistency proofs), and so of mathematical knowledge. The finitary is the ultimate (least possible) ground of mathematical knowledge, without which mathematics is impossible because any axiomatic systems rely upon finitary concepts (typically in the form of recursive definitions, without which formal systems cannot even be expressed).

Bohr's idea of classical concepts works like meta-theory, which is usually finitism in proof theory; e.g., you usually prove theorems about quantum logic over classical meta-theory. Summing up, here is a conceptual parallelism between Bohr's classicism and Hilbert's finitism, especially in terms of Neo-Kantian philosophy. And Bohr's philosophy may be given a logical interpretation in terms of the distinction between object-level and meta-level: even if the object theory is quantum, the meta theory is classical. And classical concepts are the only way we can reasonably understand the world. Bohr, as a romanticist, sought for harmony between quantum and classical physics. Hilbert, too, as a romanticist, sought for harmony between ideal and real mathematics. From the Bohr-Hilbert point of view, the Wittgensteinian Turn requires the departure from classical reality at the object-level, and yet classical realism is still kept at the meta-level as the condition of the possibility of knowledge.

Let us finally emphasize again the Neo-Kantian origin of foundational thought in the science and philosophy of the early twentieth century, especially the substantial influence of Hertz and Helmholtz, in particular their picture theory, on Hilbert, Bohr, and Wittgenstein. As noted above, Wittgenstein's picture theory stemmed from Hertz's sign theory. Hilbert's axiomatics has the same origin. What is crucial in Hertz's conception of sign or picture is its independence of reality. Patton (2014) says as follows:

> The definition of 'sign' with which Hilbert was most familiar is from the sign and depiction (picture) theories of Hermann von Helmholtz and Heinrich Hertz, respectively. In this tradition, a sign that does not copy properties of an external object can be employed to express thoughts with content within a given framework.

This nature of sign allowed Hilbert to develop the fundamental idea of axiomatics. Both Hilbert and Wittgenstein read Hertz's *Principles of Mechanics*, which led them to the axiomatic foundation of mathematics and the picture theory of meaning, respectively. Corry (2004) says:

> Hertz's Principles of Mechanics made a strong impression on Hilbert, which can be counted among the stimuli for the consolidation of his axiomatic conception.

Hertz is at the common root of Hilbert's axiomatics and the early Wittgenstein's picture theory. The influence of Hertz on Bohr was more indirect, that is, through Bohr's Neo-Kantian teacher Hoffding. Logical positivism also has a Neo-Kantian root, especially in its Marburg School, which included Cassirer in particular. It is well known that Heidegger was Heinrich Rickert's assistant in his early career, influenced by his Neo-Kantian philosophy in the Southwest School. Neo-Kantianism is philosophy just before the analytic-continental divide as well as just before the massive fragmentation of science; the Davos conference in which Heidegger, Cassirer, and Carnap met each other is considered to be the crucial point after which the analytic-continental divide got more and more serious (Friedman 2000). Neo-Kantianism may thus allow us to lay a foundation for philosophy beyond the analytic-continental divide and beyond the fragmentation of science in general.

20.4 Contextuality of Reason and Contextuality of Reality: Bell-Type Theorems beyond Physics

Wittgenstein's later philosophy (especially, *Philosophical Investigations* and *On Certainty*) counts as a form of contextualism. Roughly, contextualism is the position that truth, meaning, or existence is a function of contexts. Put another way, truth, meaning, or existence can only be determined within specific contexts, and in general, remains indeterminate without them. Wittgenstein says:

> Just as the words "I am here" have meaning only in certain contexts, and not when I say them to someone who is sitting in front of me and sees me clearly, – and not because they are superfluous, but because their meaning is not determined by the situation, yet stands in need of such determination. (Wittgenstein 1969, 348)

The more contexts are provided, the less scope of meaning is allowed. In other words, contexts allow us to narrow the possible space of meaning. In the limit of this process, meaning converges into a single point. The complete determination of meaning is not necessarily possible in practice, especially when the possibility of re-interpretation ceaselessly remains. This is actually what happens in the so-called rule-following problem or Kripkenstein's paradox (Kripke 1982):

> [N]o course of action could be determined by a rule, because any course of action can be made out to accord with the rule. (Wittgenstein 1986, p. 81)

A rule cannot determine the resulting practice even if its contexts are added more and more; for the rule and contexts can be re-interpreted again and again. Contexts, in general, allow us to restrict the scope of meaning, and in the ideal limit, meaning can be determined once and for all. Yet the meaning of a rule cannot be completely determined by any additional specification or context since the rule is always open to re-interpretation. This is the gist of the rule-following problem and Kripkenstein's paradox (Kripke 1982). An alleged solution to this sceptical paradox is what is called the sceptical solution: the meaning of a rule is determined not by any logical specification, but by collective linguistic practice. This is a sort of semantic externalism: meaning does not exist in the single human mind, but in a bunch of collective minds within a linguistic community as a coherent whole. There is still the problem of the coherency of the community, without which no single meaning may result after all. The indeterminacy of meaning is also discussed, for example, in Willard Van Oman Quine's argument on what he calls the radical interpretation and ontological relativity (Quine 1969), and even in contemporary French philosophy such as Jacques Derrida's, which refers to Kurt Gödel's incompleteness theorems in order to support the indeterminacy thesis. Hilary Putnam's model-theoretic argument (Bays 2008) may be seen as ontological relatives of indeterminacy, too. Let us come back to Wittgenstein. He says:

> The truth of certain empirical propositions belongs to our frame of reference. (On Certainty, 83)

This suggests a contextual notion of truth. Actually, a strand of epistemic contextualism is generally understood to have an Wittgensteinian origin:

> One such strand was the entertaining of the possibility of some kind of pluralism concerning epistemic standards. In one instance, this took the form of the claim, in response to skepticism, that there are "two senses of 'know"– one 'strong' or 'philosophical', the other 'weak' or 'ordinary' [...] So too, some of Wittgenstein's [...] claims about the relation between

> meaning and use and the multiplicity of "language games", each with its own set of norms, opened the way for a more thoroughgoing kind of semantic pluralism with regard to epistemic concepts and/or terms. (Rysiew 2016)

Although there are some disagreements about whether Wittgenstein's philosophy is relevant to contextualism in contemporary philosophy, it does not really matter here. Wittgenstein's strong emphasis on actual practice seems to be a driving force for his contextualism and pluralism. Human cognition or reason is inherently contextual and plural in actual practice; subtle differences in contexts can make large differences in cognition. The relationships between cognition and contexts are hotly debated in artificial intelligence, under the theme of embedded, embodied, or situated AI. Cognition does not exist in vacuum, coming to existence by being embedded in environments, embodied in actual bodies, and situated in actual situations. A similar recent trend of research in cognitive science is 4E (embodied, embedded, enactive, and extended) cognition. The extended mind thesis is a form of externalism as discussed above, telling that certain environments are part of the mind. Enactivism is the view that cognition is enabled via dynamic interactions with environments. Simply saying, they say that cognition would not exist without environments or contexts. So they are different forms of contextualism in cognition. There is also what is called Heideggerian AI (Dreyfus 2007), which too emphasizes the inseparability between cognitive agents and their environments. Contextuality of Being is a central issue in Heidegger's philosophy. Contextualism is a philosophy transgressing the boundary between the analytic and continental traditions.

In the following we shall discuss recent developments of contextuality studies across physics and cognitive science. In physics, Bell-type theorems prove contextuality of reality, a particular case of which is quantum non-locality, what Einstein called "spooky action at a distance" and what, at the end of the day, defeated Einstein's hope for local hidden variable theories, which allow us to keep the classical picture of reality, and yet turned out to be impossible. Although quantum theory is occasionally said to support indeterminism, there are actually deterministic interpretations of quantum theory, such as Bohmian mechanics and many-worlds interpretation. The non-classical feature of quantum theory is arguably best understood as contextuality and non-locality rather than indeterminism. Contextuality in quantum physics means that there is no consistent assignment of values or probability distributions on all variables in different contexts involved; put another way, measurement values or measurement statistics essentially depend upon measurement contexts. That is to say, we can measure each variable in each context, and yet the results thus obtained are not consistent as a whole when combined together. This is why contextuality is characterized as local consistency plus global inconsistency. There is no problem in measuring each variable in different local contexts, and yet it is impossible to have the global assignment of values or distributions to all variables at once that is consistent with the results in local contexts. Logically phrasing, all physical propositions cannot have bivalent truth values at once. Truth values in quantum physics are contextual, i.e., the truth values of physical propositions only exist within particular contexts.

As Wittgenstein's philosophy tells us, meaning is contextual as well as reality. Contextuality is a phenomenon observed across different sciences, including physics, linguistics, artificial intelligence, and cognitive science. Among other things, psychology is necessarily involved in contextuality of cognition or reason. Subjects in psychological experiments are readily influenced by different environments or contexts. And context-sensitivity in psychology conceptually looks analogous to that in physics. This idea has led to recent developments of quantum psychology and quantum cognitive science. In the following we shall review them, and discuss contextuality in cognition and its relationships with contextuality in reality. So what is quantum cognitive science? And what is it good for? Quantum cognitive science (aka. quantum cognition) is an emerging field; the literature on it has rapidly expanded in the last ten or fifteen years. Quantum cognitive science is highly interdisciplinary, involving psychology, linguistics, decision theory, behavioral economics, and so fourth. It is well known that quantum physics exhibits super-classical correlations such as non-local correlations, and yet there is still an upper bound on the strengths of quantum correlations such as signaling correlations and so-called PR box correlations, which are known to be super-quantum. Some researchers think that context-sensitivity in psychology results in super-quantum correlations; indeed, the no-signaling condition is violated in some psychological experiments while it holds in quantum physics (Dzhafarov 2016). Quantum cognition is a rapidly developing, vital field and yet its future is still uncertain. We do not yet know whether it is a "new kind of science" or a new kind of "fashionable nonsense." Still, quantum cognition is shedding new light on the fundamental nature of human reason, such as rationality and contextuality, as we shall see in the following. Typical topics of quantum cognition include:

- The order effect in psychology. It is non-commutativity in psychology. Let Q_1 be "Is Clinton honest and trustworthy?", and Q_2 "Is Al Gore honest and trustworthy?". Then, Q_1 and Q_2 are non-commutative. That is, it is experimentally verified (Wang-Solloway-Shiffrin-Busemeyer 2014) that if you ask Q_1 first then you have got a less probability for Q_2 (i.e., less people answer yes for Q_2), and if you ask Q_2 first you have got a higher probability for Q_1 (i.e., less people answer yes for Q_1). Cognition is non-commutative.
- The conjunction effect in cognitive biases. Succinctly saying, it shows that $\mathrm{Prob}(\varphi \wedge \psi) \leq \mathrm{Prob}(\psi)$ does not hold in certain cognitive bias experiments such as the Linda problem (Tversky and Kahneman 1982). In the Linda experiment, subjects are given a description about Linda that makes Linda look like a feminist, such as the following:

  > Linda is 31 years old, single, outspoken, and very bright. She majored in philosophy. As a student, she was deeply concerned with issues of discrimination and social justice, and also participated in anti-nuclear demonstrations.

  And then they are asked to judge which of "Linda is a bank teller" and "Linda is a bank teller and active in the feminist movement" is more probable. A significant number of people tend to choose the latter, regardless of their various backgrounds. Hence the violation of $\mathrm{Prob}(\varphi \wedge \psi) \leq \mathrm{Prob}(\psi)$.

- The disjunction effect in the prisoners' dilemma. Rational decision theory tells us that a prisoner defects regardless of whether the other prisoner defects or not. But this is experimentally violated, and information about the other prisoner's decision significantly affects the prisoner's decision (Pothos-Busemeyer 2009). Mathematically, it is violation of the sure thing principle or the law of total probability: $\mathrm{Prob}(\varphi) = \mathrm{Prob}(\psi)\mathrm{Prob}(\varphi|\psi) + \mathrm{Prob}(\neg\psi)\mathrm{Prob}(\varphi|\neg\psi)$.

Quantum cognition exploits quantum mechanical models (or general probabilistic models) to account for such non-classical features of cognition, in particular violations of the laws of classical probability; the statistics of non-classical phenomena in psychology can (often or sometimes) be explained using statistical models in quantum physics (Busemeyer-Bruza 2014). There are some physical phenomena that classical physics cannot account for, and thus we need quantum physics to do that. Likewise, there are some cognitive phenomena that classical psychology or cognitive science cannot account for, and thus we need quantum psychology or quantum cognitive science. Logically, these effects may be interpreted in terms of substructural logic. Quantum cognitive science is particularly interesting for the following reason: if there are structural mechanisms shared by both physics and cognition, it would pave the way for overcoming the Cartesian dualism of matter and mind, just as Chalmers' property dualism or double-aspect theory of information (Chalmers 1996) aims at elucidating the higher structural laws of information that govern the actual laws of matter and of cognition. It would count as a scientifically sound embodiment of Chalmers' property dualism or double-aspect theory of information.

There are some Bell-type results recently found in cognitive science. Bell-type No-Go results in physics often exhibit obstructions to classical realism (e.g., local realism). Do cognitive Bell-type results imply the end of some cognitive realism? Do violations of Bell-type inequalities in cognitive experiments refute any realism? More broadly, how do they affect the realist wordlview? We shall discuss these questions in the following, but before that, let us touch upon past research on the alleged quantum nature of the brain. Roger Penrose is known for his controversial argument for the quantum nature of the brain. In particular, Penrose (1989, 1994) argues as follows:

(i) AI or the computational theory of mind is misconceived in light of Gödelian incompleteness. This is the (notorious) Lucas-Penrose argument; the capacity of human cognition is not bound by computability. It has an origin in Gödel (1955): "the human mind (even within the realm of pure mathematics) infinitely surpasses the power of any finite machine" or "there exist absolutely unsolvable diophantine problems", which was not criticized much, though.
(ii) The mind is materially quantum; consciousness emerges via material quantum processes in microtubules. Let us call it the Material Quantum Mind Thesis (Material QMT for short).

Penrose was criticized so much for these arguments. And approximately two hundred pages of his *Shadows of the Mind* are devoted to replies to different criticisms. David Chalmers (1955) argues against Penrose as follows: "Why should quantum processes in microtubules give rise to consciousness, any more than computational processes should?"; "reader who is not convinced by Penrose's Gödelian arguments is left with little reason to accept his claims that physics is non-computable and that quantum processes are essential to cognition." Yet, ten years later, Pothos and Busemeyer (2009) say:

> [T]he success of human cognition can be partly explained by its use of quantum principles.

There is now some reason to accept Penrose's claim that "quantum processes are essential to cognition." Yet we have to be careful here; quantum effects may not necessarily arise from the material brain. Indeed, Tegmark (2000) argues as follows:

> Based on a calculation of neural decoherence rates […] the degrees of freedom of the human brain that relate to cognitive processes should be thought of as a classical rather than quantum system […] This conclusion disagrees with suggestions by Penrose and others that the brain acts as a quantum computer, and that quantum coherence is related to consciousness in a fundamental way.

If this is correct, there is no quantum effect on the material brain. And there is no problem on classical realism about the brain. In the following I assume that Tegmark is correct in this respect: the cognitive mechanism of the human brain is classical; human beings are classical macroscopic objects within the scope of classical realism. We shall later use this classical brain assumption. The classical brain assumption may be comparable with Bohr's classical observer thesis. Note that Penrose actually does not claim that the brain is a quantum computer (while Hameroff does). Indeed, Penrose says in his 1995 paper "Consciousness Involves Noncomputable Ingredients" as follows:

> When I argue that the action of the conscious brain is noncomputational, I'm not talking about quantum computers. Quantum computers are perfectly well-defined concepts, which don't involve any change in physics; they don't even perform noncomputational actions. […] I don't think it can explain the way the brain works. That's another misunderstanding of my views."

Tegmark understood the brain, but probably not Penrose's.

Now let us think about how to fix Penrose's argument. A promising way is as follows. Replacing (i) computability by complexity, and (ii) the material QMT by the structural QMT we have got:

(i) Classical AI or the classical computational theory of mind is misconceived in light of the super-classical features and effectiveness of human cognition. The mind cannot be a classical computer due to differences in complexity.
(ii) The mind is structurally quantum; the structure of cognition is homomorphic to the structure of quantum information. It is not that quantum processes are materially going on in the macroscopic physical brain.

The structure of economic systems is homomorphic to that of physical systems, and so we can apply physical models to explain economical phenomena. Yet this never means that the nature of economy is materially physical. The same goes for quantum cognitive science. We can apply quantum models to explain (the statistics of) cognitive phenomena. Yet this never means that the nature of cognition is materially quantum; rather it is structurally quantum. If the universe is a materially quantum computer, the mind might be a structurally quantum computer. The above idea is also related to functionalism, which states that mental states are characterized by what they do rather than by what they are made of. The essence of the mind is the cognitive function rather than the material substance (cf. Cassirer's *Substance and Function*). Different functions of the mind may be explained by different quantum-like models.

Contextuality in cognition involves the fundamental problem of psychology in the following sense.

- Like quantum systems, cognitive systems are sensitive to contexts of measurement.
- Unlike them, cognition is so embedded in contexts that contextual side-effects cannot adequately be controlled.

Cognitive systems are "beings-in-the-world" (cf. Dreyfus' embedded/situated AI). The life scientist often says, "Life is warm, wet, and noisy." Physical experiments, too, are subject to contextual noise, which can still mostly be controlled. There are both internal and external noise in psychological experiments, caused by uncontrollability of mental states and by uncontrollability of environments, respectively. Contextual effects make state preparation difficult in psychology, in which it is unclear what kind of cognitive states are to be measured in the first place. Due to such contextual noise, cognition gets super-quantum, and violates the no-signalling (marginal selectivity) principle as noted in Dzhafarov et al. (2016). The relationships between parts and wholes, therefore, are even more complex in psychology than in quantum physics (cf. Heisenberg, *Der Teil und das Ganze*, 1969):

- In quantum physics: wholes ($\otimes$) are not direct sums ($\times$) of parts. It is called quantum holism.
- In psychology: parts (marginals) are not direct restrictions of wholes (joint distributions). It may be called cognitive holism.

The first property is called quantum holism, which is conceptually analogous to Gestalt psychology. Note that the difference between tensor product ($\otimes$) and direct product ($\times$) gives rise to quantum phenomena such as entanglement and non-locality. When the no-signalling condition is violated as in the second case, Bell-type inequalities must be extended so as to take such additional effects into account. The violation of no-signalling is caused by contextual noise, which Dzhafarov et al. (2016) call direct contextuality as opposed to properly quantum contextuality. Dzhafarov et al. (2016) extends Bell-type inequalities so as to take direct contextuality into account, and to detect properly quantum contextuality even in the coexistence of direct contextuality. In terms of the characterization of quantum contextuality as local consistency plus global inconsisntecy, cognitive holism implies that even local consistency may not hold in the statistics of cognition. Local inconsistency and global inconsistency together imply that parts are independent of wholes; this property may be called strong holism as opposed to weak holism such as quantum holism.

It should be remarked that contextual noise is what threatens reproducibility in cognitive and life sciences while leading to the aforementioned statistical violation of no-signalling; the reproducibility crisis has been a big concern in cognitive and life sciences. According to The Reproducibility Project: Psychology, only 35% of published experimental studies were reproducible (Weir 2016). There is a similar problem in medical science (e.g., the STAP affair in Japan). The internal and external noise as hidden contextual noise threatens the reproducibility of cognitive experiments. It is relatively rare in physical experiments; contextual noise is more controllable in physics, even though it is true that cosmology or the history of the universe is not really reproducible. Cognitive systems are so complex and sensitive that subtle contextual noise cannot be controlled on them.

Summing up the points so far, No-Go theorems in quantum foundations, such as Bell's and Kochen-Specker's, tell us that fundamental reality is inherently contextual (and non-local as a particular case); from a theoretical point of view (e.g., Abramsky-Brandenburger 2011), contextuality is local consistency plus global inconsistency (of contextual information; e.g., probability distributions on measurement contexts). According to recent research in quantum cognitive science (e.g., Cervantes-Dzhafarov 2019), Bell-type inequalities may be reformulated so as to be applicable in cognitive science, and they are actually violated in certain cognitive experiments. Now several questions arise as follows. Do Bell-type results in cognitive science show that human reason is contextual as well as fundamental reality? If so, in what sense? Put another way, do cognitive systems exhibit the same kind of contextuality as quantum systems? And ultimately, do Bell-type results in cognitive science have such a massive impact on our understanding of the world as those in quantum physics indeed had? There are analogies and disanalogies, both mathematically and philosophically, between contextuality of reality and cotenxtuality of reason. In the following let us attempt to articulate the analogies and disnalogies between contextuality of reality and contextuality of reason, especially in light of the nature of probabilities involved. The point is that quantum and cognitive systems exhibit the same kind of contextuality at a mathematical level of statistical correlation (apart from the issue of violation of marginal selectivity), and yet physical contextuality differs from cognitive contextuality in terms of how relevant probabilities are interpreted therein. This disagreement about the nature of probabilities, arguably, makes the meaning of Bell-type theorems in cognitive science depart from that in quantum physics in a significant manner. This would also explicate how science is not just about the analysis of statistical correlation.

So our question is: what demarcates cognitive contextuality from quantum contextuality? And how does cognitive contextuality impact our conception of the world? Conceptually there is relevance to the Cartesian dualism. It may account for the reason why the Cartesian dualism must be a dualism. Contextuality across physics and cognitive science tells us that there are substantial commonalities between the laws of Nature and the laws of Reason; this suggests that the gap between matter and mind in the Cartesian dualism may be filled in to some extent. Yet, as we shall see below, it also tells us fundamental differences between the science of matter and the science of mind, or why the Cartesian dualism must be a dualism rather than a monism. Understanding the relationships between cognitive and physical contextuality contributes to this broader task of elucidating the Cartesian dualism.

The intrinsic difference between physical and cognitive contextuality is concerned with the origins of probabilities involved. Where are origins of probabilities in quantum physics and in psychology?

- Physics: you get statistics by repeating an experiment on a single state (e.g., entangled state). Probabilities come from the state per se. They are state-intrinsic.
- Psychology: you get statistics by repeating an experiment on different subjects in different mental states (e.g., someone on a university campus). Probabilities do not come from any specific state. They are state-extrinsic.

States are not fixed in psychological experiments while they are fixed in physics. Why? There would be both positive and negative reasons. The negative reason is that it is difficult to fix mental states of human subjects in psychological experiments. The positive reason is that the results get averaged if the number of experiments tends to infinity, and this is crucial because you want to apply the inferred law of cognition to arbitrary persons (e.g., someone who suddenly came to you at a hospital). Put another way, if you fix a cognitive state in an experiment, the law of cognition that results from it will only apply to those who are in the same cognitive state, and there is no reason that different persons can actually have the same cognitive state. In a nutshell, if cognitive science gets highly personalized, it loses its broad applicability, generality, and universality. There is a merit for not fixing cognitive states: by doing so, we can make cognitive laws applicable to arbitrary persons, not with absolute certainty, and yet with some statistical certainty. This is the positive reason.

What happens in the phenomenon of quantum contextuality or non-locality in particular is basically that you have a special state, and then you get special statistics (here we do not think about state-independent contextuality arguments). What happens in cognitive contextuality is that you have ordinary people in ordinary, yet different mental states, and yet a special experiment allows you to get special statistics. To elucidate the difference between cognitive and quantum contextuality, let us think of an experimental version of Laplace's demon. The theoretical Laplacian demon has the infinitary power of exact computation. We think of the experimental Laplacian demon as follows:

- The experimental Laplacian demon has the infinitary power of precise experimentation. In particular, the demon can fix every hidden parameter of a (cognitive) experiment in any noisy environment. The demon repeats the experiment countless times. Yet he always gets the same result since he fixed every parameter (under the classical brain assumption). Hence no probabilities in the demon's psychology experiment.
- To the demon, therefore, there is no indeterminacy or contextuality in cognition (because he fixed every parameter himself). Note that there is no statistics to be analyzed in the demon's experiment because the experimental Laplacian demon can fix every hidden parameter in the environment.
- What if the demon performs contextuality (or non-locality) experiments on quantum systems? The demon cannot erase contextuality by manipulating parameters since quantum contextuality is caused by the intrinsic properties of states and

measurements. Rather the demon would obtain a more precise detection of contextuality in the system. In a nutshell, quantum contextuality is immune to the demon.

This (Gedanken) experiment explicates the crucial difference between cognitive contextuality and quantum contextuality. And we can conclude that what causes cognitive contextuality is the nature of experimental set-up rather than the nature of reality. The brain is deterministic, and does not involve any probabilities. Still a special experimental-set up can yield contextuality by performing the experiment on different cognitive states, which are the source of probabilities. From another angle:

- Probability models in quantum physics (such as Bell's) arise by measuring a single state in different contexts.
- Probability models in cognitive science (such as Cervantes-Dzhafarov's) arise by measuring different states in different contexts.
- The demon encounters contextuality in the first case, and yet can erase contextuality in the second case by fixing parameters involved.

The difference may be conceived of as the difference between intrinsic and extrinsic probabilities in the following manner. At a mathematical level, cognitive and quantum contextuality have basically the same statistical structure. As to the nature of probabilities, quantum contextuality concerns intrinsic probabilities; probabilities come from the nature of a state measured. Cognitive contextuality concerns extrinsic probabilities; they come from the nature of different states collected together. For exact prediction in cognitive science, you have to take into account all relevant hidden contexts. This leads to the same kind of infinite regress as in the frame problem in artificial intelligence. And it is practically impossible to consider all possible relevant contexts. The demon alone can erase contextuality; the human being could not. If the demon conducts experiments on different states without fixing parameters, the demon may observe contextuality.

Let us wrap up the discussion. What particularly demarcates cognitive from quantum contextuality would be the following:

- The human mind is both classical and quantum: its hardware, the material brain, is classical and deterministic; its software, the cognitive function, can be quantum (-like) and indeterministic when measured collectively.
- Whilst physical Bell-type results refute classical realism about single state dynamics, cognitive Bell-type results refute classical realism about collective state dynamics, but not about single state dynamics.
- Physical No-Go results refute (non-contextual) hidden variable theories about single state dynamics. Cognitive No-Go results refute (non-contextual) hidden variable theories about collective state dynamics, but not about single state dynamics.
- Each single state is classical in psychology, and there is no indeterminism, contextuality, or anything quantum within the single state, as shown by the experimental Maxwell's demon thought-experiment.

Since it is practically impossible to explicate all relevant contexts, we have got probabilities in cognitive experiments, and when the experimental set-up is specially prepared, we have got contextuality as the violation of Bell-type inequalities. Yet this never means that the brain is indeterministic. Cognitive contextuality is caused by two factors: the statistical nature of collective state dynamics and the special structure of experimental set-up. Quantum contextuality is caused by two factors: the statistical nature of single state dynamics and the very existence of special states or operations. Measuring a cognitive agent in different states (e.g. via questions) yields different results (e.g., answers); it, at least, looks as if there is some free will involved here. This phenomenal free will is actually a source of cognitive contextuality as clearly observed in the Cervantes-Dzhafarov experiment, in which two correct answers are always possible, and subjects have to choose one of them by their free will.

In a broader perspective, we can still find commonalities between Bell-type No-Go results in cognitive science and in quantum physics. Quantum cognition arguably shares the same spirit as original Bell-type theorems to the following extent:

- Quantum Physics: classical models of physics do not hold any more. Because there are contextual effects in fundamental reality.
- Quantum Cognitive Science: classical models of cognition and decision do not hold any more. Because there are contextual effects in human reason. We need new models to take into account non-classical features of cognition. And several quantum models have been shown to work for this purpose.

Amartya Sen (1977) says:

> [T]he puzzle from the point of view of rational behavior lies in the fact that in actual situations people often do not follow the selfish strategy. Real life examples of this type of behavior in complex circumstances are well known, but even in controlled experiments in laboratory conditions people playing the Prisoners' Dilemma frequently do the unselfish thing.

Classical rationality is selfish; human rationality is unselfish, at least occasionally. Quantum rationality can be unselfish as there is a quantum model of the non-classical deviation from the classical model of the prisoners' dilemma. In a broader perspective, contextuality in cognition does matter for two reasons: it explicates the nature of human rationality, and it elucidates the embeddedness and situatedness of Being, which is relevant in AI as well.

Acknowledgements I am grateful to Samson Abramsky for discussions about contextuality, to Kohei Morita for discussions about Bohr and his philosophy of classical concepts, and to Takeo Uramoto for discussions about quantum physics as well as Wittgenstein's philosophy of language. I hereby acknowledge financial support from JST (grant number: JPMJPR17G9) and JSPS (grant number: 17K14231).

References

Abramsky, Samson, and Brandenburger, Adam, 2011, The Sheaf-Theoretic Structure Of Non-Locality and Contextuality, *New Journal of Physics*, vol. 13, 113036.

Baird, David, et al. (eds.), 1998, *Heinrich Hertz: Classical Physicist, Modern Philosopher*, Netherlands: Springer.
Bays, Timothy, 2008, Two arguments against realism, *Philosophical Quarterly*, vol. 58, pp. 193–213.
Bella, Mara, 2001, *Quantum Dialogue: The Making of a Revolution*, Chicago: University of Chicago Press.
Berman, Morris, 1981, *The Reenchantment of The World*, Ithaca: Cornell University Press.
Bohr, Niels, 1934, The quantum of action and the description of nature, in: *Atomic theory and the description of nature*, Cambridge: Cambridge University Press.
Bohr, Niels, 1949, Discussions with Einstein on Epistemological Problems in Atomic Physics, in: *Philosopher Scientist*, pp. 199–241.
Bourguignat, Christophe, 2014, Interpretable VS Powerful Predictive Models : Why We Need Them Both, *Medium*, 17 September 2014, retrieved on 23 June 2019.
Busemeyer, Jerome, and Bruza, Peter, 2014, *Quantum Models of Cognition and Decision*, Cambridge: Cambridge University Press.
Cass, Stephen, 2011, Unthinking Machines, *MIT Technology Review*, 4 May 2011, retrieved on 23 June 2019.
Cassirer, Ernst, 1923, *Substance and Function, and Einstein's Theory of Relativity*, Chicago: Open Court.
Cassirer, Ernst, 1945, Structuralism in Modern Linguistics, *Word*, vol. 1, pp. 99–120.
Cassirer, Ernst, 1955, *The Philosophy of Symbolic Forms*, New Haven: Yale University Press.
Chalmers, David, 1995, Minds, Machines, and Mathematics, *Psyche*, vol. 2, pp. 11–20.
Chalmers, David, 1996, *The Conscious Mind: In Search of a Fundamental Theory*, Oxford: Oxford University Press.
Cervantes, Victor, and Dzhafarov, Ethibar, Snow Queen is Evil and Beautiful: Experimental Evidence for Probabilistic Contextuality in Human Choices, *Decision*, vol. 5, 193–204.
Chomsky, Noam, 1957, *Syntactic Structures*, Berlin: Mouton & Co.
Chomsky, Noam, 1975, Recent contributions to the theory of innate ideas, in: S. Stitch (ed.), *Innate Ideas*, Berlin: California University Press.
Chomsky, Noam, and Ronat, Mitsou, 1979, *Language and Responsibility: Based on Conversations with Mitsou Ronat*, Paris: Pantheon Books.
Chomsky, Noam, 2011, Keynote Panel: The Golden Age — A Look at the Original Roots of Artificial Intelligence, Cognitive Science, and Neuroscience, *MIT Symposium on Brains, Minds, and Machines*, held in 2011, recording available at: https://mit150.mit.edu/symposia/brains-minds-machines.html, retrieved on 23 June 2019.
Christiansen, Frederik Voetmann, 2006, Heinrich Hertz's Neo-Kantian Philosophy of Science, and its Development by Harald Hoffding, *Journal for General Philosophy of Science*, vol. 37, pp. 1–20.
Corry, Leo, 2004, *David Hilbert and the Axiomatization of Physics (1898–1918)*, Netherlands: Springer.
Dreyfus, Hubert, 2007, Why Heideggerian AI failed and how fixing it would require making it more Heideggerian, *Philosophical Psychology*, vol. 20, 247–268.
Dzhafarov, Ethibar, et al., 2016, On contextuality in behavioural data, *Philos. Trans. A Math. Phys. Eng. Sci.*, vol. 374, 20150234.
Faye, Jan, 1991, *Niels Bohr His Heritage and Legacy: An Anti-Realist View of Quantum Mechanics*,
Faye, Jan, 2002, Copenhagen Interpretation of Quantum Mechanics, *Stanford Encyclopaedia of Philosophy*.
Frege, Gottlob, 1884, *Grundlagen der Arithmetik*.
Friedman, Michael, 2000, *A Parting of the Ways: Carnap, Cassirer, and Heidegger*, Chicago: Open Court Press.
Gold, Kevin, 2011, Norvig vs. Chomsky and the Fight for the Future of AI, *TOR.COM*, 21 June 2011, retrieved on 26 June 2019.

Gödel, Kurt, 1961, The modern development of the foundations of mathematics in the light of philosophy, in: *Collected Works*, vol. III, 1981, Oxford: Oxford University Press.

Gödel, Kurt, 1995, *Collected Works*, vol. III, New York: Oxford University Press.

Gray, Jeremy, 2008, *Plato's Ghost: The Modernist Transformation of Mathematics*, Princeton: Princeton University Press.

Griffin, David, et al., 1992, *Founders of Constructive Postmodern Philosophy: Peirce, James, Bergson, Whitehead, and Hartshorne*, New York: SUNY Press.

Griffin, David, 2001, *Reenchantment without Supernaturalism*, Ithaca: Cornell University Press.

Hilbert, David, 1983, "On the Infinite", in: *Philosophy of Mathematics: Selected Readings*, Cambridge: Cambridge University Press.

Katz, Yarden, 2012, Noam Chomsky on Where Artificial Intelligence Went Wrong, *The Atlantic*, 1 November 2012, retrieved on 23 June 2019.

Kamp, Hans, and Partee, B., 1995, Compositionality and Prototypes, *Cognition*, vol. 57, pp. 129–191.

Kuhn and Johnson, 2013, *Applied Predictive Modeling*, New York: Springer.

Kripke, Saul, 1982, *Wittgenstein on Rules and Private Language*, Massachusetts: Harvard University Press.

Lofts, Steve, 2000, *Cassirer: A "Repetition" of Modernity*, Albany: SUNY Press.

Markie, Peter, 2017, Rationalism vs. Empiricism, *Stanford Encyclopedia of Philosophy*

Maxwell, James Clerk, 1990, *The Scientific Letters and Papers of James Clerk Maxwell: 1846-1862*, Cambridge: Cambridge University Press.

Norvig, Peter, 2011, On Chomsky and the Two Cultures of Statistical Learning, in: W. Pietsch, J. Wernecke, M. Ott (eds.), *Berechenbarkeit der Welt?* (pp. 61–83), Wiesbaden: Springer. (made available online in 2011; published in 2017)

Parsons, Charles, 2014, *Philosophy of Mathematics in the Twentieth Century: Selected Essays*, Massachusetts: Harvard University Press.

Patton, Lydia, 2014, Hilbert's Objectivity, *Historia Mathematica*, vol. 41, pp. 188–203.

Penrose, Roger, 1989, *The Emperor's New Mind*, Oxford: Oxford University Press.

Penrose, Roger, 1994, *Shadows of the Mind*, Oxford: Oxford University Press.

Peter C. Kjaergaard, 2002, Hertz and Wittgenstein's Philosophy of Science, *Journal for General Philosophy of Science*, vol. 33, pp 121–149.

Pothos, Emmanuel, and Busemeyer, Jerome, 2009, A quantum probability explanation for violations of 'rational' decision theory, vol. 276, 1665.

Putnam, Hilary, 1983, *Realism and Reason*, Philosophical Papers vol. 3, Cambridge: Cambridge University Press.

Quine, Willard Van Oman, 1969, Ontological Relativity and Other Essays, New York: Columbia University Press.

Rorty, Richard, 1962, Realism, Categories, and the "Linguistic Turn", *International Philosophical Quarterly*, vol. 2, pp. 307–322.

Rorty, Richard, ed., 1967, *The Linguistic Turn: Recent Essays in Philosophical Method*, Chicago, The University of Chicago Press.

Rorty, Richard, 1970, *Philosophy and the Mirror of Nature*, Princeton: Princeton University Press.

Rysiew, Patrick, 2016, Epistemic Contextualism, *Stanford Encyclopedia of Philosophy*, Stanford: Stanford University.

Sen, Amartya, 1977, Rational fools: A critique of the behavioral foundations of economic theory, *Philosophy and Public Affairs*, vol. 6, pp. 317–344.

Tegmark, Max, 2000, Importance of quantum decoherence in brain processes, *Phys. Rev. E*, vol. 61, 4194.

Turney, Peter, and Pantel, Patrick, 2010, From Frequency to Meaning: Vector Space Models of Semantics, *Journal of Artificial Intelligence Research*, vol., 37, pp. 141-188.

Tversky, A., and Kahneman, D., 1982, Judgments of and by representativeness, in: *Judgment under uncertainty: heuristics and biases*, Cambridge: Cambridge University Press.

Wang, Zheng, Solloway, Tyler, Shiffrin, Richard, and Busemeyer, Jerome, 2014, Context effects produced by question orders reveal quantum nature of human judgments, *Proc. Nac. Aca. Sci.*, vol. 111, 9431–9436.
Weir, Kristen, 2016, A reproducibility crisis?, *American Psychological Association*, retrieved 23 June 2019.
Wittgenstein, Ludwig, 1961, *Tractatus Logico-Philosophicus*, New York: Humanities Press.
Wittgenstein, Ludwig, 1969, *On Certainty*, Oxford: Basil Blackwell.
Wittgenstein, Ludwig, 1970, *Zettel*, California: University of California Press.
Wittgenstein, Ludwig, 1986, *Philosophical Investigations*, Oxford: Basil Blackwell.
Zach, Richard, 2003, Hilbert's program, *Stanford Encyclopedia of Philosophy*.
Zach, Richard, 2007, Hilbert's program then and now, in: *Philosophy of Logic*, Amsterdam: North Holland. pp. 411–447.

Chapter 21
Can Robots Learn to Talk?

Lars Hertzberg

We have all seen or read about talking robots. They will utter greetings, make small talk, answer questions up to a point, and so on. The question is: how do robots acquire the capacity to talk? Can we think of robots as having *learnt* to talk, or are we to think about their coming to talk in a different way?

What I am interested in is not technology and its limits, but rather conceptual issues: what concerns me is the way reflecting on the question whether robots can learn to talk might throw light on *what is involved in learning to talk*, whether for a robot or a human being. By a robot we usually mean an entity all of whose functioning is directly or indirectly controlled by programming. (The control is indirect when part of the programme consists in instructions to receive inputs from the surroundings of the robot and apply them in its functioning—what is called "machine learning".) Beginning with Descartes, people have tended to fend off claims that robots may one day be indistinguishable from humans in their capacities by invoking the complexity of human cognitive abilities, the power of human beings to respond to novel situations in innovative ways, etc. This line of thought is often countered by the argument that there may be no limits to what robot technology may achieve, if not now, then in a distant future. I for one find the back-and-forth of this debate unsatisfactory. As I wish to argue, for a better understanding of these issues, we need to undertake a deep revision of the categories in which we tend to think about them. On the whole, I have a sense that discussions of how robots are like or unlike human beings should focus more on human beings and less on robots than they often do.[1]

I will start by discussing an experiment in which an attempt is made to model teaching a robot to speak on a child's learning to speak. The experimental design reveals the assumption that learning to speak is based on learning to use words to refer to objects, an assumption which seems to form a recurrent theme in Western thought

[1]For an essay pursuing similar lines of thought to those developed here, see Beran (2018).

L. Hertzberg (✉)
Department of Philosophy, Åbo Akademi University, 20500 Turku, Finland
e-mail: lhertzbe@abo.fi

S. Wuppuluri and N. da Costa (eds.), *WITTGENSTEINIAN (adj.)*,
The Frontiers Collection, https://doi.org/10.1007/978-3-030-27569-3_21

about language, being expressed, among others, by St Augustine in his *Confessions*. In criticizing Augustine's conception, Wittgenstein introduces the so called builders' game, which illustrates a case in which speaking is limited to referring to objects. He wishes to show how impoverished such a language would be, a critique which is later deepened by Rush Rhees, in his arguing that learning to speak is radically different from learning to play a game. What is missing in playing a game, or in simple forms of word use such as verbal routines or skills—forms of activity which could be simulated on a robot—is the dimension of words becoming an expression of the speaker, being given a place in her life. In the case of the robot, this dimension is lacking.

1. *Is it all about objects?*

The inspiration for this essay came from reading an article in *The Guardian* by Alex Beard, called "How babies learn – and why robots can't compete".[2] It was about Deb Roy, a robotics expert at MIT, married to a professor of speech pathology. In 2005, Roy initiated a research project called "The Human Speechome Project", which consisted in setting up a video recording system in their home, which recorded virtually every moment in their son's life from the moment he came home from the maternity ward until he was around three. The *Guardian* article was written 13 years after the start of the project.

This material, if treated with understanding, would seem to be an amazing source of insight into child development. However, as far as I have been able to make out, surprisingly little has come out of the project in that regard. One reason for the paucity of outcomes may be that Roy wanted to limit himself to results that could be given a quantitative treatment; that is, results that did not require trying to make sense of what went on in various interactive situations. One thing Roy studied was in which part of the house various words were the most frequently used. The words "bye bye", for instance, were most frequently heard in the hallway.[3]

In fact, the real motivation for the project was not the study of child development as such, but the possibility of using these observations in teaching robots to speak. The idea was to deal with a robot in the same way one dealt with the child in situations involving speech, in the hope that the robot like the child would pick up speech in this way. These hopes were frustrated however.

Roy used a robot he had constructed:

> Using voice recognition and pattern-analysing algorithms, Roy had painstakingly taught Toco to distinguish words and concepts within the maelstrom of everyday speech. Where previously computers learned language digitally, understanding words in relation to other words, Roy's breakthrough was to create a machine that understood their relationship to objects. Asked to pick out the red ball among a range of physical items, Toco could do it.

Now this passage raises a host of questions, some of which I will get back to later. In any case, Roy gradually gave up on his ambition. As put by the journalist,

---

[2] Also in Beard (2018).

[3] https://www.ted.com/talks/deb_roy_the_birth_of_a_word?language=en. Accessed 19 January, 2019.

> Watching his son, Roy had been blown away by 'the incredible sophistication of what a language learner in the flesh actually looks like and does'. Infant humans didn't only regurgitate; they created, made new meaning, shared feelings... He had discovered that human learning was communal and interactive. For a robot, the acquisition of language was abstract and formulaic. For us, it was embodied, emotive, subjective, quivering with life.

Well, Roy must have known to start with, as we all do, that human learning is communal and interactive. His disappointment may have lain in the fact that he found the processes of learning did not open themselves to quantitative treatment. The project dissolves in a haze of adjectives: "embodied, emotive, quivering with life". Anything that cannot be measured, it appears, is shrouded in mystery.

What I find philosophically interesting is Roy's ambition and the thinking that lies behind it. The aim of Roy's project seems to have been to get robots to connect words with objects. This is what learning to speak is, he seems to have assumed, and this is what he failed to achieve by using the methods one uses with children. Roy recounts the following episode:

> The first time his son uttered something that wasn't just babble, Roy was sitting with him looking at pictures. "He said 'fah'", Roy explained, "but he was actually clearly referring to a fish on the wall that we were both looking at. The way I knew it was not just coincidence was that right after he looked at it and said it, he turned to me. And he had this kind of look, like a cartoon lightbulb going off – an 'Ah, now I get it' kind of look.

This episode is interesting for various reasons. It is revealing that Roy regards this episode in terms of *referring*. He fails to see that applying this term to the situation described is highly problematic. Referring is something we *do* in the course of communicative interaction. It is something I mean for my interlocutor to take in a certain way; something I, as speaker, take responsibility for, for instance, in the context of saying something true or false. Furthermore, we refer to something under some *aspect*: one might refer to the fish picture, say, as a fish, or as a fish picture, or as this particular object, or as a picture, or as an object of a certain colour, or as a funny-looking object, or as an object hanging on the wall, etc. But in the situation described by Roy the context for singling out some particular aspect is missing. The idea that the child might be referring to the fish picture as *this* rather than *that* can get no foothold. For it to do so, a conversational context would be required.

In short, the gesture Roy attributes to his son is one that in itself presupposes membership in a community of speakers.

It is significant that, until what Roy could imagine as a case of *referring* presented itself, he thought of what his son produced as "just babble"—not a very sophisticated category to use for someone who is interested in studying how children learn to talk. (It should be admitted that this formulation may be that of the reporter rather than Roy's own.) Roy saw the process through a filter, in which he registered only occurrences that seemed to him to confirm certain preconceived notions of what it is to become a speaker. Along similar lines, there is talk about the robot Toco as acquiring the ability to pick out words (and, yes, concepts!) in "the maelstrom of everyday speech"—words, it seems, are what gives structure to the inchoate stream of speech sounds; what needs to be discerned against a background of "semantic noise", as it were.

Rather than impose such a highly intellectualized interpretation on this interactive episode, one could plausibly assume that the child, as children are wont to do, was picking up on the activity (game if you like) of pointing at objects and emitting sounds—different sounds for different objects. Roy does not give enough background to help us try to figure out what was actually going on in this situation; as is natural since he had a determinate notion of exactly what context was relevant for understanding the child's gesture.

Roy is far from alone in adopting this conception of what learning to speak consists in. Its philosophical interest lies exactly in its giving concrete expression to a line of thought which many of us (I for one) are inclined to embrace when thinking about learning to speak. When these matters are discussed, an episode in Helen Keller's autobiography is often cited, and Roy is no exception: the story of how she suddenly realizes that the tactile sign for water her teacher forms in her hand has something to do with the water gushing over her hands. If we scrutinize this anecdote more closely, we will note that it is hard to know how we are to understand it. What is actually involved in Helen's realization? Are we to think that she in a flash forms an idea of the entire human network of using signs to interact; does she see a connection between the repeated signings to which she has been subjected over the past (months?) and the words she had at an earlier time learned to speak; or is the whole episode a creation in the mind of her teacher, Anne Sullivan, who co-wrote the book? Unfortunately, well-known as this story is, I have never seen a critical discussion of it.

What makes an account in terms of reference seem so attractive is the fact that in the act of connecting a sound with an object we seem to have a concrete, almost bodily illustration of how words are bound up with reality, and thus, how meaning is able to seep into the learner's mind.

The idea that we learn to speak by learning to connect names with objects is of course also articulated in the famous passage from St Augustine's *Confessions* that Wittgenstein quotes at the beginning of his main work, *Philosophical Investigations*:

> When grown-ups named some object and at the same time turned towards it, I perceived this and I grasped that the thing was signified by the sound they uttered, since they meant to point *it* out. ... In this way, little by little, I learnt to understand what things the words, which I heard uttered in their respective places in various sentences, signified. And once I got my tongue around these signs, I used them to express my wishes.[4]

Wittgenstein, who greatly admired Augustine, thought this passage was an eloquent expression of a conception of what it means to learn to talk which many of us will find natural. It constitutes the backdrop to Wittgenstein's introduction of the concept of a language game.

After quoting Augustine Wittgenstein goes on to imagine the so called builders' game. A is building something, B helps him by bringing him building stones when A calls for them: "block", "pillar", "slab", "beam". Wittgenstein asks us to "[c]onceive of this as a complete primitive language." (I am not quite sure of the spirit in which this exhortation is to be taken.)

---

[4]Augustine, Confessions, I.8. Quoted in Wittgenstein (2009).

Wittgenstein's aim here is to steer his reader away from the Augustinian conception. His strategy is what has sometimes been referred to as Wittgensteinian irony. In this case, it consists in going along with the account proposed by Augustine in order to see whether it will really yield the result it was meant to yield. For one thing, even if we imagine a language restricted to referring to objects, it is hard to imagine the distinction between common names and proper names being introduced through mere gestures of pointing? And what about number words, demonstratives? Even these simple names of objects require a context of human activity in which the relation between name and thing named is constituted. Furthermore, it is evident that a language consisting only of names of objects, as imagined by Augustine, is extremely impoverished. In fact, it is not even clear what this means. The point (or at least one point) is to make us realize how little is to be achieved through the gesture of pointing and vocalizing a word.

I will get back to the discussion of the builders later.

2. *Observing communication*

Literature offers another famous account of learning to speak. This is in Mary Shelley's novel *Frankenstein.* The monster, which Frankenstein has created out of body parts glued together and enlivened by administering an electric shock, is in some ways a precursor of our latter-day robots.

When the monster comes alive, Frankenstein runs away in shock at the sight of him and leaves him to fend for himself. After roaming about, the monster happens on a family living in a cottage in the forest. Having discovered that his looks tend to give people a fright he avoids making himself known to the family; however he hides in a woodshed from which without being observed he has a clear view of the family's life through a hole in the wall, rather as if he had been watching them on a television screen. The monster's story is reminiscent of Augustine's:

> By degrees I made a discovery of still greater moment. I found that these people had a method of communicating their experience and feelings to one another by articulate sounds. I perceived that the words they spoke sometimes produced pleasure or pain, smiles or sadness, in the minds and countenances of the hearers. This was indeed a godlike science, and I ardently desired to become acquainted with it. But I was baffled by every attempt I made for this purpose. Their pronunciation was quick; and the words they uttered, not having any apparent connexion with visible objects, I was unable to discover any clue by which I could unravel the mystery of their reference. By great application, however, … I discovered the names that were given to some of the most familiar objects of discourse; I learned and applied the words *fire, milk, bread,* and *wood.* I learned also the names of the cottagers themselves… I cannot describe the delight I felt when I learned the ideas appropriate to each of these sounds… I distinguished several other words, without being able as yet to understand or apply them; such as *good, dearest, unhappy.* (pp. 129 f.)

The mind of Frankenstein's monster, like that of Deb Roy's robot, is a paradigmatic Lockean *tabula rasa.* And as was the case for Roy and Augustine, the referential connection is here taken to be the clue to language. However, the monster's account adds an important dimension. Indeed, the starting point of his discovery and his motivation for wishing to learn this skill is that people use their voice to *communicate*, for instance, in producing pleasure or pain, smiles or sadness in their hearers. In other words, unlike what was the case with Roy, an idea of *what words are for* enters here. It is just that he, or rather Mary Shelley (who had been brought up in the tradition of British empiricism) takes it for granted that the basis of communication must be a shared system of reference. Anyway, at the end of this process, it turns out, the monster has learnt enough to be able to carry on an eloquent conversation the next time he meets his creator.

This story, of course, is no less fantastic than that of Augustine. For one thing, as in the case of Augustine, the monster is credited with the possession of fairly sophisticated ideas to start with. Thus, one might ask, what notion of *communication* might one have independently of language. We would not really expect a child to become a speaker in some such way; any more than, say, by shutting her up in a room and exposing her to endless hours of Youtube. What Mary Shelley is describing, again, is a case of learning without doing. It seems a mystery how someone could acquire a competence without ever having practised it, or, alternatively having had it explained to them (the latter of course is ruled out in the case of learning to speak). Frankenstein's monster, like Roy's robot and Augustine, are made out to be passive recipients of a conventional code, created in order to enable people and other beings to make themselves understood to one another by referring to the objects surrounding them.

There is of course a famous critique of the empiricist notion of learning to speak, that primarily associated with Noam Chomsky.[5] Chomsky takes the bull by the horns and declares, in effect, that the new-born baby *must indeed* be in possession of the rudiments of language at birth. The problem with the *tabula rasa* is that it lacks the cognitive structure required to extract a grammar from the samples of speech with which the child is confronted. We can only assume that the child is equipped with a machinery for forming and testing hypotheses about the grammatical properties of the speech sounds she hears. Thus, daunted by the Scylla of the *tabula rasa* we might be tempted to address the Charybdis of innate grammar.

Chomsky is right, of course, in as far as a child must have a *readiness* for speaking of one kind or another. A chair will not start speaking, nor will a horse. For my present concerns, however, the difference between empiricism and nativism does not matter. No matter how complex the innate apparatus for learning to speak, the apparatus must gradually make contact with the outside world. The ability to form grammatically correct sentences is not enough. The question is how this can happen: under what conditions can a child grow into what we might call a linguistic agent: i.e. into someone who is capable of using words in order to express herself and make herself understood to others?

There is no reason why a *robot* could not be a Chomskian learner, i.e. have a built-in capacity to parse utterances and to construct syntactic hypotheses and test them out in practice. However, on *one* score a Chomskian robot would be in the same position as a *tabula rasa* robot. What appears to be lacking in either case is an account of how this robot comes to acquire the interactive habits which characterize speakers

[5] See Chomsky (1959).

of a language—the habits manifested in our speaking to people and responding to their speech in meaningful ways.[6]

3. *Becoming a speaker*

As a child develops, she learns to call on the attention of others. At some stage words come into this. The child does not just express her discomfort into the blue: she addresses someone with sounds, gestures or words. If she fails to elicit a fitting response, she may repeat the signal, try something different or express her frustration. Gradually, too, she will learn to recognize that others are calling on *her* attention. Here also words enter at some point. In this way, she grows into the roles of speaker and listener. To become a speaker, we might say, is to become a participant in a network of roles. The reciprocity is essential.

In fact, to speak only of *a* role of speaker and of listener is misleading. The roles we acquire are a variety. There is requesting, demanding, reporting, exclaiming (out of joy, surprise, fear, anger, etc.), asking questions, expressing affection, consoling, thanking, asking forgiveness, etc.[7] And their counterparts: complying, obeying, registering, answering, etc. The roles may often be intertwined. Thus becoming a speaker is not *one* thing but involves multiple forms of interaction; learning each is a separate process. These forms of interaction grow out of the child's relations with her elders. Her addressing others or taking herself to be addressed by them are not just random events. Their sense is bound up with the personal interconnection of which they are a part.

What I am adducing here are of course perfectly familiar facts. My purpose in reminding us of them is to turn our focus away from the sorts of concern that tend to preoccupy us when we address the issue of learning to speak in philosophical contexts. For one thing, learning to speak is not basically a cognitive venture, but a process of growing into an interactive network. The cognitive structures we invoke in characterizing people's linguistic abilities are not something that lies at the foundation of our interactive practices, but rather something we read out of (or should we say: read into?) the practices. The relevant cognitions, rather than linguistic ones ("The word 'Water' stands for such and such a substance", etc.), are matters like, "There's someone there who might give me some water", etc. For another thing, the interesting development is not something to be found by searching the dark recesses of the individual mind, but something happening out in the open, in the spaces shared by individuals. We might say: the central question when we wonder how a child comes to be able to make herself understood is not: how is she *able* to say what she says, but: *why* does she say what she says? What is she up to? What is she doing or trying to do? These are the questions that Roy apparently failed to ask in the course of his inquiry.

---

[6]For a discussion of related issues, see Shanker (2001).

[7]For a sample of the various things people may do with words, see Wittgenstein, *Philosophical Investigations*, § 22.

4. *Routines and skills*

Let us get back to the builders' game. It illustrates two simple kinds of interactive roles:

(1) The helper's role. He has to bring the stone called for by the builder. In this sense, what he has to do is fixed: within the confines of the game as described by Wittgenstein he has no alternative options. If he does not bring the stone, or if he does something different, he is no longer playing this game. To play it he must be able to recognize different kinds of stone, but since the task of recognition is a simple one, and the requisite action is straightforward, we could imagine his learning his role entirely by rote.

Reciprocally, many *uses* of words are learnt by rote: say, simple games like "This little pig went to market", saying the alphabet, counting objects, doing the multiplication tables. Let us call them, as well as the required responses, verbal routines. Such performances are of course easy to simulate on a robot. It might be programmed to say (in the right circumstances): "My batteries need recharging", "There's a leak in the nuclear plant", or "We are arriving at Brisbane Central Station"; equally it might be programmed to respond to voice commands. Anyone who knows English would understand the robot's messages or be able to give commands without needing to know anything about robots. One would simply trust the programming.

There are other uses of words that are routine-like, yet involve a higher degree of complexity than these, say greeting or thanking someone. We might call them semi-routines. Thus, in certain fairly well-defined situations saying "Thank you", or closely similar words, is the only acceptable response. Yet recognizing those situations requires a degree of cultural competence. You would not thank me if I handed you the hammer to hold while I put a mark on the wall, but you would if I handed it to you in order to help you, or as a gift. Could a robot learn to recognize these situations? Could there be such a thing as a gift to a robot—can a robot *own* things? Is it even clear what "helping" a robot might be? (Do I help my car when I push-start it?) Sometimes the help may be quite subtle, such as supplying the word I was looking for. But do robots ever look for words? I short: does a robot ever have anything to be thankful for?

(2) The builder. He is bound to a limited vocabulary—the names of building-stones—and a limited range of verbal activity—telling the helper to bring him a stone. Yet he has a range of options when it comes to choosing which stone to ask for. His use of words constitutes a low-level *skill*—he is able to use words to bring about certain specified results. The words he uses are mediators between the material circumstances and the end-result. There is no need for him to adjust his words to an individual helper: any helper who has been taught his task will respond in the same way.

In the simplest cases, the counterpart of a verbal skill is a routine, as in the case of the builder's helper. There should be no obstacle in principle to a robot being programmed to exercise certain kinds of verbal skills and respond to their exercise.

5. *There's more to speaking*

Verbal routines (and semi-routines), then, are a matter of conforming to a pattern in speaking or responding. Verbal skills, in turn, are a matter of bringing about a specific result, by addressing someone or after being addressed by someone. However, there are many dimensions of human verbal interaction that cannot be captured in these terms. (I should point out that routines and skills are not sharply demarcated categories but that aspects of them may be present to a higher or lower degree in all our speaking.)

Consider, for instance, verbal reactive expressions of pain, disgust, joy, surprise, pleasure, anger, fear, etc. As Wittgenstein suggests, we may think of them as replacements of the natural expressions of those sensations and feelings:

> Here is one possibility: words are connected with the primitive, natural, expressions of sensation and used in their place. A child has hurt himself and he cries; and then adults talk to him and teach him exclamations and, later, sentences. They teach the child new pain-behaviour.[8]

These expressions may simply be wrung from me, or they may constitute a plea for help or consolation, a shameful confession, etc. Could a robot, say, moan with pain? Of course it could be programmed to simulate moaning, in the sense that it could be caused to produce sounds that resemble, or may even be mistaken for, human moaning. But the sound by itself does not make something a case of moaning with pain. It must be a *genuine* moan. We may be tempted here to say that the moaning must be bound up with an *inner* occurrence. However, as Wittgenstein makes clear, if by something "inner" we mean a process to which no one else may have access, this "inner process" cannot be appealed to as a test for the authenticity of the expression.[9] To experience the expression as genuine *is to experience it as an expression of pain*, and vice versa. One person may credit the expression, another may not (he may have reason to suspect that I am simulating pain). There are no independent criteria by which it is to be decided who is right, although closer observation or closer acquaintance with me or a different understanding of the situation may bring one of the observers around.

How do spontaneous verbal expressions come about? Consider, say, a child learning to ask for a drink of water. This is hardly to be understood in terms of the child's recognizing that she is thirsty, then developing a technique for setting in motion a process that will ultimately lead to her having her thirst quenched; rather in learning to ask for a drink the child develops an understanding of what it means to be thirsty. This is *part* of the story: there is of course an element of reciprocity in learning to understand about thirst: I do not know what it means to be thirsty unless I realize (whether I act on it or not) that someone else's expression of thirst may involve a call on me to give him something to drink. Asking for a drink is neither a routine nor a skill. Learning to express thirst or learning to understand expressions of thirst does

---

[8] *Philosophical Investigations*, § 244.

[9] On this, cp Wittgenstein's comparison of the inner to a beetle in a box, *Philosophical Investigations*, § 293.

not consist in acquiring a technique for achieving certain ends. A more natural way of describing what goes on is to say that the word "thirst" comes to be incorporated into the child's life, into her relations to the people around her. There is no meaning kernel that can be considered in separation from the forms of social interaction in which the word "thirst" has its use. How can we tell that a child has mastered this use of words? We see it from the circumstances in which she uses them and her behaviour—whether she actually drinks the water we give her, for instance. Of course, a child may ask for a drink just to put off having to go to sleep, but that tactic will only work once her elders have come to believe that she has mastered this use of words (and it will stop working once they see through her ruse).

We are beings to whom sensations like thirst, or wishes, intentions, etc. can be attributed because we are beings who may acquire the ability to say we are thirsty, we wish for something, intend to do something, etc. (It is tempting to say we have wishes because we can say we have wishes, but that would be a simplification.) Wittgenstein writes:

> By nature and by a particular training, a particular education, we are predisposed to express wishes in certain circumstances. (A wish is, of course, not such a 'circumstance'.) (PI § 441.)

6. *Expressive robots?*

For something to constitute for me a genuine expression of pain, joy, fear, etc. is connected with its being bound up with *the life* of a being, whether human or animal. A robot, on the other hand, does not have a life. Suppose, on the other hand, someone were to insist that a robot does have a life, *what* is it that he would be attributing the life to? The more or less humanoid or animal-like entity that is moving before us and where sounds are coming from, or the software which is perhaps somewhere else, which records the inputs of the robot and which guides its functioning? The robot in front of me, we might say, is the physical instantiation of a range of programmes and a data base, which may of course be equally instantiated in innumerable other robots. So it is not a particular speaker. As a speaker it might be compared to a ventriloquist's doll: the source of the speech is not where the sounds are produced or seem to be produced.

As long as I keep in mind that what I have before me is a robot (no matter how humanoid its external design), it will be clear to me that the moaning is preprogrammed to occur when it does. This means that there is no room for raising the question whether the robot's moaning is genuine of fake. In the robot's "behavioural repertoire" (if we may call it such) there is no room for anything to which the grammatical categories—the genuine/fake distinction—of expressions of pain, or other expressions of feeling or sensation, applies. And the same goes for the verbal counterpart of those expressions.—"But suppose the moaning-like sounds are produced by some malfunctioning of the hardware—say, that the fans need lubricating." Still, unlike human expressions of pain, this would not be a manifestation of an "ability" on the part of the robot, any more than the creaking of a door displays an ability on the part of the door. For a human being to be moaning with pain may be caused, if you like, by a malfunctioning of the body, but the moaning itself is not a malfunction.

This is not to say that we may not respond to robotic expressions *as though* they were real. In care for the elderly, for instance, robots are sometimes used to provide a sense of companionship, to forestall loneliness and dejection. One of the best-known health care robots is the Japanese robotic seal Paro. Even though a client is clear that what he has before him is a piece of machinery in animal form, he may still feel some sense of encouragement from the seal's responses. This could be compared to a conjurer's trick: we may have a strong sense that the magician is picking eggs out of his assistant's ear; even so, most of us will not really believe that that is where the eggs came from. We simply let ourselves be lulled into a form of oblivion, we lean back and enjoy the paradox.

A nursing home in Pittsburgh employed a Paro to provide companionship for the residents. Dr. Marette, a clinical psychologist working in the facility

> said she initially presumed that those who responded to Paro did not realize it was a robot — or that they forgot it between visits.
>
> Yet several patients whose mental faculties are entirely intact have made special visits to her office to see the robotic harp seal.
>
> "I know that this isn't an animal," said Pierre Carter, 62, smiling down at the robot he calls Fluffy. "But it brings out natural feelings."
>
> Then Dr. Marette acknowledged an observation she had made of her own behavior: "It's hard to walk down the hall with it cooing and making noises and not start talking to it. I had a car that I used to talk to that was a lot less responsive."
>
> That effect, computer science experts said, stems from what appears to be a basic human reflex to treat objects that respond to their surroundings as alive, even when we know perfectly well that they are not.[10]

We should note that while Dr Marette and Mr Carter speak about reacting to Paro as though it were alive, they are never tempted to claim that the robot may indeed be expressing actual feelings.

7. *Spontaneity in speaking*

Rush Rhees, who was student of Wittgenstein's and who devoted much of his work in philosophy to the exposition and discussion of Wittgenstein's work (besides being one of the editors of Wittgenstein's written *Nachlaß*) was quite critical of Wittgenstein's use of the builder's game; more specifically, of his exhorting us to think of the game as a complete language. This was the theme of his essay "Wittgenstein's Builders".[11]

There were two points in particular that Rhees wanted to emphasize. First: language forms a unity. Not a formal unity in the sense of deductive logic, rather a unity in the sense that different occasions for speaking hang together, bear on one another:

> In learning to speak [the child] learns what can be said; he learns, however fumblingly, what it makes sense to say. He comes to have some sense of how different remarks have

[10] Harmon (2010).

[11] Originally in *Proceedings of the Aristotelian Society*, 1959–60. Reprinted in Rhees (1970). I quote from there—A slightly amended version occurred in Rhees (1998).

> something to do with one another. … in beginning to carry on a conversation, in trying to tell you something and trying to understand your answer, he *is* getting a sense of how different remarks have a bearing on one another. And because he learns this, he can go on speaking or go on learning. (pp. 79 f.)

Second: in her becoming a speaker, words become a way for the child to express herself. She comes to use words spontaneously, not just out of routine or because the situation demands a particular response. Words become part of who she is. They are not something external, imposed on her. Learning to speak is not

> like learning a game. We may *use* something like a game in teaching [the child]. We say his sounds back to him, and in this way we bring him to imitate other sounds we make. And this is a game. But it is not what we are trying to teach him. And if all he learns is how to play like this, he will not have learned to speak. He will never tell you anything nor ask you anything either.
>
> When he can speak we may be delighted because 'He can say things himself now – not just repeat'. But what is important is that he can *say* things: not that he can construct new sentences – as it were in an exercise. (p. 80.)
>
> If someone learns to speak, he does not just learn to make sentences and utter them. Nor can he merely have learned to react to orders. If that were all he ever did, I should not imagine that he could speak, and I should never ask him anything. When he learns to speak, he learns to tell you something; and he tries to. (p. 79.)[12]

These two points—the unity of language, and spontaneity—hang together. It is because the uses of words are not tied to specific situations that we are able to use them to express ourselves.

The dimension of spontaneity characterizes a very important part of the lives we live together, the ways we speak. For someone's use of words to be spontaneous, in the sense I am talking about, she need not be particularly lively or quick on her feet. I mean simply that she says things even though they are not required by convention or by some urgent practical need, yet not speaking randomly, but rather saying things that it makes sense for her to say at the time of speaking. A married couple over breakfast may exchange ritual expressions (how did you sleep?), they may use verbal skills (could you pass me the butter? or, on a more advanced level: will you need the car today? etc.), or express themselves (oh, I'm so tired!) But if they are never spontaneous in what they say, you might be worried about their relationship. That is, suppose they never volunteer things like: how did you like the movie we saw last night? or: did you read the latest news about the Brexit vote in parliament? or: where should we go for our holiday this year? In a certain sense, they would not be speaking to one another. They would not be related to one another as speakers.

To be a speaker is not simply to have mastered certain verbal routines or verbal skills. I believe that is what Rhees was emphasizing. But how would we go about training a child to speak spontaneously? That is the problem: for to speak

[12] Arthur Coady puts a similar point as follows in his thought-provoking essay "On the Creation of a Speaker": "For me to speak there must be an understanding of what there is to say. That understanding must be *mine*." *Mind* 77 (1968), pp. 68–76. The quotation is from p. 75.

spontaneously is precisely to speak independently of any specific requirements, and yet to speak in a way that is intelligible in the conversational context. The speaker should express herself, and in doing so make contact with those she is addressing. It seems there are no routines or drills we can go through in order to teach a child to speak spontaneously. The child will just have to latch on to speaking. There is not some separate thing that needs to be done in order to reach the stage of spontaneous conversation. We learn conversation in taking part in conversation.

This could also be expressed by saying that the child will not have a grasp of herself as learning something, of there being something she needs to learn. This seems to be what lies behind the point made by Coady, in "On the Creation of a Speaker":

> A child puts forth no effort to learn his native tongue (ignore learning grammar or big words), but he does struggle. Bewilderment, confusion, failure must be overcome, though it takes no effort. (p. 74.)

I take it that, for Coady, an effort entails striving towards a goal of which one is conscious. A child is not conscious of speaking as a goal she is trying to attain—unlike the case of someone learning a second language. Yet in a particular situation she may struggle to "get it right", that is, to put things in a way others will understand or accept.

8. *Spontaneity in robots?*

How, then, could one programme spontaneity into a robot? A robot has no life that it could learn to express. Perhaps one might programme it to make certain utterances at the occurrence of certain inner states. But that is not what we mean by spontaneity. Saying "I'm thirsty" is not a response to a state of dehydration. It is a way of verbalising a wish. The robot does not have wishes in this sense. Alternatively, the robot could be programmed to "bring up" various topics at random points in time. But neither is that what we mean by spontaneity. To use words to express oneself is not to speak at random.

The things we say may be a reflection of our temperament, our mood and the mood of the situation. They are—or we take them to be—suited to the person we are addressing, to our relation to her, to the wider context, to the history we share, or do not share. The role one assumes in a conversation will be an expression of one's life. A life, however, is precisely what a robot lacks.

A human speaker's relation to his words has depth:

> … if I know you can speak, then it makes sense for me to ask you what you mean, to try to get you to say more clearly what you want, and to ask you questions about it … The example of the builders does not seem to allow for any of these.[13] (p. 81.)

With a human speaker, the question of whether she really means what she says may arise. We may on occasion have to search our soul in order to be clear whether we really meant what we were saying. Did I only say that to be polite? was I carried

[13]Rhees, *op.cit.*, p. 81.

away by the general mood? was I putting on an act? did I speak without thinking? did my emotions get the best of me? did he cajole or frighten me out of my determination? etc. The fact that questions like these may be raised is part of what it means to be a speaker. There is no room for such worries where the speech of robots is concerned.

Maybe this was what Deb Roy realized, and why he gave up the attempt to teach his robot to speak by the method he had used to teach his son. Or better yet: maybe he realized that his son did not become a speaker because of any *methods* he used.

Acknowledgements I wish to thank Merete Mazzarella, as well as the participants in the philosophy research seminar at the University of Queensland, for helpful comments.

References

Beard, Alex, *Natural Born Learners: Our Incredible Capacity to Learn and How We can Harness it* (London: Weidenfeld & Nicolson, 2018).

Beran, Ondrej, "An Attitude towards an Artificial Soul? Responses to the 'Nazi Chatbot'", *Philosophical Investigations* 41 (2018), pp. 42–69.

Chomsky, Noam, Review of Skinner's *Verbal Behavior*. Language 1959; 35:26–58.

Harmon, Amy, "A Soft Spot for Circuitry", *The New York Times*, July 4, 2010. https://www.nytimes.com/2010/07/05/science/05robot.html?.

Rhees, Rush, "Wittgenstein's Builders", in Rhees, *Discussions of Wittgenstein* (London: Routledge and Kegan Paul, 1970).

Rhees, Rush, *Wittgenstein and the Possibility of Discourse*, ed. D. Z. Phillips (Cambridge University Press, 1998).

Shanker, Stuart, "What Children Know when they Know what a Name is" (*Current Anthropology* 42 (2001), pp. 481–497.

Wittgenstein, Ludwig, *Philosophical Investigations* (4th ed., Chichester: Wiley-Blackwell, 2009).

Chapter 22
Wittgenstein versus Zombies: An Investigation of Our Mental Concepts

Edward Witherspoon

22.1 Introduction

We human beings perceive our environment; we reflect on what we perceive; we have desires; we form intentions and act on them. All this mental activity is accompanied by what philosophers call 'consciousness', or the subjective experience of thinking. This is *what it's like* to sense, reflect, desire, intend, and act. There is a distinctive inner feeling that goes with craving chocolate ice cream, and another that goes with seeing a purple armchair, and another that goes with wondering whether the weather will clear this afternoon.

I can be sure that I have consciousness, and I naturally suppose that the people around me do too. But some philosophers have argued that it is possible for there to be a creature that looks, acts, and talks just like a human being, but that lacks consciousness. They have appropriated the term 'zombie' as the name for such a hypothetical being. There is something that it is like to be me—a way my experience *feels*—and there is something very different that it is like to be a bat, but there is nothing it is like to be a zombie. A zombie is all dark inside. To make this possibility vivid, imagine a twin who is a molecule-for-molecule duplicate of you. It is logically possible, according to these philosophers, that your twin could lack the consciousness that you possess. In similar fashion, we can imagine a 'zombie world', viz., a world that is like ours in every physical respect, where every particle of matter is just as it is in our world, but in which there is no consciousness. I will call the claim that zombies are possible the 'zombie hypothesis' and philosophers who advance this claim 'zombie theorists'.

The notion of a zombie has roots in early modern philosophy, but it came to prominence in the nineteen-seventies, when the possibility of zombies was used to argue against *physicalism*, a popular thesis in philosophy of mind. (This use of the

E. Witherspoon (✉)
Department of Philosophy, Colgate University, 13 Oak Drive, Hamilton, NY 13346, USA
e-mail: ewitherspoon@colgate.edu

S. Wuppuluri and N. da Costa (eds.), *WITTGENSTEINIAN (adj.)*,
The Frontiers Collection, https://doi.org/10.1007/978-3-030-27569-3_22

zombie hypothesis receives its most developed expression in *The Conscious Mind*, by David Chalmers, and I will largely rely on this book as my source for what zombie theorists think.) Physicalism is the thesis that the physical facts determine all the facts, including all the facts about what we think and feel. The physical facts are the facts that belong to a completed physics, whatever that turns out to be; for our purposes, we may regard physical facts as the spatio-temporal disposition of the fundamental particles of matter, together with the laws of physics. Where Descartes leaves us with a conception of the mind as metaphysically and causally independent of matter, physicalism holds that the mind and all its features are ultimately determined by matter.

The zombie hypothesis challenges physicalism. For suppose there could be a zombie world that is particle-for-particle identical with our world, but that contains no consciousness. The physical facts (including, we may assume, the laws of physics) would be the same in both worlds, but since our world has consciousness while the zombie world would not, the physical facts do not determine the facts of consciousness. In other words, physicalism is false.

If physicalism is indeed false, its demise leaves an explanatory lacuna in the philosophy of mind: how is consciousness to be explained? It cannot be explained as a function of what is happening in the brain: the brain is a physical organ, completely describable in terms of the disposition of its neurons, and the falsity of physicalism entails that no physical facts (not even the physical facts about my brain) determine the facts of consciousness. Chalmers and many other philosophers of mind who follow him in rejecting physicalism have therefore cast about for other means for explaining the existence and character of consciousness. This problem has proven difficult, and so in philosophy of mind it goes by the quaint sobriquet, 'the hard problem'.

Like Chalmers, I am no fan of physicalism. Like him, I am interested in reflecting on the nature of consciousness. But in this essay I am not going to focus on either refuting physicalism or explaining consciousness, for I have trouble with the initial step of Chalmers's approach. I find it hard to accept the claim that zombies are possible. Indeed, I believe that claim is incoherent. What Chalmers and many others think is a description of a logical possibility turns out to be a piece of nonsense. This is a negative result, but from it we will be able to draw a positive lesson for how we should understand our mental concepts. This lesson will not solve the hard problem of consciousness, but it will perhaps make it seem less of a problem.

My inspiration for thinking through the problems with the zombie hypothesis is Ludwig Wittgenstein's *Philosophical Investigations*. Wittgenstein's major themes include the connections between thought, feeling, and action, and the tendency of philosophy to yield distorted descriptions of those connections. Wittgenstein can help us recognize that the zombie hypothesis rests on such a misdescription.

In what follows, I will explain more precisely just what a zombie is supposed to be. I will argue that the zombie hypothesis entails a form of skepticism about other minds. That in itself is not a definitive argument against the zombie hypothesis; indeed, Chalmers contrives to take a dose of skepticism in stride. Drawing on Wittgenstein, I will argue that the skeptical consequences of the zombie hypothesis are more extensive than zombie theorists acknowledge. Even so, one might think that the possibility of zombies shows that skepticism is something that we have to live with. So I turn to another line of thought from Wittgenstein, about how the meaning of sensation language is established, to argue that the zombie hypothesis collapses into incoherence.[1]

22.2 The Character of the Possibility of Zombies

Let's return to the key claim, that zombies are possible. Chalmers presents it thus:

> How, for example, would one argue that a mile-high unicycle is logically possible? It just seems obvious. Although no such thing exists in the real world, the description certainly appears to be coherent. …
>
> I confess that the logical possibility of zombies seems equally obvious to me. A zombie is just something physically identical to me, but which has no conscious experience – all is dark inside. While this is probably empirically impossible, it certainly seems that a coherent situation is described; I can discern no contradiction in the description. … Almost everybody, it seems to me, is capable of conceiving of this possibility. (Chalmers 1996, p. 96)

When a zombie theorist says that zombies are possible, that does not mean that they are *empirically* possible or *physically* possible. The relevant notion of possibility instead is that of *logical* possibility. We can illustrate the differences between these notions of possibility with some examples. It is logically possible that there is an island of pure gold rising out of the seabed in the Pacific Ocean. It is physically possible as well, in that the existence of such an island would violate no laws of physics. But it is not empirically possible, in that there is not enough gold in the vicinity of earth to form such a mountain; such a mountain cannot exist in the actual world. It is logically possible that there is a unicycle one mile tall, but such a unicycle is physically impossible, because no possible material could be rigid enough yet also slender and light enough to construct one; and since it is physically impossible it is *eo ipso* empirically impossible. But a golden mountain and a mile-high unicycle are *logically* possible because the descriptions of these things entail no contradictions. So when philosophers say that a zombie is possible, they are saying that there is no contradiction in the idea of a being that is identical to a conscious human being in every physical respect but that nonetheless lacks consciousness; in other words, we can give a coherent description of such a being.

Chalmers finds that the logical possibility of zombies is obvious. I confess that I find the logical possibility of zombies far from obvious. Perhaps this is because I have been influenced by passages like this one from Wittgenstein's *Investigations*:

> But can't I imagine that people around me are automata, lack consciousness, even though they behave in the same way as usual? — If I imagine it now – alone in my room – I see people with fixed looks (as in a trance) going about their business – the idea is perhaps a

[1] Eric Marcus reaches a similar conclusion via a quite different argument in "Why Zombies are Inconceivable" (Marcus 2004).

> little uncanny. But just try to hang onto this idea in the midst of your ordinary intercourse with others – in the street, say! Say to yourself, for example: "The children over there are mere automata; all their liveliness is mere automatism" And you will either find these words becoming quite empty; or you will produce in yourself some kind of uncanny feeling, or something of the sort. …(PI, §420)

Wittgenstein tries to imagine that the people around him lack consciousness (*Bewusstsein*). He asks us to try saying, about some frolicking children, 'Their minds are all dark inside; they feel nothing.' When I try this I find, as Wittgenstein predicts, that those words start to feel empty; if I nevertheless hang onto those words, I find my attitudes towards the children become unstable. In either case, the children's being zombies is *not* a possibility I can easily imagine or conceive.

Even if others share this experience, we are not yet at a useful result, for our experience may be a merely psychological feature of me and like-minded thinkers. If zombie theorists find it easy to imagine zombies, perhaps they just have different psychological powers—greater powers of imagination, perhaps. Wittgenstein's remark does not settle the question of whether zombies are logically possible. We need to find some additional argumentative resources. We will start this search by considering the mental states zombies *do* have, according to zombie theorists.

22.3 Psychological and Phenomenal Mental Concepts

A zombie theorist holds that a being physically identical to me—or physically identical to a frolicking child—could lack consciousness. But still there has to be some sense in which even a zombie is conscious: for zombies are sometimes awake (and conscious of their surroundings) and sometimes asleep (and unconscious). Moreover, zombies have to be able to (in some sense) perceive and learn and plan, for otherwise they would behave very differently from human beings and so would fail to be physically identical to them.

Accordingly, zombie theorists allow that a zombie does have mental states. A zombie can perceive, can learn, can form intentions and act on them. In short, it can think and have experiences. When zombie theorists say that a zombie lacks consciousness, they do not mean that a zombie is unconscious in the way someone drugged or asleep is unconscious. An awake zombie possesses consciousness in that it is aware of and responds to its environment. But its awareness and responsiveness lack the subjective 'experience' of thinking; it doesn't *feel* anything. For us human beings, there is a characteristic feeling of joy, for example. When *we* see a loved one after an absence, there is a special thrill of reunion that accompanies our smiles and hugs. A zombie lacks all these feelings (despite recognizing a 'loved one' and smiling and embracing them just as you or I would).

So a zombie has experience and consciousness, in that it is aware of and thinks about its environment; but it lacks experience and consciousness in that it lacks the subjective, inner *feelings* that belong to our awareness and our thought. To forestall possible confusion arising from these ways of describing a zombie, Chalmers distinguishes two categories of mental concepts: psychological and phenomenal. The psychological concepts apply to the mind viewed as a causal nexus. These concepts are defined in terms of what produces them and what effects they have. The mind conceived in terms of psychological concepts is an information-processing organ that mediates between inputs (e.g., sensations) and outputs (actions); it infers, reflects, and learns. (A zombie's learning is the modification of its information-processing pathways so that its future behavior enables it to cope better with its environment.) Psychological concepts include concepts such as perception, inference, intention, belief, and desire. They apply to an agent's mind from a third-person perspective; that is, the attribution of these concepts is based on observation of a creature's interaction with its environment.

Phenomenal concepts belong to the mind considered as the locus of feeling. Consciousness in the phenomenal sense is *what it is like* to have experience. (I will follow the usage of zombie theorists in using the term 'consciousness'—without modification—to refer to consciousness in the phenomenal sense.) Phenomenal concepts apply to an agent from a first-person perspective, in that they concern what experience is like *for me*. Phenomenal concepts include the *qualia* of pain, the feeling of joy, the sensation of red, and the *feels* that accompany thinking, willing, desiring, and so on.

Every mental property, Chalmers argues, is either psychological, phenomenal, or partakes of both categories. For example, color perception is a psychological concept; the qualia or *look* of the color is a phenomenal concept. Color perception is a key part of the explanation of how both human beings and zombies respond differentially to (for example) red and green traffic lights. But zombies have no qualia: their experience does not extend to the *look* of the color or the *feel* of a piece of velvet or the *taste* of a pineapple. *Pain* is an example of a concept that has both psychological and phenomenal dimensions. As Chalmers puts it:

> The term [sc. 'pain'] is often used to name a particular sort of unpleasant phenomenal quality, in which case a phenomenal notion is central. But there is also a psychological notion associated with the term: roughly, the sort of state that tends to be produced by damage to the organism, tends to lead to aversion reactions, and so on. (Chalmers 1996, 17)

We will return to this idea that the word 'pain' really encompasses two distinct notions: a psychological information state whose content is that the organism is probably being damaged, and a phenomenal state whose content is the painful feeling itself. The distinction is more problematic than zombie theorists realize.

22.4 Zombies and the Problem of Other Minds

I mentioned above that the zombie hypothesis has recently been used to argue against physicalism in the philosophy of mind. But the description of zombies can also be seen as an updated expression of a much older problem, that of skepticism about other minds. (Compare the use of brain-in-a-vat scenarios to provide a vivid, contemporary portrayal of skepticism about the external world.) Skepticism about other minds has various versions, but the characteristically modern form of it holds that all the information I could ever get about other people (their words and behavior, their history, even the workings of their brain and nervous system) does not enable me to know what they are thinking or what their experiences are. Skepticism represents the other's body as a curtain that drops between my mind and hers: the body emits what sound like groans of pain or exhibits what appears to be a smile of joy, but the groans could be pretense or the smile could be a mere reflex act that expresses nothing.[2]

There are obvious affinities between the zombie hypothesis and skepticism about other minds. But it is worth making some of the connections explicit, because this will show some of the philosophical costs of accepting the zombie hypothesis. I will argue that the zombie hypothesis entails skepticism about other minds. It will turn out that, for a zombie theorist, there is no justification for believing that anyone besides oneself has consciousness in the phenomenal sense. This may come as a surprise, given the fact that the zombie hypothesis is one premise in an argument against physicalism whose other main premise is that there are facts of phenomenal consciousness, that is, that human beings (the physical twins of zombies) are conscious. But this anti-physicalist argument only requires the premise that *I* (the proponent of the argument) am conscious in the phenomenal sense. It would work even if I doubted the existence of any other consciousnesses.

The connection between the zombie hypothesis and the doubtfulness of consciousnesses other than my own can be seen in Chalmers's own discussion of skepticism about other minds. He entertains a thesis called 'eliminativism', which is the view that, really, no creatures are conscious, despite our tendency to attribute feelings to ourselves and others. He writes:

> Eliminativism about conscious experience is an unreasonable position *only* because of our own acquaintance with it. ... To put it another way, there is an *epistemic asymmetry* in our knowledge of consciousness that is not present in our knowledge of other phenomena. Our knowledge that conscious experience exists derives primarily from our own case, with external evidence playing at best a secondary role.
>
> The point can also be made by pointing to the existence of a problem of other minds. Even when we know everything physical about other creatures, we do not *know* for certain that they are conscious, or what their experiences are (although we may have good reason to believe that they are). (Chalmers 1996, 102)

Chalmers is saying that I know only that *I* have consciousness. My phenomenal experience is the only basis for my belief that consciousness exists. When it comes to other creatures, I do not know what their phenomenal experiences are; I do not even know whether they have experiences. At best, I have 'good reason' to believe that they do.

[2] One could press the question of antecedents, to ask what gave rise to skepticism about other minds and why that problem became acute in the nineteenth century and remains so today. (For a brief synopsis of the history of the problem, see Hyslop (2018). These important questions lie beyond the scope of the present essay.

This is not a merely incidental remark of Chalmers. Anyone who accepts the zombie hypothesis is committed to skepticism about other minds. To see this, reflect on the thesis that there could be two physically identical creatures, one of whom possessed consciousness while the other one lacked it. So let us assume that this is the case, and suppose further that I encounter these two identical humanoids.[3] (We may add, if you like, that the creatures have some properties that enable me tell them apart, but that are clearly irrelevant to whether they possess consciousness: let's suppose that they are in different rooms within my thought-experimental laboratory and are wearing different-colored wristbands.) Could I ever know which one of the two humanoids before me has consciousness, and which one doesn't? This is clearly a case in which whatever I know about the others will be based on my perception of them. If the range of my perceptions is limited to physical properties, then it follows straightforwardly that I cannot perceive any difference between the person (with phenomenal experience) and the zombie (without it), since *ex hypothesi* they have the same physical properties, except for whatever trivial properties allow me to distinguish one individual from the other. It follows that I cannot know whether any humanoid that I encounter is a zombie or a human being; and if I cannot know that, then I certainly cannot know anything about the character of the consciousness of the being before me.

This argument depends on the premise that my perception of others is limited to their physical properties. Could it happen instead that I perceive another's phenomenal consciousness directly, without my perception being mediated by a physical state of the other? Chalmers's distinction between psychological and phenomenal elements of the mind entails that the answer is no. My perceptions are *psychological* features of me and are therefore entirely dependent on the physical properties of my sense organs and the stimuli they receive. (I and my zombie twin have the same perceptions.) If we were to suppose that I could perceive another human's consciousness directly (without a physical intermediary), then the other's consciousness would be making a difference to my physical properties without itself being physical, and that would violate Chalmers's position that the physical realm is causally closed.[4]

Hence we may conclude that, if zombies are possible, I cannot know anything about the phenomenal character of another's mental life. We can make this conclusion vivid through the well-known idea of an 'inverted spectrum'.[5] For all I can tell, another person may have color experience (in the sense of the phenomenal *qualia* or *feel* that colored objects evoke in her) that is systematically different from mine. The objects that give her the subjective qualia of *blueness* give me the qualia of *redness*,

[3] I use the term 'humanoid' to refer to a creature that is either a human being (who possesses consciousness) or a zombie (who doesn't).

[4] 'The best evidence of contemporary science tells us that the physical world is more or less causally closed: for every physical event, there is a physical sufficient cause. If so, there is no room for a mental "ghost in the machine" to do any extra causal work.' The qualification 'more or less' is to allow for quantum indeterminacy which, Chalmers argues, 'cannot be exploited to yield a causal role for a nonphysical mind' (Chalmers 1996, p. 125).

[5] Chalmers embraces the possibility of an inverted spectrum and uses it as an additional argument against physicalism at (Chalmers 1996, pp. 99–101).

and things that look green to her look yellow to me, and so forth throughout the color spectrum. If, for each of us, our qualia are consistently aligned with colored objects, we will use the same color words for the same objects. We will both say that the setting sun is 'red', even though the person whose subjective color spectrum is inverted with respect to mine is having the experience that, if it were mine, I would call 'blue', while I of course am having the experience that I call 'red'. There is no way for me to know what another person's color experience is like, for these possible differences can never show up for us. The skeptical conclusion we have reached from the zombie hypothesis generalizes this result for all our phenomenal experiences. The zombie hypothesis implies that I can never know what experiences another person has, or even that the humanoids around me *have* experiences. For all I know, my friends and family might all be zombies—thinking, talking, reasoning zombies, but beings that lack consciousness nonetheless.

One might think that this would be a disturbing consequence that should cause philosophers to rethink their commitment to the zombie hypothesis. But zombie theorists have a way to make peace with skepticism. Their way of accommodating themselves to skepticism about other minds is hinted at in the Chalmers passage quoted above: although I cannot *know* whether others have consciousness, I can have *good reason to believe* that they do, and that their experiences align with mine. In other words, I am justified in believing that other humanoids are indeed human beings. Perhaps zombie theorists can be content to give up knowledge claims about others' experiences because they can rely on well-grounded beliefs about them.

Indeed, it is not unreasonable to hold that well-grounded beliefs—not knowledge—are sufficient for our practical needs and for sustaining our relationships with others. One could argue that *knowledge* makes demands for certainty that we don't have to satisfy in order to live a rationally ordered and meaningful life.[6] Do I have to *know* that my car is in the parking lot in order for me to act rationally in walking to the lot in order to get home? Isn't it enough for me to have a well-grounded belief that it is there? (Do I even know that my car is in the lot? Do I know that it has not been stolen? Or towed away? Or swallowed by a sinkhole?) Similarly, I could base my relationships with others on a well-grounded belief that our phenomenal experiences are similar. Such beliefs could be reassure me, for example, that my loved ones have feelings too.

But *do* I have good reasons for believing that others have phenomenal experiences similar to mine? For a zombie theorist, any attempt to provide such reasons will have to start with my own conscious experience. (As Chalmers writes, 'Our knowledge that conscious experience exists derives primarily from our own case' (Chalmers 1996, p. 102).) From the facts about my consciousness, together with any information I can obtain about the physical attributes of myself and others, I will need to construct a reasonable basis for ascribing a similar consciousness to others. The resulting

[6] I do not mean to endorse this line of argument. A contrary position is that the concept of knowledge is fundamental, and that *merely having a justified belief that p* should be understood as a privation of the fundamental relation of *knowing that p*. Investigating this issue, while important, would take us beyond the scope of this essay.

argument is usually called 'the argument from analogy'. It runs as follows: *I* am conscious, and my consciousness is associated with certain physical facts (e.g., facts about environmental impingements on my sensory organs, facts about the sounds coming out of my mouth, facts about my movements). When I observe other beings receiving relevantly similar impingements, making similar sounds, moving in similar ways, then I may infer that they are conscious too. The behavior of other humanoids is similar to mine; my behavior is accompanied by consciousness; therefore, their behavior is accompanied by similar consciousness as well. This argument appears to give me good reasons for believing what I am naturally inclined to believe anyway, viz., that others have phenomenal experiences like mine.

But now I want to bring Wittgenstein to bear to argue that the zombie hypothesis has the consequence, not only that I don't *know* about another creature's consciousness, but also that I don't have any *well-grounded beliefs* about it either. This is a much more troubling consequence than Chalmers's official position that I don't have knowledge but do have justified beliefs about others' consciousnesses.

22.5 Wittgenstein on Skepticism About Other Minds

Wittgenstein thinks that the argument from analogy is hopeless. He thinks it has two flaws. First, the inductive base (the data about my own consciousness and its connection to physical properties), from which I am supposed to infer features of the consciousness of others, does not exist. Second, even if I did have such data about myself, extrapolating that data to a conclusion about others' consciousnesses would be an inductively weak generalization.

Let's consider the data from my own case from which I am supposed to generalize to others'. This is supposed to be the set of correlations I have observed between my states of consciousness and physical facts about me and my environment. For example, I feel a burning sensation when I grab a pot on the stove, and I immediately yell and pull my hand away. According to the argument, this provides me with a data pair: the feeling of burning is associated with a certain type of stimulus and physical movement and a characteristic noise (a yelled 'ouch'). When I get enough of these data pairs in my data set, then when I observe someone else jerking her hand away from a hot pot and yelling 'ouch!', I can justifiably infer that she feels a burning pain.

The first problem with this argument is that my relation to my own behavior is not that of observer to observed. My awareness of what I am doing normally comes from my doing itself, not from my observation of my doing. And when I act, I already think of my behavior as caused by or expressive of my consciousness. I jerk my hand away because of the pain; my smile expresses my contentment, as my weeping expresses my sadness or dismay. It is a difficult mental exercise (one which most people rarely perform) to separate out the physical features of one's action from its intimate connection with consciousness and to regard these as independently describable elements; but such a separation is just what one has to do to in order to construct the required database. And while I can sometimes observe myself acting, doing so requires special circumstances and a stance towards myself quite different from my normal engaged perspective. The very specialness of these circumstances shows that we do not have the resources to construct a database of correlations between my observed behaviors and my phenomenal experience.

An especially crippling gap in this putative data set concerns facial expressions. For proponents of the argument from analogy, reading others' facial expressions would have to play a huge role in my attributions of feelings to them. But I hardly ever observe my own facial expressions. (In the days before mirrors and mobile phone selfies, we could safely say that people almost never observed their own facial expressions.) So we cannot construct the data set that, according to this argument, is required for the vast majority of attributions of feelings.

But now even if I could construct a database of the correlations between my consciousness and physical facts about my behavior, the argument from analogy would run into a second problem. For my database would contain those correlations for just one humanoid: me. (I cannot include observations regarding the correlations between consciousness and its physical manifestations from any other humanoids, for that would assume the conclusion that the argument is trying to justify.) It would be an extreme case of hasty generalization for me to use the fact of a correlation between *my* behavior and *my* consciousness to infer a correlation between other humanoids' behavior and their (alleged) consciousnesses. This idea is captured in Wittgenstein's remark in PI §293: "If I say of myself that it is only from my own case that I know what the word "pain" means—must I not say *that* of other people too? And how can I generalize the *one* case so irresponsibly?".

Finally, in addition to criticizing the argument from analogy, Wittgenstein notes that we do not actually employ any inference such as it posits:

> "You say you attend to a man who groans because experience has taught you that you yourself groan when you feel such-and-such. But as you don't in fact make any such inference, we can abandon the justification by analogy." (*Zettel* §537)

That we treat other humanoids as conscious human beings is a fundamental fact about us: our so treating them is not justified by argument. That is not a rational defect on our part, because our attitude towards them does not stand in need of justification.

Wittgenstein has shown that the cost of accepting the possibility of zombies is higher than Chalmers thinks it is, for that possibility entails skepticism about other minds, in a radical form according to which we can have no well-grounded beliefs that others are conscious. But someone who thinks it is *obvious* that zombies are possible could take this in stride. Skepticism about other minds might just be our human plight. In other words, Wittgensteinian arguments show that the zombie hypothesis comes at a very high cost. But if you think that zombies are possible, you might decide that you simply have to pay that cost: you must give up any claim to rational beliefs that there are conscious beings around you. You might have to tolerate the existence of widespread prejudices that others are conscious, for ingrained habits of thought tend to linger, even after we have recognized that they are groundless. But the idea that others are conscious will be mere prejudice, and rationally scrupulous thinkers will carry on their lives as much as possible without ascribing consciousness to other beings.

Wittgenstein offers some striking statements of how difficult it is to sustain such a stance. "Just try – in a real case – to doubt someone else's fear or pain!" (PI, §303). But one might think that these remarks merely express the *psychological* difficulty of giving up ingrained habits of thought. They don't suggest—nor does the preceding critique of the argument from analogy entail—that zombies are *not* logically possible.

To deepen my Wittgensteinian criticism of the idea of a zombie, I will argue that the zombie hypothesis is incoherent, that it only *appears* to be a philosophical thesis. This is a very different argumentative tack from the preceding one, in which I have been treating the zombie hypothesis as a thesis with logical consequences (e.g., that one can never have a justified belief about the consciousness of another). In what follows, I will use Wittgenstein's ideas to argue that zombie theorists deprive themselves of the resources required to make sense of the language of phenomenal concepts.

22.6 The Meaninglessness of the Attempt to State the Zombie Hypothesis

The problem I find in the zombie hypothesis is that it makes it impossible to account for the acquisition of a language for phenomenal concepts. To see why, we need to recall the distinction Chalmers makes between the two kinds of mental concepts, the psychological concepts that apply to both zombies and me, and the phenomenal ones that apply to me but not to zombies. Let us recall that a psychological mental concept applies to the mind considered as an information-processing organ; such concepts are grounded in the network of causal relations between environmental stimulus and the behavior of the organism. A phenomenal mental concept applies to the mind as a locus of feeling. According to Chalmers, many of our ordinary mental concepts partake of both aspects: even the all-important concept of consciousness can be subdivided into a psychological component and a phenomenal component.

Our ordinary concept of *pain* is another case of an amalgam of two distinct components, according to zombie theorists. Pain in the psychological sense is a causal intermediary between injury to the organism and aversive behavior; it is the concept that applies when we explain why an organism pulls away from a hot fire. Pain in the phenomenal sense is the distinctive feeling that (in human beings but not in zombies) accompanies injurious stimuli and aversive responses.

As a shorthand, instead of talking about pain in the phenomenal sense, I will refer to 'felt pain'. Now this expression, 'felt pain' and other phenomenal ones, like 'hurts', 'feels good', 'tastes like vanilla' are required in order to state the zombie hypothesis: a zombie is a creature that is physically just like me, but that does not have the felt pains, the felt pleasures, and the rest of the qualia that I have. Since the zombie hypothesis entails that anything I know about phenomenal experience is derived from my own case, when I learn expressions like 'felt pain' I can draw only on my own resources.

How do I learn how to use the expression 'felt pain'? The story will have to go something like this. Something happens to me: I stub my toe, say. This event is accompanied by an experience in the phenomenal sense. I decide to label this experience 'felt pain'. Later, something else happens: I twist my ankle. This event is accompanied by another experience, which I judge to be relevantly similar to the experience I remember from when I stubbed my toe. So I say that I am having another felt pain, this time in my ankle instead of my toe.

This account of learning phenomenal language sounds simple and straightforward. But Wittgenstein shows us that it is anything but. Let's focus on a key step of the above account: I have a sensation, and I focus my attention on it and dub it 'felt pain'. What is involved in this? This act of establishing the meaning of my expression 'felt pain' has to set up a rule. The rule will determine whether my future applications of those words is correct or incorrect. (It is a necessary feature of words' having a meaning that uses of them are normatively constrained, that is, that a speaker's use of them can be correct or incorrect.)

Wittgenstein asks, how do I establish this rule for myself? One might think that there is nothing to it: I mentally focus on the pain, and set up the rule that my expression 'felt pain' will refer to *that*. Now for this rule to govern my future use of the words 'felt pain', I have to have some criterion for determining that a future sensation is the same as the one I am having now. At first blush, this presents no special problem: the criterion can be my memory of the experience. When I twist my ankle some days later, I recall my earlier experience of felt pain and compare my present sensation to it so as to judge that my present sensation is relevantly similar to my earlier one.

Wittgenstein would draw our attention to two conditions that the zombie hypothesis imposes on this account. First, the criterion I employ is one which I alone possess and which I must be able to apply without any check from others. For if the humanoids around me are zombies, whatever words they come out with will not get their meaning from a connection with a felt pain (as mine does), so anything they say about the correctness (or not) of my use of my words 'felt pain' will not have any bearing on what *I* am using as the criterion of correctness. And I cannot assume that any of the humanoids around me are people, *not* zombies, because, as we have seen, it would be unreasonable for me to believe that. Moreover, what we are trying to explain is how I am able to meaningfully say, 'So-and-so is not a zombie but has felt pains and other phenomenal states'; so at this point in the argument, I cannot help myself to the conceptual resources required to formulate the assumption.

A second feature of the above account of how I learn the expression 'felt pain' is that it relies on my *memory* of the earlier pain, since that memory becomes the criterion that determines the correctness of later applications of the expression 'felt pain'. Wittgenstein points out that if I am remembering my earlier pain (from when I banged my toe) when I later say again 'I feel pain', then there is a possibility that I am *misremembering* it. Suppose I do misremember, and now use 'felt pain' for a sensation that in the past I would have labelled 'an itch'. If this happened, if my memory deceived me, I would have no means to detect the deceit. I cannot rely on anything that other humanoids say. And I cannot rely on the causal nexus between my own sensory inputs and typical behavior to reason that this is the kind of injury that typically results in *pain* and not itches; such reasoning could establish only that the psychological concept of pain applies to me, but the zombie theorist has severed the tie between the psychological concept and the phenomenal concept, so I cannot infer from the fact of my injury that I am feeling pain, not an itch.

So when I use the expression 'felt pain', I have nothing to rely on except my memory, and nothing against which I can check my memory. In other words, I am relying only on what I *seem* to remember. If my present experience *seems to me now* like a pain, then I will call it a pain. But in that case, I don't actually have a criterion for the correct use of the words 'felt pain'; for all I can tell, I might be applying those words to an itch that my memory has mistakenly brought up as the criterion of application for the words 'felt pain'.

The argument that I have been developing is expressed by Wittgenstein in a passage about an attempt to define a sign 'S' as the name of a certain sensation. Wittgenstein describes how one might try to ostensively define 'S'. (An ostensive definition in one in which I pick out what a word means by pointing to the object it describes or refers to.)

> Can I point to the sensation? – Not in the ordinary sense. But I speak, or write the sign down, and at the same time I concentrate my attention on the sensation – and so, as it were, point to it inwardly. – But what is this ceremony for? For that is all it seems to be! A definition serves to lay down the meaning of a sign, doesn't it? — Well, that is done precisely by concentrating my attention; for in this way I commit to memory the connection between the sign and the sensation. – But "I commit it to memory" can only mean: the process brings it about that I remember the connection *correctly* in the future. But in the present case, I have no criterion of correctness. One would like to say: whatever is going to seem correct to me is correct. And that only means that here we can't talk about 'correct'. (PI, §258)

The account the zombie theorist has to give of how we learn expressions for phenomenal concepts leaves us with no criterion for the correct application of those expressions: whatever is going to seem correct to me at the moment I use them is correct, which is as much as to say that the notion of correctness does not apply. And if I have no criterion of correctness for these words, they don't mean anything. The language of phenomenal experience ('felt pain', 'sensations', and all the rest) is, in the mouth of a zombie theorist, nonsense.

Here is another way to present Wittgenstein's argument. In order to establish a meaning for the expression 'felt pain', I resolve to use those words for the feeling I am having now, when I stub my toe. And I resolve to use those words *only* for this feeling, whenever it may recur. I am *not* resolving to use the words for whatever will at some future time *seem* to me then to be this feeling. For suppose that, in the moment when I resolve to use the words 'felt pain' for the feeling I am currently having, I also imagine that my memory were to go haywire in such a way that I would in the future think that the words 'felt pain' named the feeling that I now call 'an itch'. I would not regard it as acceptable to use the words 'felt pain' in the future for that itch; I would not regard that haywire use of words as correct or as in line with my intention for using 'felt pain'. My memory on its own cannot distinguish what I will then mistakenly regard as *felt pain* from what I now label 'felt pain'. So my memory cannot provide a criterion of correct application of the words, and so cannot yield a rule for the use of the expression 'felt pain'. But a consequence of the zombie hypothesis is that the only resource I have for guiding my future use of the words is my memory. And this implies that (given the commitments of a zombie theorist) I lack the resources for framing a rule for my use of the expression 'felt pain', which is to say that I cannot assign a meaning to those words.

In giving this argument, I have taken as an example of phenomenal language the expression 'felt pain', which I have stipulated to be synonymous with the phenomenal dimension of the concept of pain. But there is nothing special about *felt pain* as compared to any other phenomenal concept. So this argument establishes that any expression that zombie theorists purport to use to refer to their phenomenal consciousnesses is meaningless.

This prepares us to locate an incoherence in zombie theorists' attempt to formulate their hypothesis. This emerges most clearly when we take a first-person perspective on the attempt. I start, then, by saying that it is logically possible for there to be creatures that are like me in every physical respect but that lack phenomenal consciousness. A consequence of this is that I have no good reason for thinking that anyone else is a human being (with consciousness) rather than a zombie (without). This may be disturbing. But at least I know that *I* have consciousness, don't I? Don't I know that I, unlike zombies, have felt pains, sensations, feelings of all sorts? But what do I mean by the expression 'felt pain'? *I* cannot give it a meaning (according to the argument we have just seen), and I cannot draw on a meaning provided from elsewhere. The same goes for any other expression that purports to apply to phenomenal concepts, including the concept *phenomenal consciousness* itself: in my mouth, all those expressions are nonsense. And so when I (the would-be zombie theorist) come out with the words 'I have consciousness, but a zombie does not', my words have no meaning. What looked like a sentence expressing a hypothesis to the effect that zombies are possible turns out to be nonsense.[7]

22.7 Conclusion

This argument, if it is successful, exemplifies the method that Wittgenstein describes in PI §464: "What I want to teach is: to pass from unobvious nonsense to obvious nonsense." As I see it, the zombie theorist traffics in unobvious nonsense. What has emerged is that there are logical conditions for the meaningful employment of phenomenal language that the zombie theorist cannot meet. The zombie theorist is com-

[7] Robert Kirk asserts that using Wittgenstein's private language argument against the conceivability of zombies will amount to begging the question against those who think zombies are conceivable (Kirk 2019). I hope that the structure of my presentation shows that a Wittgensteinian argument does not have to beg the question.

mitted to establishing the meaning of phenomenal language using only the resources of his own feelings and memories. But these resources are not sufficient to enable the theorist to give a meaning to phenomenal language. And without that phenomenal language, the attempt to state the zombie hypothesis results in a nonsense utterance. The zombie theorist only appears to be stating a thesis. Consequently the zombie theorist does not present us with a possibility that refutes physicalism; the zombie theorist does not saddle us with the 'hard problem' of explaining consciousness.

This criticism of zombie theorists appears to be in tension with my earlier argument that the zombie hypothesis entails a radical form of skepticism about other minds. This cannot be an appropriate description of the earlier argument, since we have just concluded that the attempt to state the zombie hypothesis results in nonsense, and a piece of nonsense cannot have any logical implications. How then should we characterize the status of that earlier argument? Similarly, how should we characterize the intermediate steps of the argument of the preceding section, in which I have sometimes used expressions like 'a consequence of the zombie hypothesis'? These are not easy questions, but I think we can begin to address them by drawing an analogy between the zombie theorist and someone who has inconsistent beliefs without being aware of their inconsistency. So long as the subject is thinking about each belief in isolation from those it is inconsistent with, she can draw conclusions from it. She is stymied in her cognition only when she brings the inconsistent beliefs together in thought in a way that manifests their incoherence: at that point, she will not be able to retain all of them, and some of the inferences that her former beliefs supported will become unjustified. In a similar fashion, so long as we (and the zombie theorist) are taking for granted a vocabulary for phenomenal concepts, we can (seemingly) describe a zombie, entertain the (apparent) possibility that such beings exist, and draw consequences from that (apparent) possibility. But thinking about the acquisition of phenomenal language and concepts brings out latent inconsistencies: here, an inconsistency between the belief that phenomenal language gets its meaning through association with phenomenal experience and the belief that my phenomenal experience is something only I can be aware of. At this point, the thinker cannot retain all of these beliefs. If one does try to hold onto all of them, the result is that we cannot make sense of her words, just as we cannot make sense of someone who affirms p and $\sim p$. One way to describe the status of my own arguments is that, up until the denouement of the second argument, I was participating in an illusion of meaningfulness along with the zombie theorist. The second argument (if it is successful) exposes the illusion. My earlier conclusions are like the conclusions drawn from a belief that is part of an incoherent set: when the beliefs the conclusions rest on are revealed to be incoherent, the conclusions must be relinquished. So also must my earlier argument be relinquished once the zombie hypothesis is revealed to be incoherent.

I will conclude this critique of the zombie theorists' attempt to define zombies by pointing to a conclusion we should draw from the failure of their attempt. Zombie theorists assume—and in this they follow a major strand of modern philosophy that is explicit in Descartes and implicit in the empiricist tradition—that mental concepts are logically independent of physical ones. Descartes assumed that a thinker's mind could be fully stocked with ideas, including the full gamut of phenomenal experience, even if the thinker had no body; this is the possibility the meditator entertains in the Second Meditation. Even once he has established that his body exists, and that his mind is interfused with it so as to be in a certain sense one thing, the meditator argues that the relations between states of his mind and states of his body are purely causal. (Versions of these claims are presupposed by the empiricist argument from analogy.) Against this, Wittgenstein argues that our mental concepts logically presuppose the possibility of the *expression* of those concepts through behavior and action. To have the concept of pain is (in part) to have the ability to express one's pains (through groan or grimace, word or gesture) and the ability to recognize the expression of pain in others. Here, expressing a pain is making it manifest, so that others who possess the concept *pain* can recognize that you instantiate it. And recognizing an expression of pain is perceiving another's pain through her expressing it. The zombie theorist attempts to sever this link between having the concept *pain* and being able to express pain: if zombies are possible, nothing that one does could ever amount to an expression of pain (in the phenomenal sense), since *ex hypothesi* my zombie twin could do the same thing without expressing pain. Furthermore, as we have seen, for a zombie theorist no perceptual intake could ever amount to a perception of another's pain. And what goes for pain goes for any phenomenal concept. In effect, the argument I draw from Wittgenstein about the requirements for giving the words 'felt pain' a meaning is an argument for the logical connection between phenomenal consciousness and the possibility of an outward expression *cum* manifestation of that consciousness. The zombie theorists' denial of any such logical connection leaves them unable to refer to their experience and thereby reduces to nonsense their attempts to propound a philosophical thesis.

Works Cited

Chalmers, David. *The Conscious Mind*. Cambridge, MA: Harvard University Press, 1996.

Hyslop, Alec. "Other Minds", *The Stanford Encyclopedia of Philosophy* (Winter 2018 Edition), Edward N. Zalta (ed.), URL = <https://plato.stanford.edu/archives/win2018/entries/other-minds/>.

Kirk, Robert. "Zombies", *The Stanford Encyclopedia of Philosophy* (Spring 2019 Edition), Edward N. Zalta (ed.), URL = <https://plato.stanford.edu/archives/spr2019/entries/zombies/>.

Marcus, Eric. "Why Zombies Are Inconceivable", Australasian Journal of Philosophy 82 (3):477–490 (2004).

Wittgenstein, Ludwig. *Philosophical Investigations*. 4th ed. Malden, MA: Blackwell, 2009.

Wittgenstein, Ludwig. *Zettel*. Berkeley, CA: University of California Press, 1967.

Chapter 23
The Soul and the Painter's Eye

Michel ter Hark

In a visionary passage in the Philosophical Investigations, Wittgenstein asks:

> 'How does the philosophical problem about mental processes and states and behaviourism arise? - The first step is the one that altogether escapes notice. We talk of processes and states and leave their nature undecided. Sometimes perhaps we shall know more about them - we think. But that is just what commits us to a particular way of looking at the matter (The decisive movement in the conjuring trick has been made, and it was the very one we thought quite innocent.) - And now the analogy which was to make us understand our thoughts falls to pieces. So we have to deny the yet uncomprehended process in the yet unexplored medium. And now it looks as if we had denied mental processes. And naturally we don't want to deny them'. (PI§ 307)

Wittgenstein here predicts the course the philosophy of mind would take, including the responses to his own position. The passage is about what it means to understand, or coming to understand, the nature of mental states and processes. We think we understand what kind of thing a mental state or process is because we are comparing it to a physical state or process. Thus, we speak of thought processes and treat them on the model of behavioural processes, like (silent) speaking, or on the model of physiological processes, like brain processes. Rather than advancing thereby our understanding of mental states and processes, Wittgenstein warns us, the parallel between the mental and the physical *forces* the mind body problem upon us. Moreover, as long as we fail to recognise that it is a metaphor that forces the mind body problem upon us, we will accuse him who confronts the very metaphor or picture of the 'inner and the outer' with various critical questions of denying the inner realm.

The suspicion of being himself a sort of behaviorist kept him alert to the end of his life. In his writings on the philosophy of psychology one finds him writing—in 1948—'Am I saying something like, "and the soul itself is merely something about the body"? No. (I am not that hard up for categories)' (RPP2 §690). Yet it has

M. ter Hark (✉)
Faculty of Humanities, Vrije Universiteit Amsterdam, De Boelelaan 1105, VU University Amsterdam, 11th and 14th Floor, 1081 Amsterdam, The Netherlands
e-mail: m.r.m.ter.hark@vu.nl

S. Wuppuluri and N. da Costa (eds.), *WITTGENSTEINIAN (adj.)*,
The Frontiers Collection, https://doi.org/10.1007/978-3-030-27569-3_23

proven difficult to interpret Wittgenstein's writings on psychology in a way which accommodates his categorial denial of reducing the 'Seele' to the body. In this essay, I will not attempt to give a waterproof reading of Wittgenstein as a non-reductionist. Instead, my aim is merely to include in the discussion a number of his dispersed remarks about the aesthetic domain, in particular painting, which may illuminate in what sense he is not 'that hard up for categories'. To my knowledge, these remarks have not been discussed elsewhere in this specific context.

23.1 Behaviourism and Mentalism

In the period in which Wittgenstein wrote the preceding remark, various forms of behaviourism dominated the agenda in epistemology, philosophy of mind and experimental psychology.[1] For instance, Rudolf Carnap formulated a form of logical behaviourism along the following lines: 'every assertion *about* (my italics) a particular mental state of another person, e.g. "A is now happy", could be translated into an assertion which *referred* (my italics) only to physical occurrences, namely to expressive movements, actions, words, etc.' (1932, p. 65).

By its emphasis on bodily behaviour and dispositions to behave behaviourism has also been called 'peripheral-state' materialism. The identity-theory of mind and brain that replaced behaviourism is a form of 'central-state' materialism because it identifies mental states and processes with brain states.[2] The view that took the place of central-state materialism was functionalism. According to this view, mental states and processes have to be identified with computational or functional states and processes.[3] Now, both the identity theory and functionalism saw behaviourism as a too simple defence of a scientific outlook on the mind and the brain. Both conceptual and empirical arguments have led to the dismantling of behaviourism. One sort of conceptual argument concerned the theory's failure to deal with types of behavioural processes that do not match types of mental states, as in the case of lying and pretence. Empirical arguments concerned more the scientific ally of the theory, i.e. the radical behaviourism of B.F. Skinner, which could not explain the use and creativity of language. Both the identity theory and functionalism had little patience with

[1] A year later Gilbert Ryle's The Concept of Mind (1949) was published in which the notion of category and in particular category mistake would play a prominent role. I will stay away from the historical questions concerning the relationship between the two and merely note that there has been a tendency in the literature to treat both of them as behaviourists. See for instance Fodor (1975) who proposes a form of mentalism or functionalism. In this essay, I will speak of mentalism in a broad sense, encompassing both dualism as forms of functionalism. 'Materialism' is used only for the identity theory of mind and brain. Behaviourism too is used in a broad sense without distinguishing between logical behaviorism and Rylean behaviorism (see also Dennett 1978).

[2] See for an early defence of the identity theory U.T. Place (1956).

[3] An influential defence of functionalism based on the analogy between mental states and computational states is from Putnam who formulated his theory originally under the titles of 'Psychological Predicates' in 1967, but it has become more known under the title "The Nature of Mental States' printed in his collected works (1975).

conceptual problems and hence the agenda shifted towards a naturalistic philosophy of mind which took the current scientific paradigm as its point of departure. With the emphasis shifted towards central inner states either to be identified with brain states or, at a higher level of complexity, with computational states, behavioural processes became treated as a series of clues for inferring other people's thoughts and feelings. The idiom of 'processes and states' was again firmly reestablished but their nature had to be left undecided as long as scientific psychology and brain science had not arrived at an encompassing theory of the mind/brain supported by empirical evidence.

Wittgenstein's approach to the 'problem of mental states and behaviourism' is totally unlike the preceding positions in the philosophy of mind. His question is 'how does the problem arise?', rather than 'how do we solve the problem?'.[4] To be sure, the question 'how does the problem arise' is independent of an advanced scientific theory. Indeed, a scientific theory may itself rest on philosophically naive assumptions rooted in a stereotyped way of looking at psychological language.

To see how the problem of mental states and behaviourism emerges Wittgenstein invites us to study psychological language in the form of language-games, i.e. forms of linguistic practice in which language, behaviour and context are interwoven. When we study language from this perspective, it will appear that both mentalism and behaviourism operate according to the same semantical or conceptual scheme which is totally disconnected from the way psychological words are used in actual practice. To illustrate the naive semantical view, I have italicised in the quotation of Carnap above the word 'about' because it signals Carnap's semantical view that the mind body problem can be rescued from the arms of dualism by shifting the reference of psychological concepts from *mental* states and processes to *physical* states and processes, like expressive movements, actions and words. From the perspective of Wittgenstein, however, this shift of reference of psychological language does not advance our understanding of mental states and processes at all. In fact, behaviourism operates according to the same scheme as mentalism. The mentalist typically maintains that just as the language that serves the description of things in the outer world is about things in the outer world, the language that serves the description of things in the inner world is about things in the inner world. The prevailing picture therefore is of two languages, differing in subject matter, but within which language functions in parallel ways: in both areas its function is to name and describe the relevant subject matter.

It is also against this background that a rare reference to Carnap must be understood. Thus, Wittgenstein writes: 'One may also say: "He made *this* face" or "His face altered like *this*", imitating it—and again one can't describe it in any other way. (There just are many more language-games that are dreamt of in the philosophy of Carnap and others.)' (RPP1, §920).

[4]To be sure, this has also been the starting point for Carnap's brief flirtation with a strong version of behaviourism. The difference between Carnap and Wittgenstein, however, is that Carnap relies from the start upon an ideology of verificationism which is alien to Wittgenstein's careful study of language in use. Put otherwise, Carnap's attempt to dissolve philosophical problems is normative, whereas Wittgenstein's approach is descriptive.

23.2 Inner and Outer

In his later writings on the philosophy of psychology, one can see Wittgenstein exploring the richness and subtlety of psychological language-games often in the context of studying the use of psychological concepts in the artistic domain, as when we speak about portrait painting or about emotions in music. It is especially these, largely neglected, remarks that throw more light on Wittgenstein's full treatment of psychological concepts as they are applied in human interactions.

Consider as a start the following remark which occurs in the Remarks on Colour:

> A *shine*, a 'high-light', cannot be black. If I were to substitute blackness for the lightness of high-lights in a picture, I wouldn't get black lights. And that is not simply because this is the one and only form in which a high-light occurs in nature, but *also* because we react to a *light* in this spot in a certain way. A flag may be yellow and black, another yellow and white. (ROC§22)

In portraiture, the magical moment occurs when a painter brings in the highlight in the eyes. The portrait comes to life and we feel as if we can see through the face. It is now in the dimension of depth and of transparency. As Wittgenstein observes, it is not simply the use of a certain colour or mixture of colours, say yellow ochre and titanium white, which makes a dab of paint a highlight. It is above all our response to a light in the particular surrounding of the eye.

Our response to what we see is a pervasive feature of Wittgenstein's extensive discussion of a wide variety of phenomena which fall under the umbrella of aspect-seeing or seeing-as, including other people's mental states. It is at this point that his approach to the mental sharply diverges from mainstream philosophy of mind. Mainstream philosophy of mind, whether in the form of mentalism or behaviourism, treats third person statements about other people's mental states analogous to observation reports of things in our environment. Indeed, the very notion of *outer* behaviour exemplifies the tendency to see the behaviour of others as an array of stimuli that trigger our retinal and higher brain processes[5]:

> What I really *see* must surely be what is produced in me by the influence of the object" Then what is produced in me is a sort of copy, something that in its turn can be looked at, can be before one; almost something like a *materialisation*.
>
> And this materialisation is something spatial and it must be possible to describe it in spatial terms. For instance (if it is a face) it can smile; the concept of friendliness, however, has no place in an account of it, but is *foreign* to such an account.... (PI, p. 199)

What we really see, according to this view that is implicit in the philosophical views about other minds of many authors, is a complexity of colours, shapes and movements which excludes the 'perception' of mental phenomena like friendliness.

Wittgenstein's study of language games reveals the opposite. As he puts it at the occasion of emotion concepts:

---

[5] 'Outer behaviour' has become the standard phrase in contemporary philosophy of mind, especially in textbooks. See for instance Churchland (1984).

> "We *see* emotion" - As opposed to what?- We do not see facial contortions and *make the inference* that he is feeling joy, grief, boredom. We describe a face immediately as sad, radiant, bored, even when we are unable to give other description of the features. - Grief, one would like to say, is personified in the face. This is essential to what we call "emotion". (RPP2§570)

But how to interpret the emphasis in the quote above upon 'see'? Does Wittgenstein simply deny the sceptical position which argues that we can only infer emotion by reaffirming that we *do* see them? Given Wittgenstein's deep insight into the nature of philosophical problems and his careful practice of philosophical methods, it is unlikely that he would argue in this way. To understand his method of solving the question whether other people's mental states can be seen or only inferred it is fruitful to compare his position to that of the Gestaltpsychologist Wolfgang Köhler, with whose writings Wittgenstein was familiar.

At first sight Köhler's position looks similar to Wittgenstein to the extent that both reject the 'mosaic view' of introspectionism (and current physiological accounts) which maintains that all we really see are spatial items like shape and colour that are represented at the retinal image; all other items that we report as seeing, like friendliness, are the product of mental processes, like interpretation.[6] Köhler therefore is back at the common sense position which says that we do see friendliness and the like. His defense of the common sense position, however, is supported by theoretical considerations supplied by his more general explanation of the perception of wholes. Thus, according to Köhler's position, it would be wrong to say, as inferentialists do, that we only see facial features and their movements. In addition to these facial features. Köhler contends, we also see the interrelations between the various features, that is, we see organised wholes, rather than simply a number of parts. Phenomena like friendliness find their place at the level of such organised wholes. Put more strongly, we see such wholes, e.g. friendliness, *as much as* we see the colour of the lips.

Wittgenstein's approach is different, though. Thus he says: 'On the other hand I have not *admitted* anything with my way of putting it, and I contradict anyone who tells me I see they eye's glance "just as" I see its form and colour.

For "naive language", that's to say our naif, normal, way of expressing ourselves, does *not* contain any theory of seeing—it shows you, not any theory, but only a *concept* of seeing' (RPP 1, §1101)

Wittgenstein's way of arguing that we do see, e.g. friendliness is not to draw our attention to things or relations that we fail to take notice of, as Köhler does, but to the *concept* of seeing itself. It is not the objects of seeing that require further clarification (and defense), but the very concept of seeing that we use in speaking about the world around us is in need of clarification.

---

[6]Köhler (1929) deals with the subject of social perception in his chapter 'Behavior'. The similarity between Wittgenstein and Köhler can be found in passages like the following: 'Look into someone else's face and see the consciousness in it, and also a particular shade of consciousness. You see on it, in it, joy, Indifference, interest, excitement, dullness, etc. The light in the face of the other.

Do you look within yourself, in order to recognize the fury in his face? It is there as clearly as in your own breast' (RPP1, §927).

Although Kohler is correct in rejecting the classical inductive inference to other people's mental states, his reasons for doing so are not. It is not the case that proponents of the inductive (or adductive) inference have overlooked something which is really there to be seen. According to Wittgenstein, what is wrong with the inferentialist approach is that it fails to distinguish between different uses of the concept of seeing. The dominant view of perception in the inferentialist tradition is an attempt to conform to the findings of science. One of the key assumptions of psychological and physiological science is that the brain *in itself* is the seat of consciousness, and the rest of the body is a sort of mechanistic, organic vehicle in which the brain resides and by which it manoeuvres through the world. The eyes are treated as mere tools which enable us to take in sensory data, e.g. lines, shapes, colors, from the outside world. Once they have found their way into the brain, it is the task of the mind/brain to interpret the unorganized data into recognizable perception.

Applied to the perception of thoughts and feelings of other people the behavior of other people emerges as the proper object of perception. Significantly, the notion of 'outer behaviour' has been coined, thereby emphasising that behavior has nothing to do with the inner and is a mere interface of raw data on the basis of which we have to infer hidden inner mental states in the mind/brain. The category of 'outer behaviour', Wittgenstein wants to say, is as misleading as 'inner states'. The complaint that all we see is outer behaviour bypasses a more detailed investigation of what we understand by behavioural evidence. Indeed, the focus on mysterious inner states and processes in dualistic and mentalistic traditions has caused a serious neglect of what 'outer' really is, and how the notion of 'evidence' actually functions and how it relates to other uses of 'evidence', as in ordinary perception of physical objects or in scientific context.

23.3 Art, Perception and the Soul

What philosophers commonly understand by outer behavior is a myth. Their notion of outer behavior implies that we are capable of observing and describing the physical movements of the body and the face in minute detail, which then provides a basis for inferring mental states. But the truth is that in most cases we observe and describe the behavior of other people by means of *psychological* concepts. Indeed, the very assumption that the eye is a tool for absorbing data is contradicted by the fact that we see the eye totally different:

> We don't see the human eye as a receiver; it seems, not to let something in, but to send out. The ear receives; the eye looks. (It casts glances, it flashes, beams, coruscates.) With the eye one may terrify, not with the ear or the nose. When you see the eye, you see something go out from it. You see the glance of the eye. (RPP1§1100)

Moreover, in many cases we would feel embarrassed if asked to specify, say, the facial contortions 'underlying' our judgement that a person feels ashamed. As Wittgenstein notes, we may be certain that a person is glad without being able to describe what we have noted in terms of facial contortions and bodily movements, and without being able to convince another person of our report (cf. LW2, p. 86).

Let us consider a first application of Wittgenstein's conceptual approach to the problem of seeing other people's mental states.

> "What I *see* can't be the expression, because the recognition of the expression depends on my knowledge, on my general acquaintance with human behaviour.' But isn't this a merely historical observation. (RPP 1, §1073)

The idea here is that what you see is not interpreted, but given to the senses as 'raw data'. On the other hand, in order to be able to say that we see friendliness we must know (or interpret) what we see in terms of the concept of friendliness. Given Wittgenstein's conceptual approach, his reply would be that the one who says that we cannot really see friendliness has merely stipulated a certain use of 'to see'. Thus, suppose someone says: 'He looked sad'—and someone else asked: 'Have you seen this?', does it make sense if the former person replied: 'I have seen certain shapes and colours and I *know* that they are the expression of sadness'? But probed further, could one answer the question *which* shapes and colours? And in the likely case that one cannot answer this question, what is meant by the reply that one sees the facial contortions and knows that they are the expression of sadness? Is not the conclusion that all one has achieved hereby is to prioritize a certain use of the concept of 'seeing' over other uses?

Clearly, the dominant picture of seeing describes seeing as what can be taken in by the eye, the retina and the visual tract. But when we apply this use of the concept of seeing, we do not see a face as a face either, but simply an arrangement of shapes, some of them in shadow others lighted. And in this case it would hardly make sense when one says: I see only shapes and know that this pattern of light and shades expresses a face. Even a portrait painter who starts his portrait by first carving out the shapes, rather than the facial features, does so in the context of a preliminary sketch of the contours of a *face*.

Wittgenstein's approach to the ordinary way of speaking about other people's mental states therefore is neither to deny the legitimacy of using the concept of seeing nor to ground the use of this concept in terms of a theory about perceptual wholes. Rather, his approach is to remind us of distinctions between the two uses of the concept of seeing, i.e. seeing and seeing-as.

> Two uses of the word "see". The one: "I see this" - and allude to a description, or point to a picture or a copy. With this I might tell someone else: over there, where you haven't been able to see it, there is such-and-such. An example of the other use: "I see a likeness between these two faces" Let the man I tell this to be seeing the faces as clearly as I do. (LW1§431, cf. PI II, xi, p. 193a)

Wittgenstein continues by specifying the different uses of 'seeing' by noting that one man might make an accurate drawing of the faces, whereas the other man might notice in the drawing their likeness which the former did not see (cf. LW1§432; cf. PI II, xi, p. 193b). In the Philosophical Investigations, Wittgenstein does not discuss the two different uses of seeing in the context of painting and portraiture any further, but in his other writings on the philosophy of psychology we find him asking questions that directly concern the application of seeing and seeing-as in the aesthetic domain. For instance, in the following remark Wittgenstein argues that in the case of face perception seeing and seeing-as may figure independently, in the sense that it is quite imaginable that a person does not see the face as smiling, whereas his way of painting the face testifies to his accurate perception of the shapes and colours of the face:

> First and foremost, I can very well imagine someone who, while he sees a face extremely accurately, and can, e.g., make an accurate portrait of it, yet doesn't recognise its smiling expression as a smile. I should find it absurd to say that his sight was defective. And equally absurd to say that his subjective visual object just wasn't smiling, although it has all the colours and forms that mine has. (RPP1 §1103)

In fact, with beginning painters it is often the case that they may succeed in portraying the shapes and colour transitions between the various planes of the face accurately yet fail in getting the expression right. Conversely, a (skilled) painter may get the expression right without bothering much about representing the facial features accurately:

> But a painter can paint an eye so that it stares; so its staring must be describable by the distribution of colour on the surface. But the one who paints it need not be able to describe this distribution. (RPP 1 §1077)

To be sure, a painter will be able to indicate how he deals with light and shadow in order to represent the spatial form that is the eye. But this is part of the anatomy of the face. In order to paint an eye so that it stares, or throws a glance, one is concerned with the expression of the face. Now if asked whether he can explain the expressive quality of the portrait in terms of the distribution of the shapes of various planes, form and colour he typically will not be able to provide such a description.

Describing what one sees in terms of anatomical features and describing what one sees in terms of expressive features, therefore, are different linguistic practices, even though in both language-games the same word 'to see' is being used. But how is the way we respond to what we see a determinative feature of our perception of other people's thoughts and feelings? Again, Wittgenstein makes a detour via the perception of pictures and in particular portraits in order to make clear to what extent seeing-as is informed by our reactions to what we see.

Consider the following remark:

> Hold the drawing of a face upside-down and you can't tell the *e1xpression* of the face. Perhaps you can see that it is smiling, but you won't be able to say what *sort* of a smile it is. You wouldn't be able to imitate the smile or describe its character more exactly.
>
> And yet the upside-down picture may represent the object extremely accurately. (RPP1 §991)

Here we can see clearly to what extent the language games of seeing and seeing-as are different. The criterion of identity of what we mean by 'the same picture' is different in the case of the anatomical representation from the case in which we ask whether the expression is the same. In the case of an anatomical representation, possible differences in our visual impression of two pictures must be traceable to objective, i.e. anatomical changes. But when we focus upon one picture, and turn this upside down, is our impression the same as before? From the perspective of anatomy, nothing has changed, and so one should have to conclude that the upside down picture of the portrait is the same as before. At the same time, especially when a human figure is being depicted, our impression of the upside down picture is different. For even though we may still have a rough impression that the face is smiling, we no longer can see and describe its specific character. Rather than concluding that this demonstrates that friendliness does not belong to the category of what can be really seen, Wittgenstein draws our attention to the different use of 'seeing' here. For as he implies in the quote above, the upside down picture of the portrait does not *make* us smile. The picture of the Mona Lisa turned upside down has a greater impact upon us than an architectural drawing turned upside down. We no longer react to the picture in the same way, and we no longer see it as a smiling face. What counts as the same here is fixed not by what can be traced to objective features of the picture but by our (subjective) reactions to the picture.

By reminding us of different criteria of identity of concepts relating to the anatomy of behaviour and concepts relating to the expressive aspects of behaviour, Wittgenstein would reject the notion of outer behaviour as it is used both within behaviourism and mentalism as a myth, a product of a mistaken view of the way psychological concepts work in first and third person uses. Yet the distinction that has been made so far between the anatomical perspective on behaviour and the expressive perspective is more subtle than it appears. In particular, it would be wrong conclude that the expressive perspective relates to facial features that are characteristically associated with certain psychological phenomena only, e.g. the smiling mouth and gladness. In several of his remarks, Wittgenstein is explicit about it that it is not just characteristic features that define the meaning of psychological concepts but also their bodily flexibility and mobility.[7] For instance, he writes: ' "Just look at the face and the movements of a dog, and you'll see that it has a soul." But what is it about the face? Is it only the similarity with the play of features of the human face? Is it, at least among other things, the lack of stiffness?' (LW II, p. 65). More general remarks make clear that to Wittgenstein lack of stiffness or mobility of facial and bodily features are as important for our ascriptions of psychological concepts to living beings as are the characteristic expressions of joy or sadness. Put otherwise, flexibility and mobility of expressions and gestures are reckoned by Wittgenstein among the physiognomic features rather than the anatomical features of the body. Without its variability the behaviour of human beings and higher animals would not be what we call behaviour. The use of our concept of expression itself is dependent on this variability. For consider someone who has seen only one sort of expression, say a smile. For the first time he sees a face changing its expression: 'Couldn't we say here that he hadn't noticed a facial expression until now? Not until the change took place was the expression meaningful; earlier it was simply part of the anatomy of the face' (RPP II, 356). In another series of remarks Wittgenstein supposes the existence of a 'human without a soul', an organism equipped with only the biolog-

---

[7]In an earlier paper, I have argued that Wittgenstein's remarks about the flexibility of psychological phenomena has to be understood as his way of digesting Darwin's (18 work about emotional expressions of animals. See ter Hark 2004).

ical functions. Suppose the organism could be designed so that it produces signs of pain. In that case his suffering look could still be treated as mechanical 'if he always produced exactly the same suffering expression' (LW II, p. 67). It is only against the background of a *play* of expressions that the concept of expression, say of an emotion or of a cheated emotion, makes sense.

The dependency of third person ascriptions of thoughts and feelings on behaviour as sketched above also has another important implication that is often neglected or rather overruled by philosophers and psychologists who argue that our knowledge of other minds is based on a sort of theory we have internalised about the relation between inner states and outer behaviour.[8] On the basis of this theoretical knowledge we would make constantly predictions about the behaviour of other people. But as Wittgenstein correctly notices, from his descriptive point of view, 'only rarely does one predict someone else's reactions' (LW2, p. 65). And yet, the question as to the predictability of other people's behaviour is highly relevant for the question what constitutes the evidence for psychological ascriptions. Indeed, when we say that we cannot figure out what goes on in someone else, this certainly has to do with the fact that we cannot foresee his behaviour. The main difference between Wittgenstein's approach to these questions and contemporary philosophy of mind, is that he does try to answer them by means of a general theory about inner states and outer behaviour. One might even say that such a theory is based on a false analogy between a physical mechanism and a human being. When we say that we don't know what to expect from a certain mechanism, it is natural to phone an expert who is capable of opening the machine. The fact that we cannot foresee what it will do is a defect, which has to be remedied. But in the case of human beings, the situation is the reverse: 'I think unforeseeability must be *an* essential property of the mental. Just like the endless multiplicity of expression' (LW2, p. 65).[9]

23.4 Psychological Indeterminacy

The dependency of psychological concepts upon patterns of life implies that their use may lack a definite or determinate sense. In an earlier paper (ter Hark 2000), I have called this psychological indeterminacy. Psychological indeterminacy is to be distinguished from epistemological notions like uncertainty and vagueness. Psychological indeterminacy is a constitutive feature of language-games with psychological concepts. It is also a feature which once again makes clear the difference between straightforward observation reports of physical objects and third person applications of psychological concepts. Because the natural foundation of psychological concepts is formed by the flexibility and mobility of facial features the evidence for saying that

[8] An early exponent of this view is Churchland (1984).

[9] Here again the analogy between Wittgenstein's treatment of psychological concepts and the aesthetic domain is evident. Thus he says: 'Soulful expression in music - this cannot be recognised by rules' (RPP2, §695).

a person is, say, angry or acting as if he is angry, is necessarily unsharp: 'Sufficient evidence passes over into insufficient without a borderline. A natural foundation for the way this concept is formed is the complex nature and the variety of human contingencies' (RPP II, 614). Conversely, if there would have been much less variety, Wittgenstein wants to say here, sharper concepts would have to seem more natural. Immediately he adds how difficult it is to imagine a more regular life and his example clearly turns on the importance of the mobility of our physiognomy: 'Is it as if one were trying to imagine a facial expression not susceptible of gradual and subtle alterations; but which had, say, just five positions…Would this fixed smile really be a smile? And why not?—I might not be able to react as I do to a smile. Maybe it would not make me smile myself' (RPP II, 614). As these observations make clear, Wittgenstein would sharply reject an explanation of indeterminacy which is quite natural, one in terms of shortcomings in the evidence. Seeking to make the difference between, on the one hand, indeterminacy as a constitutive feature of the language games we play with psychological concepts, and on the other epistemic uncertainty, he distinguishes between subjective uncertainty and objective uncertainty. Subjective uncertainty is a form of epistemic uncertainty owing to their being insufficient information available in order to determine whether a certain term applies or not. In these cases one is not yet able to give a definite answer, but one knows reasonably well what kind of observations would lead to a positive or negative answer; it is just that, at present, one is not in a position to make such observations because of insufficient evidence, lack of time or inadequate instruments. Subjective uncertainty is typically at home in observation reports about things in our environment. Thus, although we no longer see the bird owing to whether conditions, we know that the bird is still over there where we can longer see it; there is fact of the matter. The uncertainty of psychological judgements, by contrast, is not (just) due to a lack of knowledge or evidence; rather, the uncertainty betokens a difference in the manner of judging about motives and feelings on the one hand and, say, about the length of physical objects on the other hand. As Wittgenstein puts it, it is, 'an indeterminacy in the nature of the game, in the admissible evidence' (LW I, 888).

In particular in language-games concerned with the judgement of human motives there may be no fact of the matter. The point is not only that it is sometimes unclear whether to call an expression real or feigned, but also that in some cases it is unclear whether it makes sense to insist that the evidence must be *either* genuine *or* feigned. For instance, a person smiles but the way he subsequently reacts fits neither the pattern of genuine joy nor that of feigned joy. To insist, as later defenders of materialism like H. Feigl have done, that the presence of margins of evidence is a sign of an epistemic defect due to our limited access to a person's mind or brain, is to ignore that indeterminacy is part of the very meaning of the concepts of deceit, concealment and lying (and hence to ignore that and the ways in which these concepts are based upon patterns of life). For even when one is epistemically sure about another person's feelings, and knows perfectly well what to say, the application of the relevant concepts is still different from concepts the meaning of which quite determinate.

For instance, uncontroversial applications of emotion concepts are different from simple perceptual judgements of physical objects. Wittgen-stein puts the difference thus:

> Statement: A I know that the bottle was standing there. B "How do you know that?" B "I saw it there" - If the state-ment is: "I know that he was glad", and the question is asked: "How do you know that?" - what is the answer? It is not simply the description of a physical state of affairs. Part of it is for instance that I know the person. (LW II, p. 89)

Many psychological judgements are agent-relative in a way that perceptual judgements about physical objects are not. We do not view human behaviour from a neutral stance, an optimal stance from which the best predictions can be made. Rather our descriptions vary with our own dispositions, such as, whether we like the person, trust him, or know him or her well enough. To illustrate the difference in another way, suppose a psychologist gives his subjects a verbal description of a person's behaviour in certain circumstances and asks them whether this behaviour manifests deceit. It is very likely that the subjects will protest and say that they will only be able to answer the psychologist's question when they see and even know the person.

Although those who are personally acquainted with a person can make most subtle psychological judgements, this does not mean that they are capable of specifying conclusive criteria for their judgements. In this respect Wittgenstein speaks of 'imponderable' evidence (see PI II, p. 228; LW II, pp. 65–66). Imponderable evidence is evidence which can make us certain about someone's psychological state, without our being able to specify what it is in their behaviour that makes us so sure. Imponderable evidence is, therefore, not the basis of a proof that another person is feeling such and such. And if it makes no sense to speak of proving that another person is, say, lying, it is not necessarily the case that one can convince a third person on the basis of the evidence of the insincerity of the person's words. For instance, Chamberlain visiting Hitler, September 15 1938. Hitler hides his war plans from Chamberlain and promises that peace can be preserved if the Czechs will meet his demands. Chamberlain is fooled and tries to convince the Czechs not to mobilize their armies. To his sister he writes: '…in spite of the hardness and ruthlessness I thought I saw in his face, I got the impression that here was a man who could be relied upon when he had given his word…'.[10] Even confronted with a perfect liar like Hitler one can imagine someone, unlike Chamberlain, being capable of detecting facial management. Suppose Churchill visiting Hitler too. As some historians have claimed, Churchill would not have been taken in by Hitler's lies and perhaps would have been convinced of his unreliability; perhaps he would even have noticed thinking clues that ultimately betrayed Hitler's lies. Discussing Hitler's reliability afterwards, Churchill and Chamberlain might have disagreed and the former might not have succeeded in convincing the latter of Hitler's unreliability.

The dependency of psychological concepts upon complex patterns of life, therefore, opens up a possibility of disagreement and uncertainty, which is characteristic of psychological judgements. As Wittgenstein observes:

[10]Quoted from Ekman (1985, pp. 15–16).

> Given the same evidence, one person can be completely convinced and another not be. We don't on that account exclude either one from society, as being unaccountable and incapable of judgement. (RPP II, 685)

Bibiliography

Carnap, R. (1932), 'Psychologie in physikalischer Sprache'. In: *Erkenntnis* 3, 107–142. Translated by G. Schlick as 'Psychology in Physical Language'. In Ayer (1959), pp. 165-198.

Churchland, P.M. (1984). Matter and Consciousness. The MIT Press.

Darwin, C. (1872), *The Expression of the Emotions in Man and Animals.* Penguin Classics.

Dennett, D. (1978) Brainstorms. Harvard University Press.

Ekman, P. (1985), *Telling Lies*, New York: Norton & Company.

Feigl, H. (1958), *The 'Mental' and the 'Physical'*, Minneapolis: University of Minnesota Press.

Fodor, J. (1975) The Language of Thought. Massachusetts: Harvard University Press.

Hark ter, M.R.M. (2000), 'Uncertainty, Vagueness and Psychological Indeterminacy'. In: *Synthese*, 124 (2), 193–220.

Hark ter, M.R.M. (2004) 'Patterns of Life': A Third Wittgenstein Concept, in: D. Moyal-Sharrock (ed.) The Third Wittgenstein: The Post-Investigations work. Routledge London and New York.

Köhler, W. (1929) Gestaltpsychology. Liveright: New York.

Place, U.T. (1956) 'Is Consciousness a Brain Process?' *British Journal of Psychology* 47: 44–50.

Putnam, H. (1975), 'The Nature of Mental States'. In his *Mind, Language and Reality: Philosophical Papers*, vol.2 (Cambridge: Cambridge University Press, 1975).

Wittgenstein, L. (1953) Philosophical Investigations. Oxford: Blackwell.

Wittgenstein, L. (1977) Remarks on Colour. Oxford: Blackwell.

Wittgenstein, L. (1980) Remarks on the Philosophy of Psychology, volume 1. Oxford: Blackwell.

Wittgenstein, L. (1980) Remarks on the Philosophy of Psychology, volume 2. Oxford: Blackwell.

Wittgenstein, L. (1982) Last Writings on the Philosophy of Psychology, volume 1. Oxford: Blackwell.

Wittgenstein, L. (1992) Last Writings on the Philosophy of Psychology, volume 2. Oxford: Blackwell.

Chapter 24
Wittgenstein on Seeing as; Some Issues

Paul F. Snowdon

In his middle and later periods one of Wittgenstein's concerns was perception. This is, of course, precisely what one would expect given his obvious interest then in the notion of experience and in the language we employ to describe and express our experiences.[1] However, the passage which has attracted most attention is the discussion in sec. XI of part II of *Philosophical Investigations* which is concerned with "seeing as", or "aspect seeing". In fact it is really only in the first half of that long section that Wittgenstein discusses perception. In the second half he turns to remarks about meaning, understanding, certainty, and so on, a passage which contains some famous but obscure obiter dicta of Wittgenstein's. Wittgenstein in fact says at about the halfway stage of sec. XI; 'The importance of this concept lies in the connexion between the concepts of 'seeing an aspect' and 'experiencing the meaning of a word'. For we want to ask 'What would you be missing if you did not experience the meaning of a word?'[2] The important concept that Wittgenstein seems to have in mind here is that of aspect blindness.

It is actually very difficult, I think, to come to an assessment of this section. This reflects the richness of the discussion but also, as I shall argue, it also reflects how hard it is to discern a unity in the range of cases that Wittgenstein is writing about, and also that it is not easy to extract what one might call a substantial proposal by Wittgenstein resolving the difficulties. I strongly suspect, therefore, that what I have to say here is inadequate both as an interpretation and as an assessment.

The question I shall begin with is; what problem, or problems, is Wittgenstein engaged with? To answer this we need to (i) describe the examples he discusses and

[1] I have discussed some of Wittgenstein's ideas from that period in Snowdon (2011).

[2] Wittgenstein (2001, p. 182). Since the paragraphs in Section II are not numbered I cite Wittgenstein's text by the page where the passage is located.

P. F. Snowdon (✉)
Department of Philosophy, University College London, Gower Street, WC1E 6BT London, UK
e-mail: p.snowdon@ucl.ac.uk

S. Wuppuluri and N. da Costa (eds.), *WITTGENSTEINIAN (adj.)*,
The Frontiers Collection, https://doi.org/10.1007/978-3-030-27569-3_24

say what they are in a general way, and (ii) determine what it is about them that he is trying to understand.

24.1 The Examples

In this section I want to present the types of cases that Wittgenstein is engaged with, but there are many examples and also features of some of the examples that I cannot here attend to, so my presentation of the cases is in many respects incomplete.[3]

Almost without exception the examples or cases that interest Wittgenstein are cases which we would describe (or ascribe) using the term 'see', or with some of them, cases that there is a strong but perhaps resistible temptation to describe using 'see'. So in one sense the topic is *sorts of seeing*. It is, though, surely clear that there would be parallel cases for the other senses, or some of them, for at least some of the types of case, and it is simply that Wittgenstein does not consider them. Thus one kind of example is where a subject recognizes in a flash that the thing he or she is seeing is a certain individual. Obviously this can happen with the other senses. If Wittgenstein thought there was something essentially visual about the problem cases he certainly does not say so or argue for that. We have then to concentrate on seeing.

Wittgenstein starts by distinguishing what he calls two uses of 'see'. He compares someone saying that he sees this, and linking 'this' to a description of what is seen, or a drawing of it, with someone who says that he sees a likeness between two faces. He then says that there is a 'difference of category between the two objects of sight'.[4]

Now, what are the two categories? It seems to me that in the first example the use of the description is supposed to be to convey, as one might say, how things looked. One could do that by pointing to a picture of the scene as one saw it. In the second category what is reported is some feature or aspect or perhaps property that the percipient registered—in this case the likeness. In no sense, though, are these all the uses of 'see'. Another category of object for the verb 'see', as one might put it, worth contrasting with both of these is where one simply says what object one saw. Thus I may say 'I saw the Eiffel Tower yesterday', without thereby conveying from what angle it was seen, or how much of it one saw, or what it looked like at all. What is conveyed is simply what *item* was in view. Wittgenstein is not interested in that type of object, though a lot of philosophical analysis has focused on that type of ascription.

Wittgenstein's then immediately develops the case of seeing a likeness, namely, when someone is looking at a face, and then 'suddenly notice its likeness to another'.[5] What seems important to him here is the *suddenness* of the noticing. He calls this

---

[3]My own method here, as elsewhere, is to look closely (and critically) at certain passages, but many passages cannot be given such attention, and so have to remain in the background.

[4]Wittgenstein (2001, p. 165).

[5]See Footnote 4.

experience 'noticing an aspect'.[6] And he introduces in connection with it a characterization of the example which is quite often taken as defining the whole class in which he is interested. He says; 'I *see* that it has not changed; and yet I see it differently'. I want to query this characterization shortly.

It seems that Wittgenstein is thinking of the case as one where both the faces, between which a likeness is noticed, are seen simultaneously, and so one is noticing a likeness between two perceived items. But there is surely no interesting difference between this case and another where I am simply seeing one head and then notice how alike it is to another one, not at that moment perceived? For example, I suddenly notice how alike a certain person I can see is to Wittgenstein (whose face I remember). But if this is also a case of the sort Wittgenstein is considering, we are very close to the topic simply being *the nature of recognizing some aspect of a perceived item*. Or perhaps, it is the sudden dawning of recognition of some aspect of a perceived item which one has been perceiving.

In fact latter on Wittgenstein considers cases which seem to amount to precisely, and no more, than this. He asks; 'if someone sees a smile and does not know it for a smile, does not understand it as such, does he see it differently from someone who understands it?'[7] What these examples suggest, then, is that Wittgenstein is interested in the difference between perceiving an item which is of kind K (something which could take lots of values, for example, being like that, being a smile, being Wittgenstein) and not recognizing it as being K, and perceiving it and recognizing it as being K. Wittgenstein seems to think, for some reason, that a particularly crucial case would be where a single person moves suddenly from the first to the second condition.

We can point out that there are at least two other cases of recognition. One sort of case is where the recognition is attained slowly by a process of consciously checking off signs. Thus I might be shown a picture of a castle and asked whether it is a Norman castle. Now, maybe I know that only Norman castles have feature F1, F2, and F3. I check that the depicted castle has those features, and I can be said to have thereby recognized it as a Norman castle. But recognition achieved this way seems not to be amongst Wittgenstein's concerns in the present passage. And there is also recognition which is normal and standard, for example, when I see my son I simply recognize him.

I think it is somewhat surprising to view Wittgenstein's discussion as in part concerned with recognition in this broad sense. What can be said, though, is that it is very hard to say where to draw the line around the examples to prevent the discussion spreading so wide.

I want to return, given this clarification, to Wittgenstein's remark that his cases are ones where one sees that the seen object 'has not changed, and yet I see it differently'. This can be said about the cases that Wittgenstein presents but if the cases that are

[6]One feature of Wittgenstein's discussion that is for him unusual is that in the first three pages he introduces a sequence of technical terms—namely, 'noticing an aspect', 'continuous seeing', 'the dawning of an aspect', and 'picture object'.

[7]Wittgenstein (2001, p. 169).

being focused on are cases of sudden recognition why should Wittgenstein add that one sees it has not changed? Suppose the case is one where one sees a bird eating something and so changing shape and moving, and when one is seeing it one recognizes in a flash that it is a vulture, why does the absence of seeing that it has not changed (since it *has changed* as one looks at it) mean that the nature of sudden recognition will be different?

Next, Wittgenstein's second case is what he calls later 'the schematic cube'.[8] He points out that the cube could occur a number of times in a book, at one moment being an illustration of 'a glass cube, there an inverted open box, there a wire frame of that shape, there three boards forming a solid angle'.[9] He suggests that we can '*see* the illustration as one thing, now as another'. He also wishes to say that 'we see it as we interpret it'. In fairness it needs adding that Wittgenstein displays from time to time some uncertainty that we should say this, though his dominant tendency is to affirm that we do and should speak that way.

As an example what is especially striking is that it is different from the first recognitional ones. The seeing that is going on does not correspond to recognition. Rather, this is a matter of reacting to a picture the interpretation of which is provided. In seeing the diagram as a glass cube, for example, I am not recognizing it as a depiction of a glass cube. It is fair to say, I think, that Wittgenstein is asking for trouble in setting out this example, for it in fact exhibits *two* independent features that crop up in his discussion. First, the schematic cube is a manifestly visually ambiguous or unstable figure. As we look at it, the central three-cornered angle is visually unstable. At times it looks to be projecting, and then it looks to be receding, but we can also see it as neither, for we can see it as a two-dimensional figure consisting of a rectangle and two rhombuses. It is though quite hard to keep that way of seeing it. Now, clearly, as it flips between being seen in these three ways we are seeing it differently. We tend to see it three-dimensionally, but as ambiguous between the joint receding and projecting. However, what Wittgenstein is officially interested in at this point is a supposed seeing of the item differently as, say, a solid or a frame or made of glass etc. depending on the provided interpretation. So in choosing an example which is visually ambiguous, as well as differently seeable in the sense in which he is supposed to be interested in the example, there is an invitation to confusion. I am inclined to say that Wittgenstein should have chosen a figure which is not visually ambiguous but which can, supposedly, be interpreted in different ways (as an illustration of, or depiction of, different things), and which can, perhaps, be seen in different ways. The reader can easily supply examples of this.

This example is important because Wittgenstein introduces with it a notion which plays a key role in his discussion, namely that of interpretation. Thus we are to envisage that the text, containing repeated occurrences of the cube, supplies, each time, an interpretation. The text says what the cube *represents*. This then provides some sort of anchor for subsequent uses of the notion of interpretation. Wittgenstein then repeatedly asks to what extent we can see an item in a distinctive way, corresponding

[8]See Footnote 4.

[9]See Footnote 4.

to, but in some way going beyond, the interpretation of it as representing that kind of thing. Wittgenstein's question is; do I merely interpret the diagram a certain way, or do I *see* it as I interpret it?

Wittgenstein next introduces his most famous example, the duck—rabbit.[10] It is, of course, just one example of what psychologists call ambiguous figures.[11] It is, like the second case, pictorial; however, it seems clearly different, in that as we look at it and see it first as one thing then as another, there is no dependence on an extraneous interpretation, given by the context, to guide how we see it. In this sense the duck-rabbit shift does not involve anything that might be called 'imagination' on the part of the percipient, in the way that the shifts which occur in the second example seem to demand an imaginative capacity on the percipient's part. (I mean, of course, the shifts in how the schematic cube is seen when it is seen in accordance with an interpretation).

Wittgenstein himself uses the duck-rabbit picture to introduce a distinction between the 'continuous seeing' of an aspect, and the 'dawning' of an aspect.[12] It is clear enough what the dawning of an aspect is; it is the onset of the new way of seeing the picture. What is less clear is (i) what illumination of the distinction Wittgenstein thinks he can provide, and (ii) how Wittgenstein conceives of what he calls the continuous seeing of an aspect.

As to (i), Wittgenstein introduces the idea that a sentence such as 'It is a rabbit' is ambiguous in some sense. If I always see the figure as a rabbit then, if I am asked what I see, I might reply 'It's a rabbit'. That is what Wittgenstein calls 'the report of a perception'. However, if I am suddenly struck by it as a rabbit, then if I were to say, 'It's a rabbit' then that would not be a report of a perception. He links this to the fact that in the second case it would be appropriate to have said, 'Now it's a rabbit', but that would not have been appropriate in the first case. And Wittgenstein says slightly later; 'Both things, both the report and the exclamation, are expressions of perception and of visual experience. But the exclamation is so in a different sense from the report; it is forced from us. - It is related to the experience as a cry is to pain'.[13]

The question which these remarks raise is how helpful are they? It is clear, I think, that part of the significance is that for Wittgenstein talk of 'report' or 'descriptions' is always of deep importance. He always regards it as significant to point out that a certain linguistic occurrence is not a report. Actually in this case its significance is unclear. It is true that the dawning of an aspect may elicit an exclamation, which would be an exclamatory expression of the event. Though, once we are familiar with such an occurrence there is no temptation at all to exclaim when it occurs. It is also unlikely that that the continuous seeing of an aspect will elicit an exclamation. However, it may elicit one. If I want a change of aspect but it refuses to happen, as is sometimes the case, I may exclaim 'It's (still) a rabbit'. (I assume that that

[10] Wittgenstein (2001, p. 166).

[11] For other examples see Rock (1975), and Frisby (1980).

[12] See Footnote 10.

[13] Wittgenstein (2001, p. 168).

can count as an exclamation.) So, an exclamation can express continuous seeing. In fact, the self-same words could express in an exclamatory way both a change and a non-change. So, as far as I can tell, there is no distinctive expressive role for such a sentence linked to changes of aspect. Further, the difference between saying S non-exclamatorily and saying it as an exclamation does not represent an ambiguity in any normal sense of that term. Thus, when the doctor after studying the body says 'He's dead', and the spouse exclaims 'He's dead!', they have said the same thing. Finally, there is what is clearly a non-exclamatory way to report the dawning of an aspect, and that is to describe what has happened in these words; 'Before I saw it as an X, and now I see it as a Y'. I conclude that there is no sense in which the dawning of an aspect is *alien* to report or description and no important sense in which it is distinctively tied to an exclamation.

As to (ii) Wittgenstein clearly allows, and in this he is surely right, that another would say of me that I am seeing the figure as a picture—rabbit, even if I had not seen it any other way, and would not myself report the situation by saying, 'Now I am seeing it as a picture-rabbit'. Wittgenstein seems interested in the first person/non-first person contrast but there are problems with the way he handles it. For, he compares my saying 'Now I see it as …', which would be inappropriate, with another saying, 'He sees it as …', which is true. But the inclusion of 'now' in the first person report means that the contrast between what it is appropriate to say need have nothing to do with the contrast between speaker perspectives. If the 'now' is removed from the first person sentence then I believe that there is no contrast worth remarking on between what it is appropriate to say from the two perspectives.

However, Wittgenstein follows the paragraph I have just commented on with two in which he discusses seeing cutlery. He says that there is little sense in saying 'Now I am seeing them as a knife and fork', and further, that 'One doesn't 'take' what one knows as the cutlery at a meal for the cutlery'. Neither of these amount to Wittgenstein literally claiming that one does not see what is manifestly cutlery as cutlery, which is, I believe, how these paragraphs are normally interpreted. In fact, it seems true to say that any normal person does see the cutlery as cutlery, though such a fact belongs in the realm of facts that are not worth expressing in normal contexts.

Thus Wittgenstein leaves his view as to when we can say that S sees O as Y unclear, and he therefore leaves unclear when someone can count as continuously seeing an aspect, assuming, that is, that to see an aspect one must be seeing something *as* something.

24.2 The Problem

I have argued that it is not easy to characterize the cases in which Wittgenstein is interested. The next question, however, is what the problem is (or problems are) which Wittgenstein thinks such cases raise.

When he is discussing the dawning of an aspect, Wittgenstein repeatedly says that he is not interested in the causes of aspect-dawning.[14] He also says that if someone could give a physiological account of what is happening, that would not solve the problem.[15] Wittgenstein's attitude is, therefore, in sharp contrast to the response to such phenomenon which is characteristic of psychologists. Thus, Professor I. Rock, after discussing ambiguous figures, says;"The more interesting question, however, is what determines the change in organization after the observer has been perceiving the figure in a particular way? Why should a change occur at all?"[16] Wittgenstein is not saying that this question is a bad one but he is claiming that it is not *his* question.

What, then, is his question? Budd expresses what he sees as the question in these words; 'What does this change (viz. noticing an aspect) consist in? What constitutes the change from seeing the item one way to seeing it the other way?".[17] Now, it is easy to sympathize with this formulation of Wittgenstein's question. For Wittgenstein does describe his issue as 'conceptual', and currently an enquiry into the concept of F would characteristically be formulated as an attempt to specify what the presence of F consists in, that is, what it essentially requires. Moreover, most of us would hold that Budd's question, as well as being 'conceptual', is a reasonable and interesting question to ask. Finally, Budd's interpretation fits the evident fact that Wittgenstein certainly *discusses* what seem to be constitutive hypotheses. One such hypothesis is that the change is just in interpretation, another that it is a sensory change like a change in apparent colour involved in a perception.

It has to be said, though, that it is not obvious that Wittgenstein thinks of a 'conceptual' enquiry in this way which is congenial to us. One reason for saying this is that Wittgenstein does not offer anything like a constitutive hypothesis himself. This strongly suggests that he himself is not aiming to give one. This absence can be brought out by asking what, according to Wittgenstein, a change of aspect does consist in? Now, it is clear that Wittgenstein holds that it is genuinely a difference in how the object is *seen*, which is not a constitutive claim so much as an emphatic description of what the case is. And, beyond that it would be impossible to extract any more informative constitutive thesis. I think, then, that this makes it very difficult to accept Budd's specification of Wittgenstein's question is.

A second reason is that there are indications in the passage, but also in Wittgenstein's discussion generally, that his aim is not to provide a constitutive thesis. The general point is that usually Wittgenstein has to be understood as countering mistakes; his general conception is that we are prone to adopt mistaken accounts of certain psychological states. His goal is to characterize the conceptions and to reveal them as mistakes. Viewing his discussion this way, the central interpretative task is

[14]Wittgenstein (2001 pp. 165, 174).

[15]This is probably a reference to the significance for Wittgenstein's problem, as he saw it, of the kind of physiological speculation which Kohler engages in. See for example, Kohler (Kohler 1992, pp. 217–219). However, it would not be correct to identify causal hypotheses with physiological hypotheses, since evidently the cause need not be identified physiologically.

[16]See I. Rock, op. cit. p. 265.

[17]See M. Budd, op. cit. p. 78.

not exactly to specify a question that he is attempting to answer which the cases raise, but rather to say what the supposed mistaken conceptions are, and to understand why, according to Wittgenstein, they are mistaken. Unless we can say what the basic mistake is that Wittgenstein is combatting we have not properly understood his discussion. Of course this is a general suggestion, and it would be excessively rigid to insist that we must in every case understand Wittgenstein's discussion in this way.[18] However, there is something amounting to a presumption in favour of this, and it seems to me to fit how in general Wittgenstein proceeds. Wittgenstein comes close to representing the present section in this way. He says;" What we have rather to do is to accept the every-day language game, and to note false accounts of the matter as false. The primitive language-game which children are taught needs no justification; attempts at justification need to be rejected".[19] Again, he says; '(In giving all these examples I am not aiming at some kind of completeness, ... They are only meant to enable the reader to shift for himself when he encounters conceptual difficulties.)'[20] These passages suggest that Wittgenstein is interested in primarily noting mistaken accounts, in enabling difficulties to be avoided. The way we speak has to be accepted.

There is something odd about Wittgenstein's remark in the first passage above, which is worth noting. People who offer analyses of what is involved in, say, seeing as, need not be aiming at justifying the description that we offer. They may have no suspicion that anything needs a defence, any more than someone aiming to study what is involved in digestion or in memory is trying to justify calling things by those terms. If there is any justifiable animus against justifications for the ways we speak it does not carry with it any implication that we should restrict ourselves to rejecting false accounts.

His aim is to determine, then, how we describe such cases, and to combat mistaken accounts of what our descriptions commit us to. That seems to be what a conceptual enquiry must be, or at least is, for Wittgenstein here. Now, I have criticized Budd's account of Wittgenstein's question, but the issue still remains as to why Wittgenstein does not pursue the problem which Budd picks out. What was Wittgenstein's attitude to Budd's question? What, if anything, were Wittgenstein's reasons for taking that attitude?

This is, I think, the most fundamental methodological question raised by this passage, as by many others. It is clear, however, that any justification Wittgenstein has for his attitude must be general, and so nothing particularly to do with perception.

24.3 Wittgenstein's Attitude

What is it, then, that Wittgenstein is wanting to say about the cases? If we try to fit his account into the framework set out in the last section then, I think, we have to say

[18] I have tried in Snowdon (2018) to argue that Wittgenstein is not totally a-theoretical (or quietist).

[19] Wittgenstein (2001, p. 171).

[20] Wittgenstein (2001, p. 175).

that the major mistake which Wittgenstein is opposed to is the thesis that when we notice an aspect (when an aspect changes) we do not *see* it differently. Wittgenstein repeatedly affirms, or affirms his temptation to say, that it is seen differently.

It seems to me that in Wittgenstein's view there are two major errors to which we are prone. One error is that already mentioned of denying we see things differently. Wittgenstein tends to think that anyone saying this will affirm that what is different is an interpretation (not the seeing). But the second major error is made by people who agree with Wittgenstein's claim that there is a difference of seeing but who have a restricted conception of seeing (or the sensory) and who therefore have to treat the difference, because it is in their view also a difference in seeing, *in a mistaken way*. Such thinkers want to identify some feature which is akin to color (or apparent color) and treat the change as a change like that.

This represents a very neat conception of the situation for we can regard both errors as flowing from a single error, namely *a restricted conception of what seeing is*. The first group realize that on this conception changes of aspect cannot really be changes of seeing, and so draw the conclusion that it must be some other sort of change. The other group wish to cling to the truth that it is a change in seeing, but are forced by their restricted conception of seeing to give a mistaken account of the change.

Now, although this interpretation of Wittgenstein's view fits very neatly with the conception already endorsed of what a Wittgensteinian approach typically is, it is so far totally schematic. What is the restricted conception of seeing? How does Wittgenstein think that he can show it to be restricted?

One way that helps to flesh out the interpretation is to identify people whom Wittgenstein might have thought of as occupying the mistaken positions. I think (as do many) that there is considerable evidence that Wittgenstein would have regarded the Gestalt psychologist Wolfgang Kohler as someone who is right in acknowledging that there is a change in how things are seen, but who had too restricted a conception of experience to accommodate this truth.

It is tempting to suggest that the second erroneous view would be part of the standard theory of perception which has dominated thought about perception, the received wisdom running from Locke to sense-datum theorists even in the present century. Anyway, I shall discuss this possibility.

Why is Wittgenstein so sure that there is a difference in how the item is seen? It is clear that Wittgenstein would not provide a justification for his claim by articulating a conception of seeing and then arguing that there is a change in seeing *on that conception*. For that would be to provide a *justification* for what we say when we say that there is a difference in how it is seen. His claim is that what we say needs no justification.[21] Further, to argue in the rejected way would require an analysis of experience to show that it conforms to the conception, and that is also something that we should not do. Wittgenstein's approach is, simply, that we do report the change as a change in what is seen. That is our language game. Now, there are other concepts of seeing in which what is seen is not different, is the same. But that simply is the

[21] See Footnote 20.

structure of our concepts. This view is what Wittgenstein is expressing, as I interpret it, when he says; '"Is it a genuine visual experience?" The question is in what sense is it one? … A concept forces itself on one.'[22] Here Wittgenstein means that there is no pre-existing and pure conception of genuine experience; it is a genuine change in experience because that is how we describe it.

24.4 On the Interpretational View

We can approach what Wittgenstein has to say about the interpretational account by asking who might be expected to take such a view. It seems likely that it, or something like it, would be held by people who accept the traditional conception of perception in which perceptual occurrences are thought of as two-sided, involving a sensory, sensation like, event, which has an intrinsic phenomenal character, the reporting of which requires in principle no reference to the perceiver's environment, plus a belief or belief-like state resulting from the sensory occurrence, where the belief can have an external content, which can be assessed as accurate or inaccurate in respect of the perceiver's environment. This seems to be the view of, for example, Locke, Berkeley, and Hume. How might they handle the duck-rabbit example? They have three options; it is either a change in the sensory experience, or a change in the resulting belief-like state, or a combination of the two. Now, Wittgenstein has an observation which may be an argument he would see as both correct and as pushing against the view that the change is, either totally or in part, a purely sensory change. He points out that if that were what happens it should be possible to describe the change without invoking the concept of duck or rabbit.[23] It ought to be able to describe the change in a 'direct' way. Now, Wittgenstein implies that this is not possible. It would follow that the change has to be placed on the belief-like side.

There is an interesting question which can be raised once this point has been exhibited. It is; what concepts exactly *are* required by the so-called duck-rabbit experience? Is there any reason for thinking that the shift between perceptions could not occur if the subject lacked the concepts of a duck and a rabbit? There is, after all, a more basic description of the change. Let us assume the picture is first of all seen as the duck. Now, to see it as a duck, given how it is drawn, is to see it as animal facing to the left, with a beak-like, to the front of the face, protrusion. After the shift it is seen as facing to the right, with rear ear-like protrusions. Now, a shift described this way, is quite possible for someone not acquainted with ducks or rabbits. It concerns concepts like the front of the face. It follows that there is no necessity, concerning this famous example, that a description of the change must utilize the concept of ducks and rabbits. (There is merely the empirical question, of which I know of no investigation, whether such shifts do occur to people who do not have the concepts of duck and rabbit).

---

[22] Wittgenstein (2001, p. 174).

[23] Wittgenstein (2001, p. 165).

In fact, Wittgenstein's argument or point is not decisive. For a supporter of the bi-partite account might not accept straight off that these examples are not exhausted by a change reportable without use of the concepts which, according to Wittgenstein, are essential, whatever exactly those concepts are. However, Wittgenstein clearly does accept the claim that reference to some such concept is indispensible in a description of them, and so would hold, presumably, that a theorist who views perception as having these two components must place the change we are interested in on the belief-like side. He seems to think that they are then forced to describe the change as a change in interpretation.

Against such an account he has three main objections. The first is simply that we can see the items in accordance with the way we interpret them. Wittgenstein is in effect saying that we can distinguish between merely adopting a new interpretation and adopting an interpretation in accordance with which we see the figure. I think that this is correct. It can be illustrated by the example of puzzle pictures. I might be told that a puzzle picture is the picture of a face, and adopt that as an interpretation of it, that is, accept the claim and try to confirm it, without being able to see it that way.

Wittgenstein has two other points though.[24] He says; 'To interpret is to think, to do something; seeing is a state'. Does this show anything? The claim that interpreting is thinking should be accepted, of course, but can do no work, since the dispute is precisely whether what we call seeing O as ... is a sort of thinking. So it seems that the point is that thinking is a doing, whereas our phenomenon is a state. However, the bi-partite account of perception must have available to it a notion of interpretation which allows interpretation to be a state, where this is contrasted with being a doing. For, clearly, what we are disposed to think while engaging in perception is usually not a matter of our doing anything. The thinking need not be recited nor controlled. Thinking and belief are states.

Wittgenstein's third point is this; 'When we interpret we form hypotheses, which may prove false.-"I am seeing this figure as a ..." can be verified as little as (or in the same sense as) 'I am seeing bright red'. So there is a similarity in the use of 'seeing' in the two contexts." '[25] I am not sure what force this has. Presumably, 'verifying' links with 'proving false', but, in that case, the interpretationalist will not say that what is verified (or falsified) is that 'I am seeing this figure as a ...', but, rather, it can be verified that this figure is ... or that it is not .... Thus, one would not verify that I am interpreting this figure as ..., although the interpretation is verifiable.

24.5 The Non-interpretational View

Fairly early on in his discussion, Wittgenstein makes a series of remarks which he concludes in the words; 'And this by itself wrecks the comparison of 'organization'

[24]Wittgenstein (2001, p. 181).

[25]Wittgenstein (2001, p. 212).

with color and shape in visual impressions'.[26] Wittgenstein is using the term 'organization' to stand for that aspect which changes when there is a change of aspect. The example he is taking is the schematic cube. So, he is saying that when how we see it changes, it can be described as a change in organization. We cannot sensibly compare the change in organization to a change in color or shape, because if we are asked to represent to someone what colors and shapes we are seeing we could do so by making a drawing of the scene. If there were a change in features at the level of color (etc.) then a subsequent drawing would be different. However, when an aspect changes, whatever picture would have been accurate at the start will also be accurate after the change. At the level depicted by a drawing it remains the same. It follows, therefore, that it cannot be right to think of organization as akin to color or shape.

Now, Wittgenstein makes a distinction between a drawing, which is a two-dimensional representation, and a model, which is a three dimensional representation. In the case of the schematic cube this is of significance, because when it is seen differently the difference lies in apparent three-dimensional lay out. So, a three-dimensional model for the first experience would not be the same as a three-dimensional model of the second experience. But the contrast between model and picture is not appropriate for all cases. Thus in the duck-rabbit shift since we are seeing a picture in the first place we could not make a model of it, and yet there is a difference on organization. In fact this is simply an aspect of the point that Wittgenstein is talking about a wide range of cases.

Wittgenstein's point here seems persuasive and important. But against whom is he making it? There is some reason for thinking that Wittgenstein is addressing his remarks here to Kohler. Thus, Ray Monk points out that when Wittgenstein lectured on aspect perception he would often quote passages form Kohler to begin the discussion. It is quite clear, both from what is known about Wittgenstein's activities at the time he was thinking about aspect perception, and from the content of the text, that Kohler acted as a major stimulus. What is not entirely clear is which bits of Wittgenstein's discussion are supposed to relate to Kohler's views.

However, it is significant that Wittgenstein uses the term 'organization' to stand for the general notion of seeing an aspect precisely at the point where he is criticizing the comparison of organization and color. For, 'sensory organization' is the title of chapter V of Kohler's book *Gestalt Psychology*. There is surely a temptation to see a link in that. Second, Kohler's point, the theme of Gestalt psychology, is that organization belongs to the sensory. It is quite clear that he thinks also color belongs to the sensory, and his arguments sometimes proceed by denying differences between color and 'organization'. Thus he says;' When I see a green object, I can tell the name of the color immediately; yet I know that green is used as s signal on rail-road tracks and also as a symbol of hope. But I do no believe that, therefore, the color green as such must be explained by meaning. .... In exactly the same manner, sensory units have acquired names and may have become richly symbolic in the context of our knowledge, while existing, nevertheless, as segregated units in the sensory

[26]Wittgenstein (2001, p. 168).

field prior to such accretions'[27] We have then that Kohler uses 'organization' and compares organization to color. Further, thinkers interest Wittgenstein because of errors they make, and I assume that Kohler is no exception, and he certainly cannot be accused of the errors of interpretationalism, since Kohler's point is to expose that error himself.

Although this point does not provide any extra justification for the interpretation I am considering, a very striking parallel between Kohler and Wittgenstein is that Kohler talks of 'a sort of 'blindness' in person's, who at the same time are not totally blind'.[28] Kohler is thinking of people who perceive but who do not 'group' elements in their sensory fields. And, of course, at the end of his discussion Wittgenstein asks about aspect-blindness.

However, if this interpretation is accepted, it needs to be asked whether Wittgenstein's criticisms are sound. The crucial point, it seems to me, is that for Kohler 'sensory' basically means—in the psychological system prior to the operation of thought or memory. He argues against theorists who treat 'organization' as imposed by thought. This means that when Kohler describes organization as sensory he is primarily making a negative point about it. He is not committing himself to its being exactly the same type of thing as color, or present in one's experience in the same way that color is. I think, therefore, that it cannot be said that Kohler did compare organization with color in a way that Wittgenstein's argument brings out is wrong.

There is another aspect of Gestalt theory which is relevant. Kohler emphasises Wertheimer's work on the principles of grouping (principles which are normally presented whenever Gestalt theory is described). But he was attempting to formulate principles of grouping which rely on the presence in the sensory field of more primitive features, e.g., spatial features, to which grouping is sensitive. This approach seems to involve the idea that grouping is different from color and shape features, although still remaining sensory, in the negative sense.

This establishes quite a strong case for saying that Wittgenstein's remarks are not totally accurate criticisms of Kohler's conception of organization. However, it is not simply the comparison of organization with color that Wittgenstein objects to at this point, he also links it to the idea of a visual impression as an 'inner picture'. We need to ask, therefore, whether Kohler can be accused of doing that. Did Kohler accept an 'inner picture' model of vision?[29] It seems fairly clear that Kohler accepted a model of perception in which it involves an inner experience, produced in the subject by his environment. He also thought of this inner experience as a physical process or

[27] Kohler (1992, p. 151).

[28] Kohler (1992, p. 169).

[29] In Ray Monk's discussion of aspect perception he obviously implies that Wittgenstein did see Kohler's theory in these terms. Thus he says; "Kohler's treatment of them (ie. these problems) falls foul of the very conceptual confusions that Wittgenstein had been trying to dispel in his 'Private Language Argument'. These confusions begin with Kohler's description of a Gestalt as 'a concrete individual and characteristic entity, existing as something detached and having a shape or form as one of its attributes'. This already makes it sound as if what was being described was an object, a private object". See Monk (1990, p. 512). One reason that his is unhelpful is that it is not actually clear what confusion Wittgenstein is dispelling in his argument against a private language.

condition, and since organization relates to it overall, that is different parts of the experience are united in the organization, he was keen to find acceptable analogies to this kind of holistic property in physical systems. He cites the distribution of 'differences of potential and the densities of current' along a series of conductors. What, however, there seems no reason to accept is that he also treated this inner experience as like the presentation of an inner picture. The notion of an inner picture and an inner experience are obviously different. The model of an inner picture is a model or analysis at some level of inner experience, the act/object analysis, but there are other models of inner experience, and the act/object form is not compulsory.

24.6 Wittgenstein's Solution

Wittgenstein's basic point seems to be that when an aspect shift occurs we *do see things differently*. That is how we report such cases, and there are no good reasons to modify our way of speaking. What else does he say? A few more things need to be added.

(1) Wittgenstein's strategy in dealing with expressions of doubt that how we see is different amounts to claiming that to say that amounts to simply assuming that we have one concept of seeing. Thus, it cannot be right to object to it as a case of seeing by arguing that it fails to fulfill some condition which has been selected as necessary for seeing, because what that really shows is that it simply is a distinctive concept of seeing. It is, of course wrong to demand that it satisfies some constraint appropriate only to other sorts of seeing.

Here is an example. Wittgenstein points out that mastery of certain concepts is necessary for seeing O as F (say, as an apex), and then considers the remark; 'But how queer for this to be the logical condition of someone's having such-and-such an experience!' His reply is; 'From this it follows that we cannot be dealing with the same concept of experience here.'[30]

(2) Wittgenstein alludes to behavioural consequences of a person's seeing O as F. Thus he says;"What tells us that someone is seeing the drawing three-dimensionally is a certain kind of 'knowing one's way about'. Certain gestures, for instance, which indicate the three-dimensional relations: fine shades of behaviour".[31] These are what ground our judgements about another, but also what hearing and accepting such judgements prepare us to expect. I do not want to examine this suggestion except to remark that Wittgenstein's stress on behavior amounts to some sort of justification for describing things in terms of seeing, and so marks a little retreat from his insistence that justification is not needed.

(3) Much of Wittgenstein's discussion concerns the perception of pictures. This is a natural topic given that many of the examples are cases of pictorial, or

[30] Wittgenstein (2001, p. 178).

[31] See Footnote 8.

representational, experience. It is, however, clear that when an aspect is seen, or when it 'dawns', the aspect need not be pictorial or representational. Thus, seeing the likeness between two faces is not pictorial experience. This means that pictorial experience does not lie at the heart of the phenomenon Wittgenstein is analyzing. I wish, however, to make a few remarks about picture perception and Wittgenstein's discussion of it.

I have given one reason for supposing that picture perception is not central to Wittgenstein's approach. There is also another reason, and that is that picture perception is a very special sort of perception. One just needs to remind oneself of the non-pictorial character of most of our perception, to realise that an account of it cannot be central in the theory of perception. It seems to me that Wittgenstein gives the impression that aspect perception is of more importance than it would be if it were regarded as primarily evidenced in representational experience.

Despite the fact that it is a very special sort of perception (representational experience being primarily restricted to vision) there is a way in which it is of great significance to the theory of perception, and that is that the perception of pictures has been taken as a general model of perception. There has been a tendency to think of perception as involving an inner picture, apprehended by an inner eye. This means, I think, that although seeing as has nothing essentially to so with pictorial experience Wittgenstein is right to look hard at the nature of picture perception.

(4) Since it is Wittgenstein's central point that the dawning of an aspect is a change in seeing, (or perception), you would expect him to consider whether the sort or level of perception he has isolated can be absent where other levels are present. This is the question whether there can be aspect blindness, discussed by Wittgenstein.[32] Interestingly, Kohler himself introduces the concept of organisational blindness. Unfortunately, one of the central problems with Wittgenstein's discussion prevents Wittgenstein from getting any distance with the issue. The problem which I have in mind is that it is not clear what aspect perception is. This unclarity was noted earlier. Since it is unclear what aspect perception is, it is automatically unclear what aspect blindness is. Thus, Wittgenstein says that the aspect-blind man is not supposed to see 'the aspects A change'. Wittgenstein has given this curious label to the two ways of seeing the 'double cross'. So he is understood not to have aspect shifts in his experience. Hence, he will not undergo the kind of experience change which will lead him to exclaim, 'Now it's a black cross on a white background'. However, this amounts, *not* to aspect blindness, but rather to stability of experience once it is underway. He is allowed to recognise that the figure contains a black cross and a white cross. So, in what sense is he 'aspect blind'? What, even in this example, is he blind to? Further, no account has yet been given of a general condition of aspect blindness. So, it is not surprising that Wittgenstein should add; 'Is he supposed to be blind to the similarity between faces?.... I do not want to settle this.'[33] I think that

[32]Wittgenstein (2001, p. 182).

[33]Wittgenstein (2001, p. 181).

it has to be concluded that Wittgenstein does not properly get to grips with aspect-blindness.

(5) One question to which we have not as yet given a satisfying answer is why exactly Wittgenstein was so interested in 'seeing an aspect'. He provides an answer himself when he says, straight after his discussion, "the importance of this concept lies in the connexion between the concepts of 'seeing an aspect' and 'experiencing the meaning of a word'".[34] What exactly are we supposed to understand by this? Now I assume that this concept is 'aspect blindness'

However, the examples Wittgenstein next gives and which are presumably supposed to be examples of not experiencing meaning are, first, not understanding a request to pronounce an ambiguous word meaning one thing by it, and, second, not loosing a sense of the meaning of a word which is repeated ten times over. Now, the first of these seems to be unconnected to aspect perception, since it concerns some sort of active meaning. That is, if we label the missing phenomenon 'experiencing meaning', it is a different sort of 'experiencing meaning' from experiencing meaning in the perception of language. The second example does indeed resemble the cases Wittgenstein has been studying, but what is absent is, rather, a case of, what we might call, 'aspect vanishing', rather than 'aspect dawning'. I think that Wittgenstein does not mention what I have called 'aspect vanishing', but it is a real occurrence which certainly belongs with the kinds of cases he does mention. Thus, with puzzle pictures one can have the experience of loosing a sense of what it represents, just as one can have the experience of gaining such a sense. But, it would be very odd to think, however, that there would be any significant deficit, of interest to the study of language, in the case of a person whose sense of meaning is not disturbed by repetition.

I suspect that the oddity of Wittgenstein's examples again flows from the general unclarity of what 'aspect perception', and hence, 'aspect blindness' is. Thus, if by 'aspect perception' Wittgenstein fundamentally means perceptual recognition, then it is obvious that normal language understanding just is a case of aspect perception. We hear the sounds and recognise them as saying certain things. Since Wittgenstein's basic point about aspect perception is that it is 'perceptual' the application of that claim to language leads to the thesis that understanding is perceptual. This is a claim which is emphasised by many theorists of language. However, since Wittgenstein defends his thesis by claiming that we have here a distinct concept of perception little is conveyed as to the significance of labelling it perceptual. The negative element in Wittgenstein is, primarily, that all reasons for denying the cases are perceptual depend on over restrictive concepts of perception. But Wittgenstein's treatment attaches no significance to calling it perceptual other than it being our practice to attach that label to it.

At other times Wittgenstein seems most interested in the experience of aspect shifting. That is, he is interested in the experienced instability of recognition, or the experienced dawning of recognition. It seems to me that our perception of language exhibits this. One kind of example is where one hears a remark, but only understands it after a brief lapse of time. This is slightly different from the standard case considered

[34] See Footnote 33.

by Wittgenstein because the heard utterance is not continuously perceived, whereas, for example, the ambiguous figures are seen for a period. With a dawning of linguistic understanding it is usually the significance of a just heard remark that dawns. This reflects, however, a difference between hearing and sight, in that what we might call the significance of a sound typically lies in a temporally extended structure, with only a part presented at a time, whereas with vision we can discern a significance in the complex structure presented at a single time. Still, the dawning of understanding seems to be the same sort of thing as the dawning of an aspect.

It remains quite obscure, though, both what according to Wittgenstein is the significance for our account of perception that there are experienced aspect shifts, and what, according to him, is the significance for our account of language that there is experienced dawning of understanding. It is also obscure what the actual significance of both things is as well.

I conclude that the one reason which Wittgenstein gives for the importance of the phenomenon, its link to language, is hard to make anything of.

24.7 Some Reactions

I have tried to provide an account of Wittgenstein's discussion, in the course of which some criticisms have been expressed. I want now to give what as I see as a partial assessment of Wittgenstein's treatment.

The most serious problem is that there is no clear boundary around the phenomenon with which Wittgenstein is concerned. It is very hard to say what the dawning of an aspect is. It is not clear, therefore, what the scope of Wittgenstein's claims are.

Since the most obvious suggestion he defends is that the item is seen differently, this means that we do not know what Wittgenstein is committing himself to. On the one hand there is a temptation to suppose that there is a dawning of an aspect whenever a subject recognises something about a perceived item without a change in the appearance of the item leading the recognition. However, if it is understood so broadly then Wittgenstein's claim is implausible. Consider this example. I am listening to the performance of a song which I know well, with which I am completely familiar. I cannot, though, remember who the composer is. Then it dawns on me that it is, say, Mahler. In this case I cannot see that it is true to say that I now hear it differently. Now, there may or may not be generalisations relating what aspect is recognised and whether the item, recognised as falling under that aspect, is thereby seen differently, but Wittgenstein does not really explore whether there are or not.

The obvious response to this comment is to say that Wittgenstein's main point is just that it is *possible* to see an item differently as recognition of it occurs. It can be seen differently although in terms of colours and shapes it is not seen differently.

If this is treated as the central point then it represents Wittgenstein as endorsing the main claim (or an implication of the main claim) about perception advanced by the Gestalt theorists. What significance, though, does this claim have for Wittgenstein?

It seems that it has basically a two-fold significance. The first is that the difference, which we describe as a difference in seeing, cannot be analysed as a difference in interpretation. It is not that kind of difference. This is a negative significance. The second is that we describe the shift, and when undergoing it, express the shift, by using the term 'see'. This seems to be a positive significance.

Now, the second claim is not insignificant, but it has no deep significance for the theory of perception. It hardly matters how we express the phenomenon, since someone who wanted to say that the proper account was that there had merely been a difference in interpretation could happily agree that we use the word 'see' to report it, or express it. A proponent of such a view does no have to criticise our linguistic practice since he can say precisely what Wittgenstein says when he claims that this is just one of our concepts of seeing. If so, the real significance of Wittgenstein's thesis is the negative claim about interpretation.

We have to enquire whether Wittgenstein successfully argues the case against interpretation. When discussing that question I claimed that two of Wittgenstein's points are not strong, although it has to be allowed that there is a difference between what we would call holding as an interpretation of, say, a picture, that it is of an F, and what we would call seeing it as an F. The problem is that this does not really settle the question whether the approach which would receive expression in the talk of interpretation is wrong. The approach which is supposed to be captured by talking of the interpretative view is one in which what we call perceiving is thought of as involving two components. The first component is a sensory experience the character of which can be captured by specifying the color present at a minimum point in a visual field, and the second component is a thought, as to how the environment is. If someone taking this approach accepts that there can be aspect shifting while the basic sensory state remains constant, he must regard the aspect shift as being a change in the thought component. He may talk of a change in interpretation. But it is quite consistent to recognise that there are *differences between different kinds of thought component changes*. So, this account is not immediately defeated by acknowledging a contrast between so called perceptual aspect shifts and another sort of change, labelled changes in interpretation. That is to say, a proponent of this dual component account need not hold that all thought-shifts are the same. So, at most, the contrast which should be agreed to amounts to a challenge to the theorist to give a plausible account of the difference between different thought shifts.

For example, it might be suggested that what we standardly call 'interpretation' is sensitive to reasons, whereas the thought shift in a change of aspect is autonomous of such reasons.

I believe, then, that the traditional account which, we are treating it as Wittgenstein's main point to deny, is not decisively criticised in any argument he adduces. Furthermore, Wittgenstein is in a weaker position to criticise that view than Kohler, and the Gestalt theorists, were, because he accepts that the phenomenon of aspect shift requires concepts, whereas the Gestalt thesis was that organisation and grouping is pre-conceptual. It is, in their view, part of the pre-conceptual bedrock on which our concepts rest, out of which they grow. They could argue that the evidence is that grouping is pre-conceptual and that, therefore, the thought based account must be wrong. Wittgenstein himself cannot argue that way.

24.8 Conclusion

Wittgenstein's discussion of seeing as is dense and very rich, and packed with remarks that illuminate aspects of perception and illuminate too our theoretical thought about perception. Many of these have had to go unmentioned here. Wittgenstein's thinking was, for its time, highly original in philosophy, where thought about perception tended to consist in a re-engagement with a limited number of well-known arguments about what is directly perceived. Wittgenstein escaped from that trap, in a way that has had a lasting effect. Further, and somewhat paradoxically, part of this escape came from his engagement with psychology. What I have tried to argue, though, is that there are features of his discussion which are unclear, the most important one being what aspect seeing is supposed to be, and the direction of some of his arguments is uncompelling. But, no doubt, there is far more to be unearthed in this section that I have managed to bring to light.

Bibliography

Budd M (1989) *Wittgenstein's Philosophy of Psychology* London, Routledge.

Cahill K & Raleigh T (eds) (2018) *Wittgenstein and Naturalism* London Routledge.

J.P. Frisby, Seeing, (Oxford, Oxford University Press), 1980 pp. 18–19.

Kohler W (1992) *Gestalt Psychology* New York Liveright.

Kuusela O and McGinn M (eds) *Oxford Handbook of Wittgenstein* Oxford, OUP.

Monk R (1990) *Ludwig Wittgenstein* London Jonathan Cape.

I.Rock, An Introduction to Perception, (N,Y., Macmillan), 1975, pp. 264–266.

Snowdon P F (2011) 'Private Experience and Sense-Data' in Kuusela O and MeGinn M. (2011) pp. 402–428.

Snowdon P F (2018) 'Wittgenstein and Naturalism' in (eds) Cahill K and Raleigh T (2018) pp 15–33.

Wittgenstein L (2001) *Philosophical Investigations* Oxford Blackwell.

Chapter 25
Essay in Formal Biology

Nikolay Milkov

> The philosopher must twist and turn about so as to pass by the mathematical problems, and not run up against one,—which would have to be solved before he could go further….
> Even 500 years ago a philosophy of mathematics was possible, a philosophy of what mathematics was then.
> —Ludwig Wittgenstein (1956, IV, § 52).

25.1 Opening

The task of this essay is to put biological individuals in formal terms. This approach is not directly interested in matters of time (for example, in evolution), but rather in the formal shape of biological objects.[1] So it is radically different from natural science. In his later years, Wittgenstein made similar investigations in psychology and mathematics.[2] Unfortunately, he found no time to make extensive remarks on philosophy of biology. This is what we are going to advance here.

The approach followed in this essay opposes, above all, the intensionalism in the philosophy of biology. A typical case of biologically motivated intensionalism is the theory of "organic wholes" adopted, among others, by one of the founding fathers of analytic philosophy, G. E. Moore. The most radical form of intensionalism, however, is vitalism. Vitalism, which is as old as biology is, contends that compared

[1]An ontological (structuralist) concept of the biological kinds, which doesn't necessarily oppose Darwin's theory of evolution, was suggested by Thompson (1917); see also Rescher (2011) and Chap. 2 in this volume. On the compatibility of the formal biology with the mainstream theory of evolution see § 22.

[2]See the motto for this essay.

N. Milkov (✉)
Philosophy Department, Universität Paderborn, Warburger Straße 100, 33098 Paderborn, Germany
e-mail: nikolay.milkov@upb.de

S. Wuppuluri and N. da Costa (eds.), *WITTGENSTEINIAN (adj.)*,
The Frontiers Collection, https://doi.org/10.1007/978-3-030-27569-3_25

to the elements of matter, biological individuals contain elements which are of a fundamentally different kind. This conception was already embraced by Aristotle who introduced the concept of "final cause" which is to be kept apart from the material cause.

The most radical form of vitalism was developed in German Philosophy, first, in the eighteenth century by Leibniz and his followers. In the nineteenth century, Nietzsche maintained that life is much stronger than knowledge (1874, § 44). At the beginning of the twentieth century, Hans Driesch still explained the life of biological individuals with the help of the Aristotelian term of *entelechy*. This is a substantial entity, controlling all biological processes (1905).

Our objective in this essay will be to tentatively estimate the possibilities of applying formal approach in biology, not to develop an explicit and strict formal biology. To this purpose, we shall briefly discuss three different approaches in formal biology: (i) Wittgenstein's formal biology; (ii) mereology of biological individuals; (iii) topology in biology.

25.2 Wittgenstein's Formal Biology

25.2.1 Wittgenstein's Material a Priori

As cases of formal biology can be cited some ideas or, more precisely, hints of Bertrand Russell and Ludwig Wittgenstein. In "The Philosophy of Logical Atomism" (1918) Russell compared philosophical logic to zoology. According to him, logic investigates the "'zoo' containing all the different forms that facts may have" (1956, p. 216). In fact, however, this conception of Russell has also its flip side: Zoology can be seen as logic. The task of the logic of zoology—and of biology in general—is to describe zoological/biological individuals in formal terms—in terms of their forms.

The conception of logical forms as biological individuals was implicitly developed, but in a rudimentary form, by Russell's pupil Wittgenstein. In *Remarks on the Foundations of Mathematics* the latter raised the question:

> But could one not study transformations of (e.g.) the forms of animals? But *how* 'study'? I mean: might it not be useful to pass transformations of animal shapes in review? And yet this would not be a branch of zoology.
>
> It would then be a mathematical proposition (e.g.), that this shape is derived from *this* one by way of *this* transformation. (The shapes and transformations being recognizable.) (1956, III, § 13)

According to this idea, geometrical figures, numbers, but also biological individuals, can be seen as shapes (forms) which can be studied by respective formal disciplines that describe their transformations. Importantly enough, Wittgenstein now is reluctant to call these formal disciplines *logic* because is indignant at the all-embracing claims of this discipline. He made these claims when he wrote the *Tractatus*.

Formal biology, in particular, describes qualitative forms (*Gestalten*) of biological objects. Hence it is a type of morphology. Not a morphology of Goethe's type, however. Wittgenstein's philosophical morphology tries to find the forms of the individuals with the only aim to analyze, resp. to discuss them. In contrast, Goethe's morphology is not analytical (not formal) but complementary. It is a kind of natural science—a suggested alternative to the science of Isaac Newton.

Here is a brief description of Wittgenstein's approach:

> Wittgenstein sees the task of elucidating here, above all, as to put of the data together [*zusammenstellen*]; the data are to be so put together that to suggests an understanding. … In this way he suggests a 'general picture' of the data, which presents them perspicuously, without to pay attention to their temporal order. (Schulte 1990, p. 32)

Now, it is exactly this seeking of new formal orders of biological individuals, that suggests new *patterns* of individuals that are the proper subject of this type of formal biology.

This approach is based on the understanding that biological individuals are mosaics of forms, so that at every instance—in every biological individual—we see the object in only one of its many forms; its other forms remain implicit.[3] The task of formal biology is to find out these forms. Its method is the comparison of the different forms of biological individuals (including that of "folk biology") that it collects in formal patterns of most variegated order. In contrast, mainstream biology is only interested in the spatial–temporal (in the real) orders of biological individuals.

Wittgenstein's formal biology is clearly a rich research program. It contrasts, the conventional taxonomies in biology, introduced into it by Carl von Linnaeus, are linear and so monotone. Also today, the mainstream taxonomies in biology are usually built on an inclusive relation of genus to an art. Their ideal is the *arbor Porphyriana* by which every new classification relates a genus to an art. Wittgenstein was sharply critical to it. In "Conversations with M. O'Drury", he noted: "I have always thought that Darwin was wrong: his theory doesn't account for all this variety of species. It hasn't the necessary multiplicity (1981, p. 174).

Wittgenstein's description of biological individuals compares forms (shapes) of objects in order to pick out new patterns in which new forms of theirs are presented. An instructive example of this method is provided by Wittgenstein in *Remarks on the Philosophy of Psychology*:

> But might it not be that plants had been described in full detail, and then for the first time someone realized the analogies in their structure, analogies which had never been seen before? And so, that he establishes a new order among the descriptions. He says, e.g. 'compare this part, not with this one, but rather with that' … and in so doing he is not necessarily speaking of *derivation [Abstammung]*. (1980, § 950)

Despite the fact that Wittgenstein did not elucidate what this new discipline exactly is, he gave hints at how to develop parts of it when he introduced the concept of *forms of life* (*Lebensformen*). He uses this expression in an idiosyncratic way at that. For

[3]This conception is based on an idiosyncratic interpretation of the Wittgenstein's Tractarian objects, developed in Milkov (2001).

Wittgenstein, this term is a biological, not a social category. Thus he held that there is a man's form of life, lion's form of life, etc. In short, every biological species has its own form of life. In contrast, in ordinary German, this concept usually means social, or spiritual *ways* of life, not biological forms of life. One maintains, for example, that different social classes in the nineteenth-century Germany had different *Lebensformen* (ways of life). This was the sense of the concept *Lebensformen* also according to the man who first introduced it in the theoretical humanities—Spranger (1914).

Comparing different forms of life is only one type of the patterns in formal biology. As already noticed, however, biological individuals have data of different order. So we can investigate different orders of biological forms (shapes):

(i) The outer form of the animals. "Are tomatoes fruits of vegetables?" "Is a zebra without strips a zebra?"
(ii) Animals' behavior (*habitus*) understood as a form;
(iii) The animals' psychology: can animals strive for, hope for, dream for, etc.

Etc.

These different kinds of investigations give rise of different kinds of formal biology. At the end we have descriptions of different patterns of biological forms which can be ordered in separate chapters.

Importantly enough, these investigations are similar to that Wittgenstein made of colors in *Remarks on Colour* (1977). Historically, it followed the approach of "pattern analysis," introduced in humanities by Oswald Spengler who compared the archetypes of different cultures (Haller 1988, pp. 74–89). These are analyses of data, of indefinables, of objects (faces, shapes) that cannot be analyzed further.[4] In *Philosophical Investigations* (1953) Wittgenstein analyzed this way the forms of: the fly (§§ 284, 309), goose (p. 221), hare (§§ 520, 521, 524, 542), dog (§§ 250, 357, 650, pp. 174, 229), beetle (§ 293), cat (§ 647), cow (§§ 120, 449, p. 220), lion (p. 223), mouse (§ 52), parrot (§§ 344, 346), cat (§ 647). "They constitute an autonomous, natural order that is not capable of being invented or constructed through acts of man" (Smith 1992). In this sense, they are cases of *material a priori*.

25.2.2 Formal Ontology of Biological Objects

Formal ontology of biological individuals was also advanced by Nicolai Hartmann who held that the world is a united system, which nevertheless has many *strata*. The strata are ordered one onto another, so that each one "is not isolated in itself, but rather sets out the conditions and regularities for the next strata" (1940, p. 182).[5] Every stratum of a higher order is a superstructure on the grounding stratum. Perhaps we can better understand this ontological dependence in the light of John Searle's

---

[4]See Milkov (1999).

[5]Compare with the Wittgensteinian modal ontology explicated in Chap. 2

conception that strata are not causally but ontologically dependent. The strata of higher order are only a trait of the strata of lower order (1983, pp. 20–1).

This position was also embraced by Karl Marx according to whom the subject is parasitic upon the world: it does not have anything else to do than to imitate reality.[6,7] The subject makes this in an attempt to "grasp" the world, to *assimilate* it, becoming in this way identical with it.[8] As Marx put it—following Aristotle, via Hegel,—the subject makes this in an act of "appropriating of alienated, objectivated essence" (1974, p. 242).

As for their uniformity, in our formal ontology we hold that philogenetically, biological individuals are different developments of one and the same form of nature (see Thompson 1917). We follow in this a long tradition in searching for a uniform ontology in biology and philosophy of mind. Two examples: (i) Nietzsche claimed that thinking has its primitive stages in the pre-organic world. "'Thinking' in primitive states (pre-organic) is forming on—so as it is by the crystals" (1901, § 499). (ii) The same did Bertrand Russell who claimed that "lifeless instruments, such as photographic plates and gramophones ... have something closely analogous to perception" (1927, p. 270).

The acceptance of the uniformity of the nature of biological individuals paves the way for elaborating a powerful formal ontology in biology. We, more precisely, can see different ontological individuals as constructed out of simple forms, through their modifications and different arrangements.

According to this conception, different kinds of organic individuals, plants, animals, *homini sapienti* and their thinking, are results of different *ways* of composing complexes which can be described by logical means. In order to demonstrate this, we will refer to our investigations in Wittgenstein's Tractarian ontology.[9] According to them:

(i) The objects are arranged in complexes, fitting one to another in certain *way*.
(ii) By picturing, the forms of facts are delivered (*expressed*) by other facts (pictures)—in a certain way (style). Importantly enough, the form of projecting *hinges* on the form of facts.
(iii) The copying (deliverance, expressing) of facts by thinking is the way of projecting the facts through language.
(iv) Creating (gestures, postures, etc.) is also an act of delivering of forms in a certain way. This time, however, the act of projecting (expressing) changes the form completely so that a new form is created.

---

[6]The alternative positions, which accept that the subject can achieve something of her/his own, are necessarily circular. Take, for example, the suggestion that the subject's aim is to be "capable of making molecules swerve from their paths" (Searle 1984, p. 92). It again assumes an action *upon* the world and not independent of it.

[7]A similar position was maintained by Wittgenstein. We are going to discuss it in § 25.3.1.

[8]In this sense, Konrad Lorenz states that biological individuals make pictures (*Abbildungen*) of the real world, which are similar to a plaster copy of a coin (1973, p. 39).

[9]As developed in Milkov (1997, i, pp. 372–4; 2003: Chap. 3, § 3, (vi)); and in the Chap. "Wittgenstein's Ways" in this volume.

Our final thesis is that not only the different strata of the world, but also the strata of biological objects, can be logically constructed and also explained in this form. Accordingly, our formal ontology of biological individuals describes the different ways in which biological individuals are composed.

Our hope is that the formal ontology of biological objects we are suggesting in reference to Wittgenstein puts many points of both contemporary biology and contemporary philosophy in a new light. Take as an example biological evolution. The formal ontology describes it as accomplished in accordance with the laws of composition of complexes. This conception is supported by the mainstream evolutionary biology today, which maintains that the change in the life-world is a result of gene mutations. In fact, the genes can be seen as nothing but forms (from a logical point of view, they can be seen as the identification numbers of the biological objects) that are quasi "attached" to some of their building-blocks-chromosomes. On this view, the natural selection consists in that only those individuals survive that have more developed forms. Importantly enough, the "development" here is understood in terms of information.[10] In other words, these types of individuals survive that have better information-gain (Lorenz 1973, p. 38).[11]

This conception of the species-evolution maintains that there is no place for teleology in it but for abundance of forms and their subsequent combination. In this variety, the winner is contingent. So Daniel Dennett: "Replication is not necessarily for the good of anything. Replicators flourish that are good at… replicating" (1990, p. 129). The selection among the forms takes place in a process of ceaseless experiments, through a constant confrontation with the forms of reality. The method is that of experimenting and correcting. The neo-Darwinian Richard Dawkins sets out as necessary conditions for the biological evolution via natural selection three factors:

(i) Abundance of most variegated elements;
(ii) The elements must be able to copy themselves;
(iii) Interaction between the form of the elements and the form of environments (1976, p. 127).

25.3 Mereology of Biological Individuals

25.3.1 Wittgenstein's Biological Mereology

Mereology is a well-established approach in philosophy of biology, already used by Aristotle in *De partibus animalium* and also by the Stoics. In modern times, Joachim Jungius, Gottfried Leibniz and Bernard Bolzano made important contributions to mereology. But mereology was also widely explored in early analytic philosophy,

[10] Karl Popper's "evolutionary epistemology" (1984) maintains virtually the same.

[11] Here it is to remember that the etymology of the concept of information is connected with the concept of *form*.

for example in Wittgenstein's conceptual analyses as developed in the last pages of the *Notebooks 1914–1916*.

According to Wittgenstein,[12] the subject and the world are separated one from another; they are independent one from another and are self-contained. In another sense, however, the subject is compelled to imitate the world in attempts to grasp it—to become identical with it. The aim is to quasi abolish the difference between the person and the world. This relation is made either in the form of *contemplation*, or in a form of *creation*.

The very impulse to know the world comes out of the consciousness that we are unique subjects. "Only from the consciousness of the uniqueness of my life arise religion–science–art" (1979, p. 79). At that, the person and the world lie in two different planes. It is true that the impulse to know the world comes from the person. However, she/he doesn't exist in the plane of this knowledge: we cannot find it amongst the pictures made by him. "The subject does not belong to the world: rather, it is a limit of the world" (1922, 5.632). This explains why "solipsism, when its implications are followed out strictly, coincides with pure realism" (5.64).

All this means that there is in the world a thinking subject and also another willing subject, which are situated in two different planes. This means that they cannot be seen together in one and the same time. The volitional subject exists only as the cause of our knowledge. Unfortunately, very often it is extrapolated onto the world of ideas, thus producing confusions with important consequences. This mistake is easy to commit since the volitional subject is embedded in a body, which is a part of the material world.

25.3.2 Hegel's Mereology of Biological Individuals

In *Philosophy of Nature* (Volume 3 of his *Encyclopedia of Philosophical Sciences*), Section Three ("Organics"), Hegel has made a bunch of interesting suggestions in the "logic" of biology. Hegel's "science of logic", however, like all classical logic before Frege, was built "within the bounds of a mereological idea of logic" (Stekeler-Weithofer 1992: 100). That is why his logic of biology is more of a mereology of biological objects. Unfortunately, so far this part of Hegel's philosophy remained in the shadow.

Hegel's biological mereology makes analyses similar to these of Wittgenstein as discussed above: it describes biological individuals in terms of wholes, their parts and their location. In what follows, we are going to set it out—indeed, only in rudimentary form,—but recovered from Hegel's speculations.

(a) *Biological Individuals*. To Hegel, only animals and humans are biological individuals. This is already shown in the fact that usually, the flowers in our private garden have no proper names, whereas our pets have names. In order to mark off

[12]In the next paragraphs we follow Milkov (1997 i. pp. 381 ff.); see also Milkov (2004).

this difference, we can call, for convenience, the plants *particulars* and the animals *individuals*.

The fact that animals are individuals while plants are only particulars, explains why the former have, while the latter haven't:

- Self-expression: Animals express themselves through their voices; in persons this characteristic is further developed in language.
- Locomotion: Animals are not connected with one point in space as plants are connected; but they are nevertheless connected with some spatial point which they chose and which they occupy now.
- Senses: Animals have feelings: pain, joy, etc.; plants don't have feelings.[13]
- Souls: animals have souls, plants have not.

How are all these—and other—characteristics to be formally interpreted? Following Hegel, in the next sections we are going to give a tentative answer to this question.

(a) *Shape* (*Gestalt*). An organism has characteristically a *form* (*Gestalt*), which it reproduces in its biological successors. But while the plant "is still closely related to geometrical forms and crystalline regularity" (§ 345),[14] the form of the animal is something radically new. It is a biological form.

Another formal difference between plant and animal is that whereas plants mature through a change of form, animal's maturation involves a change in size, the animal's shape remains the same (§ 343). This point is another argument to hold that the plant is not an individual but a particular: it ceaselessly changes its outer individuality (see (e), below).

(b) *Mereology of Biological Individuals*. Organic individuals can be also seen as conventional unities that contain parts—the members (limbs) of the individuals. The difference between them can be well described as a difference in the set-up of unities:

- The members of plants are not related one to another.
- In animals, the members are related; they are elements of their unity—elements of their form. The members of animals are interchangeably means and ends (§ 355); indeed, they are perpetually negating their independence (§ 350).
- What is new with the human person is that he concentrates his power into a single point of his body, deploying it into certain direction, and for particular purposes.

(c) *Sense* . Most important characteristic of animals is their sentience: they have feelings. In fact, feeling is nothing but a form of connection of the biological individual with the world made from the perspective of its self-preservation (§ 344, *Add.*). This connection is realized through the stimuli produced by the individual that recurrently encounters matter. In other words, perception is stimulation by matter, and a reaction of self-preservation in the face of that stimulation (§ 354).

Senses are spatio–temporally determined. In sight, the physical reality manifests itself spatially; in listening, it does so temporally. This explains why in hearing, the object ceases to be a thing—it becomes a sound.

[13]*Pace* Gustav Theodor Fechner, who believed that plants have souls (1848).

[14]Here and in what follows the quotations are from Hegel (1970).

(d) *Soul as self-reflexing*. As already mentioned, a soul is characteristic of animals only. Plants do not have a soul. The soul can be defined as the self-reflexive sentience of a logically-defined individuality. This logical characteristic of the soul explains why it is not composite: it is an expression of the unity of the individual (§ 350, *Add.*); of "the incomposite existence of the unity of selfhood" (§ 344, *Add.*).

The animal in general is self-relating (self-referring)—but not self-reflexive—in the sense that it develops in two logical directions that have one crossing point. (In contrast, the chemical particles are one-dimensional). On the one hand, it follows the reality and in this sense it is part of it. On the other hand, it is only one moving point of reality which is its center as an individual.[15,16]

(e) *Relation to the External World*. Animals and men are oriented towards externality in the same extent in which they are internally strung in opposition to it (§ 357). This means that individuals' opposing to reality is not absolute. Time and again, they realize that they are not self-sufficient. This happens in the relation of individuals' *reproduction* through the environment, from which they are split off. There are two such relations: of assimilation of nutrition, liquids, smoke; of sex. Both processes are cases of *consuming* (see Milkov 2011). The process begins with the awareness of deficiency, and the drive to overcome it (§ 359).

(i) Assimilation:

- The external nature to which the plant—the particular—relates is not individualized, but consists of elements which are mass particulars (§ 347). Its relation to the external nature is a transformation of one external material to another (§ 345).
- In contrast, when it interacts with the environment, the animal preserves its selfhood. It *feels* an unpleasant need (hunger) which is nothing but "a connectedness with the universal mechanism and abstract powers of the world" (§ 361). The aim is a satiation (*Sättigung*),[17] by which no deficiency is felt anymore. Now "the organism is objective" (§ 366).

(ii) Sex-relation: It is characteristic for animals only. In it, the animals are related one to another through their senses. This is so since "the genus is present in individual as an urge to attain its sentience in the other of its genus" (§ 368). In contrast, the plant is sexless; its sexual parts (organs) are not an integral part of its individuality (§ 347).

---

[15]Helmut Plessner has expressed this point as follows. Organic individuals have a double reference, realized in two planes, determined by the boundaries of the individuals:

(i) They relate themselves to this boundary and so to the environment. This relation is reciprocal.
(ii) They relate self-referentially to themselves (1928, p. 181; see § 25.4.1).

[16]This point suggests a logical explanation of why the personal life is intrinsically paradoxical (see Milkov 2004).

[17]"Satiation" (*Sättigung*) is also a term of Frege's logic which comes to denote the satisfying of a function through its argument. It was often translated as "saturation". For criticism see Milkov (2015).

(f) *Genus as One Whole*. Seen mereologically, the biological natural kind *homo sapiens* is one whole. Its different exemplars, the individuals, are only its elements. So in a sense, they are one thing; what differentiates them one from another are accidental factors: geographical, biographical, etc.

This means that the individuals of the genus *homo sapiens* have souls that are intrinsically identical one with another: they are numerically different, but qualitatively identical. To be more exact they all are exemplars of the one World Soul (*Weltseele*).[18] This explains why biological organism is a *microcosm* (§ 362, *Add.*).[19]

25.4 Topological Biology

25.4.1 A Reconstruction of Helmuth Plessner's Scheme

Another variant of formal biology was advanced by the German philosopher Helmuth Plessner (1892–1982). Plessner was a student of Edmund Husserl, Wilhelm Windelband and Max Weber and an admirer of Immanuel Kant and also of Hermann von Helmholz. Plessner was the first philosopher who pursued a formal biology in topological form in Germany, thus opposing the conception of his teacher Hans Driesch, the father of the neo-vitalism. In this section, we shall try to put his theory into new terminology. Our motive for doing so is that Plessner advanced his theory in 1928 in a German cultural environment. As a result, his terminology appears to the readers of today rather obscure. This partly explains why his topological biology is rather neglected today.

To put it briefly, the main idea of Plessner's topological biology was to apprehend life in terms of the relation of the individual to its boundary. He assumed the *differentia specifica* of biological objects to be the fact that they have "positionality"—we shall call it a *location*—in space. Plessner thus presented biological objects as topological entities that are marked off from their environment through their boundary: the membrane that encloses the cell, the hide of the animals, the skin of man, etc. The important point is that the biological boundary belongs to the body. In contrast, mineral bodies merely have edges; the latter do not belong to the objects themselves.

Different strata of biological objects have quite different topologies:

(i) Plants' location (positionality) is open. This means that, despite being sharply determined by their boundaries, plants directly explore their environment.
(ii) In contrast, the location (positionality) of animals is closed: they make their location a constitutional principle of their being. This means that they are organized around one point—more precisely, around one *figure*: the body of the animal that Plessner calls *center*. Accordingly, Plessner calls animals *centric* biological individuals. Animals live towards, and through, their individual

[18] For this term used also in Wittgenstein (1979, p. 49e).

[19] Also according to Wittgenstein, "I am my world. (The microcosm.)" (1922, 5.63).

form. But they are not this form: they are not individuals. This explains why despite the fact that they *experience* contents of the outer world (they have perceptions and experience), they do not experience their *personality*.

(iii) Only man has personality. He does not only live out of its location (body); he also refers to his location.[20] That is why Plessner calls him *eccentric*. The most important consequence of this newly explored ontological dimension—developed into a new stratum of ontology—is that man experiences his experience. (This makes him self-reflexive, and also free.) In this way he trespasses his boundary and exactly in cultivating it, becomes what he is.[21] To be more exact, the man *makes* himself through his "life-acts":[22] the man constructs himself.[23] He becomes himself only referring to the objects of the external world (1928, p. 17). It is this reflexive making of himself through reference to the external things that constitutes his *existence*.

To put this point in other words, man is incomplete; he needs supplementation (is *ergänzungs-bedürftig*) (ibid., p. 18);[24]and he finds his finite form only in culture. In fact, what we call *culture* is a result of just this characteristic of the human person (ibid., pp. 383 ff.). Ultimately, this is neither a result of his will to power (Nietzsche); nor a result of the *libido* (Freud). Rather, culture arises out of the incompleteness of the man from topological point of view.

(iv) It is the existence that makes the common spiritual life of the humanity possible: the social life; the mind (*Geist*) as distinct from the individual consciousness or the personal soul (Plessner 1982, p. 15).

25.4.2 Alternative Biological Topologies

The topological approach was often used in philosophy of biology in the past. Already Aristotle has noted that "in some animals we find all such distinctions of parts as

[20]An alternative topology of man and animal is set out by Karl Marx. According to him, while the animal forms its product according to the form and the need of the species to which it pertains, the man produces according to any species. The man knows how to lay out the form of any object (1974, p. 158).

[21]This theory is close to Aristotle's conception that the "thought thinks on itself because it ... becomes an object of thought in coming into contact with and thinking its objects" (*Met.*, 1072b19–20). Among other things, this conception was warmly embraced by Hegel who quoted this passage in a more extended form at the very end of his *Encyclopedia*. This point explains why some authors find Hegel's *Phenomenology of Spirit* a precedent for Plessner's project of formal biology (see Gerhardt 1998, p. 43b).

[22]This conception was initially developed by Wilhelm Dilthey in his book *Die Lebensakten der Menschen* (1907, p. 164). Cf. with the concept of speech-acts, developed around 1910 by Adolf Reinach and after the Second World War by J. L. Austin and John Searle.

[23]This conception strongly reminds John Searle's theory that social reality is constructed by the way of social acts (1995).

[24]Another term also used by Gottlob Frege (see n. 16) but in quite different context. To him *ergänzungsbedürftig* are the function and its argument in logic.

this of right and left clearly present, and in other some; but in plant only above and below" (*De caelo*, 284b15–17).

In a recent article Berit Brogaard and Barry Smith, following some ideas of Roman Ingarden, try to precisely define the beginning of human life (an issue of first-rate importance by discussing such issues of applied ethics as abortion) using the same topological approach. They fixed the beginning of human life at day 16 after the fertilization of the egg-cell by a sperm. Indeed, it is first the embryo, developed in the period of gastrulation, that is topologically a self-sufficient entity and which can be seen as an ontological individual. This is determined through: (i) its boundary; (ii) it being a unified causal system which is relatively isolated from its surroundings (2003, p. 15).

In contrast, the opposing conventional view that the threshold event in the making of a human being is to be located towards the end of the first trimester after the fertilizing of the egg-cell, refers to such a vague concept as the visible *resemblance* of the embryo to a human being.

25.4.3 Ecological Topology of Biological Individuals

A case for ontology of environments (*Umwelten)* or for ecological topology was first developed by the constructivist biology of Jakob von Uexküll. Uexküll claimed that every animal creates his own environmental reality. There are as many enclosing realities as there are animals; these worlds are, practically, as separated as Leibniz' monads are. Their "environments surround them like solid but invisible glass houses" (1928, p. 62). Since all organisms are embedded in their own niches, their forms (shapes) can be differentiated according to the forms of the niches: according to their exterior boundary. In this way Uexküll suggests an alternative—ecological—topology of biological individuals. In this section we are going to set out some of its points.[25]

(i) At one extreme we find the niches that are fully bounded by a retainer. This class can be subdivided into:
   - Those whose retainers offer full protection without an access point: an example here is a larval cocoon;
   - Those which have access point. Examples are: a closed oyster shell, which must actively keep its shell closed to prevent predators from prying it open. The degree of protection depends on the physical properties of the retainer. The walls of the crocodile's egg, for example, have a higher protective value that the thin membrane enclosing the mammalian zygote.

(ii) Most biological niches, however, are not fully bounded. Examples are: a kangaroo-pouch, a nest, a hive. All of these are niches which do not involve

[25]In this we follow Smith and Varzi (2001, pp. 235 f.).

closed cavities but rather natural or artificial hollows within their respective environments. Nevertheless, from the perspective of protection it is fairly robust.

(iii) There are also niches which are bounded by a partial retainer which offers a low degree of protection. Example is the niche of the oxpecker removing ticks from the back of the African rhinoceros's hide, a niche bounded by a part of the rhinoceros's hide.

(iv) Finally, at the other extreme of the continuum between bound and free niches we find biological niches which lack a retainer altogether. Such niches are bubble-like zones in some region of space, as in the case of the niche of the fish of the open ocean. This class of niche may manifest a range of different topologies. So, for example, when a falcon is flying in the sky circling above the area where its prey is to be found, the niche of the falcon is its orbit.

References

Barry Smith, "An Essay on Material Necessity", in P. Hanson and B. Hunter (eds.), *Return of the A Priori*, *Canadian Journal of Philosophy*, Supplementary Volume 18 (1992).

Barry Smith, and Achille Varzi, "Environmental Metaphysics," in *Metaphysics in the Post-Metaphysical Age*, eds. P. M. Simons and Uwe Meixner (Vienna: Hölder–Pilcher–Tempsky, 2001), pp. 231–40.

Berit Brogaard, and Barry Smith, "Sixteen Days," *Journal of Medicine and Philosophy*, 28 (2003), pp. 45–78.

Bertrand Russell, *The Analysis of Matter* (New York: Dover, 1954, 1st ed. 1927).

Bertrand Russell, *Logic and Knowledge*, ed. R. C. Marsch (London: George Allen & Unwin, 1956).

Daniel Dennett, "Memes and the Exploration of Imagination," *The Journal of Aesthetics and Art Criticism*, 48 (1990), pp. 127–35.

D'Arcy Thompson, *On Growth and Form* (Cambridge: Cambridge University Press, 1917).

Eduard Spranger, *Lebensformen* (Halle: Niemeyer 1914).

Friedrich Nietzsche, *Vom Nutzen und Nachteil der Historie für das Leben*, Bd. 2 von *Unzeitgemäße Betrachtungen* (Leipzig: Fritzsch, 1874).

Friedrich Nietzsche, *Der Wille zur Macht* (Leipzig: Fritzsch, 1901).

Georg W. F. Hegel, *Philosophy of Nature*, Vol. III, ed. and trans. M. J. Petry (London: George Allen and Unwin, 1970).

Gustav Theodor Fechner, *Nanna oder über das Seelenleben der Pflanzen*, Leipzig: Voss, 1848.

Hans Driesch, *Der Vitalismus als Geschichte und als Lehre* (Leipzig: Barth, 1905).

Helmuth Plessner, *Die Stufen des Organischen und der Mensch* (Berlin: de Gruyter, 1928).

Helmuth Plessner. *Mit anderen Augen* (Stuttgart: Reclam, 1982).

Jakob von Uexküll, *Theoretische Biologie* (Berlin: Springer, 1928).

Joachim Schulte, *Chor und Gesetz: Wittgenstein in Kontext* (Frankfurt: Suhrkamp 1990).

John Searle, *Intentionality* (Cambridge: Cambridge University Press, 1983).

John Searle. *Minds, Brains, and Science* (Cambridge MA: Harvard University Press, 1984).

John Searle. *The Construction of Social Reality* (Allen Lane: The Penguin Press, 1995).

Karl Marx, *Ökonomisch-philosophische Manuskripte aus dem Jahre 1844* (Leipzig: Reclam, 1974).

Karl Popper, "Evolutionary Epistemology", in *Evolutionary Theory: Paths into the Future*, ed. J. W. Pollard (London: John Wiley & Sons Ltd., 1984).

Konrad Lorenz, *Die Rückseite des Spiegels* (München: dtv, 1977; 1st ed. 1973).

Ludwig Wittgenstein, *Remarks on Colour*, ed. G. E. M. Anscombe (Oxford: Blackwell).
Ludwig Wittgenstein, *Tractatus Logico-Philosophicus*, trans. C. K. Ogden and F. P. Ramsey (London: George Allen & Unwin, 1922).
Ludwig Wittgenstein. *Philosophical Investigations*, eds. G. E. M. Anscombe *et al.*, trans. G. E. M. Anscombe (Oxford: Blackwell 1953).
Ludwig Wittgenstein. *Remarks on the Foundations of Mathematics*, eds. G. E. M. Anscombe *et al.*, trans. G. E. M. Anscombe (Oxford: Blackwell 1956).
Ludwig Wittgenstein. *Notebooks 1914–1916*, 2nd edn., eds. G. E. M. Anscombe *et al.*, trans. G. E. M. Anscombe (Oxford: Blackwell 1979).
Ludwig Wittgenstein. *Remarks on the Philosophy of Psychology*, vol. I, eds. G. E. M. Anscombe *et al.*, trans. G. E. M. Anscombe (Oxford: Blackwell 1980).
Ludwig Wittgenstein. *Ludwig Wittgenstein. Personal Recollections*, ed. Rush Rhees (Oxford: Blackwell 1981).
Nicolai Hartmann, *Der Aufbau der realen Welt* (Berlin: de Gruyter, 1940).
Nicholas Rescher, *Productive Evolution: On Reconciling Evolution with Intelligent Design* (Frankfurt: Ontos, 2011).
Nikolay M. Milkov, *The Varieties of Understanding. English Philosophy Since 1898*, 2 vols. (Frankfurt: Peter Lang, 1997).
Nikolay M. Milkov. "Tractarian Topology: A New Interpretation of the Objects in Wittgenstein's *Tractatus*", *Beiträge der Öster-reichischen Ludwig-Wittgenstein-Gesellschaft* 6 (1997), ii. pp. 653–8.
Nikolay M. Milkov. "What is Analytical Phenomenology?", *Analecta Husserliana* 59 (1999), pp. 491–8.
Nikolay M. Milkov. "Tractarian Scaffoldings", *prima philosophia* 14 (2001), pp. 399–414.
Nikolay M. Milkov. *A Hundred Years of English Philosophy* (Dordrecht: Kluwer, 2003).
Nikolay M. Milkov. "The Meaning of Life: A Topological Approach", *Analecta Husserliana* 84 (2004), pp. 217–34.
Nikolay M. Milkov. "Sexual Experience," in: Adrienne McEnvoy (ed.), *Sex, Love, and Friendship*, vol. 2 (Amsterdam: Rodopi, 2011), pp. 155–66.
Nikolay M. Milkov. "Frege and the German Philosophical Idealism", in: Dieter Schott (ed.), *Frege: Freund(e) und Feind(e)* (Berlin: Logos Verlag, 2015), pp. 88–104.
Pirmin Stekeler-Weithofer, *Hegels analytische Philosophie* (Paderborn: Schöningh, 1992).
Richard Dawkins, *The Selfish Gene* (Oxford: Oxford University Press, 1976).
Rudolf Haller, *Questions on Wittgenstein* (Lincoln: University of Nebraska Press, 1988).
Volker Gerhardt, "Die rationale Wendung zum Leben. Für die Zukunft ein Blick zurück: Helmut Plessners, *Die Stufen des Organischen und der Menschen*", *Die Zeit*, Dezember 30, 1998, p. 43.
Wilhelm Dilthey, *Das Wesen der Philosophie* (Stuttgart: Reclam, 1984; 1st German edition 1907).

Chapter 26
The Ecological Economics Revolution: Looking at Economics from the Vantage-Point of Wittgenstein's and Kuhn's Philosophies

Rupert Read

26.1 Overview

Is there a scientific revolution taking place in economics? This piece seeks to apply the thinking of Wittgenstein and of the major philosopher of science who was, I have argued elsewhere,[1] most influenced by him—Kuhn—to the emergence of 'ecological economics'. (This is a topic of not-inconsiderable importance, given the power of economics in our world today, and the role of conventional economics in degrading the Earth, a degrading which is now threatening to put not just myriads of other species but humanity itself 'out of business'.)

Due to the financial/economic crisis of the last decade, Economics itself ought to be in profound crisis and rebirth. It does not seem to be, despite the apparent flux due to 'behaviourial economics' etc. But, at a slower pace, there might be a paradigm-shift taking place in economics: the (slower, much deadlier) ecological crisis seems gradually to be yielding a green or ecological economics.

This chapter will investigate what Wittgenstein's philosophy has, implicitly, to say about this situation. Is the fantasy of 'progress', as critiqued by Wittgenstein,[2] a key motivation for the dream of a scientific economics? Is scientism, as also critiqued by Wittgenstein? I will (also) investigate what Kuhn's (broadly Wittgensteinian) philosophy has to say about this situation. Is economics the kind of thing that Kuhn famously terms a 'discipline without a paradigm'? Does it/could it have a scientific status? (Insofar as it does, or could, does the emergence of ecological economics

[1] See e.g. my Kuhn (co-authored with Wes Sharrock), Polity Press, 2002.

[2] See my "Wittgenstein and the illusion of progress", *Philosophy* 78 (July 2016), 265–284.

R. Read (✉)
School of Politics, Philosophy, Language and Communication Studies, University of East Anglia, Norwich Research Park, Norwich, Norfolk NR4 7TJ, UK
e-mail: R.Read@uea.ac.uk

S. Wuppuluri and N. da Costa (eds.), *WITTGENSTEINIAN (adj.)*,
The Frontiers Collection, https://doi.org/10.1007/978-3-030-27569-3_26

around Herman Daly et al. relevantly resemble (say) the Chemical revolution around Lavoisier et al.?)

My conclusion is double: That, insofar as economics does or could have scientific status, then one might justly say that the ecological economics revolution is or could be a real scientific revolution. *But,* and *superordinate* to this point, that there remain significant (broadly Wittgensteinian and Kuhnian), probably unovercomeable reasons to believe that it is misleading even to aspire to have a scientific economics—'even' from within the standpoint of a scientifically-respectable ecology. Economics does not and probably never will have scientific status. This means that attempts to 'be scientific' within economics are liable to be pointless. Rather, we should seek to understand economics principally as something else, and something needful and well worth having: especially, as politics and as philosophy, operating within the *limits* of (natural) science (the limits imposed by ecology; i.e. by the Earth).

26.2 How to Make Economics into a Science…?

Economics, as Philip Mirowski has taught us, was 'made' into a 'science' by the crude expedient of directly copying the paradigm of (the frontiers of) science in the mid-nineteenth century, mechanics. In the twentieth century, there have been numerous efforts to 'correct' this model, to make it *more* scientific. Perhaps the most impressive and plausible has been the effort by Robert Costanza to integrate economics with real science, in particular with the environmental sciences—and in particular to value economically the biosphere. My suggestion will be that, on balance, *this* move nevertheless points in precisely the *wrong* direction.

However, I will further suggest below that if we rotate the axis of our examination of economics, we can come to see the contribution of Costanza et al. in a quite different light. In this light, it appears, not as a way of making economics genuinely scientific at last, but as a way rather of showing economics to be what, I contend, it really is or at least ought to be: a *subset* of Ecology and of Politics/Labour Studies (or—perhaps better—simply a subset of ecology, *human and non-human*), a subset with in fact a peculiarly *philosophical* face. This takes economics back to its roots in the great works of such economic *philosophers* as Marx, Smith, the Physiocrats, and so on. Or even, as one might put it, *before* those roots: to political economy as a craft, an art, and a way of thinking about life. On this understanding, we are beyond the idea of economics as a science. Rather, economics *gives way* to environmental science and to political economy, and what insights if any it offers 'of its own' become primarily *non-quantitative*, transitional, broadly philosophical insights. (Just as, as we will see, Kuhn at his best would lead us to expect.)

For: it is an unjustly-neglected fact, and one of potentially-immense philosophical and political and ultimately practical significance, that neo-classical economics originated, in the work of Jevons and Walras, by means of an appropriation of nineteenth century physics. Specifically, that the creation of neoclassicism involved the transformation of Helmholtzian energy-conservation principles into the 'general equilibrium' principles of the conservation of utility.[3]This creation of allegedly-scientific economics was accomplished in remarkably direct fashion: the mechanics equations in question were kept intact, and what the variables stood for was simply substituted, economic variables for physics variables. Thus economics seemingly attained the legitimacy of a science: it mathematicised itself by presenting itself as a strict analogue for physics. The 'laws of economics' were then supposed to apply just as rigorously to 'social atoms' as the laws of physics did to physical atoms. At each stage in the growth of the economy, a new equilibrium was supposed to have been attained: no limits were present, then, to the figures one could plug into these equations, representing now not energy but utility.

My own view, already argued elsewhere,[4] is that the self-identification of economics as a science has proven disastrous to academic and intellectual inquiry, to the world, and (in the very long run) to economics itself. My view, following Kuhn, would be in particular that the leap to numbers, to quantification, prior to a serious effort *qualitatively* truly to understand the economic world, is deeply suspect. (And this is already a worrying precedent for Costanzian ecological economics.)

Let me now explain this.

26.3 Social and Economic 'Science'?

Drawing extensively on Wittgenstein, I've argued over the years in a series of books[5] that the trouble with the categories of 'social/human/economic science' is that they are an *apeing* of science. Acting as if you have a paradigm in Kuhn's sense, when you actually don't.

The most common way in which this apeing occurs is through a numbers fetish. A premature effort to quantify one's discipline. With garbage in, and so garbage out….

Here is the opening of perhaps the most seminal of all Kuhn's critiques of the urge to go forth and quantify:

"At the University of Chicago, the façade of the Social Science Research Building bears Lord Kelvin's famous dictum: "If you cannot measure, your knowledge is meager and unsatisfactory." Would that statement be there if it had been written, not by a physicist, but by a sociologist, political scientist, or *economist*? [I suspect that] the answer to [this question] is no… Because physical science is so often seen as *the* paradigm of sound knowledge and because quantitative techniques seem to

---

[3]Or, more strictly, of the conservation of the sum of utility and income. Never mind that the former is unmeasurable, and that the idea of summing the two is simply bizarre and meaningless. Never mind, that is, unless you mind about the foundations of the 'science' of economics. For detailed argument, see for instance Mirowski (1988, 1989, 2002), Ingrao and Israel (1990).

[4]See for instance my paper "Economics is philosophy", and my critique of Friedman in my Wittgenstein among the sciences (Ashgate, 2012).

[5]Including There is no such thing as a social science (co-authored with Phil Hutchinson and Wes Sharrock, 2008, Ashgate). This 'series' reached reached its apogee in my Wittgenstein among the sciences.

provide an essential clue to its success, the question how measurement has actually functioned for the past three centuries in physical science arouses more than its natural and intrinsic interest. Let me therefore make my general position clear at the start. Both as an ex-physicist and as an historian of physical science I feel sure that, for at least a century and a half, quantitative methods have indeed been central to the development of the fields I study. On the other hand, I feel equally convinced that our most prevalent notions both about the function of measurement and about the source of its special efficacy are derived largely from *myth*."[6]

Kuhn goes on to suggest that the key to understanding why this is, and overcoming the myth that has long had 'social scientists' in its thrall, is to understand that it is only if one achieves a new *qualitative* understanding, of the kind arrived at when a scientific revolution occurs and institutes a new paradigm (exemplar) at the heart of its fresh puzzle-solving normal scientific activity, that a quantitative understanding can develop and be of any use. (A paradigm of this—to which we will return—was for Kuhn the Copernican Revolution in astronomy. Copernicus gave us a new, perspective-revolutionising, simple *qualitative* understanding of our place in the universe. His *calculations* by contrast were for a long time inferior to Ptolemaic calculations (mainly because he assumed that the paths of the planets were circular, rather than elliptical).)

The dread temptation that has lured economics down the path of scientism is of course that it has a *ready-made* quantification to hand. Money preceded economics!

Economics typically aims at a scientificity that perpetually eludes it because it quantifies before it has arrived at a fundamental qualitative understanding.

With regard then to economics and science, then, once one has admitted the stark almost uncontroversial obviousness of economics's scientific inefficacy to date (exemplified in the utter failure of mainstream economics to predict the financial crisis, and in fact rather to *facilitate* it), two opposite options present themselves:

(1) We could seek to make economics *more* scientific; genuinely scientific.
(2) We could admit that economics is *not* a science. That it is value-saturated, and needs to be pluralistic, in ways that biology is not, even once we have (as we should) re-understood biology in the kind of way recommended by John Dupre.[7]

Now, I respect the impetus behind (1). And I am fully in favour, as I will explain, of making economics more *compatible* with science.

Moreover, I am 100% against the impression that might be given by (2) that there is nothing to be said for a change of pre-analytic vision in economics. On the contrary, I believe that economics desperately needs the new Earth-centred pre-analytic vision that Herman Daly has repeatedly proposed as a new foundation for economics. That is the ecological economics 'revolution'. But the question I am raising is whether that revolution would genuinely be a scientific revolution in Kuhn's sense. And whether the new Earth-centred pre-analytic vision of planetary limits etc. should be

[6]The opening of "The function of measurement in modern physical science", *Isis,* 52 (2) June 1961, 161–193. Emphases mine.

[7]See his "The disunity of science", *Mind* 92, No. 367 (Jul., 1983), pp. 321–346.

considered therefore a paradigm. I am suspicious of that idea, and a key aim of the present piece is to convey the reasons for that suspicion.

Wouldn't it be simpler for me simply to urge that economics undergoes a Copernican revolution? One that decentres humanity, as we were decentred from our place at the centre of the Universe by Copernicus. Wouldn't this be exactly the kind of qualitative improvement stressed by Kuhn? If, instead of the diagrams by way of which students are introduced to economics, diagrams constructed around human needs, picturing an endlessly expanding economy to which the ecosystem is a kind of externality or addendum (see e.g. Figure 26.1), we moved to a richer picture which sees the ecosystem as the limit and encompasser of a much 'smaller' subsystem, the economy (see e.g. Figure 26.2):

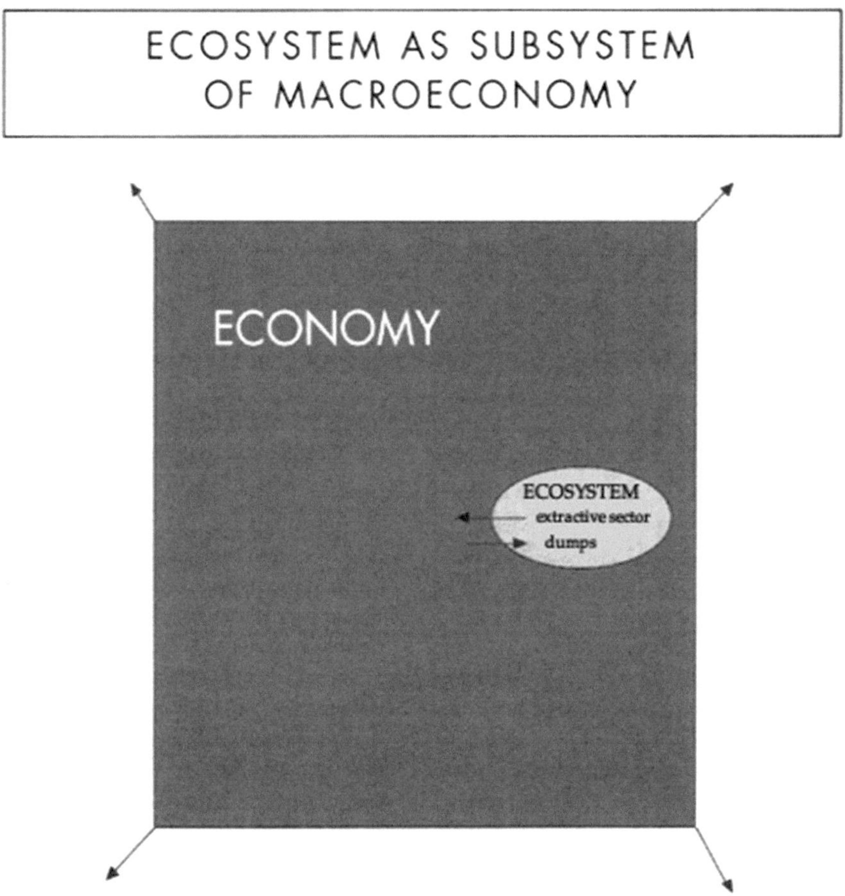

Fig. 26.1 Sourced from Feasta under a creative commons attribution-noncommercial-ShareAlike 3.0 unported license

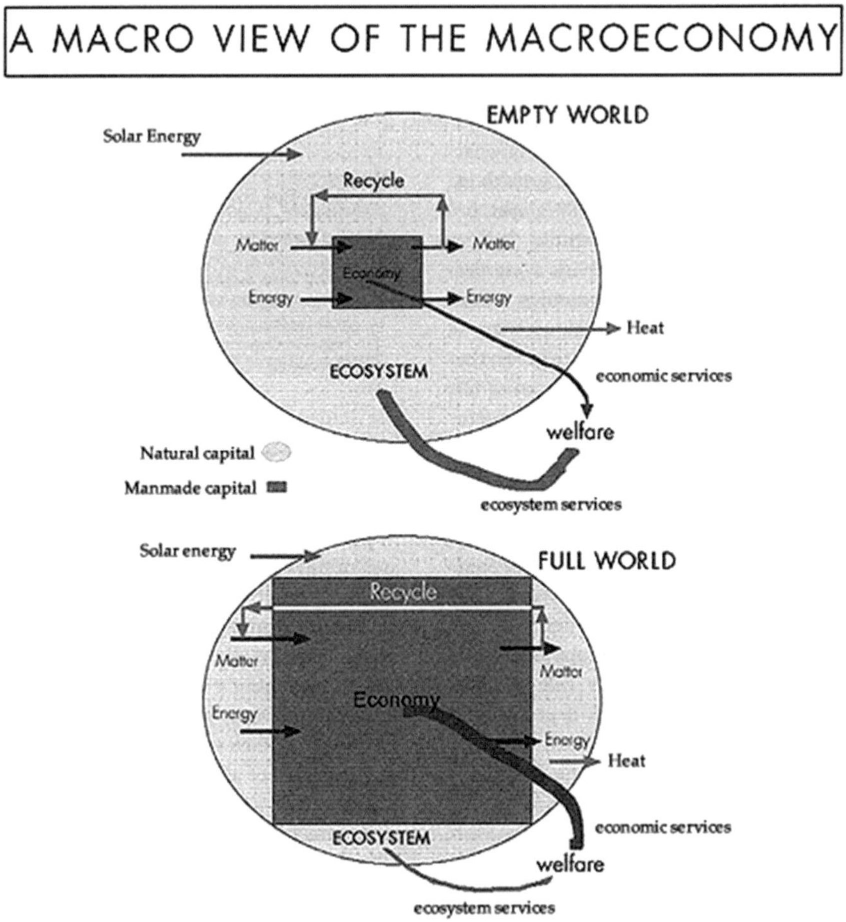

Fig. 26.2 Sourced from Feasta under a creative commons attribution-noncommercial-ShareAlike 3.0 unported license

My reluctance to cave into the temptation to see the ecological economics 'revolution' as a scientific revolution pure and simple—one which, like the Chemical Revolution did for chemistry, establishes economics on the secure path of scientificity for the first time, by moving from Figs. 26.1 and 26.2,—can only be fully understood once we go further into the hazards of the endeavor to arrive in the kind of way that Costanza did at a quantitative ecological economics, and picture instead a way of seeing economics through the lens of energy (the lens Costanza recommends) which we will be *less* tempted to regard scientistically.

26.4 Quantitative Ecological Economics?

The modelling of economics on mechanics, specifically on energy-conservation principles, a modelling which has been of incalculable rhetorical significance and has had vast intellectual and real-world effects, as can be seen writ large in the works of such luminaries as Samuelson, and as exposed to the cold light of day in the work Mirowski and others, can actually shed a good deal of light on the nature of ecological economics.

The particular apeing of physics that we find in neoclassical economics has proven so effective (I mean in terms of being adopted; I don't mean scientifically effective) in part because it actually does key into one deep similarity. Go back from the idea of utility to the idea of energy—and then an environmentally-centered economics makes broad sense. There is a real sense—*if* one stretches the concept of energy as far as it can possibly go (and so empties it of *scientific* content—see below)—in which the founding equations of neoclassicism get something very important *right*. But only because what economics needs to understand—and what neoclassicism *systematically* shields from view—is that it, economics, is (or rather, *must* be, *ought* to be) nothing other than the study of all the producers of or vehicles of energy, in the broadest sense of that word: encompassing for instance not only fossil fuel energy and renewable energy, but also the energy provided by the labour of animals, including crucially of course the labour of probably the most vital economic animal of all, humankind. That is, of energy (in the ordinary sense of the word) and ecology (i.e. both the finitude of non-renewable energy and the finitude of Earth's natural resources and of the capacity of the Earth to absorb pollution), *and* of the 'energy'-contributions of human and non-human-animal actors *and* of technological change. And so on.

This would of course require a complete revolution in the practice of economics.

Is what I am saying this: That if economics were *to take seriously* the origins of the equations which birthed neoclassicism, then, maybe it might be scientific after all? Possibly: But I think not quite. For what would it be? It's *environmental science* (and perhaps '*ecological* economics'—see below) which take seriously the study of energy sources and the hazards attending their mass use: economics would then in effect be a human- and societally- oriented *branch* of a widened environmental science, *and would be exposed to the* limits *which the latter centres (or ought to centre) its investigations on.* I.e. The planetary limits (and social limits to growth, and so forth). It would no longer be able to hold out the mad prospect of indefinitely-prolonged 'growth'; it would have to give up the equations which founded it altogether *except insofar as they were radically reconceived to include* all *that had been previously excluded as mere 'externalities'*. And this inclusion would have to be done not in monetary terms, but in terms of the manifold of energy, impact, sustainability, and so on. I.e. It would be economics that was subsumed into ecology, *not* vice versa.

The way that the economic valuation of nature has proceeded (in Costanza and his many successors in 'natural capital' valuation) has been *to try to assimilate nature to (economic) capital,* rather than vice versa. This, I am suggesting, is very troubling.

Furthermore: economics which took seriously that it is *about* labour and the use of animals and so on too would have to recognise not only their limits but also their sufferings and tragedies (and joys etc.). It would be open to the downsides for its 'objects' of for instance growth—such as, perhaps, greater stress, greater exploitation, less free time, more sickness. It would have to consider the problematic *consequences* of increasing 'use of human (or other animal) energy'. For, rather than abstracting from these, as one does when one simply *assumes* that utility is growing as the economy grows, it would have to *look and see* whether that was in fact the case. (The Index of Sustainable Welfare, the Genuine Progress Indicator etc. are perhaps a first tentative step in this direction. But even these, we should be suspicious of. We should be deeply suspicious of any effort, however well-intentioned, however 'complete', however 'scientific', that seeks to reduce the entire economy, let alone the broader society, with all its incommensurables, to *one number*.)

What this all amounts to is that, from the point of view for instance of labour or of ecosystems—and actually these combine, in the concept of ecology as including *human* ecology[8]—we *don't* have (as of course we do in physics) a conservation of energy, except in the exceptional (desirable) case when we practice (human) ecology in the normative sense of the word, behaving with ecosystemic rectitude and treat people aright, etc. When for instance we use almost exclusively genuinely renewable energy, rather than fossil fuels or nuclear (for the latter, just like fossil fuels, is non-renewable (depending on finite fuel sources) and dangerously polluting); and when we don't abuse labour or animals. We might then put the point this way: for (post-neo-classical) economics to really be true, it would have to *become* the study of and respect for labour (and, as Polanyi reminded us in his great work *The great transformation*, labourers are also of course actually citizens, people) and for the entirety of our ecosystems. Economics must decisively lose its absurd sense of itself as (the study of) a Pareto-optimal pseudo-closed system. A closed circuit of exchange of goods and services. For it is completely open to (the implications of) the finitude of our world—and all that is in it. When we construe labour as energy, for instance, then, as implied above, its increase does not yet register as the cost it probably is to the *quality of life* of the labourer(s). When we construe fossil fuel energy only as energy, too, its increased use does not register as the cost it certainly is to future generations. The physical universe (though *not* the Earth, because of the Sun), from whence the conservation principles came, is (roughly) a closed system. For the economic system to characterize itself similarly is for it to ignore its through-and-through embeddedness in and co-constitutiveness with the environment—with the entire planetary ecosystem—which only works effectively as a semi-closed system open to the flow of solar energy when it is both lucky and is

---

[8]For argument to this effect and in-depth further discussion, see pp. 90-1ff. of Kovel's (2002). See also p. 165–6 and p. 132 of that work, on the interlinkedness of capital's exploitation of nature and of labour.

run well by its denizens/citizens. 'Run well', in the sense *not* necessarily (or rather: almost necessarily *not*) of the logic of actually-existing neoclassical economics, nor even of any souped-up quantitatively-improved version of it, but rather in the far subtler and wider and still-less measurable logic of the happiness and integrity in inter-relatedness of the creatures of the world, including of future generations. For if we think of the world as a finite whole, as ecosystemic thinking—but not of course neoclassical economic thinking—demands, then included in that whole are all the generations as yet unborn, too….[9]

When one thinks through then what it would take to be able to take seriously an understanding of economics as potentially involving a kind of conservation of 'energy', one sees that it will among other things involve recapturing lost greatness in the economics that came before the triumph of neoclassicism. For example, a return to what insights were present in 'the labour theory of value' (which reached its apogee in Marx[10]) and in land theories of value (e.g. the forgotten work of the Physiocrats). But it is not, I submit, a *theory* of value that we really require, not even one which intelligently 'combines' the (value of and in the) labour of humans and animals, the energy extracted from the Earth in fuel-form, and so on, into an ecological synthesis.

To see why I say this, it is worth looking further at some of the most important work done by the influential founding 'ecological' economist: Costanza. Costanza tried in 1980 to develop an 'energetic' theory of value. That is, he tried to convert not only natural resources but also labour and even government services into energy equivalents (Here we see why any assumption that total literal quantification of energy *in the economic sense* might be possible by commensurating it as in for example the old concept/unit of 'house power' would be in vain). He claimed that his results showed that inputs of energy values *are extremely closely correlated to outputs of monetary values.*

One way this effort might be seen is as an (extremely-valuable and much-needed?) effort to *internalize* the ecosystem, labour etc. into economics… But *is* that actually what is needed? Or should things rather, as I intimated above, move the other way: *should economics be subsumed into the study and understanding of ecology, of labour, and indeed of political understanding and action*?

Of course, Costanza's work that attempted to reduce everything to energy should be distinguished from his even more famous work that attempted to reduce everything to money, to price the 'natural capital' of the entire world. But he is hardly in a good place from which to object to a way that I may seem here to have run these two things together; because, as just outlined, *he tends to run them together, himself!*

---

[9]For their sake, we *must not* pollute more than the Earth (with our help) can sustainably absorb, and we *must not* use up non-renewable resources unless we are confident that future generations will not need them. For argument as to the policy implications of this, which are I think 'Contraction and Convergence' and a far more egalitarian distribution of resources than that which at present we have, see my paper "Contractarian liberalism cannot take future generations seriously".

[10]Or arguably, in fact, in Wittgenstein's great colleague and influencer, Sraffa (See his <u>The production of commodities by means of commodities</u>, 1960).

26.5 On the Roots of the Dangerous Delusion of Measuring 'Natural Capital'

Is Costanza's analysis an *alternative* to neoclassicism, showing up its limitations or blindspots? Or is it a *correction* to it; does it merely put neoclassicism right where it had gone a little wrong? Or is it even *the bringing of neoclassicism to full self-consciousness*? Is Costanza's achievement actually to show what it would be to take *seriously* the founding intuition of neoclassicism—that economic activity IS energy—, while retaining neoclassical pricing methodology and the concept of equilibrium?

My own suspicion is that the latter is very close to being the truth. I think that Costanza's 'energetic' theory of value is precisely of a piece with his pricings of the atmosphere, of the sea, of the entire ecosystem of our planet. His 'Promethean' would-be valuation of the totality of natural capital. I regard the latter as an obscene *triumph* of the logic of neoclassical economics, not as any kind of alternative to it.[11] It seems to me that ecological economics has yet to decide whether it is really ready to put ecology and human life first, or whether it is merely a kind of adjunct to—in fact, a *completer* of, a more efficient and grander, more total *version* of—neoclassicism.

In a way, then, Costanza is deepening our difficulties, not solving them. He inappropriately tries to make economics *more* scientific, by including factors that it has excluded.

The effort to price 'natural capital' (or 'ecosystem services') in effect contends that we can scientifically analyse the entire planetary ecosystem *and convert it into money*. That the category of capital is superordinate to the category of nature. That money is more important than life. This is a bizarre destiny for what was meant to be an *ecological* economics.

But, if Costanza saw what he was wanting to do in the alternative philosophical light that I am developing here, he would see that there is another possible way of following up the 'energetic' insight with which he began. For when one takes seriously the sense in which energy (in a broad enough sense of the word) just is identical with economic activity, without attempting to theorize or scientise this insight, then one sees that actually one's insight is not and could not be best characterised as a piece of science nor even theory. It is rather 'just' *a way of seeing*—a 'point of view'[12]—a perspicuous, helpful and insightful more or less philosophical (in Wittgenstein's sense) way of recovering a way of seeing that economic ideology can get in the way of. *Of course* economic activity is ((describable/readable as) meetings of and various forms of) energy—one might even justly wonder what else it could possibly be? And now one will be much more able—better primed—to look at what counts so much for us, at this point in history—the *impact* of these employments of energy, upon our quality of life, upon the liveability of the planet for future generations, and so on. The things which neoclassicism has little or no conceptual space for. Which it treats,

---

[11]For argument to this conclusion, see Chap. 6 of Beder's (2002). And several of the other essays in this collection.

[12]See Gavin Kitching's books on Marx and Wittgenstein for development of this idea.

ludicrously as 'externalities'. Bringing those 'externalities' inside a system that is set up fundamentally on the wrong basis is not progress, it is the making seemingly coherent of something monstrous.

It won't make sense to do what Costanza wanted to do: literally to *quantify* these things. (The wide bringing together under the heading of 'energy' that I've briefly and speculatively carried out is not a commensuration, let alone a mensuration, let alone the basis of a quantification. Because life is incommensurate with non-life, for starters.) But that's just as well: for that attempt at quantification was pretty absurd or just repellent to begin with. To quantify the value of what sustains life is as nonsensical as it is abominable: infinite value is of an altogether different order to finite (genuinely quantifiable/measurable) value; infinity is not a number. Likewise: Costanza's 'result' that inputs of energy value are closely tied to outputs of monetary value is a fake piece of (pseudo-)'science': for when *everything* is reduced to energy, *of course* there is a virtually constant relationship between energy and output! The contribution to growth of labour, government etc. is computed *by reference to the output*—so there is only a *fake* appearance of something having been explained here.[13] Costanza's famous pricing of the world's ecosystem services is in this way based in a dangerous piece of metaphysical pseudo-science. Because it is only the illusion that we can assimilate everything natural into one thing whether that be energy or money, that enables his pricing to seem possible.

Economic activity 'is' energy: but this is not a basis for any kind of science, but only a reminder of something[14]—or, rather, of a whole lot of things—that only a scientistic ideology could have caused us to forget, in the first place. Value *can't* truly be theorized—this deep truth, Costanza doesn't seem to have realized, any more than Marx did. But their ideas have the germ in them of profound philosophic *reminders* to us.[15] Being reminded of the ineradicable contributions of labour, of (now in the ordinary sense) energy, etc., to the economy is a portion of the study of—and the realization of—who we are, what we really want and need etc.; with us (humans) understood as utterly embedded in and co-constitutive with nature.

---

[13]My point here is similar to a central point concerning 'the capital controversy'. You can't *measure* capital by means of capital, by means ultimately of money. Similarly: you can't measure the contribution of energy to output by means of energy, if everything is in any case 'turned into' energy. In each case, the measuring 'device' then lacks the necessary independence needed for true measurement to occur.

For detailed development of this point, see my "The natural capital controversy", co-authored with Prof. Molly Scott Cato MEP, *Journal of Human Rights and the Environment*, 5(2), (2014): 153–167.

[14]I use the word "reminder' in Wittgenstein's sense—see *PI* Sect. 127.

[15]See my "Marx and Wittgenstein on vampires and parasites: a critique of capital and metaphysics", in G. N. Kitching & Nigel Pleasants (eds.), *Marx and Wittgenstein: Knowledge, Morality and Politics*. Routledge. pp. 35–254 (2002) for discussion of the sense in which Marx demands to be re-read as a philosopher of political economy, rather than as an economist. I suggest there that 'the labour theory of value' is hopeless as scientific economics—and so much the better for it, as philosophy, as a reorientation, that we desperately need, of Marx's readers and followers toward the world of exploitation and inequality that we inhabit.

Where do otherwise promising 'ecological economists/philosophers' such as John Gowdy[16] and Robert Costanza most deeply go wrong? In hoping for quantifications of the value of nature and/or scientific appraisals or theorizations of its value. For these hopes embody incoherent goals. *Science* cannot tell us anything about nature's value. (And attempts to quantify it will always underestimate nature's value for unknowably-distant future generations, who are typically 'discounted' to some extent in the calculations even of ecological economists. A true steady-state economics is needed. With—contra Costanza's famous, catastrophic allowance of a discount rate of approx. 2% p.a.[17]—no discounting.)

When we think of economic activity/growth etc. as *essentially* dependent upon energy in the expanded sense of that term which I outlined above, then it doesn't mean anything any more to characterise that concept, "energy", as the focus of a *science*. Far from it. The concept of "energy" no longer has quantifiable content: it cannot be measured as Costanza wished to measure it, nor used in equations as standard neoclassicism wish in effect to use it (to 'index' utility or its positivistic successor-concepts). It is conceptually unmeasurable. It is empty, from the point of view of any self-respecting science. Rather, what we have in our hands, in this expanded concept of "energy", is the focus of—the nub of—a philosophical insight. That *returns* us to the land, to labour, and so on. To this world, where we find the real—and perilously open-systemed—economy. An economy that is a subset of society, and that depends on *societal* (and ultimately, on philosophical) valuation.

We have seen then that economics was made into a 'science' by the crude expedient of directly copying the paradigm of (the frontiers of) science in the mid-nineteenth century, mechanics. This copying was well-contested by real scientists at the time, but to no avail.[18] Nor was it of any avail when physics moved on, as it swiftly did—e.g. the Second Law of thermodynamics, letalone quantum mechanics, have seemingly barely been noticed by the 'scientific' economists who have utterly dominated their discipline, to date. In the twentieth century, various efforts from within economics to contest this 'scientific' model have been fought in the trenches, or co-opted, (or arguably both, in the important case for instance of Keynes). There have also been very numerous efforts to correct the model, to make it *more* scientific. One of the most interesting of recent times has been the influential effort by Robert Costanza et al. to integrate economics with real science, in particular with ecology and the environmental sciences in general. My suggestion above however is that this move, at least as it has been attempted thus far, points in precisely the wrong direction.

---

[16]Take for instance Gowdy's useful https://ideas.repec.org/p/rpi/rpiwpe/0402.html. Gowdy has long nursed the hope of rendering economics truly scientific by basing it in environmental science. I have applauded such hopes—but *only* if one does not hope to understand or explain love of life or even quality of life through such science alone. "Ecosystem" is in part *a value-concept,* a normative concept, not reducible to science. That is why in the present paper I have stressed the importance of a role for *philosophy*, too. Philosophy can reflect values back to us without loss; science, no matter how 'environmental', cannot.

[17]At p.71 of his major piece, "Natural capital and sustainable development", co-authored with Herman Daly, pp.65-74 of his Frontiers in ecological economics (Lyme: Elgar: 1997).

[18]See again Mirowski's vital *oeuvre*.

No effort to save the scientific status of economics which does not *entirely* give up the starting point of neoclassicism will have any lasting effect, and such efforts, *especially* if intelligently carried out, with attention to what does indeed need to be attended to (e.g. the ecological crisis), are merely particularly-dangerous obscurers of this truth. Such is my assessment of the 'movement' for economic valuation of natural capital and ecosystem services.

However, I suggested also that if we rotate the axis of our examination of economics, we can come to see the contribution of Costanza in a quite different light. In this light, it appears, not as a way of making economics genuinely scientific at last, but as a way rather of showing economics to be what, I contend, it really is or at least ought to be: a subset of Ecology and of Politics/Labour Studies (or—better—simply a subset of ecology, *human and non-human*) a subset with in fact a peculiarly *philosophical* face. This takes economics back to its roots, in the great works of such moral *philosophers* as Marx, Smith, the Physiocrats, and so on. On this understanding, we are beyond the idea of economics as a science. Economics must be *compatible* with science; and at present it is not. But if we try to make it into a science, we end up with the absurdities of Costanzianism. I've suggested that the way out of those absurdities lies via philosophical rethinking in a broadly Wittgensteinian mold.

Rather than itself becoming a science, then, the optimum solution is for economics as we currently know it to *give way* to environmental science and to political economy, and what insights it offers 'of its own' become chiefly non-quantitative, transitional, philosophical insights: like the insight, which is really just a way of reminding ourselves of what is so obvious, so constantly in view, that we are inclined to look *through* it and forget it,[19] that there can be no economic activity without one kind or another of 'energy', and *a fortiori* that economic growth tends of course to requires an increase in such 'energy'. And then the only truth in neoclassicism is this, seen through a glass darkly. The equations that neoclassicism borrowed from physics then become bearers of insight after all: but they point precisely, now, *not* to economics as a free-standing science, but quite beyond this idea. They point, in fact, to the absurdity of the growthist hegemony and to the beginnings of an alternative: to the world that we all always already know (so long as ideology doesn't delude us about it); to the improvement of our knowledge that comes from real, natural science (especially, ecology); and to philosophical understanding itself. Understanding of the kind essayed, for example, in the present chapter. And thus to the possibility of us deciding together what kind of economy to have, within the limits that nature sets for us.

[19]My remarks hereabouts are of course strongly influenced by Wittgenstein's conception of philosophy.

26.6 Wittgensteinian Conclusions

As I understand him,[20] Wittgenstein is above all a philosopher of intellectual freedom. He teaches us that our thinking is constantly constrained by the scribblings of dead philosophers, and that, to escape such influences, we have to make them conscious. Once we do so, as I have sought to do in this piece, then we can choose our path adequately at last.

I propose that we exercise our philosophical freedom not by creating new theoretical edifices—such as a genuine science of economics would supposedly be—but by taking the bold step of actually making explicit and then questioning the *hegemony* of science (i.e. the step of questioning scientism).

We saw in this chapter that there are two rival visions of ecological economics available:

(1) On the Costanzian vision, we seek to internalize externalities in full. This is the most superficially obvious way for a Kuhnian to go, in relation to economics. It would be exemplified by Costanza's famous valuation of the total value of the world's 'ecosystem services'.

(2) If, by contrast, we take Herman Daly's vision of ecological economics, as based in a quite distinct pre-analytic vision, seriously, and truly place Earth (and the Sun!) at the centre of our thinking, while recognizing that Earth has *simultaneously* to be thought as far more than a rock—i.e. recognizing that the reason the Earth *matters* is because of the beings (not just humans!) inhabiting it—then a different *philosophical* path opens up before us. One leaving *intact* the sociality and politicality of economics. And constraining economics tightly by way of natural science, and particularly of ecology (which, as Teddy Goldsmith labored to show us,[21] is actually far *more* than a natural science!). This is a subtler way to think Kuhn in relation to economics, a way not presuming that economics 'really is' a science. And rather, taking inspiration from Kuhn's legitimation of many ways of knowing that are non-scientific.

Route (1), I have suggested, fails, for Kuhnian reasons: because there is an inadequate *qualitative* understanding of the world, in the 'paradigm' of neoclassical economics; and all that Costanza ultimately does is to seek to correct that paradigm. Not to transcend it.

The terrible irony of the neoclassical system of economics is that if models its subject metaphorically on the conservation of energy but ignores the literal truth of the laws of thermodynamics. Entropy is a fatality for endless growth. Costanza tries to correct this failing. But it would be better to junk the whole failed vision. Substituting a better one; but *without* leaping to the conclusion that this better green ecological vision of economics is a vision literally of a *science*.

---

[20]See my Liberatory philosophy, forthcoming, and my Wittgenstein among the sciences, op.cit.

[21]In his great work The way.

26.7 Envoi: A Policy-Alternative to the Scientism of 'Natural Capital' and 'Ecosystem Services'?

What, then, is the alternative to a fake scientisation of economics? If the now-widely-influential Costanza-inspired idea of measuring the economic value of 'ecosystem services' is based in pseudo-science, and ought to be replaced by a deeper effort to politically and ecologically philosophise a world that could last, what form should the latter effort take?

That is a task for another paper. But let me make one concrete proposal. One of the gravest dangers of the economic valuation of natural capital or ecosystem services is that it encourages us to think that we now *know* the value of any 'chunk' (sic) of nature. In relation to 'natural capital', a *precautionary* approach by contrast clarifies that it is reckless ever to assume that the purported worth in monetary terms of some chunk of nature/farmland actually exhausts its worth. Rather, we should always err on the side of thinking that there is likely to be much that is valuable that is not yet evident or understood by us. Consider in this context biodiversity loss. Conservation biologists and ecologists agree the rate of species loss is at least 100 and possibly 1000 times faster than background—the fastest mass extinction in the history of the world. But, as the wide margin of uncertainty in the previous sentence reveals, there is so much we do not yet know about the process. Meanwhile, biologists estimate that we may be aware as yet of as few as 1 in 10 species that exist in the world[22]; a far far higher level of ignorance than is present in other sciences such as chemistry. Any financial estimation of the value of biodiversity thus risks direly and systematically understating what is at risk of being lost. The Precautionary Principle would, rightly, force us to lean far more toward safety.

Our vast ignorance of ecology and biology is a serious limit even upon any serious effort to turn economics into science by integrating it with ecology and biology. Given that context, a far better basis for guarding nature at this fateful moment in human history than the fashionable fad for financialisation of nature would be: a serious entrenchment of the Precautionary Principle. And this would bring with it a concomitant great advantage: that precaution simultaneously warns against the reckless pursuit of economic growth. For of course it is still in reality, economic-growthism, and not environmentalism, that actually motivationally underlies most of the dangerous drive toward the economic valuation of so-called 'ecosystem services'.[23]

---

[22]See E O Wilson's book 'Half-Earth'.

[23]This paper was made possible by two grants from the AHRC, for the projects on Debating Nature's Value on which I was Principal Investigator, in 2016–18 and in 2019–2020. Thanks, for some comments, to Victor Anderson, and to audiences across Europe, across the years.

References

Beder, Sharon (2002). *Global spin: the corporate assault on environmentalism*. Totnes: Green Books.

Daly, Herman (1996). *Beyond Growth*. Boston: Beacon.

Ingrao, Bruno and Israel, Giorgio (1990). *The invisible hand: Economic Equilibrium in the history of science*. Cambridge, Mass: MIT.

Kovel, Joel (2002). *The enemy of nature: The end of capitalism or the end of the world*? London: Zed.

Mirowski, Philip (1988). *Against Mechanism: Protecting economics from science*. Totowa: Rowman and Littlefield.

Mirowski, Philip (1989). *More heat than light: Economics as social physics*. Cambridge: Cambridge University Press.

Mirowski, Philip (2002). *Machine Dreams: Economics becomes a cyborg science*. Cambridge: Cambridge University Press.

Read, Rupert (2007). "Economics is philosophy, economics is not science", *International Journal of Green Economics* 1: 3/4.

Part III
Miscellaneous

Chapter 27
Aesthetic Gestures: Elements of a Philosophy of Art in Frege and Wittgenstein

Nikolay Milkov

27.1 Frege's Aesthetics: Works of Art as One and Many

To grasp the sense and philosophical purport of Frege's aesthetics presupposes a sound understanding of Frege's philosophy of language. Frege held that only complete sentences express thoughts and thereby deliver information that enables us to derive truths from them. Sentences employed to command, to express a wish, to request, and so on are, to Frege's way of thinking, 'exclamations in which one vents feelings, groans, sighs, laughs' (Frege 1918, 355). Such sentences are consequently but 'mock thoughts' (*Scheingedanken*), and one acquires no knowledge from them.

Starting from this position, Frege advanced an idiosyncratic theory of art. According to Frege the poet, for example, 'does not really depict (*malt*) anything: he provides only the impetus for others to do so, furnishing hints (*winke*) to this end, and leaving it to the hearer to give his words body and shape' (Frege 1897, 140). In other words, works of art only stimulate one's imagination so that one constructs one's own, private work of art. 'The real work of art', states Frege, 'is a structure of ideas within us. … The external thing—the painting, the statue—is only a means for producing the real work of art in us' (Ibid., 132). On Frege's view, those who are genuinely taken with a work of art compose it—'in their head', as it were—for themselves.[1]

Given the seminal influence that Frege's thinking had on Wittgenstein, it is most illumining to consider Frege's conception of how art 'works' by recasting his view in the somewhat better-known Wittgenstinian terms. The artist's task as one apprehends it from this standpoint is not to depict facts (states of affairs) for us, but rather to move

[1]Quite recently, Kendall Walton made a similar contention: pieces of art express *our* thoughts, whereas we usually believe that they express the thought of their author (Walton 2015). See also Stambovsky (2004), Chap. 1.

N. Milkov (✉)
Philosophy Department, Universität Paderborn, Warburger Straße 100, 33098 Paderborn, Germany
e-mail: nmilkov@mail.uni-paderborn.de

S. Wuppuluri and N. da Costa (eds.), *WITTGENSTEINIAN (adj.)*,
The Frontiers Collection, https://doi.org/10.1007/978-3-030-27569-3_27

us to believe a narrative, for instance, or to 'imagine' for ourselves a musical theme.[2] Quite different from this is the semantics of mathematics and science, which is by contrast a-subjective, *objective*. Numbers, for example, inhere in the 'third world'—waiting, as it were, to be brought to light like unexplored continents by adventurous explorers. This holds as well for the propositions of everyday life when we speak 'seriously', i.e. with occurrent actualities in mind. Thus if someone *seriously* asks about the weather at the present moment and one replies, 'It's raining', one obviously enough aims to state the truth. What this comes down to is that propositions of mathematics and propositions about facts (states of affairs) have cognitive import. One can grasp, learn, and communicate their sense. Works of art, on the other hand, manifest, as Frege saw it, no cognitive signification, but instead evince emotive import.

Frege argues, in addition, that precisely because those engaged by a work of art construct it in different ways, no truth-in-general is to be found in this realm: every art aficionado is preoccupied with his or her own self-constructed aesthetic entity. This amounts to a Fregean ground for inferring the old adage, *de gustibus non est disputandum*. Frege's 'public' domain, by contrast, includes the world of logic, mathematics, language and science, along with the individuals and events of practical life.

We have just seen that concomitantly with this doctrine of the objective character of mathematical and scientific truth, Frege had formulated a persuasive aesthetic philosophy. But his aesthetic conception can face serious critical challenges. One problem is that individuals who make up an audience, for all their private constructs of a given artwork, nonetheless unerringly identify*one* particular artistic product to which all their personal constructive engagements with it collectively refer. Another point that would seem to challenge the Fregean view is that usually one is able at least potentially to distinguish an original work of art from its counterfeit. Further, and with respect specifically to music, we normally can discern when a new melody to which we happen to be listening is stylistically similar to a different one. One way to meet the foregoing difficulties is to assume that even if individuals do in some manner construct for themselves each work of art they encounter, different minds can unmistakably identify the source of these personal constructions as the same artwork—this in a manner analogous to how we can have many different photos of one individual but unmistakably recognize them all as pictures of the same person.

Another challenge facing Frege's theory is that it fails satisfactorily to account for the fact that some successful works of art are celebrated as masterpieces in radically different historical periods and across diverse cultures. This suggests that the seminal artworks are in some sense aesthetic universals. The problem is how to square this with Frege's doctrine, given his position that aesthetic experience is essentially private.

---

[2] A similar conception was developed a century later by Walton (1990), without any reference to Frege, though, and apparently independently of him.

27.2 That and How Wittgenstein Initially Assimilated Frege's Position, then Moved Beyond It

The early Wittgenstein adopted Frege's position on works of art, holding that we acquire information from other persons only through propositional language. In its propositional form, according to Wittgenstein, language arranges things 'experimentally', its role being logically to model—'depict', 'picture'—facts, or states of affairs (1961, 13). What distinguishes works of art, musical compositions in particular, is that they do not, as he sees it, model reality in this way and consequently have no cognitive import.

Moreover, while two propositions—true and false—correspond to every fact, or state of affairs, note-scripts of music do not. Unlike note-scripts, 'seriously articulated' propositions model facts and on that score one must recognize them as proxies for possible facts. (Those propositions which lack cogency are 'senseless' and consequently neither true nor false.) No such model/original relation obtains, however, vis-à-vis the work of art, which is no model of facts. It is on this count that the early Wittgenstein would not dismiss artworks as senseless.

Starting from the early 1930s, Wittgenstein began radically to modify his position on this theme. He slowly grew convinced that we have a command of language—we can produce and understand sentences—only because it is embedded in life practices. This novel consideration clearly looms large when one considers Wittgenstein's later thinking on how humans master languages. He argued that we learn to speak and understand sentences by training ('drill'), by examining acts of speaking and understanding the language employed by those who teach us language, and by being attentive more generally to the linguistic practices of others. Possessing a command of language does not mean, though, that we learn to follow *rules* that govern linguistic praxis. It indicates, rather, that we have the capacity to speak and understand language so far as it is embedded in our lived experience.

Wittgenstein ultimately concluded that even if not identical with language, art (again, music in particular) is more akin to language than he had earlier thought. This evolved perspective is adumbrated in the following, typically gnomic Wittgenstinian declaration: 'In art it is hard to say anything [is] as good as: saying nothing' (1980, 23). In other words, art is not a language as such, yet it is 'almost' a language. This position led Wittgenstein to delineate some pregnant distinctions applicable to the relation between art and language. He opined, for instance, that 'music, *some* music at least, makes us want to call it language; but some music of course doesn't' (62).

Wittgenstein moved on from this to assert that while composers and writers can 'think', they do not 'think' equally well or consistently. In this regard, he pronounced Beethoven and Brahms 'good thinkers' but judged Gustav Mahler less accomplished by this measure. This evaluative approach correlates with Wittgenstein's belief that like the thoughts that language expresses, those that a musical composition can express bear messages. The cardinal point for Wittgenstein is that such messages, the content of musical thought, are what the composer aims at bottom to communicate. It is true that musical thoughts cannot be put into words: indeed, artists turn to the work of music, as they do to other genres of art, precisely in order to express with often incomparably compelling precision messages that discursive language simply cannot convey.

We turn, next, to the nature of the messages that Wittgenstein associated with the work of art. All the while we shall keep in mind that as he thought through this development in his philosophy of art, Wittgenstein did not abandon but rather supplemented his earlier, Fregean aesthetic in ways that expanded its scope and depth, opening in light of it new horizons. We shall return to this point in § 27.7.

It is worth summarizing at this juncture the key findings of the two foregoing sections: (i) A work of art, we saw, does not to Frege's mind and also to the early Wittgenstein refer to individuals or events that one can measure, which is to say to individuals or events with a strictly fixed space-time determination. That is why it has no cognitive import. (ii) To the later Wittgenstein, an artwork can express thoughts and not merely amorphous feelings, the latter view being the position of such improbable bedfellows as Leo Tolstoy[3] and the logical positivists. This point is supported by the fact that the work of art is segmented and articulated. It has, that is, a specific structure that is embedded in the context of a particular art-form, one that crystallizes at some definite present moment in time.[4]

27.3 Works of Art as Sources of Information

To further clarify the Frege-inspired Wittgenstinian philosophy of art, and in particular how from that standpoint works of art have life-transforming information that can teach us, we turn to Wittgenstein's claim that 'good art' (1961, p. 83) 'insinuates itself [*schleicht sich ein*] into my life', having the character of something that 'I adopt ... as my own' (1980, 73). The artist's creation thereby comes to be intimately bound up with my inner life. Indeed, artistic creations speak to defining problems of personal life. More specifically, a poem, a musical composition, even a masterwork of architecture can inform one of ways of being that enable him or her, imbued with that aesthetically communicated 'message', to resolve problems of living (27). In the process the work of art can also profoundly deepen one's self-understanding. It is in just this respect that Wittgenstein champions works of art as sources of information, sources from which we often feel that we have important things to learn. On this Wittgenstein was insistent: 'poets, musicians, etc. ... *have something to teach*' (36). He was convinced that in their different ways the arts thus communicate knowledge.[5]

---

[3]See Tolstoy (1897). We are going to say more about Tolstoy's philosophy of art in § 27.6, below.

[4]Cf, § 27.4, below.

[5]Some authors today hold that works of art have cognitive import. They discriminate two types of cognition: knowledge (*Erkenntnis*) and acquaintance (*Kenntnis*). The acquaintance is an immediate type of cognition and can be conveyed by works of art. Knowledge, in contest, can be conveyed by science and also by facts and events of ordinary life. Cf. e.g. Gabriel (2015), p. 68. That emotions have cognitive import was clearly understood by Max Scheler. To be more exact, according to him

Referring to the musical compositions of the American folk singer/song writer Woody Guthrie, the pop artist Bob Dylan unwittingly echoed the Wittgenstinian position when he remarked that 'you could listen to his songs and actually learn how to live' (Corliss 2005). To be sure, Wittgenstein was more explicit when he argued that a work of art can 'show' a novel and fully determinate life-direction—a hitherto untraveled path that can lead one to a new way of living. Works of art can thus enhance our ability to discover what we genuinely wish for in life and how to achieve it, or the social group to which we truly belong.[6] It can implicitly school us in how to live our life and how to transform it. Given this understanding of art one can appreciate why a cohort of influential thinkers have taken works of art to be embodiments of philosophy. Friedrich Nietzsche, for one, declared that 'music is the true philosophy: it can namely intimate us a higher form of knowledge, or wisdom' (1888, pp. 2–3).

Significantly, it is difficult to predict whether, having been literally 'informed' in this manner by a novel work of art, one ought to expect a revolution in the configuration of one's life-concerns (in Wilhelm Dilthey's *Lebensbezugen*). Generally, such a fundamental transformation can occur only after a gradual inner development in the same direction, and when some, often unexpected spark triggers it.[7] We can look to the late Wittgenstein for a hint as to how it is that a successful work of art can transform one's way of pursuing a life. 'The fact that life is problematic', he avers, 'shows that the shape of your life does not fit into life's mold. So, you must change the way you live' (1980, 27). Wittgenstein's idea is that a successful work of art can prove of value in such cases by transforming the perspective from which we see our life; and this along lines that enable that life better to fit into its proper vital mold.[8]

Wittgenstein perhaps most tellingly signals the change in his conception of the work of art when he introduces in his philosophy the notion of the 'riddle', something that in the *Tractatus* he dismissed as non-existent (6.5). Wittgenstein invokes the 'riddle' as an explicit observation or question that has no equally explicit solution or explanation. He finds, for example, that colors in particular 'seem to present us with a riddle, a riddle that stimulates us—not one that disturbs us' (1980, 67). Wittgenstein takes this to be true of all qualia. He finds a similarity here with the fact that 'art *shows* us the miracles of nature. It is based on the *concept* of the miracles of nature.' (56)

---

emotions are not really acts of knowing but rather cognitive acts of grasping values of, among other things, works of arts (Scheler 1979).

[6]On how we join social groups see Milkov (2012). A similar conception was developed in Bourdieu (1979).

[7]Similar processes take place by falling in love. We have several acquaintances for whom we feel sympathy. Then all of a sudden, and for unclear reasons, we fall in love with one of them (Cf. Eastwick and Hunt 2014). This similarity is, of course, not surprising: falling in love is a paradigm cause of changing lifestyle. It is also noteworthy that decisive changes in the way of life take place only a couple of times in life.

[8]According to Wittgenstein, the role of religion is similar. 'One of the things Christianity says is that sound doctrines are all useless. That you have to change your *life*. (Or the *direction* of your life.)' (1980, 53).

The problems of life are immersed in these deep waters. To be sure, 'the problems of life', as Wittgenstein understands them, 'are insoluble on the surface and can be only solved in depth. They are insoluble in surface dimensions' (74). It comes therefore as no surprise that he regards works of art as bound up with the 'problems of life'. As reflected in music, such issues get represented 'not [in] the *melody*' as such, but rather in 'what gives the melody its *depth* and power' (37–8). It is in this connection that Wittgenstein revisited and emended his early theory of 'showing', along the complex and novel lines that distinguish his mature position in which he advances the view that 'good' works of art can show the miracle of nature and can also communicate reconfigured life-concerns in ways open for us to adopt.

The late innovations in his philosophy of art put Wittgenstein in the company of Friedrich Schiller, who famously taught that a work of art can teach things to those who genuinely encounter it on the work's own terms. That said, Wittgenstein radically diverged from Schiller on the matter of exactly what it is that artwork communicates and precisely how it effects spiritual (*geistige*) instruction. Wittgenstein argued that works of art do not teach us in any explicit manner, as Schiller thought, but rather that they tutor us in ways that are tacit or implicit. By the same token Wittgenstein sharply criticized those who reject the notion that art can educate us in life-transforming ways. He singles out for censure in this connection French poet Charles Baudelaire's pronouncement that the most one stands to learn when one goes to the theatre is the form of its newly installed chandeliers; to Baudelaire's way of thinking, art does not instruct, it simply entertains (cf. Bohrer 2015).

27.4 An Ontology of Aesthetic Gestures

While he assimilated core ingredients of Frege's philosophy of art, the later Wittgenstein dropped Frege's reference to *hints* (*Winke*), advancing instead the not-unrelated notion of aesthetic *gestures* (*Gebärde*).[9] Wittgenstein came to adopt the view that successful art posits (*setzt*) gestures in this sense—at the level of the complete work or of its component elements. Hence 'architecture', to his mind, 'is a *gesture*' (Wittgenstein 1980, 42), as is so such an aesthetic component as a 'musical phrase' (73).[10] Aesthetic gestures, as conceived by Wittgenstein, indicate in their different ways various 'styles' of life. Although this notion cannot be pursued here, there is a compelling case to be made for the idea of an entire universe of aesthetic gestures, each of which 'shows' (as a quasi 'vector') a distinctive direction in which one might lead one's life.

Significantly, this doctrine of gestures as pointing to new ways of life can be applied to non-aesthetic domains as well. (i) In the domain of religion a miracle, observed Wittgenstein, can count as a '*gesture* which God makes' that directs the devotee's life (45). (ii) In the realms, further, of oneiric experience and psychoanalytic

[9] Semantically, Frege's German term *Wink(e)* is closely related to the extension of the word *Gebärde*.
[10] See more about the works of art as gestures in Gall (2014), and Maddalena (2015).

psychology one of Wittgenstein's close followers, and his successor in the philosophy chair at Cambridge University, John Wisdom noted that not only dreams but also psychoanalytic sessions can change the light in which one sees the facts and events of one's life, and as a result can change its direction (Wisdom 1973, p. 43). (iii) More generally, an event in life too can change the path, in essence the 'style', of one's life. An incalculably influential cultural-historical instance of this occurred as the Buddha's epiphanic awakening from the lethargy of the mundane life upon encountering sick, aged, and suffering persons. One sees the same thing play out as well within the much more narrowly circumscribed events. Somerset Maugham's *Summing Up* (1938), for instance, depicts how the author's witnessing a seven-year-old boy dying of meningitis motivated a change in his life direction. (iv) By way of their rhetorical and ideational gestures, political leaders and sages typically 'put' to their communities alternative ways of life. The breathtaking range of such alternatives is manifest in the different types of gestures of, for example, political life-direction as promulgated by Lenin, Saddam Hussein, and Yasser Arafat. Every one of them points to very specific way of life.

In all of the above instances, 'gesture', as Wittgenstein understood it, denotes, similarly to artworks,[11] a mode of expression. He asserted moreover that

> what we regard as expression *consists* in incalculability. … I can after all listen again and again to a piece of music that I know (completely) by heart; and it might even be played on a music box. Its gestures would still be gestures for me, even though I knew all the time what was going to come next. Indeed, I might even keep being surprised. (1980, p. 73)

Apparently Wittgenstein reads aesthetic gestures as potential actions or, better, as 'germs' of actions—something substantiated by the circumstance that like gestures, actions can be only performed or copied (mimicked). Their intrinsic content, however, cannot be comprehensively conveyed in verbal form (Aristotle, *Poetic* 1779b25). Being germs in this sense, gestures may intimate in various ways the form of actions presently unrealized.

27.5 Style

The Wittgenstinian answer as to how a work of art alters the direction of a way of life closely connected with his notion of 'style'. Wittgenstein was convinced that virtuosic works of art actualize 'good' 'aesthetic gestures' in the form dictated by the particular style that the artist employs. It is style, in other words, that is formally constitutive of artwork. One might say, that style is the 'bridge' connecting the work of art and the form of life it depicts or expresses. It follows that the style of an artistic masterwork is operative in ways that can transformatively redirect the life practices of those who engage it on its own terms: love for freedom, strength of character, grace, etc. In other words, artistic style as a mode of expressing the way of experiencing

---

[11] In his *Notebooks 1914–1916* Wittgenstein asks: 'Art is a kind of expression.' (1961, p. 83).

the world can radically influence one's way of life, one's style of living and acting in the world.

What the modal verb 'can' signals here is that this influence is a function exclusively of what Wittgenstein would identify as successful or 'good' works of art.[12] Tellingly revealing of his aesthetic taste, Wittgenstein dismissed the music of Felix Mendelssohn as deficient by this criterion, since to his ear Mendelssohn's compositions display a lack of depth and originality, exemplifying the work of the merely '"*reproductive*" artist' (38). From early on, Wittgenstein held to the doctrine that the 'good' work of art is wholly originary as opposed to being mimetic ('reproductive'). Exemplary art of this sort achieves, he declared, a *vision* of a single object *sub specie aeternitatis*,[13] and it is as founded upon such an achievement that artwork can not only reconfigure one's perspective on a particular object or theme but can transformatively enlarge one's view of life (*Lebensanschauung*) in general. Compared to 'great' artistic creations—again, creations classified by Wittgenstein as 'good'—mediocre paintings, poetry and music are, in his terminology, aesthetic gestures that are in different respects and to different degrees anemic, insipid. Their main role is to convey entertainment, to bring their consumers away from the 'necessities' of the external world (from what Sigmund Freud had called, following the Greek mythology, ἀνάνκη[14])—and nothing beyond that.[15] One need only call to mind the contrast between the lifeworld set up in the classic novels of Fielding, George Eliot, D. H. Lawrence, Proust, and Thomas Mann with those of formulaic pulp fiction to appreciate Wittgenstein's point.[16]

Beyond its formative role as a factor of the cogency of artistic creations apprehended as aesthetic gestures, style is what typically distinguishes more generally the character of any activity that is a product of craft ($\tau\varepsilon\chi\eta\varepsilon$), of know how (Ryle 1949, 48).[17] To be sure, while there is no 'style' in the proper sense of coughing, for example, or sneezing, there are styles of such practices as chess play, a soccer play, painting, and musical and theatrical performance. In line with this wider ranging function, style in Wittgenstein's sense understandably has different implications for art, sports, politics, indeed for life at large. Wittgenstein's idea of style arguably

[12]We have already referred to Wittgenstein's thought that there are qualitatively different musical phrases: 'music, *some* music at least, makes us want to call it language; but some music of course doesn't' (1980, 62).

[13]Cf. Wittgenstein (1961), p. 83. In a similar key, in his *Aesthetics* (1835), Hegel maintained that the successful work of art 'touches' the Absolute. Unfortunately, maintains Hegel further, it cannot fuse with the Absolute since art is not the appropriate form to this purpose in principle: art only works in the field of appearances. We can reach fusion with the Absolute only in philosophy.

[14]See Freud 1916/17, 410.

[15]On this score Wittgenstein wrote: 'Is it the essence of the artistic way of looking at things, that it looks the world with a happy eye?' (1961, p. 86).

[16]Similarly, Kant maintained that there are works of art with spirit (*Geist*) and works of art that only entertain. Works of art with spirit possess the 'living up principle of the soul (*Gemüt*)' (AA. 5. 303).

[17]Merleau-Ponty, in contrast, meant (1964, 131) that style also characterizes the way we perceive the world.

applies this broadly because we express ourselves through cultural (often artefact-producing) practices[18] whereby we freely objectify the human world of will and ideas.[19] Where free will plays only a subordinate role, or none at all, namely when we produce artefacts determined by rigid norms, 'style' as Wittgenstein understands it is less significant. One sees familiar examples of this in textbooks of grammar and formal logic, wherein style is rarely if ever a substantive factor.

The philosophical bottom line of Wittgenstein's understanding is that it is the *style* of the 'good' work of art that underwrites one's being able to see (or hear or contemplate at large) what the artwork is about, its 'object', *sub specie aeternitatis*. As he put it in 1934, evidently referring expressly to works of art (as opposed to logic or grammar), style is 'the general necessity seen *sub specie aeternitatis*' (1997, 27). Apparently, Wittgenstein saw the style as 'general necessity' in the sense that the artist produces the synthetic a priori—the norm—of the work of art. It follows that the norm or 'law' of an artwork is what implies its 'necessity', and that the alternative interpretations of a given portrait or novel or symphony or sculpture are merely different ways of following its law, its 'general necessity'.[20]

27.6 Mood

Given that Wittgenstein's primary interest in the arts centered on the philosophy of music, it stands to reason that one may best appreciate the fruitfulness of his later philosophy of art by considering how it applies to music.

In his essay *What Is Art?*, Leo Tolstoy, an author Wittgenstein highly esteemed, maintained that music communicates emotions. This view, interestingly enough, can be easily seen as linked to Frege's and the early Wittgenstein's insistence that propositional language communicates thoughts. Indeed, what Tolstoy understood by 'emotions' and their communication is clearly correlated with what the two logical philosophers understood by 'thoughts' and their communication.

To make Tolstoy's definition of works of art more clear, one might reformulate it in an analogy to the Frege-inspired Wittgenstinian dictum that a proposition is nothing but a 'thought … expressed perceptibly through the senses' (1922, 3.1), which is to say a sense-wrought expression of thought. Thus we can conceive works of art, and for Wittgenstein this applies especially with musical compositions in mind, as emotions likewise 'expressed perceptibly through the senses'; so that those who 'tune in' to them may be imbued with the same emotions that animated the creative artist.

A Wittgenstinian refinement can be added here by suggesting that music expresses not simply an emotion in some general sense, but more precisely what it expresses

[18]Cf. n. 11.

[19]We are going to develop our conception of the freedom of the will in another paper.

[20]Nelson Goodman extensively discussed the problem of the relations between the original work of art, which works following a priori laws, and its interpretations (cf. Goodman and Elgin 1986).

is the species of cognitively informed affect that we term a 'mood'.[21] This follows Wittgenstein's appropriation of Schiller's notion of 'poetic mood' (1980, 65), in view of which one could regard mood as the 'fuel' of artistic expression, or even its soul.

This line of development finds support in the thinking of Gilbert Ryle, a close student of Wittgenstein and a distinguished philosopher in his own right. Ryle held that moods are neither identical with feelings, nor with emotions, for while feelings are short-lived, moods 'monopolize'—they penetrate our whole nature. 'Somewhat as the entire ship is cruising south-east, rolling, or vibrating, so the entire person is nervous, serene, or gloomy. His own corresponding inclination is to describe the whole world as menacing, congenial, or grey.' (Ryle 1949, 96) So far as works of art function this way, they can insinuate themselves 'into my life' (cf. § 27.3), and eventually change my view of life and consequently my way of life.

Among other things, Ryle's remark on moods comports with Wittgenstein's early Tractarian position that the message of the work of art lies on the 'boundary' of the world of facts, a boundary determined by an affectively keyed attitude (cf. Milkov 2004), to be more precise a boundary determined by way of a mood.[22] This 'message' can inspire us to contemplate objects of the world of facts in a new way: *sub specie aeternitatis*. Objects thus contemplated can acquire powerful, even sublimely novel meaning that can enable us to better orient ourselves in the world. Further, acquiring a new view of life in this way is tantamount to appropriating a newly redrawn boundary of the world of facts. Finally, in order to be 'perceptible through the senses', the expression of the mood that the work of art communicates is embodied, 'encapsulated' in a particular aesthetic construction, embedded in the context of a specific art-form.

Three observations on the work of art, music in particular, follow from the foregoing reflections:

(i) Once an aesthetic mood is encapsulated—i.e., created and preserved in the work of art—the individual who accesses the work of art can acquire the mood it expresses at any point in time. The mood is fully and communicatively conserved in it.
(ii) The composition of a successful art work (Rachmaninov's third piano concerto, say) goes hand in hand with the development (synthesis) of the specific structure of an artistic genre (e.g., piano concertos) as the work takes its place in the history of art (here, that of the piano repertory). This makes it clear why the craft to create 'good' works of art generally plays such a shaping role in the history of art.
(iii) At the same time, however, a defining element of genuinely accomplished, 'good' works of art is that they achieve, encapsulated in artistic compositions, an intense mood. In other words, the preeminently 'good' work of art presupposes not only masterly creative technique but also an ability to attain and encapsulate an intense mood. In a sense, it is a fusion of these two. Unfortunately, since

[21] See more about philosophy of moods Bollnow (1941).

[22] Cf. § 27.5.

moods are difficult to control and change at will, and consequently difficult to preserve, quite often the ability to compose successful works of art irreversibly wanes.[23] Interestingly enough, one finds no comparable rate of substantial decline in creative scientific and mathematical work, nor in the arts of painting and poetry—perhaps because in them mood plays a lesser role.

27.7 Frege-Inspired Wittgenstinian Aesthetic Theory Vindicated

As indicated at the outset, the aim of this essay is not just to rehearse and move beyond Frege's theory of art and the related doctrine of the early Wittgenstein. The intention is also to spell out and build upon the aesthetic thinking of the later Wittgenstein, and this in ways that shed novel light upon the enduring contributions of his own earlier and Frege's views (both now unwarrantably little-known or appreciated), and in the process to enhance our understanding of them.

Consistent with this essay's aim of at least indicating how one might build upon Frege and the early Wittgenstein, we conclude with a word about the sense in which artists are explorers of human psychology, adventurers who examine the human inner landscape in ways analogous to how geographers explore the earth and astronomers are explorers of celestial objects.[24] And this notwithstanding the all-important difference that unlike those who discover and investigate previously unknown objects of the external world, artists *construct* the objects they depict in their artwork; which is to say that the epistemology of the work of art is anti-realistic *per definitionem*. Employing their art in aesthetic 'studies' of the human mind, creative artists introduce ever-new aesthetic forms that elicit 'aesthetic pleasure' in the sense of Kant.

An epistemological aporia that would seem to threaten the plausibility of this account, one adumbrated earlier (at the end of § 27.1), has to do with the circumstance that while individual human minds react to a given aesthetic form differently, 'good' works of art commonly communicate across a diversity of human cultures what, in Wittgenstein's sense, they have to 'teach' us. This, if anything, attests to the reality of aesthetic universals (in the sense of 'eternal works of art') that cannot be simply circumvented for theoretical reasons. Of course, aesthetic universals are not the sort that populate of Frege's third world. Rather, they are relative, and on that count vague. Be this as it may, they exist, are *real*—but in what sense?

---

[23]In the realm of popular music, for instance, the Beatle Paul McCartney was in his youth a highly successful composer of pop tunes, producing a series of top-rated popular hits. But McCartney proved unable to compose equally successful pop tunes in his mature and late years. One could attribute this decline in musical creativity to McCartney's inability to sustain the level of mood he had once achieved.

[24]According to the pop artist Robert Irwin, 'to be an artist is not a matter of making paintings or objects at all. What we are really dealing with is our state of consciousness and the shape of our perceptions.' (Schoenholz Bee and Heliczer 2004, p. 269).

A singularly disclosive means, till now overlooked, of addressing this issue is by way of exploiting depth-psychologist Carl Jung's signature idea of the 'collective unconscious.' The factor of the collective unconscious as a formative ingredient of the human mind can go far toward clarifying how it is that large and diverse human communities concur on aesthetic universals. The principal divergence from Jungian psychology here is that our approach to the works of art is attuned not to religious or mythological archetypes but instead to the aesthetic archetypes that are activated when one encounters the work of art in its own terms. Hence, works of art that are 'good' in the philosophically pregnant Wittgenstinian sense we've reviewed realize the power to transform one's view of life by operating with the currency of the aesthetic archetypes that orchestrate the aesthetic of human mindfulness.

This is compellingly borne out in, for example, the most celebrated productions of the visual artists of the late nineteenth and early twentieth centuries—Seurat, Degas, Renoir, Van Gogh, Matisse. Every one of them discovered (Degas in his *Ballet*, for example) some new recurs—a new physiognomy, a new aspect of acting, a new gesture that stimulates human mind in a new way. In other words, each of these masters owed his success, in effect the 'cogency' of his most celebrated creations, to the new perspectives—new aesthetic objects—that perennially impel the serious viewer to construct it as his or her own, 'private' work of art.

Considerable historical and philosophical significance attaches to the fact that this account substantiates the unduly neglected theoretical value of a Fregean aesthetic, particularly in the ways we've seen that it was developed by the early Wittgenstein. The latter maintained, once again, that those who share an encounter with a single artistic production construct the artwork on their own, individually, 'privately'. At that point in time Wittgenstein denied that there exist any objective, generally valid aesthetic objects. As we saw at the outset, in § 27.1, this element of Frege's and Wittgenstein's aesthetic conception is sound. Nonetheless, we have found that there *are* general aesthetic archetypes that inform human mindfulness. While these cannot establish the general validity of any particular 'good' work of art—cannot justify 'eternal aesthetic truths'—they do substantiate archetypal validities of art in relative terms. Consequently, good works of art are general enough to secure the warrant of human aesthetic practice at large, this insofar as human cultures and civilizations have aesthetic universals (in what is cross-culturally esteemed as their 'great' works of art) that communicate novel and transformatively enriching visions of aesthetic life.

27.8 Concluding Remark

The upshot of the foregoing discussion is that from a genuinely philosophical standpoint the work of art is most effectively approached by recourse to psychology, not logic. Having established this along the lines pursued in the preceding sections, we are in a position to understand why Frege, the fierce warrior of anti-psychologism, ultimately turned his back on aesthetic theory. At the same time, however, by having made explicit the grounds of the authority of this approach to philosophical aesthetics we can appreciate why, in contrast with Frege, the later Wittgenstein, who nurtured a progressively deepening interest in psychology, came to treat aesthetic theory with ever-greater seriousness and penetration.[25]

References

Bohrer, Karl Heinz. 2015. *Ist Kunst Illusion?*, München: Hanser Verlag.

Bollnow, Otto Friedrich. 1941. *Das Wesen der Stimmungen*, Frankfurt: Klostermann.

Bourdieu, Pierre. 1979. *La Distinction. Critique sociale du jugement*, Paris: Minuit, 1979.

Corliss, Richard. 2005. "When He Was on His Own?", *Time Magazine*, September 19, p. 55.

Eastwich, Paul, and Hunt, Lucy. 2014. "Relational Mate Value: Consensus and Uniqueness in Romantic Evaluations", *Journal of Personality and Social Psychology* 106: 728–751.

Frege, Gottlob. 1897. "Logic", in: *Posthumous Writings*, Oxford: Blackwell, 1979, pp. 126– 51.

Frege, Gottlob. 1918. "Thoughts", in: *Collected Papers*, Oxford: Blackwell, 1980, pp. 351–72.

Freud, Sigmund. 1916/17. *Vorlesungen zur Einführung in die Psychoanalyse*, Frankfurt: Fischer Verlag, 1991.

Gabriel, Gottfried. 2015. *Erkenntnis*, Berlin: de Gryuter.

Gall, Michalle. 2014. "The Art as Gesture", *Paragrana* 23: 15–22.

Goodman, Nelson, and Elgin, Catherine. 1986. "Interpretation and Identity—Can the Work Survive the World?", *Critical Inquiry* 12: 564–75.

Hegel, G. F. W. 1835. *Vorlesungen über die Ästhetik*, in *Werke in 20 Bände*, 13.–15. Band. Frankfurt: Suhrkamp. 1986.

Maddalena, Giovanni. 2015. *The philosophy of gesture: completing pragmatists' incomplete revolution*, Montreal & Kingston: McGill-Queen's University Press.

Maugham, Somerset. 1938. *Summing Up*, New York: Doubleday, Doran & Company.

Merleau-Ponty, Maurice. 1964. *L'Oeil et l'Esprit*, Paris: Gallimard.

Milkov, Nikolay. 2004. "The Meaning of Life: A Topological Approach", *Analecta Husserliana* 84 (2004), pp. 217–34.

Milkov, Nikolay. 2012. "The Cement of Social Alchemy: Philosophical Analysis of Mind Group and Personal Identity", in: O. J. Schulz and B. E. Koch (eds.), *New Developments in Anthropology Research*, New York: Nova Science Publishers, pp. 53–70.

Nietzsche, Friedrich. 1888. *The Case of Wagner*, New York: Random House, 1974.

Ryle, Gilbert. 1949. *The Concept of Mind*, Harmondsworth: Penguin Books, 1963.

Scheler, Max. 1979. *Erkenntnislehre und Metaphysik*, 2 vol. of idem, *Gesammelte Werke in 15 Bände*, Bern: Francke.

Schoenholz Bee, Harriet, and Heliczer, Cassandra (eds.). 2004. *MoMA Highlights*, New York: The Museum of Modern Art (1st published 1999).

Stambovsky, Phillip. 2004. *Philosophical Conceptualization and Literary Art: Inference, Ereignis, and Conceptual Attunement to the Work of Poetic Genius*, Vancouver (BC): Fairleigh Dickinson University Press.

Tolstoy, Leo. 1897. *What is Art?*, transl. by Larissa Volokhonsky and Richard Pevear, London: Penguin Books, 1995.

Walton, Kendall. 1990. *Mimesis as Make-Believe: On the Foundation of the Representational Arts*, Cambridge (Mass.), Harvard University Press.

[25]First variant of this paper was delivered under the title "Aesthetic Gestures: Essay in Frege–Wittgenstein Theory of Art" at the 39th International Wittgenstein Symposium at Kirchberg am Wechsel, Austria, on August 10, 2016. The author is indebted to stimulating remarks of the audience.

Walton, Kendall. 2015. *On Other Shoes: Music, Metaphor, Empathy, Existence*, Oxford: Oxford University Press.
Wisdom, John. 1973. "Mr Köllerström's dream: enlightenment and happiness—a few remarks", in: I. Dilman (ed.), *Philosophy and Life*, The Hague: Nijhoff, pp. 35–48.
Wittgenstein, Ludwig. 1922. *Tractatus logico-philosophicus*, trans. by C. K. Ogden, London: Kegan Paul.
Wittgenstein, Ludwig. 1961. *Notebooks 1914–1916*, 2^{d} ed., ed. by G. E. M. Anscombe, Oxford: Blackwell. 1979.
Wittgenstein, Ludwig. 1980. *Culture and Value*, ed. by G. H. von Wright, Oxford: Blackwell.
Wittgenstein, Ludwig. 1997. *Denkbewegungen. Tagebücher 1930–1932, 1936–1937*, ed. by Hg. von Wilhelm Baum, Innsbruck: Haymon Verlag.

Chapter 28
The Lens of Architecture: Wittgenstein, Vision, Space, and Language

Nana Last

Some forty pages into *Philosophical Investigations*, the central work of Ludwig Wittgenstein's later philosophy, the author tells us: "A rule stands there like a sign-post" (PI §85).[1] One thing that makes this analogy so intriguing is the placement of its status as linguistic rule, visual image, and spatialized artefact—that is, language, vision, and space. The association Wittgenstein makes between a *rule* and a *sign-post* concerns the function of each as something we *follow*. As one of the myriad visual, spatial, and architectural images the *Investigations* employs, the sign-post is a conglomeration of the logics that appear throughout the *Investigations*. Following the text's discussion of the various ways to interpret the sign-post leads one both further into the text—as the sign-post circles back on itself to ponder how the *Investigations*, itself, should be followed—and beyond the text to Wittgenstein's practice of architecture as demonstrated by the house in Vienna he designed for his sister, Margarethe Stonborough-Wittgenstein.[2]

Historically and conceptually, Wittgenstein's practice of architecture occupies a unique slot between the early philosophy of language realized in his *Tractatus Logico-Philosophicus* (1922) and the late philosophy of language developed in

[1] All references to the *Philosophical Investigations* will be denoted "PI" followed by the number of the relevant section (if not explicitly delineated in the main text).

[2] The story of Wittgenstein's involvement in the design of his sister's house is well known. While his sister Margarethe Stonborough-Wittgenstein had wanted her brother involved in its design from the start, she began the project by engaging architect Paul Engelmann. In the spring of 1926, Engelmann outlined the massing of the building and basic layout of the rooms on the ground floor. Over time, Wittgenstein became increasingly involved in the process, first officially becoming co-architect for the house by September of that year, and then taking control over the project completely. Wittgenstein thus was the sole architect during the later stages of the house's conception, during which he designed the house's interiors, windows, and finishes, and made all final determinations on the house's massing.

N. Last (✉)
The University of Virginia, Campbell Hall, PO Box 400122, Charlottesville, VA 22904, USA
e-mail: ndl5g@virginia.edu

S. Wuppuluri and N. da Costa (eds.), *WITTGENSTEINIAN (adj.)*,
The Frontiers Collection, https://doi.org/10.1007/978-3-030-27569-3_28

the *Philosophical Investigations* (1953). Wittgenstein's work on the Stonborough-Wittgenstein house began in the autumn of 1926 and lasted two-and-a-half years, constituting the final portion of a ten-year abandonment of philosophy separating his early and late phases. The distinction between early and late work, however, is more than temporal; it registers a decided shift in both his understanding of language and his approach to philosophy. Given that distinction, questions emerge on either of the two sides of that ten-year period. On one side lies Wittgenstein's declaration that, with his completion of the *Tractatus* in 1919, all philosophical problems were solved and his involvement with philosophy was over. The other side coincides with the house's completion in 1929 and Wittgenstein's concurrent proclamation that he was returning to philosophy to "work on visual space [Gesichtsraum] and other things" (Nedo 1993).

In apparent disregard for this stated goal, the central work of Wittgenstein's later stage (as with his early one) continues to focus on the philosophy of language. Yet the remark is more accurate than it might initially appear. As the *Investigations* constitutes itself in relation to the early philosophy so as to be deeply invested in maintaining a discussion with it, another way to approach this quandary is to ask how visual space might form a medium of discussion between Wittgenstein's early and late philosophical views. With any such discussion necessarily spanning the ten-year period capped by the practice of architecture, one possibility is to consider what role Wittgenstein's practice of architecture played in this philosophical advancement. How does the philosophy of language engage the practice of architecture? If we follow the sign-post to Wittgenstein's practice of architecture, it points to how architecture does not merely precede his return to philosophy—rather, it enters the late philosophy as the lens through which his view of language is cast. In this way it becomes that through which philosophy looks, providing both the impetus for Wittgenstein's return to philosophy and the means by which he establishes a new approach within it.

The lens of architecture, as with the ten-year intervening period between Wittgenstein's phases of work, has two facets. It acts across the two philosophical sides of this equation by not only connecting these sides, but also forming a way of viewing the connection between the early and late work. One side of the lens consists of looking forward through it, that is, looking through architecture to the *Investigations*'s construction of a reimagined philosophy of language. In this view, language becomes fully spatialized, summed up by Wittgenstein's declaration in the *Investigations* that he is "talking about the spatial and temporal phenomenon of language, not about some non-spatial, non-temporal phantasm" (PI § 108). Equally, however, the lens can be looked through in the opposite direction, by looking back at the *Tractatus* from the standpoint of the later philosophy and practice of architecture. Looking in this direction becomes a means of assessing the spatiality of the *Tractatus,* showing it to operate with a flattened and restrictive understanding of space that, although concerned with spatial constructs, attempts to maintain itself outside of both space and time. This view understands language as (following the *Investigations*) a "non-spatial, non-temporal phantasm" (*ibid.*). Transforming the non-spatial and non-temporal into their spatial and temporal other is Wittgenstein's practice of architecture, allowing the latter to serve as a particular kind of intermediary between his two philosophies of language. Architecture provided a forum or venue for bringing to the surface the critical, underlying spatial aspects of both early and late philosophy.

Central to the workings of the lens is the understanding that the practice of architecture was impactful not because it introduced visuality and spatiality into Wittgenstein's understanding of language, but because it exposes the *Tractatus*'s idealized view of language to be just that—a *view*. It highlights the *Tractatus*'s underlying (and unacknowledged) fundamental linguistic-spatial supposition. With this understanding, the post-practice of architecture philosophy is not simply the product of Wittgenstein having been a philosopher who *then* designed a house, but rather one whose philosophy and methods *already contained* related ideas. The persistent concern with spatial topics appears on both sides of the lens, in the insistent focus on boundaries, limits, showing, and spatial ordering of propositions, as well as in the literal form of each text.[3] The view across the lens reveals that these constructs exist on both sides, but act differently, showing how such fundamentally spatial propositions cannot be conceived of without committing to certain conceptions of space. What appears through the lens of architecture are thus two competing premises of the relation of language to space, two competing underlying spatial logics.

Wittgenstein's engagement with architecture was thereby able to have the impact it did precisely because his thinking did not suddenly turn to spatial issues after practicing architecture. Instead, the mode of spatiality in relation to language was altered, its view changed. His designing of the Vienna house played a critical role in this by creating an arena in which the flattened and constrained spatial logic of the *Tractatus* had to contend with that of architectural practice's fully spatial one. Transpiring through Wittgenstein's work on the house, in its literalization, manifestation, and materialization of the architectural process, the restricted spatiality of the *Tractatus* was challenged, following which it emerges transformed in the later work. This shift allowed Wittgenstein to address language and philosophy anew, metamorphosing his philosophy of language from the logic-based approach advanced in the *Tractatus* to the practice-based one indicative of the *Investigations*.

Revealing the *Tractatus*'s debt to a particular mode of spatial thinking plays a critical part in grasping spatiality in the *Investigations*. By showing this, the lens of architecture does more than provide a way of seeing: it manifests how each philosophy's fundamental constitution is indebted to a particular visual-spatial-linguistic association. Apprehending how philosophy and view are formed together makes it apparent that early and late work alike rely upon specific spatially located views of language to form their philosophical positions: the *Tractatus* is the product of the view from *outside* and *above* language (what might be thought of as the "view from the ladder") while the post-practice of architecture (given by the *Philosophical Investigations*) develops the view from *within* language.

[3]The *Tractatus* is a series of discrete propositions, logically ordered and numbered to the fifth decimal place. By contrast, the *Investigations* is an open-ended sequence of sections, numbered 1–693, each composed of one or a few short sentences or paragraphs.

28.1 The View from Without

The view of language that the *Tractatus* creates is one in which language behaves in accordance with the rules of logic. To do so, it fundamentally constructs a picture of language as if viewing it from outside and above. The view exhibits a series of specific characteristics: it lies below, is framed, and is presented as a horizontal surface. From its vantage, language appears to be a flat, ideal, delimited construct. The hallmark of this view is its removal from ordinary language. In lieu of engaging with ordinary language, it oversees it, hovering above the fray of everyday linguistic practice. Situated as it is, this view allows for establishing a clearly discernible relation among language, logic, and philosophy, proposing that all three share a coincident series of boundaries. The ability to grasp these boundaries, to capture language's limits in its sights, is what lends the view from without its sense of being comprehensive. As such, it purports to give a complete picture of meaningful or sensibly determinate language, one which fulfills the dictates of logic.

The view from above is both formed by the text and produces it. It appears most succinctly in the *Tractatus*'s declaration of a series of coincident spatial limits that frame the text, define the relation of the subject to language and the world, and encapsulate the realms of language and logic. These limits define its critical relations with subjectivity, the world, and language, as well as bracketing the text by forming the literal beginning and end of the book.

> 1. The world is all that is the case.
>
> 5.6. *The limits of my language* mean the limits of my world.
>
> 5.61. Logic pervades the world: the limits of the world are also its limits. So we cannot say in logic, "The world has this in it, and this, but not that."
>
> 5.632. The subject does not belong to the world: rather it is a limit of the world.
>
> 7. What we cannot speak about we must pass over in silence. (TLP)[4]

The *Tractatus*'s ability to locate limits to language, the subject, and the world suggests it has both a complete and clear view of language, yet its viewpoint is as restrictive as it is omniscient. The very propositions designating neatly aligned limits are both the product and the enabler of these restrictions. These circumscribed limits demarcate language into that which is logically determinate (meaningful) and logically indeterminate (nonsense). In varying ways, the text draws these same limits again and again, tracing and retracing them, separating meaning from nonsense—distinguishing what language is able to say sensibly from what it can only show. This stance, however, constrains language as much as it does vision. From across their shared boundary, in the continual reinscribing of these divisions, language and vision (philosophy and view) are bound together in mutual formation. View and limits go hand in hand. This state of being locked together is what the *Investigations* will react against.

---

[4]All references to the *Tractatus Logico-Philosophicus* will be denoted "TLP" followed by their associated decimal reference number (if not explicitly delineated in the main text).

The *Tractatus*, in attempting to define the limits of the thinkable, is itself unthinkable outside the severely restricted understanding of space that I am calling the "view from without." This view constructs an understanding of language literally described as the "picture theory." The picture theory sets forth a parallel between language and the world in which linguistic propositions depict possible states of affairs in the world, proposing a logical isomorphism between representation and that which is represented, or comparing the concept of a logical picture with a spatial one. Accordingly, the particular components in a proposition mirror the things they stand for in a direct correlation between words and facts. The picture the *Tractatus* develops of language, however, is a product of the view from above, a fixed view challenged by Wittgenstein's subsequent practice of architecture. Against the *Tractatus*'s singular image, the practice of architecture introduced an array of complex spatial constructs irreducible to a single view or representation—views, that is, at times incompatible with the view from above.

The text reaches the extent of the view from above at the same moment its closing lines acknowledge the limitations of both view and text—at the very moment the *Tractatus* tries to see itself. A paradox ensues as its closing lines try to contend with these limits. The result places the claims of the *Tractatus* against its ability to see them. It acknowledges this in its closing propositions, declaring that:

> 6.54. My propositions serve as elucidations in the following way: anyone who understands me eventually recognizes them as nonsensical, when he has used them—as steps—to climb up beyond them. (He must, so to speak, throw away the ladder after he has climbed up it.) (TLP)

If the primary aim of the *Tractatus* is to set limits to what can be said or thought sensibly, it violates this dictum by trying to say what it acknowledges can only be shown: the limits of language and the world. Wittgenstein concludes that the propositions must be transcended to "see the world aright" (TLP 6.54). The penultimate proposition makes clear that this view of language leaves no place from within it for self-reflection. It has literally reached its prescribed limits, after which the final proposition can offer only silence. Transgression of the text's rules and limits becomes the only other possibility. Wittgenstein follows this rather literally at first by abandoning philosophy. It is only after his work on the Vienna house that Wittgenstein finds his way to act otherwise, to inhabit a fully spatial philosophy disallowed in the view from above.

28.2 Without to Within

If the view from *without* language characterizes the *Tractatus*'s depiction of and engagement with language, the view from *within* it defines the space of the *Investigations*. As if to acknowledge its role in this reformation, Wittgenstein's practice of architecture announces itself almost immediately. It enters the text as a response to a quoted passage from the *Confessions* relaying Augustine's memory of the "essence of [learning] human language" (PI § 1). Wittgenstein then subjects Augustine's characterization to a practical test. He transposes it to an everyday linguistic interchange between a shopper and shopkeeper centering around the words "red," "five," and "apples." Wittgenstein is here proposing a model that fulfills Augustine's description yet exaggerates its limitations. In his example, the shopkeeper checks these words against preset, fixed tables that show what "apple" or "red" refer to. The result is a stilted way of being, clearly incommensurate with ordinary language. Wittgenstein caps this by saying, "Well, I assume that he *acts* as I have described" (*ibid.*).

The Augustine-based example, while lacking the *Tractatus*'s emphasis on logic, evinces a comparably narrow description of language's relation to the world. It might be thought of as how the *Tractatus*'s approach to language appears through the lens of architecture: the lens shows how this narrow view is entirely at odds with regular uses of language. Wittgenstein shortly makes that association explicit, stating that Augustine's description

> does describe a system of communication; only not everything that we call language is this system. And one has to say this in many cases where the question arises "Is this an appropriate description or not?" The answer is: "Yes, it is appropriate, but only for this narrowly circumscribed region, not for the whole of what you are claiming to describe." (PI § 3)

The section immediately preceding introduces the constructive side of the lens, appearing as Wittgenstein's example of the builders.

> Let us imagine a language for which the description given by Augustine is right. The language is meant to serve for communication between builder A and an assistant B. A is building with building-stones: there are blocks, pillars, slabs and beams. B has to pass the stones, and that in the order in which A needs them. For this purpose they use a language consisting of the words "block", "pillar", "slab", "beam". A calls them out;—B brings the stones which he has learnt to bring at such-and-such a call.—Conceive this as a complete primitive language. (PI § 2)

If Wittgenstein's acting out of Augustine's description is designed to show how it constrains language's functioning, the builders' example is meant to show language's potential to exceed itself. Slab, pillar, etc., unlike the words in the shopkeeper example, are expressly shown to do more than designate objects in a one-to-one correspondence. They also function as calls, as shorthand for the phrases "bring me a pillar" or "bring me a slab." The words far exceed the awkwardness exhibited by the shopkeeper's list of word-meaning correlations. Although the builders' words fall into the same category as the shopkeeper's—object nouns—the example shows how they become meaningful in relation to their context, namely, the particular language-game of building with building stones. More so, this expansive ability emanates from an explicitly architectural practice. That constructive aspect underscores the language-game's implicit potential to evolve beyond its initial four words. It is an example that, by design, exceeds its own parameters. From that point forward the text expands both this particular language-game and this model of how language is meaningful.

Beginning with the builders and the idea of practice, and continuing with myriad architectural and spatial examples, architecture emerges throughout the pages of the

Investigations in any number of ways. From the start, the builders example shows how language comes to be meaningful, not in isolation, but by virtue of its basis within practice-based language-games. The text then returns periodically both directly and indirectly to this example, to show how it and other language-games are inherently expansive as they encounter various new needs, run into problems, gain new tools, develop new purposes, and so on. In such ways, the builders' language-game grows from its original four words by accruing new tools and aims. This culminates in Wittgenstein positing an amazing list of some of the endless "kinds of sentence" (PI §23). These span from the commonplace "assertion, question, and command" to "*countless* kinds" that include the specifically architectural, e.g., "[d]escribing the appearance of an object, or giving its measurements" and "[c]onstructing an object from a description (a drawing)" (*ibid.*). Furthermore, he notes that this "multiplicity is not something fixed, given once for all; but new types of language, new language-games, as we may say, come into existence, and others become obsolete and get forgotten" (*ibid.*). He concludes the section by directly comparing this understanding to his earlier thinking:

> It is interesting to compare the multiplicity of the tools in language and of the ways they are used, the multiplicity of kinds of word and sentence, with what logicians have said about the structure of language. (Including the author of the *Tractatus Logico-Philosophicus*.) (*ibid.*)

Examples such as these build the *Investigations*'s view from within language. This continually expanding, open-ended approach contrasts greatly with the *Tractatus*'s regulative one. While the view from above pronounces itself definitive, the view from within is decidedly multidirectional. In contrast to the single hierarchical viewpoint of the early philosophy's view from above, the view from within is partial, incomplete, active, engaged, and subject-laden. It results from a multiplicity of occasionally conflicting and ambiguous views. If the *Tractatus* aimed to find the limits of language, the *Investigations* turns its vision to the myriad ways language functions. As with the sign-post situated in space, it points, among other things, to the place within itself it has allowed for self-reflection.

By manifesting human subjectivity through a linguistic practice, it is the builders who set the *Investigations* in motion in ways that allow for the inhabitation of language. They are simultaneously "subjects" and "subjects engaged in an architectural practice." This allows the view from within language to become a linguistic place of inhabitation and practice in which language is apprehended as a social, temporospatial entity. This perspective is evident in myriad ways throughout the *Investigations*, beginning with its opening sections devoted to the builders, developing through the understanding of language as a practice (which includes such things as measured drawings), and continuing by finding new purposes for philosophy, including uniting the philosophical and everyday uses of language. Time and again, the view from within actively defies the fundamental tenets instilled by the view from without.

The view from within is based in everyday practices through which language becomes bound to the subject. This entanglement disallows the possibility of an absolute clarity and complete view of language only achievable from without. It cannot see or fully apprehend its own limits, although it may at times bump into them. This transposition changes the entire workings of philosophy, its aims and methods. From within, vision is no longer idealized and isolated, but becomes one of the many possible ways of interacting with the world. Another way of understanding this distinction is to see the aerial view of the *Tractatus* as having been replaced by the view from the streets, a view that would later form the "everyday" of de Certeau, who turned to Wittgenstein's thinking as a model.

Nowhere is this repositioning more evident than in the *Investigations*'s rejection of the *Tractatus*'s logical method and aim to rid language of contradictions as its way of resolving philosophical problems. Rather than seek to resolve the linguistic contradictions found in everyday language, the *Investigations* understands the ambiguity inherent in language as informative. Consequently, it dissolves the *Tractatus*'s aim to eradicate contradiction and replaces it with a radically different proposition: understanding the social status of a contradiction. Making this explicit, he writes:

> 125. It is the business of philosophy, not to resolve a contradiction by means of a mathematical or logico-mathematical discovery, but to make it possible for us to get a clear view of the state of mathematics that troubles us: the state of affairs *before* the contradiction is resolved. (And this does not mean that one is sidestepping a difficulty.)
>
> The fundamental fact here is that we lay down rules, a technique, for a game, and that then when we follow the rules, things do not turn out as we had assumed. That we are therefore as it were entangled in our own rules.
>
> This entanglement in our rules is what we want to understand (i.e. get a clear view of).
>
> It throws light on our concept of *meaning* something. For in those cases things turn out otherwise than we had meant, foreseen. That is just what we say when, for example, a contradiction appears: "I didn't mean it like that."
>
> The civil status of a contradiction, or its status in civil life: there is the philosophical problem. (PI)

Similarly, many critical moments of the *Investigations* transpire across the two sides of this lens as a continually changing vision confronts and dismantles the singular, fixed image instituted by the *Tractatus*. The *Investigations* employs this lens to build itself on the one hand and to see the philosophy of the *Tractatus* anew on the other. As these are bound together, the process entails first revealing the *Tractatus*'s limited perspective and then disputing its tenets; to contest, that is, its way of looking at language which presupposes—impresses, even—a particular view upon it. This pattern reappears throughout the text, as example after example shift from a removed, constrained, or idealized view of language to an integrated, everyday, evolving linguistic practice. With this in mind, a major task of the *Investigations* is to show how the *Tractatus*'s view of language is spatially delimiting in its attempt to maintain sensible language within the realm of logic. Together, two passages most succinctly describe how the *Tractatus* instills a literally circumscribed view whose effects Wittgenstein associates with a pair of eyeglasses which direct how we see and yield a singular picture of language.

> 103. The ideal, as we think of it, is unshakable.... Where does this idea come from? It is like a pair of glasses on our nose through which we see whatever we look at. It never occurs to us to take them off. (PI)

> 115. A *picture* held us captive. And we could not get outside it, for it lay in our language and language seemed to repeat it to us inexorably. (PI)

These and similar admissions act as parallel statements to the *Tractatus*'s list of spatial limits. Those in the *Investigations*, however, act to identify and remove preset constraints, rather than impose them. Becoming aware of the presence of the glasses leads to the realization that the *Tractatus*'s view of language is literal and, hence, has an associated viewer situated in space. This understanding inaugurates the process of prying apart ordinary, unaided vision from logic's "improved" one as manifest in the glasses. The existence of a spatially and temporally situated viewer jeopardizes the *Tractatus*'s ability to maintain ideal, fixed relations among the limits of the subject, language, and vision. Spatial and temporal bodies (as with those of the builders) are at odds with a fixed limit holding together thought, language, and view. Wittgenstein addresses the irreality of his earlier thinking when he writes:

> These concepts: proposition, language, thought, world, stand in line one behind the other, each equivalent to each. (But what are these words to be used for now? The language-game in which they are to be applied is missing.) (PI § 96)

Wittgenstein is suggesting here that, rather than logic, language-games provide the circumstances for meaning.

Wittgenstein's acknowledgement that the glasses obscure ordinary vision brings about an ironic reversal. The *Tractatus* held that ordinary language obscured the logical form of a proposition. That belief underlies its recourse to logic and the correlative belief that philosophical problems are pseudo-problems—linguistic chimera resulting from the ambiguous nature of ordinary language. That is to say, the *Tractatus* saw ordinary language as confusing and logic as clear. This view split philosophical language from ordinary language—a split the *Investigations* aimed to bridge. If the *Tractatus* sought a hidden clarity beneath language's ambiguous surface, the *Investigations* reverses this process and diagnoses the distorting agent as the *Tractatus*'s requirement that language behave as logic. This turnaround alters not only *where* we look, but *how*. Clarity remains important, but where and how to look and the criteria for clarity are brought into question. Ramifications follow. The glasses previously thought to aid vision are shown to have usurped it. Logic is revealed not to *be* language, but rather an imposition *on* language, both a limited view and a limitation on viewing and viewer alike. As the *Investigations* systematically undoes the view from above (the *Tractatus*'s primary undertaking), logical clarity, completeness, sharp boundaries, absolute distinctions, and the general form of a proposition all fall by the wayside as the arbiters of the entirety of language. In lieu of the coherence the overview afforded, the *Investigations*'s view from within must develop new measures of linguistic sense. It achieves this not by positing preset limits to language's sensible realm, but instead by "looking" at language across a wide range of practices, including uses that may contradict one another. If logic previously provided the standard of clarity, the *Investigations* must, after its removal, search for other criteria. To do this, Wittgenstein returns vision to everyday language, famously demanding of the reader to not *think* but rather to *look* at how language-games operate.

> 66. Consider for example the proceedings that we call "games". I mean board-games, card-games, ball-games, Olympic games, and so on. What is common to them all?—Don't say: "There *must* be something common, or they would not be called 'games' "—but *look and see* whether there is anything common to all.—For if you look at them you will not see something that is common to *all*, but similarities, relationships, and a whole series of them at that. To repeat: don't think, but look! (PI)

Examples of Wittgenstein's survey of the everyday use of language populate the text, most notably with this examination of games, culminating in the concept of family resemblance and its explicit abolition of any single shared aspect underlying all usages of a word. Comparably, no one gauge of perspicuity pervades the open-ended extent of language's myriad modes of being. As it does throughout the text, the *Investigations* here confronts the restricted view of language with spatial and temporal examples. This yields an unfolding terrain of examples and varying viewpoints that bridge philosophical and spatial constructs, from boundaries to methods of viewing, forms of representation, visualizations of rules, and sign-posts amidst myriad paths.

The issues of clarity and boundaries are tied together. If the *Tractatus* understood boundaries as logical functions, the *Investigations* understands boundaries and distinctions as fundamentally spatial entities. As spatial practices continue to delaminate the *Tractatus*'s reinscribed and reified limits, they unleash an array of behaviors, practices, modes of resemblance, spectrums of precision, and (perhaps most critically) a rethinking of the manner in which boundaries may function. As the *Investigations* dismantles the limits that the *Tractatus* imposed on language by undoing its view from above, it rethinks spatial boundaries themselves. As the *Investigations* questions what measure of exactitude even philosophy requires of language, it turns to spatial boundaries to provide a model. Spatial boundaries provide the locus to consider what criteria or degree of exactitude is necessary for language to make sense, both in philosophy and everyday uses. The examples he gives begin with spatial demarcations and move on to the rules of a game and then to the efficacy of language. As examples migrate from the spatial to rule-following to the linguistic (the three realms of the sign-post), the underlying thinking ties those realms together without equating them to a single way of functioning.

> 88. If I tell someone "Stand roughly here"—may not this explanation work perfectly? And cannot every other one fail too?
>
> But isn't it an inexact explanation?—Yes; why shouldn't we call it "inexact"? Only let us understand what "inexact" means. For it does not mean "unusable". And let us consider what we call an "exact" explanation in contrast with this one. Perhaps something like drawing a chalk line round an area? Here it strikes us at once that the line has breadth. So a colour-edge would be more exact. But has this exactness still got a function here… ?
>
> … Thus the point here is what we call "the goal". Am I inexact when I do not give our distance from the sun to the nearest foot, or tell a joiner the width of a table to the nearest thousandth of an inch?
>
> No *single* ideal of exactness has been laid down …. But you will find it difficult to hit upon such a convention; at least any that satisfies you. (PI)

Although by this point in the text, it is clear that neither games, boundaries, nor rules need to be ideal to be meaningful, the question remains whether the use of words like "game," "boundary," or "rule" can also function in this manner and maintain their respective senses. In so doing, Wittgenstein is asking whether the same thinking applies to words and to practices. His answer is yes—something he shows by transposing the spatial consideration of boundaries to those of sense. This requires rethinking where and how we look. Picking up on the sign-post's directional gesture, Wittgenstein posits the need for a spatial rotation to change our perspective.

> 108. We see that what we call "sentence" and "language" has not the formal unity that I imagined, but is the family of structures more or less related to one another.——But what becomes of logic now? Its rigour seems to be giving way here.—But in that case doesn't logic altogether disappear?—For how can it lose its rigour? Of course not by our bargaining any of its rigour out of it.—The *preconceived idea* of crystalline purity can only be removed by turning our whole examination round. (One might say: the axis of reference of our examination must be rotated, but about the fixed point of our real need.) (PI)

The rotation about our real need redirects the focal point away from the preconceived ideal image and toward a multiplicity of potential views in any number of directions. This leads directly to Wittgenstein's resounding conclusion of the section: "We are talking about the spatial and temporal phenomenon of language, not about some non-spatial, non-temporal phantasm" (*ibid.*).

If the *Tractatus* sought to wrench philosophical problems from their material and practical substrates in order to see their core logical form, the *Investigations* repeatedly returns philosophical problems to the material and practical firmaments in which we largely encounter them. To this end, Wittgenstein's practice of architecture provided a spatial forum for the examination of spatio-linguistic philosophical constructs central to his philosophy, including rule-following, boundaries, limits, practice, and resemblance. The lens it produces consolidates a collection of what Wittgenstein later defines as language-games: venues or instances of language's practice-based functioning. As he writes in the *Investigations*, "I shall also call the whole, consisting of language and the actions into which it is woven, the 'language-game' " (PI §7). This is what the house does as well.

Broadly speaking, the house's inter-workings of vision, space, and movement allow for exactly what the *Tractatus* sought to disallow: the interchange between philosophical and spatial constructs. As such, the house's design inserts a place where the *Tractatus*'s restricted spatiality necessarily intersects with a three-dimensional, inhabitable space. Evidence for such sites in the Stonborough-Wittgenstein house include its situated perspectives, use of various degrees of transparency, and competing centers that defy singular logics or rules. In providing such a forum, the practice of architecture initiates the *Investigations*'s formation of a practice-based, spatial understanding of language that again and again turns to this shared territory between architecture and philosophy as the means to examine the meaningful workings of language.

The house thus forms a practical arena situated between the *Tractatus*'s limited spatiality of the view from without and that of the *Investigations*'s wider and far more varied one from within. This affords it a unique condition in which to consider philosophical constructs in relation to their spatial counterparts. Similar to the *Investigations*'s insistence on a multiplicity of linguistic practices, the Stonborough-Wittgenstein house acts as a compilation of concepts and constructs that are unique to neither philosophy nor architecture but prevalent in both. And although Wittgenstein's example, itself, is unique, this thinking readily extends well beyond its specifics to the wider state of relations between architecture and philosophy, space and language.

Turning to the Stonborough-Wittgenstein house[5] itself, it appears from the exterior as an unadorned, white, modernist, asymmetric cubic building with a main central section and a largely regular grid of vertical windows which bestow on the house its particular character. On the main floor are found a salon, a library, a dining room, breakfast room, and Margarethe Stonborough's private rooms. The house's upper floors contain the private rooms for other family members and household workers. Without doubt, however, the quintessential space of the house is its central hall, which connects the rooms on the main floor. The manner in which it does this constitutes the central hall as the paramount locus of the house's inter-workings. Emanating from its complex spatial and visual configurations, it is the preeminent spatial collector and orchestrator of the house, creating contexts and consolidating spatial issues that will later appear in the *Investigations*.

To enter the house, visitors first pass through a pair of glass doors and enter a small vestibule. They then proceed through a second set of paired glass doors and ascend dark stone stairs into the brightly lit space of the central hall. Engulfed by a series of eight paired glass and steel doors on six different surfaces and three sides, the central hall is luminous. To the left of the stairs, a glass wall and doors leading to the southwest terrace admit light into the hall. On the right of the stairs, paired metal doors lead to the main salon. On the wall directly opposite the entry are two sets of paired translucent glass doors. Those on the left lead to the dining room; to the right, the pair opens to a hallway and private rooms. Turning around one hundred eighty degrees to face the entry, visitors are presented with four sets of glass doors: the two center pairs on axis with the entry, a left-hand pair that provide entry to the library, and on the right, elevated by two steps, doors connecting to the breakfast room.

The central hall thus sits poised between the simplicity and austerity of the exterior and the almost mazelike series of reflections produced by paired glass doors that erupt on nearly all sides. Set into both solid wall and glass plane, the glass doors are tall, structured by thin metal frames with one vertical division in each glass panel and no horizontal divisions. The one exception to this is the pair of doors connecting vestibule and hall in which the glass is not sub-divided, allowing for an unobstructed view of the main entry doors beyond. These like-designed doors each reflect and reiterate the others; their repeated image absorbs the focus of the house, inside and out, yet they do not exactly repeat, but are subtly transformed from one instance to the next. They mutate according to context and use, yielding an array of related cases that enact and presage Wittgenstein's subsequent formulation of his notion of family resemblance.

[5]For a more in-depth discussion of many of these issues, see Last (2008, 2012).

In tackling such problems as rule-following, boundaries, and precision, the disposition of the doors in the central hall serve as an amalgam of the spatial- philosophical-linguistic issues Wittgenstein contends with throughout his philosophy. Across the central space, the doors' continual changes, relocations, recalibrations of light, transparency, opacity, proportion, and spatial and social function speak to a range of issues. They connect both interior and exterior, one room to another, and owing to their construction, create spaces between one another, yielding a dialogue of possibilities and purpose. They play out many of the issues of Wittgenstein's philosophy by recasting, consolidating, and redistributing the house's spatial-philosophical-linguistic issues spatially, materially, and functionally.

Largely occurring in pairs, the doors create complex boundaries between the spaces they connect. Because of the climate, double exterior windows were common in Austria. But Wittgenstein extended this doubling to the interior, throughout the central hall and into the salon. With the exception of the doors to the breakfast room, and those connecting vestibule and hall, all of the glass and steel doors in the interior are bi-paneled, double doors that always open out into the rooms they connect in both directions. On the exterior where these double glass and steel doors occur, both sides are transparent glass. That is not always the case with the interior's more multifarious conditions and relations. While the doors connecting living room and hall are clear glass on both sides, those between dining room and central hall or staircase and central hall are transparent on the hall-side and translucent glass on the other. This allows for varying degrees of separation and privacy dependent upon where one is located, which way one is looking, and which panels are open or which closed.

In each of the instances in the hall where two materials or degrees of transparency are brought together, the less transparent material is placed on the more private side of the doors and the more transparent on the more public side. This would seem to form a rather straightforward approach, even constituting a rule. Concerns about rules and rule-following play a substantial role in the house as they do in the *Investigations*. Repeatedly, the book's discussions are played out around questions of visuality and spatiality. While the text grapples with these issues, related concerns arose with the house, including questions of what constitutes a rule and what it means to follow a rule. In terms of the house's use of transparent or translucent glass, what seems to be a simple declarative principle confronts in spatial practices a complex series of relationships that disrupt the direct implementation of a preset or fixed rule. Examples occur both in the dining room and salon. In the dining room, the doors to the hall are on a wall with three similar sets of paired glass and steel doors that lead directly outside to the southwest terrace. This sets up a series of four such pairs. The situation creates a dilemma as to whether the dining room-to-hall doors should exactly match the other three sets along the same wall, as they do in size and detail, or whether they should also mark what is on their other side. The question that emerges is a spatial one: To which room and which wall do the doors belong? What boundaries do they define? These questions in many ways arise as the double boundaries produced by the doors that create a sandwich composed of glass doors and space between. They lay the ground for many of the complex boundary questions posed by the *Investigations*. In this, they multiply possibilities and disallow any single boundary between spaces.

In the dining room, Wittgenstein chose to place translucent glass on the dining room side and clear glass on the hall side. This distinguishes the two spaces even as it connects them, presenting distinct faces as approached from opposite sides. The resulting situation will play a defining role in the *Investigations* where the understanding of language takes on a decidedly spatial slant wherein the approach from one direction is quite different than the approach from another. In the *Investigations*, Wittgenstein equates "knowing" and "not knowing" one's way about a place to our relationship to philosophical problems.

> 123. A philosophical problem has the form: "I don't know my way about". (PI)
>
> 203. Language is a labyrinth of paths. You approach from *one* side and know your way about; you approach the same place from another side and no longer know your way about. (PI)

A similar situation to that in the dining room/terrace/central hall connection occurs with the double doors leading from the salon to Margarethe's private living room. The twist here is that all of the interior doors leading to the salon are metal, so that these doors combine not two types of glass to create the transparent/translucent pairing, but rather glass and metal to form a transparent/opaque combination. As with the dining room, the doors in question occur along a plane that also contains a series of exterior doors opening onto a terrace. The choice of material becomes even more significant in the salon, as it is the only room in the house to possess metal doors. The connection between salon and Margarethe's private living room thus had to navigate a series of three conflicting rules: the continuation of the glass doors along the exterior wall, the placing of the more opaque material on the more private side, and the rule that all interior doors to the salon, and only to the salon, are metal. All of these could not be satisfied at once. Ultimately, Wittgenstein opted to place metal on the salon side and clear glass on Margarethe's private living room side. This allowed the metal doors to remain solely associated with the salon, but it disrupted the series of glass doors along the same wall and left the private living space designated by the more transparent material.

In this and other related design decisions, what initially suggests a singular and repeated image gives way to a multitude of possibilities arising from the specifics of site and the complex demands of use and occupation. The continual differentiation and specification of the doors ultimately yields a series of complex situations whose conflicts add as much to the defining of the nature of these boundaries as any fixed rule. What the analysis of the doors on the main floor points to is how the central hall produces no stasis nor exact repetitions of elements, neither does it execute absolute scripted rules. The spatial complexity of the hall instead multiplies effects and situations, and in so doing enacts philosophical issues as they manifest visual and spatial concepts. These include boundary, correspondence, limits, and what will become known as "family resemblance." It creates a field of interactions through which these concerns emerge repeatedly, a situation that fosters Wittgenstein's encounters with fully wrought spatio-philosophical issues.

The one-to-one correlation of word to meaning treating all language as if it functioned in a restricted way was central to the *Tractatus*'s picture theory and its delineation of absolute and clear limits to language. This thinking cannot help but be challenged by spatial practices that approach an issue from many points. Movement through space destroys singular, fixed images and idealized conditions as they give way to commonplace vicissitudes of movement, space, and inhabitation. This finding—though in many ways very ordinary—is an important product of the design of the central hall, as in the *Investigations* Wittgenstein distinctly associates movement through space with everyday language, in contrast to the reified fixity that results from the *Tractatus*'s collapse of language and logic.

Similar spatial conundrums as those of the glass doors occur throughout the Stonborough-Wittgenstein house with the location and alignment of the floor joints, location of window and door openings on inside and out, and problems of localized symmetry within the overall asymmetric plan, all of which provide spatial forums to engage related linguistic and philosophical concepts as they play out in space. It is this set of problems, introduced to Wittgenstein in the practice of architecture, which in spatializing limits, boundaries, relations, practice, and correspondence, literally demands that Wittgenstein reconsider the functioning of language and philosophy. Their spatial manifestations make clear that spatial concepts cannot be discarded and leave philosophical problems intact. The *Investigations* acknowledges this in regularly turning to related examples to show that their insistent spatiality pervades language.

By the same token, the continual transformation or mutation of the glass doors produces what will emerge as a major theme of the *Investigations*, the repeated locating and relocating of an element as a way of understanding a word or practice. By examining these in a range of positions and in a variety of places, Wittgenstein focuses on the activities he later associated with the formation of a perspicuous representation, one that acts by "'seeing connexions'" and the "finding and inventing [of] *intermediate cases*" (PI § 122). The central hall of the Stonborough-Wittgenstein house—as a distributor of both space and spatial concepts—acts to allow Wittgenstein to engage an unfolding series of such intermediary connections and cases later defined in the *Investigations* as the necessary activity by which philosophy may achieve a clear view. In each of these, view and viewer, as subjects in space, are intertwined. Clarity does not emerge outside of subjectivity, but in relation to it.

In the *Investigations*, concepts corresponding to those in the house appear in a variety of ways, frequently with multiple, even contradictory, meanings or ways of viewing. For Wittgenstein, this meant reconsidering not only the limits within which we place and define our concepts, but also how those limits function. The *Investigations*'s methods arise from the need to define new standards of clarity within the view from within, ultimately introducing an embedded form of perspicuity not previously thought possible. In identical parenthetical comments at the end of two passages discussing the importance of context in understanding language, Wittgenstein tells us this, writing: "A multitude of familiar paths lead off from these words in every direction" (PI § 525; PI § 534). In PI § 426, however, Wittgenstein explicitly employs a spatial terrain to contrast ideal views from the *Tractatus* with ordinary ones: "In the actual use of expressions we make detours, we go by side-roads. We see the straight highway before us, but of course we cannot use it, because it is permanently closed."

In all this, acting as its starting point and model, the practice of architecture is integral to the *Investigations*'s rethinking philosophy. The success of this endeavor yields its own spatial inversion, as what seems to be Wittgenstein's movement outside philosophy into practices such as architecture ultimately serves to form the basis for both his re-engagement with, and re-conceptualization of, the discipline. This reversal underscores the relation between Wittgenstein's philosophy of language and spatial constructs. While his search for visual space within language may initially seem to be outside the philosophy of language, in fact it lies at its core. It is the spatial issues and problems that the *Investigations* offers as an inseparable part of language itself that provide some of the most important and iconic examples of linguistic functioning. Tellingly, these examples are offered in exactly the same way that all meanings of a word, or examples of a practice, are, with each contributing to our understanding of a word, concept, or spatial construct. As Wittgenstein tells us when he offers these examples, he does not offer less than he knows himself. More so, in using the spatial and architectural examples as one of the ways in which he knows language to operate, Wittgenstein, through these discussions and examples, is offering not just what he himself knows, but the ways in which he came to understand given concepts. He is showing us the lens of architecture as integral to his philosophy of language.

References

Last, Nana. (2008) *Wittgenstein's house: language, space and architecture*. Reprint, New York: Fordham University Press, 2012.

Nedo, Michael. (1993) Introduction. In: *Ludwig wittgenstein: wiener ausgabe*. Vienna and New York: Springer-Verlag.

Wittgenstein, Ludwig. (1922) *Tractatus logico-philosophicus*. Trans. C.K. Ogden. Reprint, London: Routledge, 1974.

Wittgenstein, Ludwig. (1953) *Philosophical investigations: the English text of the third edition*. Trans. G.E.M. Anscombe. Reprint, New York: Prentice Hall, 1968.

Chapter 29
Wittgenstein's Torments of the Mind

Ilse Somavilla

29.1 Introduction

Reading Wittgenstein's philosophical manuscripts and personal diaries, one can observe a continuous process of searching for perfection in content and in form—a process leading up to his "torments of the mind".

His obsessive search for the adequate expression—the "redeeming word"—in order to articulate his movements of thought—reveals his notion of philosophizing not only as a way to bring clarity in propositions, but furthermore to lead to a state of peace of mind by having found "the word which finally allows the philosopher to grasp what before—unreachably—has weighed heavily upon our mind" (cf. MS 110, 17).

In this article, I will analyze Wittgenstein's way of writing philosophy according to his high ethical standards in view of the limits of language and of science.

Moreover, I will show how his struggling in view of his ethical standards is not only linked to his notion of philosophy, but also to religion. As he himself noted: "To get rid of the torments of the mind, that is to get rid of religion" (February 21, 1937, DB, 192).

29.2 The Connection Between Life and Work

For decades the reception of Wittgenstein's philosophy was focussed on the linguistic-analytical aspects of his work, as can be seen in numerous articles, books etc.

I. Somavilla (✉)
Universität Innsbruck, Innrain 52, 6020 Innsbruck, Austria
e-mail: Ilse.Somavilla@uibk.ac.at

S. Wuppuluri and N. da Costa (eds.), *WITTGENSTEINIAN (adj.)*,
The Frontiers Collection, https://doi.org/10.1007/978-3-030-27569-3_29

Even today his work is often treated and discussed from an analytical perspective, although other forms of reception have also started to become more common.

Paul Engelmann was one of the first commentators to emphasize the ethical significance of the *Tractatus*, and his *Memories of Wittgenstein* led to a reconsideration in the Wittgenstein literature, i.e. to the awareness of the ethical aspect in Wittgenstein's philosophizing and to a gradual understanding of his personality.

And after the publication of Wittgenstein's so-called *Secret Diaries*, written during the First World War, not only the connection between his life and his philosophy became obvious, but also his approach towards a religious belief.[1]

Finally, when Wittgenstein's diary from the 1930s was discovered, later published under the title *Denkbewegungen*, Wittgenstein's approach toward religious questions as well as the connection between personal and philosophical problems became even more evident.[2]

Although I do not intend to maintain that Wittgenstein's philosophy could be explained from his biography, several personal entries hint at his philosophical movements of thought in a remarkable way. This applies first of all to ethical and religious aspects, which are closely connected with one another, as I will discuss in the following.

However, before taking up this topic, I want to speak about Wittgenstein's concept of "spirit" (Geist) in his writings, while suggesting that the translation "torments of the mind" could perhaps also been translated by "spiritual torments".

Nevertheless, a proper translation for the German term "Geist" is difficult—sometimes "mind" would fit, and in other contexts "spirit" is more appropriate.

29.3 The Significance of "Geist" (Mind/Spirit) for Wittgenstein

This concept first appears in Wittgenstein's coded diaries written during the First World War, and it has to be understood in light of Leo Tolstoy's *Gospel in Brief*, which played a decisive role in Wittgenstein's approach towards a religious belief.

Again and again he turns to the "spirit" in a passionate tone—on the one hand, in his search for illumination in his philosophical movements of thought, and on the other, with regard to moral problems, which above all, is connected with his struggle to conquer his own sensual drives. Besides, in his devotion to the spiritual, he finds inner strength to endure the cruel atmosphere of the war:

"In times of outer wellbeing we do not think about the weakness of the flesh; in times of distress, though, we become aware of it. And we turn to the spirit. – " (MS 101, 33v; 5.10.14).

---

[1]Cf. Wittgenstein (1990).

[2]Cf. Wittgenstein (1997).

In order to fight sensuality viz. the weakness of the flesh, Tolstoy's booklet meant a great help to him; in fact, it literally helped him to survive, as he wrote to Ludwig von Ficker.[3]

In his coded diary entries Wittgenstein notes: "Again and again I tell myself the words of Tolstoy: 'Man is weak in the flesh, but free in the spirit…' may the spirit be inside of me!" (MS 101, 21v, 12.9.1914).

For Wittgenstein, the strive toward inner freedom was essential: Freedom from the compulsion of sensual drives and wishes, also from sensations like fear, anger and despair. In order to remain alive whilst enduring the brutality of the war, he needed more than ever the inner poise and strength that his devotion to the spiritual could provide. The spirit enables him to stand beyond the mundane, to feel free amidst the vessels of the war: "I am spirit & therefore I am free", he notes on October 13, 1914 in MS 101.

And on December 13, 1914: "May only the spirit live! It is the safe harbour protected from the endless gray sea of happenings …" (MS 102, 44v).

In his striving for spirituality and struggle with sensuality, he meticulously reports on the progress or stagnation of his philosophical work—in opposition to progress viz. improvement or failure in his moral endeavours. Notes like "Little worked", "Worked a lot", "Not worked" etc. alternate with notes like "Sensual", "Very sensual", "Again sensual" etc.

Wittgenstein considers it his "duty" to "dedicate himself whole and completely to a life in the spirit" (cf. MS 101, 35v; 7.10.14). He regards spiritual work as blessing: "Remember how great is the blessing of work!" he notes, or, "The blessing of work" (MS 102, 72v).

After having read Nietzsche's Volume 8, he is moved by Nietzsche's animosity toward Christianity, but also writes about "some truth" in his writings and concludes:

> Certainly, Christianity is the only path to happiness. But what, if someone disdains this happiness?! Might it not be better, to die unhappy, in a hopeless fight against the outer world? However, such a life has no meaning. But why not lead a meaningless life? Is it unworthy? Is it compatible with a strict solipsistic viewpoint? What do I have to do, so that I don't lose my life? I have always to be conscious of the mind (MS 102, 39vf., 8.12.14).

Thus, in the Wartime Diaries, the spirit is, in many ways, of invaluable significance to Wittgenstein: in philosophizing as part of the process of writing, in ethical-moral matters over and against sensuality and sinfulness, as well as in supporting and strengthening his belief in a personal God, whom Wittgenstein—presumably also under the influence of Tolstoy—identifies with the "spirit".

To live in sensuality and sin, means to be unhappy, unreasonable, and to be separated from the good and right life—i.e. from a life in the spirit.

Therefore, Wittgenstein alternatively calls upon "God" or the "Spirit" for help and "illumination"—the close connection between philosophical and religious significance is obvious.

[3]Cf. Wittgenstein's letter to Ludwig von Ficker (July 24, 1915), in *Briefe an Ludwig von Ficker*, 28.

The spirit primarily stands for reason—with regard to philosophy it stands for "Ratio" necessary for epistemology, and with regard to his personality it stands for reason and insight in order to lead a life of virtue, connected with the notion of God as the symbol for the good and ethical.

In Tolstoy's booklet "Reflections about God", one finds a similar emphasis on reason: God is seen as the primary source of reason; therefore, God is addressed to as "God-Spirit", not als "Godfather"; Tolstoy even goes so far as to negate the notion of God-Creator.

It is in his fear of the dangers surrounding him—in fact, in face of the constant threat of death—that Wittgenstein searches for comfort in the spiritual as well as in his confidence in God's plans: "Live only for your spirit! And leave everything to God!", he notes on November 29 in 1914 (MS 102, 33v), or: "May the spirit protect me whatever happens" (MS 102, 34v; 1.12.14).

Writing philosophy is a blessing for him.

However, the spirit cannot only be identified with his philosophical work. There are days when he does not get on in his work, yet the spirit stays with him, which is of great importance to Wittgenstein. Whereas he compares his philosophical reflections with a flame that has expired, he notes: "But my spirit is alive. I think…" (MS 102, 50vf.; 13.1.15).

Years later, suffering again of personal problems—this time due to his unhappy love with Marguerite Respinger—he tries to find comfort in the spiritual—in philosophy, especially in logic, which, he notes, makes him feel "at home", "in the warmth" (cf. DB/MT, 51). Eventually, though, this "spiritual warmth" is no longer enough for him, and thus he notes:

> It is true that one may be able to live also on the fields of rubble from the houses in which one was once accustomed to live. But it is difficult. One had derived one's joy from the warmth & coziness of the rooms, after all, even if one didn't know it. But now, as one wanders aimlessly on the rubble, one knows it.
>
> One knows that only the mind can provide warmth now & that one is not at all accustomed to being warmed by the mind. (When one is chilled, it hurts to wash & when one is sick in the mind it hurts to think.) (DB/MT, 18f.).

Aside from personal entries, the spirit/mind continues to be of decisive significance for Wittgenstein and—as is the case in the Age of Enlightement—is identified with "light" in the sense of reason viz. "ratio"—a metaphor Wittgenstein frequently uses for his philosophizing. "The idea working its way towards the light" (MS 131, 19; 11.8.1946, C&V, 54e).

Here again, a continuous struggle can be observed—be it in order to grasp a philosophical problem, or the adequate expression or word. While usually a painful process, at times it also fills him with happiness in his search for philosophical truth. All too often he finds himself in the dark, without escape to the light that only the spirit could offer him. Thus he notes in 1931: "Mind, don't abandon me! That is, may the weak flame of the spirit lamp that is my mind not expire!" (DB/MT, 122).

And he increasingly comes to see this light as given from above, i.e. from a God who still means the ethical for him. Thus it becomes obvious that Wittgenstein's preoccupation with philosophical problems is closely connected with religious questions, and that he suffers from both—the impossibility to come to a solution, as well as his dilemma between philosophical scepsis and the search for religious certainty.

During his longer stay in Norway in 1937, he suffered due to the darkness of winter, and when the first signs of spring gradually came closer, every hour of sunlight appeared as something precious to him. Light and darkness turn to a metaphor for his inner state—for his struggle for a religious belief. On March 23, 1937, he notes:

> Help & illuminate! But if I believe something tomorrow that I don't believe today, I was not therefore in an error today. For this 'believing' is not holding an opinion, after all! But my belief tomorrow can be lighter (or darker) than my belief today. Help & illuminate! & would that no darkness descend upon me! (DB/MT, 223)

The day after, he notes: "I am petitioning & I already have it as I want to have it: namely half heaven, half hell!" (DB, 223).

"Half heaven, half hell"—these words reveal the state of Wittgenstein's philosophizing and his struggle with respect to ethical-religious issues. It is a situation that leads him several times to the verge of madness, to a "terrible abyss", as he describes it.

He draws a comparison between being alone with himself—or with God—and being confronted with a wild animal that could attack him at any moment (cf. DB/MT, April 17, 1937).

God as a threat to him and reason for his pain and suffering,—God as source of light, support and strength in times of distress,—Wittgenstein's attitude is ambiguous, even contradictory, and his access to a religious belief reveals his lifelong inner turmoil.

Apart from his ever-changing and diverging remarks about religious questions in the Nachlass, there are two texts in particular that show the contradictory notions of God he held: a sheet of notes obviously cut from a diary (dated 1922) and a letter-fragment written in 1925, presumably addressed to his sister Hermine or to his brother Paul.[4]

In the diary-sheet he describes his enormous fear of God's demands that he feels compelled to obey. Were he not to obey, he would be for ever lost. This text is reminiscent of the Old Testament, where Abraham is tempted and willing to sacrifice his only son. And Wittgenstein also seems to allude to Kierkegaard's preoccupation with that topic and consequent discussion of the problem of the paradox of religious belief consisting in the contradiction between absolute love for God and ethical commitment to a human being.[5] In fact, Wittgenstein explicitly refers to Kiekegaard while confessing that his own state is "fear and trembling" (cf. LS, 21).

Until the end of his life, Wittgenstein was haunted by the fear of a frightening "Supreme Judge". On March 15, 1951, about a month and a half before his death, he noted:

---

[4]Cf. Wittgenstein (2004).

[5]Cf. Kierkegaard (1954).

"God may say to me: 'I am judging you out of your own mouth. You have shuddered with disgust at your own actions when you have seen them in other people.'" (C&V, 99e)

Wittgenstein's meticulous style of writing, and his striving for utter truthfulness and perfection, seem to be a result of his notion of a fearful judge.

In his book *Logic and Sin in the Writings of Ludwig Wittgenstein*, Philip Shields examines Wittgenstein's writings through the lens of the connection between logic and sin, and he concludes that Wittgenstein's philosophical remarks are conceived in accordance with his ethical and religious convictions. Though Shields does make some interesting points, he tends to exaggerate in so far as he tries to explain more or less every single statement and proposition Wittgenstein made—his grammar, syntax and so on—in terms of his moral-religious perspectives.

In contrast to Wittgenstein's apparently Hebrew notion of the punishing God of the Old Testament and therefore his source of fear, the other text mentioned above, displays a completely different idea of God and religion. Here religiousness is identified with actual spirituality—with the "pure" and "white light". In fact, Wittgenstein goes as far as to maintain that in view of this light viz. religious ideal, all cultural achievements appear merely as tinted lights and that only in preoccupation with religious questions man is able to create works of genius. And these works are moving and timeless: "The preoccupation with the spirit, with the light, is what seizes" [„Die Auseinandersetzung mit dem Geist, mit dem Licht, ergreift"] (LS, 45).

Thus, again, light stands for the spirit. However, while in the Age of Enlightement, light is a metaphor for "ratio", viz. reason or intellect, for Wittgenstein the concepts "spirit" and "light" above all stand for spirituality and religiousness—in contrast to secularity, art and science.

29.4 The Significance of Ethics for Wittgenstein's Life and Philosophizing

To begin with, I want to point out a remark by Wittgenstein that demonstrates the connection between ethics and religion—a connection he emphasized many years earlier, in conversation with his sister Hermine.[6]

"What is Good is Divine too. That, strangely enough, sums up my ethics" (10.11.1929, C&V, 5e).

[6]Cf. Hermine Wittgenstein's note in her diary from autumn 1917: "Ludwig sagt: Religion u. Ethik hängen absolut zusammen. Wäre Religion etwas anderes als die Begründung der Ethik, etwas das allein existiert, das man haben kann oder auch nicht, so würde es sich nicht lohnen darüber zu reden. Ich habe ein Gewissen, ich weiss was ich an mir für gut oder schlecht halte, aber ich habe keine Religion. Ich weiss auch nicht warum ich etwas an mir für gut oder schlecht halte und ich frage nicht danach weil diese Frage nicht in mir liegt.

Nach Ludwig's Meinung muss Jeder der diese Frage oder die Frage nach dem Sinn u. Zweck des Lebens stellt, auf Begriffe wie Gott, göttlich etc. kommen" (see "Aufzeichnungen von Hermine Wittgenstein", Nachlass Margaret Stonborough-Wittgenstein, Brenner-Archiv).

Numerous further remarks in Wittgenstein's Nachlass—of personal and philosophical nature—show how closely ethical and religious questions are linked together. The remark quoted above is relevant for his personal way of life. He tried to live according to very high ethical standards, all the while searching for answers to religious questions. Above all, his *Lecture on Ethics* is an excellent example of the interconnection between these two topoi, and at the same time, a kind of personal avowal of his approach towards ethics.

29.4.1 The Lecture on Ethics

Contrary to his refusal to discuss ethical and religious issues in philosophy, in the *Lecture on Ethics*, Wittgenstein dared to speak about this problematic topic—even if only by using examples from his personal experiences, thus from a totally subjective viewpoint.

His choice of topic as well as his choice of words—his use of language—was conceived for a so-called broader audience (and not for a specific philosophical one); his "tone" differs from the tone of his philosophical manuscripts and from his personal diaries.

Thus he manages to present philosophical problems of ethics by means of vivid examples involving his own life and personal experiences via a narrative style of speech. He succeeds in conveying his great concern—ethics—in a hermeneutic way, that is, by evoking a real praxis of language that he considered essential for his notion of ethics.

For him, this was only possible by way of presenting his own personal situation, in other words, by stepping forward as an individual and, as previously mentioned, by speaking from a subjective viewpoint,—by admitting, so to speak, his personal and philosophical dilemma—that is, his suffering in view of the limits of language.

Ethical problems are discussed in immediate connection with language problems—the preoccupation with ethics turns into a preoccupation with language, i.e. with our incorrect use of language, our misuse of words. In order to exemplify these problems, Wittgenstein presents examples from everyday life, and similes to seize the audience in a direct way.

In posing the question as to the nature of ethics, he differentiates between relative and absolute value by means of three personal experiences: As examples of absolute value and thus for *his* understanding of ethics, he mentions the feeling of absolute safety, the feeling of guilt and the experience of wonder at the existence of the world, while the latter he considers as his first and foremost example—his "experience par excellence".

In order to explain the difference between relative and absolute value, Wittgenstein offers the example of safety: whereas the sentence "I feel safe in the house when it is raining outside", is understandable in everyday language and thus an example of relative value, the sentence "I feel safe whatever happens" sounds nonsensical and is thus an example of absolute value. For Wittgenstein, nonsensicality is characteristic of all ethical and religious expressions, because they transcend the world of facts and thus the limits of language. The same applies to the example of wonder: to be astonished at something extraordinary is an example of relative value and completely different to the example of absolute value, expressed in the proposition "I wonder at the existence of the world". For, to be astonished at the very existence of the world, implies an ethical attitude and corresponds to Wittgenstein's remark in the *Tractatus* 6.44, "Not *how* the world is, is the mystical, but *that* it is".

Moreover, all three examples are of a religious nature, and express Wittgenstein's appreciation for the "supernatural", that is, that which cannot be expressed in ordinary words. To his conviction, there are no propositions capable of expressing the "higher" and thus there are no propositions for ethics (cf. TLP, 6.42).

29.4.2 Ethical Aspects in Wittgenstein's Writings

Wittgenstein's pursuit of absolute decency ["Anständigkeit"], as well as his striving for purity and perfection effects his use of language and his style of writing, and this interconnection has to be seen as a process of personal development and self-realization.

From the very beginning of his philosophical work, his style of writing accords with the high ethical standards he set both in content and in form. Regarding content, he distanced himself from merely babbling ["schwefeln"] about matters transcending the world of facts and thus scientific explanation; regarding form, he wrote according to Wilhelm Ockham's expression "Simplex sigillum veri", or—to quote Wittgenstein's remark about style: a style *sub specie aeterinitatis*, which he demanded of any style—be it writing or building style, as "general human necessity" (cf. DB, 28). For Wittgenstein, this meant to restrict himself to a minimum of words and precise formulation, to avoid empty phrases etc. And this is tied to his striving for truthfulness and to his condemnation of vanity. Years later, in 1936, he wrote in his diary:

"My work (my philosophical work) is also lacking in seriousness & love of thruth.—Just as in my lectures I have also cheated often by pretending to already understand something while I was still hoping that it would become clear to me" (DB/MT, 145).

Two days before he noted: "I always take joy in my own good similes; were it not such vain joy" (ibid).

His fight against vanity took him as far as to say that he did not want to flee madness, for madness was the most severe judge of whether his life was right or wrong (cf. DB/MT, 185).

To this extent, he considered madness the "guardian of vanity", for it was via his close proximity to it that he came to realize his personal inferiority, his nothingness—a possible avenue to attain modesty. He considered painful self-introspection absolutely necessary in order to write truthfully. At the same time, this way of self-recognition leads to more clarity in thought—clarity about himself, as well as with regard to philosphical problems. As he had already written in the *Tractatus*:

"The purpose of philosophy is the logical clarification of thoughts. Philosophy is not a theory but an activity. The result of philosophy is not a number of 'philosophical propositions', but to make propositions clear. Philosophy should make clear and delimit sharply the thoughts which otherwise are, as it were, opaque and blurred" (TLP, 4.112).

29.5 Self-recognition, Movement and Change

Self-recognition means the readiness to be willing to change—a most important aspect of Wittgenstein's life and philosophizing. His tendency to continuously change things in order to become better or even perfect, can be observed throughout his writings.

He desperately fights to move forward in his process of philosophizing and thus alters, again and again, not only expressions and formulations, but also his position viz. view of a problem. In his passionate search for finding the adequate expression, he again and again rejects or deletes what he has just written and tries to find ever new ways and possibilities.

"In philosophizing it is important for me to keep changing my position, not to stand too long on *one* leg, so as not to get stiff. Like someone on a long up-hill climb who walks backwards for a while to revive himself, stretch some different muscles" (MS 118, 56r, 4.9.1937; C&V, 32e).

As Friedrich Waismann recalls, Wittgenstein never rested on the same point, but constantly resumed anew his thoughts which he revised and re-revised in order to gain new insights (cf. WVC, 26). This particular trait in Wittgenstein's philosophizing finds its expression above all in his examples of "aspect-seeing" in his later writings. He examines the objects of his inquiry from ever new and changing perspectives, thus discovering (via so-called "changes of aspects") new features of the objects of which he had not been previously aware. His tendency to continuously change and rearrange his manuscripts in pursuance of his ethical and aesthetic ideals, can also be seen from a religious point of view. As he himself remarked:

"Christianity says: Here (in this world)—so to speak—you should not be sitting but going. [...] The question is: How do you go through this life?—(Or: Let this be your question!)—Since my work, for example, is only a sitting in the world, after all. But I am supposed to go & not just sit." (DB/MT, 207f.)

"To go" means an untiring process of searching for new insights in philosophy—for clarity, for truthfulness. And this process involves pain, as we see in Wittgenstein's writings, where he searches, almost obsessed, for the right or ideal expression. His pursuit exhibits a restlessness that often led him to the feeling of having failed. The same applies to his life, i.e. to his striving for perfection which he tries to attain by changing his way of life—similar to a process of catharsis. As early as in 1912, he wrote to Bertrand Russell:

> Whenever I have time I now read James's "Varieties of Religious Experience". This book does me a lot of good. I don't mean to say that I will be a saint soon, but I am not sure that it does not improve me very much: namely I think that it helps me to get rid of the Sorge (in the sense in which Goethe used the word in the 2nd part of Faust). [...] (CL, 14).

And around Christmas in 1913, he wrote to Russell:

> Here I feel different every day. Sometimes things inside me are in such a ferment that I think I'm going mad: then the next day I am totally apathetic again. But deep inside me there's a perpetual seething, like the bottom of a geyser, and I keep on hoping that things will come to an eruption once and for all, so that I can turn into a different person.
>
> I can't write you anything about logic today. Perhaps you regard this thinking about myself as a waste of time—but can I be a logician before I'm a human being! Far the most important thing is to settle accounts with myself!" (mit mir selbst in's Reine kommen) (CL, 66).

In order to come to terms with himself, he felt it necessary to plump the depths of his personality with brutal frankness in order to face his deficiencies and weaknesses. Only then he would be able to create something genuine, something borne of modesty and humility; this would stand in stark contrast to the pretending and deceiving of others and himself, tied to vanity. "Self-recognition & humility is one. (These are cheap remarks.)", he noted in 1931 (DB/MT, 97).

Radical self-introspection by confronting and admitting his deficiencies and sinfulness was the only way to escape vanity and thus attain salvation. Only then he would conquer vanity, regardless of how agonizing it might be.

In connection with his remarks about madness, Wittgenstein's torments of the mind become particularly obvious and they reveal his thoughts about intellectual vanity and the glorification of science—similar to Lenau's *Faust*. Apart from the danger of isolation, intellectual vanity is bound to lead to madness, as Weininger saw it in connection with logical thought.[7]

In these torments of the mind, one can see just how intimately connected they are with religiosity. After the remark "To get rid of the torments of the mind, that is to get rid of religion" (DB/MT, 192), Wittgenstein notes in code: "Have you not been somehow tormented in your whole life (just not in this way) & would you rather return to those torments now?!"

It seems as if his personal and wordly sufferings have now been transferred into spiritual-religious sufferings: the following entries circle around his relationship to God, more precisely, to Jesus Christ—always in connection with an ethical way of life and harboring an intensity rarely matched. In his efforts to meet his ethical standard of perfection, he notes:

> What I believe now: I believe that I should not fear people and their opinions when I want to do what I consider right.
>
> I believe that I should not lie; that I should be good to people; that I should see myself as I really am; that I should sacrifice my comfort when something higher is at stake; That I should be cheerful in a good way when it is granted to me and when not, that I should then endure the gloom with patience & steadfastness; that the condition which demands everything from me is not taken care of by the words "sickness" or "madness", that is: that in this condition

[7]Cf. Weininger (1918, 125).

> I am just as responsible as out of it, that it belongs to my life like any other and that it thus deserves full attention. I don't have a belief in a salvation through the death of Christ; or at least not yet. I also don't feel that I am on the way to such a belief, but I consider it possible that one day I will understand here of which I understand nothing now; which means nothing to me now & that I will then have a belief that I don't have now. (DB/MT, 192–194).

The connection between Jesus Christ as the personification of the ethical (cf. WVC, 118) and madness as the most severe judge about right and wrong seems evident.

Both are the origin of Wittgenstein's sufferings—on the one hand with respect to his personality, on the other, with respect to his philosophizing as a rational sceptic with all his doubts. Moreover, he becomes aware that intellectual vanity and mere addiction to science would result in total isolation (as Lenau describes in his *Faust*). As late as in 1946, Wittgenstein refers to this issue:

> I often fear madness. Have I any reason to assume that this fear does not spring from, so to speak, an optical illusion: of seeing something as an abyss that is close by when it isn't? The only experience I know of that speaks for it not being an illusion, is the case of Lenau. For in his Faust there are thoughts of a kind I too am familiar with. Lenau puts them into Faust's mouth, but they are no doubt his own about himself. What is important is what Faust says about his loneliness or isolation. (C&V, 61e).

Reading Lenau's Faust, the similarity to Wittgenstein's situation, i.e. to his thoughts and feelings—his wavering between rational considerations verging on intellectual vanity and, at the same time, being aware of his worthlessness—is obvious. In contrast to Faust, though, Wittgenstein is willing to admit and accept his inferiority, regardless of how painful it might be. For only then would he attain the kind of modesty necessary for his inner freedom, that is, the prerequisite for truthfulness in his writing as well as for a religious belief through the very consciousness of sinfulness.

"Genuine modesty is a religious matter", he notes on October 18, 1930 (DB/MT, 52).

Similarly to his time during the First World War, when Wittgenstein, in the face of death, came to realize the meaning of life, his proximity of madness in the 1930s seems to have given his life a meaning, insofar as he came to realize the difference between a right and wrong way of life. And, just as—when reflecting on the happy and thus ethical good life—he declared the fear of death as the best sign of a wrong, i.e. a bad life (cf. NB, 8.7.16), he now does not want to flee madness. For, only by confronting it, can he perceive himself fully. Just like through his preoccupation with religion, the value of which he sees in the recognition of his personal depravity, as well:

"People are religious to the extent that they believe themselves to be not so much *imperfect but sick*" (C&V, 51e).

Insofar the attempt to flee from madness would mean disgraceful behaviour, because to feel wretched is the prerequisite of religiousness. The consciousness of his failure to live up to the ideal he has in mind, can only result either in madness or lead to a total change with regard to an ethical way of life.

> To have an ideal is alright. But how difficult not to playact one's ideal. Instead to see it at that distance from oneself at which it is! Yes, is this even possible—or would one either have to become good or go mad over it? Wouldn't this tension, if it were fully grasped either open the person to everything or destroy him. Is it a way out here to cast oneself into the arms of grace? (DB/MT, 155).

To take refuge in the "arms of grace" as an alternative to inner destruction means to expose oneself in an attitude of humility, i.e. to conquer the pride and vanity that are part of madness, because the proud person cannot accept his inferiority and will thus go mad. Conversely, insight and acceptance of one's imperfection would lead man to modesty and piousness, and thus to the attainment of mental sanity. Here the religious aspect plays a decisive role, as can be seen from various remarks in the *Movements of Thought*, where Wittgenstein speaks about the storm of outrage the sinful man would feel when confronted with the perfection of Jesus Christ. To accept this contrast between him and Jesus Christ not only means to take refuge in the arms of grace, but it furthermore requires to believe in resurrection. However, neither the sceptic(al) nor the proud person will be able to believe. It is in this sense that Wittgenstein's remark, "for only religion could destroy vanity" (C&V, 54e) has to be understood, not to mention his conviction that the Christian faith is "only for the one who needs infinite help, only for the one who suffers infinite distress" (C&V, 52e).[8]

Therefore, Christianity is not a doctrine, "not a theory about what has happened & will happen to the human soul, but a description of something that actually takes place in human life. For 'recognition of sin' is an actual occurrence & so is despair & so is redemption through faith. Those who speak of it (like Bunyan), are simply describing what has happened in them; whatever gloss someone may want to put on it!" (4.9.37, C&V, 32e)

In the same year, roughly seven months earlier, Wittgenstein wrote:

"A religious question is either a question of life or it is (empty) chatter. This language game—one could say—gets played only with questions of life. Much like the word 'ouch' does not have any meaning—except as a scream of pain" (DB/MT, 203).

Thus, just as Wittgenstein defined philosophy as an activity—not as theory—, and ethics can only be shown by one's way of life, religiousness can only be experienced as an inner process, viz. in a state of pain.

29.6 The Method of Philosophy as a Method of Madness

Not only with regard to his personality, but also with regard to his philosophy, Wittgenstein considered madness an important theme. In fact, he declared the method of philosophy to be a method of madness in order to heal madness.

---

[8]Cf., too: "Religion is as it were the calm sea bottom at its deepest, remaining calm however high the waves rise on the surface" (16.10.1946, C&V, 61e).

This of course applies to the limits of language again; more precisely, it applies to our misuse of language, to the "misunderstanding of our logic of language" (MS 104, 119), and thus to the origin of philosophical problems. Therefore we have to treat the problems in a new way; they need to be approached from the bottom, at their roots, and not be treated on the surface. This means calling into question the apparently "normal" and unimportant issues in a way that would appear as mad to men of so-called common sense.

However, "the task of philosophy is to soothe the mind about meaningless questions. Whoever doesn't tend to such questions doesn't need philosophy" (DB/MT, 65).

Only if one thinks crazier than the philosophers, one can solve their problems (cf. MS 167, 19). Therefore the "method of philosophy is to drive oneself mad, and then to heal again one's madness" (cf. MS 109, 84).

29.7 The Limits of Language: Saying and Showing Viz. the Dimension of Silence

"It is strange what relief it is for me to write something in a secret script what I would not like to be written in a legible way"(MS 106,4; 1929).

Although Wittgenstein—apart from the *Lecture on Ethics*—avoided treating ethical and religious questions in his philosophical discussions, he nevertheless could not resist immersing himself in these issues, and his personal diaries reveal how he desperately sought to find other ways of treating them.

By encoding his reflections concerning ethics and religion—so to speak, by hiding or disguising them—, he saw the possibility of separating them from the realm of the sayable and thus from what can be grasped by scientific propositions.

For, as he noted: "There is a big difference between the effects of a script that is easily & fluently legible, & one which one can write but not easily decipher. In it one locks one's thoughts as in a jewelry box"(MS 157a, 59r).

This remark also suggests that the encoded notes meant something precious to him, something he wanted to protect from the superficial reader.

It was above all during the First World War that Wittgenstein started to put his philosophical reflections in normal writing on the right-hand pages and his personal movements of thought in coded script on the left-hand pages. Even though style and tone differ from one another—here speaks the sober arguing philosopher, there an irritated man—there are nonetheless similarities.

Wittgenstein's well-known distinction between saying and showing, along with his awareness of the limits of the sayable, which finds its expression in the famous final sentence of the *Tractatus*, already appears in the coded part on July 7, 1916: "What cannot be said, cannot be said!"

In the philosophical part this thought first turns up in a slightly altered form on the 15th of October in 1916: "What one cannot think, one cannot speak about".[9]

Wittgenstein's endeavours to express the inexpressible, his search for the adequate word in order to cast light on philosophical problems, can be observed throughout his writings—as a sometimes desperate process, as a struggle with language, its possibilities as well as its limits.

His painful creative process—in the coded part with respect to the ethical, in the philosophical part written in normal writing with respect to philosophical, mainly analytical problems—, becomes visible in the concept of the "redeeming word". This concept first emerges in the coded part, however, in reference to philosophical problems. On November 21, 1914, he notes in MS 102, 23v: "Rather worked. Still I cannot express the one redeeming word. I circle around it and quite close but I have not been able to grasp it yet!!"

The next day, he writes in the philosophical part that he tried to express something "which cannot be expressed". On the same day, he notes in the coded part: "Not spoken the redeeming word. Yesterday it was laying on my tongue for a while. Then again it disappears…" (MS 102, 24v; 22.11.2014).

In the philosophical part the concept of the redeeming word first appears on January 20 in 1915 and in the form of a question after a series of logical movements of thought: "The redeeming word -?!" On June 3, 1915, while preoccupied with the tautology of propositions, he remarks in brackets: "The redeeming word is by the way not spoken".

This expression would turn up again in later years and it has to be seen in connection with his search for solving philosophical problems, which is, at the same time, also a kind of "salvation" in his desperate efforts for the right expression, the adequate "redeeming" word. Still, if one considers the concept in the coded remarks of his diaries, where Wittgenstein alternatively calls upon "God" or the "Spirit" for enlightenment, the religious connotation concerning his personal situation cannot be denied. This applies to a number of further aspects originating in his personal experience and thus being treated first in the coded part.

Even though this painstaking process was necessary for Wittgenstein's life, it was also accompanied by suffering. Sometimes he had the feeling that his mind was like "a glass-rod apt to break any moment" (cf. (DB/MT, 142), or, that he "feared his brain would no longer be able to sustain its mental strain", as he noted on April 4, 1930 (cf. DB/MT, 4).

Despite his often painful process of philosophizing, Wittgenstein experienced joyous moments, as well: "The joy I have in my thoughts (philosophical thoughts) is joy in my own strange life. Is that the joy of life?" (DB/MT, 108)—a remark which, just like several other remarks, also occurs in other manuscripts (cf. C&V, 20).

Another example revealing the joy he finds in philosophizing and sign of creative activity, is the following: "O, why do I feel as if I wrote a poem when writing phi-

[9]Cf. TLP, 4.116: "Everything that can be thought at all can be thought clearly, everything that can be said, can be said clearly". Cf. also TLP, 5.61: "What we cannot think, that we cannot think: we cannot therefore *say* what we cannot think."

losophy? It is here, as if there was something small which has a wonderful meaning. Like a leaf, or a flower" (MS 133, 1946).

This remark is written amidst philosophical discourse and stands out as something quite different from his examples of logical investigations.—Different not only because it is written in code, but also with regard to its tone and wording. It seems to interrupt his strict philosophical argumentation, as if Wittgenstein suddenly paused for a moment to muse in a poetic manner, before again taking up his philosophical discourse. Immediately afterwards, he writes: "The craving for explanation is a hindrance to the full apprehension of facts, i.e. the description. The hypothesis is like a sieve which allows to see us only a small part of the facts."

Thus, what is essential, remains hidden; it is unattainable via hypothesis, viz. a scientific theory. Still, Wittgenstein finds joy in the process of writing down his philosophical movements of thought—just as he finds joy in looking at a flower or a leaf while being content with its impression instead of searching for analytical explanation.

29.8 Ratio–Reason–Explanation Versus Intuitive Approach

The dichotomy between striving for rational explanation and intuitive approach could be seen as paralleling Wittgenstein's wavering between philosophical scepticism and search for religious certainty.

As early as in the *Tractatus*, one can observe a mystical rather than analytical and scientific approach towards religion—an approach leading to silence instead of metaphysical speculations. Only in a few cryptic remarks does Wittgenstein dare to approach to the sphere outside the world of facts—remarks frequently labelled in the reception as mystic-pantheistic, presumably influenced by Schopenhauer and Spinoza.

Wittgenstein's verdict of silence with respect to ethics and religion is still relevant in later years. In his conversations with members of the Viennese Circle, he clearly maintains his position, when he—despite his rejection of metaphysics within philosophy—emphasizes the significance of ethical and religious problems, regarding them as something essential and precious, about which one need not waste words (cf. WVC, 117).

His attitude towards rational grounds [Begründung] is especially radical: While discussing with Moritz Schlick the essence of the good according to the two conceptions of theological ethics, Wittgenstein maintains that the proposition "The good is good because it is what God wants" is more profound, because it cuts off, from the very beginning, any attempt at explanation or offering reasons. In contrast, Wittgenstein holds the sentence "God wants the good because it is good"—considered the more profounder interpretation in theological ethics—to be shallow and rationalistic, because it assumes one could provide reasons for what is good. Wittgenstein contends that "What God demands, that is good" for more profound, because this conception says clearly that the essence of the good has nothing to do with facts and hence cannot be explained by any proposition (cf. WVC, 115).

In the *Movements of Thoughts*, Wittgenstein again emphasizes his view on this point: "'It is good because God has thus commanded' is the right expression for groundlessness" (DB/MT, 75).

Wittgenstein was not only against providing reasons for religious or ethical questions, but also against any kind of theory about these issues. Even if the theory were true, it would never be what he was looking for. The ethical cannot be taught. If he could explain the essence of the ethical only by means of a theory, then what is ethical would be of no value whatsoever (cf. WVC, 116f.).

29.9 "Other" Ways Toward the Ineffable: The Significance of Literature and Art

Aside from an ethical and religious way of life, Wittgenstein appreciated literature and art as further means to convey an understanding of ethical and religious problems. Above all, he considered Tolstoy and Dostoevsky as authors "who really had something important to say about religion", as they did not theorize it (cf. his remark to Drury in Rhees, 101).

However, it was a play written by Ludwig Anzengruber that revealed the significance of religion to him for the first time. In the play *Die Kreuzelschreiber*, the protagonist has a mystical experience in the surroundings of nature and the feeling of being absolutely safe in the hands of God regardless of what might happen. Wittgenstein, at that time about 21 years old, was so impressed that he later said to Norman Malcolm that he now was able to realize the possibility of religion (cf. Malcolm 1984, 58).

And, in his *Lecture on Ethics* held years later, he presents the feeling of absolute safety as one of the key-examples for his understanding of ethics, as already mentioned.

Dostoevsky was one of Wittgenstein's most favourite authors, and he was particularly taken with the novel *The Brothers Karamasov*. In this novel, according to Wittgenstein, ethical and religious problems are presented in a way much more touching and illuminating than any abstract theories. In his reflections on the justification of a happy life, Wittgenstein explicitly refers to Dostoevsky, who saw the happy life as the "purpose of existence" (NB, 6.7.1916).

This happy life is guaranteed by a life in the spirit—in knowledge—a life renouncing of the so-called "amenities of life", which Wittgenstein considers merely as "so many graces of fate" (NB, 13.8.16).

The idea of giving himself over to fate and accepting everything in a stoic manner, turns up again 20 years later: In his reflections on God, Wittgenstein writes that if one wants to quarrel with God, one has a false concept of God, i.e. one would be in a state of superstition. Thus he has to alter his concepts and regard "contentment with his fate" as the first command of wisdom (cf. DB/MT, 217).

However, Wittgenstein himself is not quite clear about the use of the word God. Even though he would agree with many passages of the Bible, he could not share the views of pious men. However, he is not able to directly fight this part of his thinking, but only try to behave in a decent—that is, not in mean, cowardly, ungood—way. If he were to succeed in doing so, his way of thinking and his use of the word God might also change.

As far as Kierkegaard is concerned, I refer to Wittgenstein's Hebrew notion of God as a frightening Judge, who could demand from him the most dreadful thing at any moment. This becomes obvious not only in the diary sheet from 1922, mentioned earlier, but also in the 1930s when Wittgenstein saw himself compelled to obey God's demands, even if it meant burning his manuscripts (cf. DB/MT, 178f.).

Moreover, Kierkegaard's radical notion of a truly Christian life with regard to a total change as well as the idea of "the leap of faith", plays a role in Wittgenstein's thought.

Just like Kierkegaard, he is convinced that "this life with all sorts of pleasure & pain is nothing, after all! It must strive toward the absolute. And the only absolute is, to battle through life toward death, like a fighting, a charging soldier. Everything else is wavering, cowardice, sloth, thus wretchedness." (DB/MT, 188). He regards the striving toward the absolute – instead of turning our gaze "level, toward the things"–, as something "glorious, sublime", but he himself admits that his gaze often turns toward the wordly things.

29.10 Art and Religiosity

Not only literature, but any form of art may help man to attain a better understanding of religious issues. Wittgenstein refers to architecture als a "gesture of eternity", when viewed *sub specie aeternitatis*, or to music, where again he emphasizes truthfulness as a prerequiste of genius.

Georg Trakl and Ludwig von Beethoven belonged to the persons he most appreciated. Trakl, who, much like Wittgenstein, suffered in his painful search for perfection in his creative process, noted: "Your poem an imperfect redemption".[10] Thus, the tone of his poems struck Wittgenstein—a tone that filled him with happiness and that he regarded as the tone of true genius.[11]

Wittgenstein considered Beethoven's music to be "totally true", "totally religion & not religious poetry. Therefore, "Beethoven can console in real pain, while others fail. He doesn't lull one into a beautiful dream but redeems the world by viewing it like a hero, as it is" (DB/MT, 72).

[10] In this context one might refer to Henrik Ibsen, as well, who saw writing poetry as judging oneself (cf. his poem EIN VERS, in *Gedichte* 1871, in Ibsen 1908, 117).

[11] Wittgenstein (1969, 22).

Both Beethoven and Trakl created their works from the experience of suffering and by considering ultimate questions, namely the religious. Therefore, their "tone" moved Wittgenstein, who suffered and fought as they did. However, in his suffering, he might have been able to find comfort in Trakl's poetry and in Beethoven's music.

Despite his appreciation for non-scientific approaches toward ultimate questions, and his longing to find an answer, Wittgenstein was aware of his inability to find a way of approach himself. "I am a thinker", he once wrote to Arvid Sjögren, and "I approach the problems of life only by thinking".[12] By thinking one could gain some kind of religious knowledge or understanding, however, in order to "possess religion", other ways are required, i.e. helpfulness, altruism, modest insight etc.—in short, a way of life that eventually would lead to a real understanding of religion.

29.11 The Aspect of Love Viz. Heart Versus Mind

As can be seen in Wittgenstein's later remarks, an other aspect with respect to the limits of rational knowledge became more and more important for him: the aspect of love.

Again and again, he speaks about the passions of the heart in contrast to the coldness of reason; he identifies religion with colours, while regarding wisdom as gray, cold and passionless. "Wisdom cannot be used for setting one's life as one can forge iron when it is cold. For a sound doctrine need not seize you; you can follow it, like a doctor's prescription. But here you have to be seized & turned around by something" (C&V, p. 61e).

In his reflections on the belief in resurrection in the course of musing about his spiritual torments of the mind, this belief only seems possible by means of the experience of love:

> But if I am to be REALLY redeemed,—I need <u>certainty</u>—not wisdom, dreams, speculation—and this certainty is faith. And faith is faith in what my <u>heart</u>, my <u>soul</u>, needs, not my speculative intellect. For my soul, with its passions, as it were with its flesh & blood, must be redeemed, not my abstract mind. Perhaps one may say: Only <u>love</u> can believe the Resurrection. Or: it is <u>love</u> that believes the Resurrection. [...] (MS 120, 108; 12.12.1937; C&V, 38e, 39e).

Thus Wittgenstein, similar to Pascal, gradually came to realize that a religious belief could only be attained through love, viz. the logic of the heart – le "logique du Coeur". In other words, and in the sense of Kierkegaard, through passion.

According to Paul Engelmann, Wittgenstein was the most passionate person he had ever met, and it was via this acquaintance that he came to understand Bettina von Arnim's words: "Passion is the only key to the world".[13]

And von Wright remarked:

---

[12]Cf. Wittgenstein's letter to Arvid Sjögren, written on the 9th of October, 1947.

[13]Cf. Wittgenstein-Engelmann (2006, 150).

> It seems to me that there are two forms of seriousness of character. One is fixed in 'strong principles'; the other springs from a passionate heart. The former has to do with morality, and the latter, I believe, is closer to religion. Wittgenstein was acutely and even painfully sensitive to considerations of duty, but the earnestness and severity of his personality were more of the second kind. Yet I do not know whether he can be said to be religious in any but a trivial sense of the word. Certainly he did not have a Christian faith. But neither was his view of life un-Christian, pagan, as was Goethe's. To say that Wittgenstein was not a pantheist is to say something important.[14]

While I do agree with von Wright that Wittgenstein's seriousness sprang from his passionate heart, I do not quite agree with the remark that Wittgenstein was not a religious person. From a narrowly dogmatic perspective, he was certainly not. However, he was religious in the sense of a passionate seeker, one who considered himself sinful and thus "wretched", and who was torn between moments of doubt and moments of believing.

Ludwig Hänsel wrote that Wittgenstein found himself in an "enormous tension between his seriousness of life i.e. his religiousness, and his analytical way of thinking—a tension only a few men would have the strength to endure."[15] To me, this tension ought to be seen in connection with Wittgenstein's conflict between head and heart, which several remarks of him confirm.

In his struggle with a religious belief, Wittgenstein experienced torments similar to those he encountered while engaging in his philosophical work, in which he found himself like a "cloud driven by the wind", exposed to hours of success and failure (cf. MS 118, 93v).

In those times of utmost despair, his tone appears humble—in contrast to his sober and critical argumentations such as in his *Lectures on Religious Belief*. In his personal diary notes, we find the voice of a human being who suffers.

> Now furthermore this is true: that I cannot through reflections make something right that appears as antics in my heart. No reasons of the world could prove, for example, that my work is important & something that I may & should do, when my heart says—without any reason—that I have to stop it (DB/MT, 179f.).

In this state of utmost despair, he asks himself what attitude he should assume to escape his suffering. That of outrage? That would be the death of him, for in rage he would only beat up on himself. Therefore he ought to surrender, i.e. not his hand, but his heart (cf. DB/MT, 183). If he were a believer, then his inner voice would tell him what to do and his suffering would come to an end (cf. ibid.).

He gives advice to himself and to others; sometimes the differences between the you and I blur into one another; a third voice or entity seems to be involved like a shadow. Just as he turned in his philosophical method from explaining to describing, he notes in his diary:

"Don't explain!—Describe! Submit your heart & don't be angry that you must suffer so! This is the advice I should be giving myself. When you are sick, accommodate yourself to the sickness; don't be angry that you are sick" (DB/MT, 183f.).

---

[14] von Wright, "A Biographical Sketch", in Malcolm 1989, 17f.

[15] Cf. Hänsel (1994, 246).

For, to get rid of the "torments of the mind" would mean to get rid of religion and both were indispensable for Wittgenstein and closely connected to one another. He was neither capable of existing without his spiritual struggles nor without religion.

With reference to Wittgenstein's words—"My ideal is a certain coolness. A temple providing a setting for the passions without meddling with them" (C&V,4e)–, it seems obvious that despite his aspirations to achieve this ideal, he did not really succeed in coming to terms with his passions, viz. the emotional side and priority of heart versus mind.

However, his passionate nature was indispensable for his way of doing philosophy. In an important sense it was the motor of his fervent way of treating philosophical problems—be they language problems or human problems in general—along with his passionate involvement with life as such—with his approach to the "existence of the world", expressed in an attitude of wonder and thus ethical commitment.

However fulfilling this passionate preoccupation with his surroundings might have been, Wittgenstein was incapable of achieving the sort of detachment necessary to gain sufficient distance to escape his suffering. Instead, he was a life-long tortured seeker, a person often feeling sinful, wretched. His suffering in spite of the futility of love—his dilemma between the spiritual and the emotional—might be compared to Ingeborg Bachmann's situation, who in a similar way suffered not only with respect to the limits of language, but also concerning love. In this connection I want to refer to her poem "Erklär mir Liebe, was ich nicht erklären kann…."—a poem that could be applied to Wittgenstein, as well.

> Erklär mir, Liebe, was ich nicht erklären kann:
> sollt ich die kurze schauerliche Zeit
> nur mit Gedanken Umgang haben und allein
> nichts Liebes kennen und nichts Liebes tun?
> Muß einer denken? Wird er nicht vermißt?
> Du sagst: es zählt ein andrer Geist auf ihn…
> Erklär mir nichts. Ich seh den Salamander
> durch jedes Feuer gehen.
> Kein Schauer jagt ihn, und es schmerzt ihn nichts.[16]

[16]Bachmann 1978, 48f.
Explain to me, love, what I cannot explain:
should I this short and dreadful time
have only contact with my thoughts, and alone,
know nothing loving and do nothing loving?
Have I to think? Won' t I be missed?
You say: another spirit counts on me
Do not explain to me.
I see the salamander go through every fire,
No shiver haunts him
and nothing gives him pain. (translation: I. Somavilla).

In their devotion to the spiritual, both Bachmann and Wittgenstein seem to have experienced the feeling of having failed in life and in love. Still, they took comfort in the salamander, who bravely managed to go through every fire without feeling pain.

For both of them, writing was of an existential necessity and their suffering and loneliness involved,—according to Bachmann, a "secret pain" [geheimer Schmerz]—was indispensable for approaching the truth and thus implicitly help others to open their eyes for it.

29.12 Conclusion

To live in the spirit, means freedom, happiness in its actual sense, but also pain. Being preoccupied with the problems of the mind, thus with philosophical problems, all too often does not lead to a solution, but ends in a *regressus infinitus*. This also applies to questions of religion that cannot be grasped, as "on the ground of grounded belief lies belief ungrounded" (MS 175, 20).

Freedom and peace of mind can perhaps be attained by accepting the limits of language—the limits of one's world–, about which Wittgenstein wrote in the course of his analytical investigations as early as in 1915.

Or through love. However, similar to passion, love, related to it, is bound to lead to suffering; and as can be observed in Wittgenstein's writings, his suffering cannot be separated from questions of ethics and religion, as the preoccupation with philosophical questions inevitably leads to "ultimate" questions transcending the world of facts and therefore involving religious questions that elude rational understanding and explanation. The impossibility of satisfying answers causes suffering to the person who reflects and philosophizes, as he cannot bridge the gap between philosophical scepsis and the search for an answer beyond scientific explanation.

Acknowledgement My thanks go to James M. Thompson for correcting my English.

References

Ariso, J. M. 2012, *Wahnsinn und Wissen. Zu Wittgensteins Lage und Denkbewegung*. Würzburg: Königshausen & Neumann.

Aurelius Augustinus 1921, *Des heiligen Augustinus Bekenntnisse*. Übertragen und eingeleitet von Herman Hefele. Jena: Eugen Diederichs.

Bachmann, I. 1978, *Gedichte und Erzählungen*. Bibliothek des 20. Jahrhunderts. Hg. von Walter Jens und Marcel Reich-Ranicki. Ausgewählt von Helmut Koopmann. Stuttgart, München: Piper.

Dostoyevsky, F. 2009, *The Brothers Karamasov*. Translated from the Russian by Constance Garnett. New York: The Lowell Press. The Project Gutenberg EBook. See http://www.gutenberg.org/files/28054/28054-h/28054-h.html.

Drury, M. O'C. 1981, "Some Notes on Conversations with Wittgenstein". In: *Ludwig Wittgenstein. Personal Recollections*. Ed. by Rush Rhees. Totowa, New Jersey: Rowman and Littlefield, p. 90–111.
Hänsel, L. 1994, „Ludwig Wittgenstein". In: Somavilla, Unterkircher, Berger (eds.), *Ludwig Hänsel – Ludwig Wittgenstein. Eine Freundschaft. Briefe. Aufsätze. Kommentare*. Innsbruck: Haymon.
Kierkegaard, S. 1954, *Fear And Trembling. The Sickness Unto Death*. New York: Doubleday & Company.
Lenau, N. 1986, *Faust*. Stuttgart: Reclam.
Malcolm, Norman 1984, *Ludwig Wittgenstein. A Memoir with a Biographical Sketch by G.H. von Wrigth and Wittgenstein's letters to Malcolm*. Oxford: Oxford University Press.
Pascal, B. 1980, Gedanken. Stuttgart: Philipp Reclam.
Schopenhauer, A. 1958, *The World as Will and Representation. Volume I*. Transl. from the German by Payne E.F.J. New York: Dover Publications.
Shields, P.R.1993, *Logic and Sin in the Writings of Ludwig Wittgenstein*. Chicago: The University of Chicago Press.
Somavilla, I. 2013, "'Und den Wahnsinn sollst Du nicht fliehen!' Wittgensteins Leiden des Geistes". In: *Geist, Sprache und Handlung / Mind, Language and Action*. Beiträge des 36. Internationalen Wittgenstein Symposiums in Kirchberg am Wechsel. Hg. von Danièle Moyrack-Sharrock, Annalisa Coliva und Volker Munz. Kirchberg a. W.: Österreichische Ludwig Wittgenstein Gesellschaft, S. 390–393.
Spinoza, B. 1976, *Die Ethik. Schriften. Briefe*. Hg. von Friedrich Bülow. Stuttgart: Kröner Verlag.
Tolstoi, L.1892, *Kurze Darlegung des Evangelium*. Aus dem Russischen von Paul Lauterbach. Leipzig: Reclam.
Tolstoi, L 1901, *Über Gott und Christentum*. Deutsch von Dr. M. Syrkin. Berlin: Steinitz Verlag.
Von Wright, G.H. 1982, *Wittgenstein*. Oxford: Basil Blackwell.
Weininger, O. 1912, *Geschlecht und Charakter*. Eine prinzipielle Untersuchung. Wien und Leipzig: Wilhelm Braumüller.
Weininger, O. 1918, *Über die letzten Dinge*. Mit einem biographischen Nachwort von Dr. Moriz Rappaport. Wien und Leipzig: Wilhelm Braumüller.
Wittgenstein, H. 1916–1919, *Aufzeichnungen*. Nachlass Margaret Stonborough-Wittgenstein. Brenner-Archiv, Universität Innsbruck.
Wittgenstein, L. 1969, *Briefe an Ludwig von Ficker*. Ed. by von Wright, G.H. and Methlagl, W. Salzburg: Otto Müller.
Wittgenstein, L. 1995, *Cambridge Letters*. Ed. by von Wright, G.H. and McGuinness, B. Oxford: Blackwell Publishers.
Wittgenstein. L. 1998, *Culture and Value*. Revised edition. Ed. by von Wright, G.H. in collaboration with Nyman, H. Transl. by Winch, P. Revised Edition of the Text by Pichler, A. Oxford: Basil Blackwell. (= C&V).
Wittgenstein, L. 1997, *Denkbewegungen. Tagebücher 1930–1932/1936–1937*. Ed. by Somavilla, I. Innsbruck: Haymon. (= DB).
Wittgenstein, L. 1990, *Geheime Tagebücher 1914–1916*. Ed. by Baum, W. Wien: Turia & Kant.
Wittgenstein, L. 2001, *Gesamtbriefwechsel. Innsbrucker Elektronische Ausgabe*. Universität Innsbruck: Forschungsinstitut Brenner-Archiv.
Wittgenstein, L. 1969, *The Blue and the Brown Books*. 2nd edition. Transl. by Kenny, A. Oxford: Basil Blackwell.
Wittgenstein, L. 1982, *Last Writings on the Philosophy of Psychology*. Vol I. Ed. by von Wright, G.H., Nyman, H. Oxford: Basil Blackwell.
Wittgenstein, L. 1992, *Last Writings on the Philosophy of Psychology*. Vol II. Ed. by von Wright, G.H., Nyman, H. Oxford: Basil Blackwell.
Wittgenstein, L. 2007, *Lecture on Ethics*. Ed. by Zamuner, E., Di Lascio E.V. and Levy, D. With Notes by Somavilla, I. Macerata: Quodlibet.
Wittgenstein, L. 2004, *Licht und Schatten*. Ein nächtliches (Traum-) Erlebnis und ein Brief-Fragment. Ed. by Somavilla, I. Innsbruck: Haymon. (= LS).

Wittgenstein, L. 2003, „Movements of Thought. Diaries 1930–1932/1936–1937", transl. by Alfred Nordmann. In: *Ludwig Wittgenstein. Public and Private Occasions*, ed. by Klagge, J. and Nordmann, A. Rowman & Littlefield Publishers. (= MT).

Wittgenstein, L. 1984, *Notebooks 1914–1916*. Ed. by Anscombe, G.E.M., von Wright, G.H. Chicago: The University of Chicago Press. (= NB).

Wittgenstein, L.1974, *On Certainty*. Ed. by Anscombe, G.E.M., von Wright, G.H. Oxford. Basil Blackwell.

Wittgenstein, L. 1974, *Philosophical Grammar*. Transl. by Kenny, A. Ed. By Rhees, R. Oxford: Basil Blackwell.

Wittgenstein, L. 1958, *Philosophical Investigations*. Transl. by Anscombe, G.E.M. Oxford: Basil Blackwell.

Wittgenstein, L. 1947, *Tractatus-Logico-Philosophicus*. With an Introduction by Bertrand Russell, F.R.S. London: Kegan Paul, Trench, Trubner & CO., LTD. (= TLP).

Wittgenstein, L. 1989, *Vortrag über Ethik und andere kleine Schriften*. Ed. by Schulte, J. Frankfurt: Suhrkamp.

Wittgenstein and the Vienna Circle 1979, Conversations recorded by Friedrich Waismann. Transl. by Schulte, J., McGuinness, B. Totowa: Barnes & Noble Books. (= WVC).

Wittgenstein – Engelmann. Briefe, Begegnungen, Erinnerungen 2006, ed. by Somavilla, I. in collaboration with McGuinness, B. Innsbruck: Haymon.

Wittgenstein, L. 1981, *Zettel*. Ed. by Anscombe, G.E.M., von Wright, G.H. Oxford: Basil Blackwell.

Wittgenstein's Nachlass 2000, *The Bergen Electronic Edition*. Bergen, Oxford: The Oxford University Press. (= MS for manuscripts, TS for typescripts).

Chapter 30
Wittgenstein and the Inexpressible

Freeman Dyson

Wittgenstein, unlike Heidegger, did not establish an ism. He wrote very little, and everything that he wrote was simple and clear. The only book that he published during his lifetime was Tractatus Logico-Philosophicus, written in Vienna in 1918 and published in England with a long introduction by Bertrand Russell in 1922. It fills less than two hundred small pages, even though the original German and the English translation are printed side by side. I was lucky to be given a copy of the Tractatus as a prize when I was in high school. I read it through in one night, in an ecstasy of adolescent enthusiasm. Most of it is about mathematical logic. Only the last five pages deal with human problems. The text is divided into numbered sections, each consisting of one or two sentences. For example, section 6.521 says: 'The solution of the problem of life is seen in the vanishing of this problem. Is not this the reason why men, to whom after long doubting the sense of life became clear, could not then say wherein this sense consisted?' The most famous sentence in the book is the final section 7: 'Whereof one cannot speak, thereof one must be silent.'

I found the book enlightening and liberating. It said that philosophy is simple and has limited scope. Philosophy is concerned with logic and the correct use of language. All speculations outside this limited area are mysticism. Section 6.522 says: 'There is indeed the inexpressible. This shows itself. It is the mystical.' Since the mystical is inexpressible, there is nothing more to be said.

Wittgenstein's intellectual asceticism had a great influence on the philosophers of the English-speaking world. It narrowed the scope of philosophy by excluding ethics and aesthetics. At the same time, his personal asceticism enhanced his credibility. During World War II, he wanted to serve his adopted country in a practical way. Being too old for military service, he took a leave of absence from his academic position in Cambridge and served in a menial job, as a hospital orderly taking care of patients. When I arrived at Cambridge University in 1946, Wittgenstein had just returned from his six years of duty at the hospital. I held him in the highest respect

F. Dyson (✉)
Institute for Advanced Study, School of Natural Sciences, Einstein Drive,
Princeton, NJ 08540, USA

S. Wuppuluri and N. da Costa (eds.), *WITTGENSTEINIAN (adj.)*,
The Frontiers Collection, https://doi.org/10.1007/978-3-030-27569-3_30

and was delighted to find him living in a room above mine on the same staircase. I frequently met him walking up or down the stairs, but I was too shy to start a conversation. Several times I heard him muttering to himself: 'I get stupider and stupider every day.'

Finally, toward the end of my time in Cambridge, I ventured to speak to him. I told him I had enjoyed reading the Tractatus, and I asked him whether he still held the same views that he had expressed twenty-eight years earlier. He remained silent for a long time and then said, 'Which newspaper do you represent?' I told him I was a student and not a journalist, but he never answered my question.

Wittgenstein's response to me was humiliating, and his response to female students who tried to attend his lectures was even worse. If a woman appeared in the audience, he would remain standing silent until she left the room. I decided that he was a charlatan using outrageous behavior to attract attention. I hated him for his rudeness. Fifty years later, walking through a churchyard on the outskirts of Cambridge on a sunny morning in winter, I came by chance upon his tombstone, a massive block of stone lightly covered with fresh snow. On the stone was written the single word, 'WITTGENSTEIN.' To my surprise, I found that the old hatred was gone, replaced by a deeper understanding. He was at peace, and I was at peace too, in the white silence. He was no longer an ill-tempered charlatan. He was a tortured soul, the last survivor of a family with a tragic history, living a lonely life among strangers, trying until the end to express the inexpressible.[1]

[1]From Freeman Dyson, 'What can you really know?', The New York Review of Books (8 November 2012).